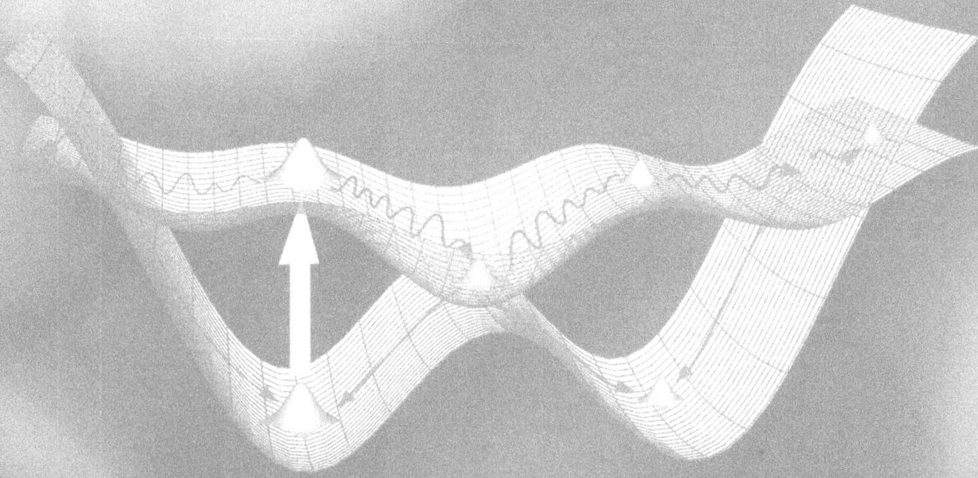

Time-Dependent Density Functional Theory

Time-Dependent Density Functional Theory

Nonadiabatic Molecular Dynamics

edited by

Chaoyuan Zhu

JENNY STANFORD
PUBLISHING

Published by

Jenny Stanford Publishing Pte. Ltd.
101 Thomson Road
#06-01, United Square
Singapore 307591

Email: editorial@jennystanford.com
Web: www.jennystanford.com

British Library Cataloguing-in-Publication Data
A catalogue record for this book is available from the British Library.

Time-Dependent Density Functional Theory: Nonadiabatic Molecular Dynamics

ISBN 978-981-4968-42-3 (Hardcover)
ISBN 978-1-003-31921-4 (eBook)

Contents

Preface

Time-dependent density functional theory (TDDFT) has emerged as an accurate and efficient quantum chemistry method for excited-state electronic structure and dynamic simulation in photochemical and photophysical processes, and especially for large complex systems in gaseous and condensed phases, TDDFT is almost unique choice for ab initio nonadiabatic molecular dynamic simulation. Many nonadiabatic dynamic processes involve internal conversion associated with conical intersections and intersystem crossing associated with spin orbital couplings among two or more electronic states. The nonadiabatic molecular dynamic simulation based on on-the-fly two or more TDDFT potential energy surfaces can reveal intermediate physical insights of photoisomerization, radiationless relaxation, and photoinduced decay processes, and its dynamic simulation can provide a nice interpretation for a chemical reaction related to electron transfer and charge transfer. In the recent years, TDDFT has been developed for computing excited-state properties of large-scale systems to high accuracy in biomolecules and nanomaterials in the sense of ab initio nonadiabatic molecular dynamic simulations.

Linear-response time-dependent density functional theory (LR-TDDFT) is one of the most popular methods traditionally for calculating excited-state properties of molecular systems such as the excitation spectra. The applicability of LR-TDDFT to conical intersection between electronic ground and first excited states might be questionable, and thus the spin-flip TDDFT (SF-TDDFT) is developed, in which the lowest triplet state is chosen as the reference state, and the ground and first excited states are treated as response states following a single electron spin flip. However, LR-TDDFT can work reasonably well for dynamics simulation involving conical intersection between the ground and first excited states if interesting simulation is involved with highly averaged quantities such as quantum yields, lifetimes, and spectral profiles. There are various ab initio nonadiabatic dynamic simulation methods among which the on-the-fly trajectory surface hopping method, Ehrenfest semiclassical method, phase space formulation method treating electronic and nuclear freedoms on an equal footing, and the multi-configuration time-dependent Hartree method are all most popular methods for dealing with large systems in photochemistry. The nonadiabatic molecular dynamics simulation methods along with TDDFT electronic structure calculations are demonstrated as accurate and efficient ab initio simulation methods for excited-state spectroscopy and excited-state reaction dynamics for large systems.

This book compiles and details cutting-edge theoretical research in quantum chemistry and chemical physics from the interdisciplinary groups around the world, including Japan, China, South Korea, the USA, Hong Kong, and Taiwan, that are developing excited-state dynamics methods involving conical intersections and intersystem crossings for large complex systems along with LR-TDDFT and SF-TDDFT electronic structure studies. Various applications can be also found on radiation chemiluminescence decay processes, photo-branching reaction mechanisms, carbon-related material photoluminescence, conformationally controlled photoexcitation, silicon quantum dot–induced nonradiative processes, excited carrier dynamics in condensed matter systems, and optical cavity–modified chemical dynamics. The book will appeal to graduate students and research scientists involved in excited-state dynamics and spectroscopy in the fields of photochemistry, biochemistry, materials chemistry, and chemical physics.

Chapter 1

Intersystem Crossing Reaction for Fluorescent 10-Methyl-9(10H)-Acridone via Dioxetanone Intermediates: On-the-Fly Nonadiabatic ONIOM Molecular Dynamics with Particle Mesh Ewald Method and Thermodynamics Simulations

Tatsuhiro Murakami and Shinkoh Nanbu
Department of Materials & Life Sciences, Faculty of Science & Technology, Sophia University, 7-1 Kioicho, Chiyoda-ku, Tokyo 102-8554, Japan
shinkoh.nanbu@sophia.ac.jp

The dissociation mechanism from the transition state of dioxetanone in DMSO solvate has been studied. The solution system has been equilibrated by isothermal-isobaric ensemble under 1 bar pressure and room temperature conditions with eight temperature replica exchange molecular dynamics (T-REMD) simulations. On-the-fly ONIOM molecular dynamics has been carried out with particle mesh Ewald method to consider the long-range Coulomb interaction from DMSO. Ab initio potential energies of S_0 and T_1

Time-Dependent Density Functional Theory: Nonadiabatic Molecular Dynamics
Edited by Chaoyuan Zhu

ISBN 978-981-4968-42-3 (Hardcover), 978-1-003-31921-4 (eBook)
www.jennystanford.com

states for dioxetanone have been calculated by B3LYP/aug-cc-pVDZ. The intersystem crossing has been occurred from the S_0 to the T_1 state at 15 out of 55 trajectories. All trajectories on the T_1 state have been dissociated to chemiluminescent 10-metyl-9(10 H)-acridone.

1.1 Introduction

Chemiluminescence is the radiation involved in the decay process of electronically excited state during a thermochemical reaction without an external a light source. The design and synthetization for various chemiluminescent compounds such as the luminol [1], firefly luciferin [2], hydroxycoumarin [3], 2-hydroxymethylbenzoate [4] and acridinium ester [5] derivatives have been performed to apply for bioimaging. Chemiluminescence reaction generates light through three sequential reactions. First a thermally activated molecule reacts at the electronically ground state. Second, the molecule produces the electronically excited species thorough nonadiabatic transition also known as a chemiexcitation process. Finally, the excess energy releases by emission of the light. At the chemiluminescence reactions [6–8], the oxidation reactions by oxidizing agent produce high-energy species and then the decay of the species is accompanied by the radiation. In the chemiluminescence of luminol [6, 9], for example, hydrogen peroxide is used as an oxidant; then the cyclic peroxide intermediate is produced before the emission. The chemiluminescent compounds used luciferin, hydroxycoumarin, 2-hydroxymethylbenzoate and acridinium ester have the four-membered ring peroxide namely 1,2-dioxetane intermediate [2–4, 10].

The 1,2-dioxetane structure appeared in the various chemiluminescence reaction has been interesting to understand the mechanism of the reaction. Adam and Baader measured experimentally the activation energies and excitation yields and studied the effect of methylation on the thermal stability of the 1,2-dioxetane [11]. It was observed that the excitation yield to spin triplet state is larger than the singlet one, but both were small. It was also clarified that the excited triplet and singlet quantum yields increase by the methylation. The trans-3,4-dimethyl-1,2-dioxetane (see Chart 1 in

Ref. 12) by Lindh et al. [13] were researched the reaction path by using multi-configurational reference perturbation theory (CASPT2) level potential energy surfaces (PESs). It has been estimated that the molecule reaches a flat biradical region (entropic trap) by O–O bond-breaking first, then the C–C bond is cleaved, and two acetaldehyde (carbonyl compounds) generates eventually. Yue and Liu et al. analyzed the time-evolved population of electronic states about the dissociation process of the trans-3,4-dimethyl-1,2-dioxetane by on-the-fly ab initio trajectory surface hopping simulation [12]. The quantum yield of the lowest triplet state (T_1) agrees with the experimental measurement [11, 12].

The 1,2-dioxetanone (see Chart 1 in Ref. 14) which is the intermediate of firefly luciferin chemiluminescence reaction [15] has also been studied to clarify the chemiluminescence mechanism. The dissociation process from 1,2-dioxetanone to formaldehyde and carbon dioxide was researched by the exploration of the minimum energy path on PESs of CASPT2 level [14] and the computation of the strength of the spin orbit coupling (SOC) [16], which gets heavily involved in intersystem crossing (ISC) process, by Lindh et al. The strength of the SOC between lowest singlet state (S_0) and the lowest triplet state (T_1) at the geometry of the transition state on S_0 was 59.4 cm^{-1}. These works suggest that the intersystem crossing from ground sate is feasible from the energy degeneracy at the transition state structure, the entropic trap at the biradical region and the SOC value.

The chemiluminescence mechanism of the acridinium ester has been studied. The nucleophilic additional reaction by hydrogen peroxide anion to carbon 9-position first occurs, then the chemiexcitation by activated molecule happens, finally the molecule decomposes to a fluorescent 10-methyl-9(10 H)-acridone. Nelson et al. [17] have suggested and McCapra et al. [18–20], White et al. [21], Dodeigne et al. [22] and Pieńkos et al. [23] have implied the existence of the 1,2-dioxetanone as an intermediate. On the other hand, Błażejowski et al. have indicated that the light-emitting acridone are formed as a result of the elimination of the phenyl carbonate anion, and not a carbon dioxide by the computation of the enthalpies and Gibbs's free energies at 298 K temperature condition under DFT/B3LYP level [24, 25]. Furthermore, Boużyk and Błażejowski et al. have measured the emission peak at 451 nm (2.75 eV) as well [26]. Recently, Nakazono et al. [27] suggested the reacted acridinium ester with hydrogen peroxide has both dioxetane and dioxetanone as intermediates and generates the excited acridone (see Scheme 1 in Ref. 27). The dioxetanone as an intermediate is still controversy up to now.

Nakazono et al. have produced a fluorescent 10-methyl-9(10 H)-acridone by making the acridinium ester derivatives react with hydrogen peroxide under pH neutral condition [5, 27, 28]. The substitution of electron withdrawing group to the 4-position phenyl moiety has shown strong chemiluminescence [28]. The possible chemiluminescence reaction path of the acridinium ester derivative substituted for methoxycarbonyl group has been shown in Fig. 5 of Ref. 5. The high-performance liquid chromatography (HPLC) peaks of 10-metyl-9(10 H)-acridone and methyl 4-hydroxybenzoate were observed, which indicates the existence of the dioxetanone intermediate.

In this study, we focus on the exploration of the chemiexcitation and dissociation mechanisms from the dioxetanone structure of the acridinium ester by theoretical approaches. Molecular structures of the dioxetanone, acridinium ester and 10-methyl-9(10H)-acridone are shown in Fig. 1.1.

The on-the-fly nonadiabatic ab initio trajectory surface hopping method is employed. This method is one of ideal tools to understand

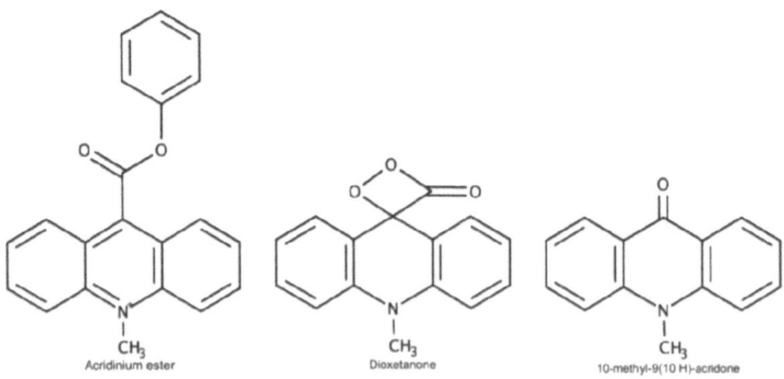

Figure 1.1 Molecular structures of acridinium ester (left), the dioxetanone (center) and 10-methyl-9(10H)-acridone (right).

the chemical reaction mechanism of multidimensional systems by pursing the classical nuclear motion on the potential energy surfaces obtained by quantum calculations. Molecular dynamics of various systems, which are the fluorescent switching molecule [29], the dissociation process of the atmospheric molecule [30] and the photoisomerization reaction of photochromic molecule [31] were studied, and the reaction mechanisms were clarified by means of the on-the-fly nonadiabatic ab initio trajectory surface hopping method. Additionally, the global switching algorithm [32] developed by Zhu et al. is embraced in the present study. The global switching algorithm does not need the calculation of nonadiabatic coupling vectors and is able to handle the transition to the different spin state in the same manner as the transition between the identical spin states. The nonadiabatic transition probability is calculated by the improved the Zhu–Nakamura formula [33–35]. The coulomb interactions of the target system from the surrounding solvent molecules are considered by the ONIOM (our Own N-Layered Integrated molecular Orbitals and molecular Mechanics) [36] by Gaussian 09 program package [37] with the electrostatic embedding (EE) method [38]. The particle-mesh Ewald ONIOM (PME-ONIOM) method developed by Kobayashi and Nanbu [39] is utilized to incorporate the periodic boundary condition. The estimation of the photoisomerization processes influenced by different solvents effects has been studied by using the PME-ONIOM method [40]. The equilibration of the solution system of the dioxetanone solute and the solvent molecule is performed by canonical and isothermal-isobaric ensembles with the temperature replica exchange molecular dynamics simulation of GROMACS 2020.6 [41].

In Section 1.2, the methodology used is presented. In Section 1.3, the results are shown and discussed. A summary is presented in Section 1.4.

1.2 Methodology

The solution model, which is the compound of the dioxetanone intermediate of acridinium ester and a phenol as solute, dimethyl sulfoxide (DMSO) as solvent, has been built to perform on-the-fly nonadiabatic ONIOM molecular dynamics simulation and to estimate intersystem crossing. The motion of nuclei has been evaluated by the expansion of the classical trajectories.

1.2.1 Electrostatic Potential from Quantum Mechanics

The geometry optimization of the dioxetanone intermediate (MIN) was carried out by the density functional theory (DFT) at B3LYP/aug-cc-pVDZ level by means of Gaussian 16 [42]. The transition state (TS) also was searched at same calculation level. The phenol, which is consider as the adjacent molecule with the acridinium ester, acridone and CO_2 which are the final products were also optimized by B3LYP/aug-cc-pVDZ level. The optimized geometry of DMSO was determined by MP2/aug-cc-VTZ level.

The electrostatic potential (ESP) charges of all of them were obtained by same calculation level for each species as mentioned above, with Merz–Singh–Kollman analysis by Gaussian 16.

1.2.2 System Setup

Each 1 MIN with a phenol and 1 TS with a phenol was packed with 512 DMSO in a 50 Å each side cubical box by Packmol [43]. The phenol molecule was arranged near the dioxetanone.

The acridone, CO_2 and 1 phenol, respectively, with 512 DMSO in a cubical box of same size for acridinium ester were also packed.

General AMBER Force Field 2 (GAFF2) has been used to generate the parameters for MIN, TS, phenol, acridone, carbon dioxide and DMSO with AmberTools18 [44]. Their ESP charges were adjusted to keep the net charge of the whole system as zero. The topology files of the systems of (**a**) MIN/phenol/DMSO, (**b**) TS/phenol/DMSO and (**c**) acridone/CO_2/phenol/DMSO were generated by the LEaP program of Amber 18 [44] with the force fields and the ESP charges as mentioned above and then were converted to the Gromacs [41] format by means of the ParmEd python module.

1.2.3 Equilibration

The temperature replica exchange molecular dynamics (T-REMD) simulation has been carried out to obtain equilibrium structure at room temperature. All simulations have been done by Gromacs 2020.6 [41].

Energy minimization for each solvated dioxetanone intermediate of acridinium ester was performed using the steepest decent method until the maximum force of atoms (F_{max}) was smaller than 1000 kJ/mol nm^{-1} or reached 50,000 steps. Initial minimization step size has been set as 0.01 nm for the systems (**a**) MIN/phenol/DMSO, (**b**) TS/phenol/DMSO and (**c**) acridone/CO_2/phenol/DMSO. The $F_{max} < 1000$ kJ mol^{-1}/nm reached 313, 85 and 93 steps for the systems (**a**), (**b**) and (**c**), respectively. For the systems (**b**) and (**c**), the compounds except DMSO were fixed during the minimization.

For NVT ensemble, NPT ensemble and T-REMD simulations, the equations of motion were integrated with a leap-frog algorithm. The time step used was 2 fs. The particle mesh Ewald method [45] was used for computing long-range electrostatic interactions with a periodic boundary condition. The cut-off for van der Waals (VDW) (nonbonded Lennard–Jones) interactions was chosen to be 10 Å. During equilibration in order to avoid drastic rearrangements of target parts which are the acridinium ester and phenol, the position restraints procedure has been employed. By setting a second-order harmonic potential against the displacement from a fixed reference position, the position movement was drastically suppressed [41]. The force constants for the position restraints of acridinium ester and phenol were 200×10^3kJ mol^{-1}/nm^2. Furthermore, for the transition state structure of the system (**b**), the bonded parameter of O3–O5 was set as 600×10^3 kJ mol^{-1}/nm^2. Note that the bonded parameter of O1–C2 from GAFF2 for TS was 5,46,095.68 kJ mol^{-1}/nm^2. The canonical (NVT) ensemble calculation was performed from the energy minimization structure for eight temperatures, which were 280.00, 286.66, 293.42, 300.33, 307.34, 314.50, 321.79 and 329.22 K individually until reaching the simulation time of 200 ps (1,00,000 steps) for all of the systems. The Bussi–Donadio–Parrinello (Canonical sampling through velocity rescaling) thermostat [46] with time dumping constant (τ_t) of 100 fs was used. The initial velocities were generated based on Maxwell–Boltzmann velocity distribution.

The isothermal-isobaric (NPT) ensemble simulations, against the eight temperatures as mentioned above, were implemented with the final structure of the canonical ensemble as the initial coordinates. A Parrinello–Rahaman barostat [47] with the pressure time constant (τ_p) of 2 ps was used to maintain a pressure of 1 bar with Bussi–Donadio–Parrinello (canonical sampling through velocity rescaling) thermostat. The simulations also were executed until leading to 200 ps for the systems (**a**), (**b**) and (**c**).

In order to carry out T-REMD simulation, the initial configuration of the systems (**a**), (**b**) and (**c**) was chosen from the last steps of the trajectories in NPT ensemble for each of the eight temperatures. The simulations were repeated until 100 ns under the same conditions of the NPT ensemble as explained above. Eight replicas with temperatures ranging from 280.00 to 329.22 K were used. The replica exchanges with the temperature neighbor were attempted every 500 steps (1 ps).

The convergence of the T-REMD simulations was confirmed by computing the replica mixing parameter $m(T, t)$ [48–50],

$$m(T, t) = 1 - \frac{\sqrt{\sum_i n_i(T, t)^2}}{\sum_i n_i(T, t)^2}, \tag{1.1}$$

where $n_i(T, t)$ is the total number of T-REMD steps of the replica i spent at temperature T. If the total number of replicas is $N_{\text{rep}} = 8$ and all replicas equally stay at temperature T, $m(T) = 1 - \sqrt{N_{\text{rep}}}/N_{\text{rep}} \simeq 0.646$.

1.2.4 ONIOM Potential Energies with Particle Mesh Ewald Method (PME-ONIOM) under a Periodic Boundary Condition (PBC)

ONIOM [36] of Gaussian09 [37] package has been used to obtain the potential energies of the solvated dioxetanone intermediate of acridinium ester. The reactive parts of a chemical system have been evaluated by a quantum mechanical (QM) calculation, also known as an ab initio calculation while the remaining nonreactive parts have been treated with an empirical molecular mechanics (MM) potential. The ONIOM potential energy is expressed as

$$E^{\text{ONIOM}} = E^{\text{MM}}_{\text{real}} + E^{\text{QM}}_{\text{model}} - E^{\text{MM}}_{\text{model}}, \tag{1.2}$$

where the *real* system contains all the atoms and is calculated only at the MM level and the *model* system includes the part of the system is evaluated at the QM level. In order to consider the Coulomb interaction between the QM region and MM region more appropriately, the electrostatic embedding (EE) method [38] has been employed. The ONIOM-EE potential energy expression is as follows:

$$E^{\text{ONIOM-EE}} = E^{\text{MM}}_{\text{real}} + E^{\text{QM}}_{\text{model}} + E^{\text{QM}}_{\text{int}} - E^{\text{MM}}_{\text{model}} - E^{\text{MM}}_{\text{int}}, \tag{1.3}$$

where $E^{\text{QM}}_{\text{int}}$ represents the Coulomb interaction energy of the electrons and nuclei of the quantum system with the embedding charges and $E^{\text{MM}}_{\text{int}}$ indicates the empirical Coulomb interaction energy between the model layer and the layer which is the real part excluded the model part.

Periodic boundary condition (PBC) enables a simulation to be performed using a relatively small number of particles to minimize edge effects in a finite system.

The PME-ONIOM method implemented by Kobayashi and Nanbu [39] has been embraced in this study. The long-range electrostatic interaction has been taken into account by a particle-mesh Ewald [45] and the minimum image convention has been employed for the short-range VDW interaction in the PME-ONIOM method. Here, the image layer is defined as the replica of the real layer with the real layer (see Fig. 2 of Ref. 39). The abbreviation of the model, real and image layers are set as the set of M, R and I. The set I is a superset of R; then R is indeed a superset of M, i.e., $I \supset R \supset M$. For example, the only replica area represents as the intersection between the set I and the complement set $\bar{R}$, i.e., $I \cap \bar{R}$. Under the definition of the particles A which indicates the point charges of the model layer and the particles B which represents the embedding charges locates the real part outside of the model part, the ONIOM-EE potential energy expression is re-expressed as

$$E^{\text{ONIOM-EE}} = E^{\text{MM}}_{\text{R}} + E^{\text{QM}}_{\text{M}} + E^{\text{QM,int}}_{M,\, B \in R \cap \overline{M}} - E^{\text{MM}}_{\text{M}} - E^{\text{MM,int}}_{A \in M,\, B \in R \cap \overline{M}}. \tag{1.4}$$

The five terms of the right-hand side (RHS) mention the same. The third term of the RHS represents the Coulomb interaction; the electrons and nuclei of the quantum system with the embedding charge particles B belong to the intersection between the set R and the complement set $\bar{M}$; $B \in R \cap \bar{M}$; the fifth term of RHS represents the empirical Coulomb interaction energy between the particles $A \in M$ and $B \in R \cap \bar{M}$. The number of particles A and B indeed correspond to the number of atoms of QM and MM layers.

Here, the PME-ONIOM potential energy expression is denoted:

$$E^{\text{PME-ONIOM}} = E^{\text{PME}}_{I \cap \bar{R}} + E^{\text{MM}}_{\text{R,MIC}} + E^{\text{QM}}_{\text{M}} + E^{\text{QM,int}}_{M,\,B \in R \cap \bar{M}} - E^{\text{MM}}_{\text{M}} - E^{\text{MM,int}}_{A \in M,\,B \in R \cap \overline{M}}, \tag{1.5}$$

where $E^{\text{PME}}_{I \cap \bar{R}}$ represents the long-range Coulomb interactions by particle-mesh Ewald (PME) method of the only replica area and $E^{MM}_{\text{R,MIC}}$ describes the bonded and VDW interactions by minimum image convention (MIC) and the Coulomb interactions inside the real layer by molecular mechanics scheme.

In this study, the dioxetanone intermediate and acridone and carbon dioxide, which are the final products as a solute, have been considered as a model layer, while the rest of DMSO as a solvent was set as a real layer. One phenol was basically set as a real layer to reduce computational costs. Based on the original ONIOM scheme, the PME-ONIOM energy has been evaluated via the external option of Gaussian 16 [42]. The ab initio calculation was carried out to obtain indeed the QM part's potential energies by the density functional theory (DFT) at B3LYP/aug-cc-pVDZ level with density fitting spin-restricted Kohn–Sham (DF-RKS) method [51] for lowest singlet (S_0) and triplet (T_1) states by using the Molpro2021 [52] program package. The electrostatic potential (ESP) and general AMBER force field 2 (GAFF2) stated in Sections 1.2.1 and 1.2.2 were used for the empirical parameters. The cut-off for VDW interactions was chosen to be 10 Å.

1.2.5 Spin–Orbit Coupling Calculation

The spin–orbit coupling (SOC) has been calculated by the multireference configuration interaction (MRCI) method [53] with the full Breit–Pauli Hamiltonian by means of Molpro2021 [52] package. The strength of SOC $\left(V^{\text{SOC}}\right)$ between singlet and triplet electronic states has been considered as the length of the spin–orbit coupling vector

[12, 16, 32, 54]

$$V_{\mathrm{jk}}^{\mathrm{SOC}}=\sqrt{\sum_{M_s=-1,0,1}\left|\langle S_j|\,\hat{H}_{\mathrm{SO}}\,|T_k,\,M_s\rangle\right|^2} \tag{1.6}$$

where j and k are the numbering for the singlet and triplet states.

The molecular orbitals have been calculated as the reference wave function by the four-state (i.e., two lowest singlet and two lowest triplet states) averaged multireference self-consistent field program [55, 56], which involved the complete active space of 18 electrons and 14 orbitals (4SA-CASSCF(18e,14o)) with aug-cc-pVDZ basis set; $V_{\mathrm{jk}}^{\mathrm{SOC}}$ between S_0 and T_1 state has been obtained in this study.

1.2.6 Transition Probability for Intersystem Crossing (ISC)

In the semiclassical picture based on the two-state linear diabatic potential curve crossing model, the probability of ISC $\left(p^{\mathrm{ISC}}\right)$ for passing through between the singlet and triplet crossing point has been given by

$$p^{\mathrm{ISC}}=1-p^{\mathrm{ZN}}, \tag{1.7}$$

where p^{ZN} is the nonadiabatic transition probability between one-dimensional two-state adiabatic potential energy surfaces (PESs) estimated by the Zhu–Nakamura formula [33–35]. In the two-state curve non-crossing case [12, 35, 57, 58], p^{ISC} becomes the following formula:

$$p^{\mathrm{ISC}}=1-\frac{p^{\mathrm{ZN}}}{p^{\mathrm{ZN}}+1}. \tag{1.8}$$

The nonadiabatic transition probability expressions $\left(p^{\mathrm{ZN}}\right)$ are as follows:

$$p^{\mathrm{ZN}}=\exp\left[-\frac{\pi}{4\sqrt{a^2b^2}}\sqrt{\frac{2}{1+\sqrt{1+b^{-4}}}}\right]. \tag{1.9}$$

The probability is used where the two diabatic forces (F_1 and F_2) have same direction. The other probability where the slopes of the two diabatic potentials (V_1 and V_2) have the opposite sign, has been represented as

$$p^{\mathrm{ZN}}=\exp\left[-\frac{\pi}{4\sqrt{a^2b^2}}\sqrt{\frac{2}{1+\sqrt{1-b^{-4}}}}\right]. \tag{1.10}$$

The two unitless parameters (a^2 and b^2), namely, the effective coupling parameter and the effective collision energy,

$$a^2 = \frac{\hbar^2}{2\mu} \frac{\sqrt{|F_2 F_1|}\,|F_2 - F_1|}{(2V_{12})^3} \tag{1.11}$$

and

$$b^2 = (E_t - E_X) \frac{|F_2 - F_1|}{\sqrt{|F_2 F_1|}\,(2V_{12})}, \tag{1.12}$$

where $\hbar$ and μ represent a reduced Planck constant and a reduced mass for the diatomic molecule, E_X is an energy at the crossing point, E_t is the potential energy plus kinetic energy component in direction of hopping direction and V_{12} is the diabatic coupling. The numbering 1 and 2 represents the electronic surface 1 and the surface 2, respectively.

In order to obtain the parameters a^2 and b^2, the multidimensional forces have been converted to one-dimensional forces with the procedure by Yu and Zhu et al. [32]. The multidimensional diabatic forces are described as

$$F_1^{i\alpha} = -\frac{\partial V_1}{\partial R_{i\alpha}} \text{ and } F_2^{i\alpha} = -\frac{\partial V_2}{\partial R_{i\alpha}}, \tag{1.13}$$

where $R_{i\alpha}$ stands for x, y and z component of Cartesian coordinates for the i-th nucleus. The reduced force and the reduced mass [32] have been generalized as

$$\frac{|F_2 - F_1|}{\sqrt{\mu}} = \sqrt{\sum_{i=1}^{N} \frac{1}{m_i} \sum_{\alpha=x,y,z} \left(F_2^{i\alpha} - F_1^{i\alpha}\right)^2} \tag{1.14}$$

and

$$\frac{|F_2 F_1|}{\sqrt{\mu}} = \sqrt{\left|\sum_{i=1}^{N} \frac{1}{m_i} \sum_{\alpha=x,y,z} F_2^{i\alpha} F_1^{i\alpha}\right|}, \tag{1.15}$$

where N is the number of nuclei in QM layer with its mass m_i ($i = 1, 2, \ldots, N$). The diabatic coupling V_{12} becomes $V_{12} = V_{12}^{SOC}$ under the consideration of the transition probability between the diabatic states with the different spin-multiplicity i.e., singlet (triplet) for the surface 1 and triplet (singlet) for surface 2. The QM adiabatic PESs are identical to the diabatic PESs, i.e., $U\ (S_0) = V\ (S_0)$ and $U\ (T_1) = V\ (T_1)$.

1.2.7 A Global Switching Algorithm

The classical trajectories have been applied for the estimation of the nuclear motions on the electronic surfaces. The transition between the surfaces has been evaluated by the trajectory surface hopping method [59]. For the multidimensional systems, the hopping direction has been defined as a following vector,

$$s_{i\alpha} = \frac{F_2^{i\alpha} - F_1^{i\alpha}}{\sqrt{m_i}} \tag{1.16}$$

with a normalized vector as

$$\mathbf{n}_{\mathrm{i}} = \frac{1}{\sqrt{s_{\mathrm{ix}}^2 + s_{\mathrm{iy}}^2 + s_{\mathrm{iz}}^2}} \left(s_{\mathrm{ix}},\ s_{\mathrm{iy}},\ s_{\mathrm{iz}}\right). \tag{1.17}$$

The surface hopping with energy conservation has been carried out. The momentum of i-th nuclei $\mathbf{P}_{\mathrm{i}}$ has been decomposed into the momentum of the hopping direction $\mathbf{P}_{i\parallel}$ and the rest component $\mathbf{P}_{i\perp}$. The description is as follows:

$$\mathbf{P}_{i\parallel} = (\mathbf{P}_i \cdot \mathbf{n}_i)\,\mathbf{n}_i \text{ and } \mathbf{P}_{i\perp} = \mathbf{P}_i - (\mathbf{P}_i \cdot \mathbf{n}_i)\,\mathbf{n}_i\,. \tag{1.18}$$

The component $\mathbf{P}_{i\perp}$ does not change during the transition; the component of the hopping direction $\mathbf{P}_{i\parallel}$ is adjusted for the energy conservation. Thus,

$$\mathbf{P}_{i\perp}(+, t) = \mathbf{P}_{i\perp}(-, t) \tag{1.19}$$

and

$$\begin{aligned} E_t &= U_+(t) + \sum_{i=1}^{N} \frac{\mathbf{P}_{i\parallel}(+, t) \cdot \mathbf{P}_{i\parallel}(+, t)}{2m_i} \\ &= U_-(t) + \sum_{i=1}^{N} \frac{\mathbf{P}_{i\parallel}(-, t) \cdot \mathbf{P}_{i\parallel}(-, t)}{2m_i}, \end{aligned} \tag{1.20}$$

where $(+, t)$ and $(-, t)$ stand for the momentum components on the upper and lower PESs, respectively, at time t.

1.2.8 On-the-Fly PME-ONIOM Molecular Dynamics

The on-the-fly ab initio molecular dynamics method [29, 30] has been employed. This method calculates the trajectories by solving the Newtonian equations of motion for the nuclei on potential energy surfaces.

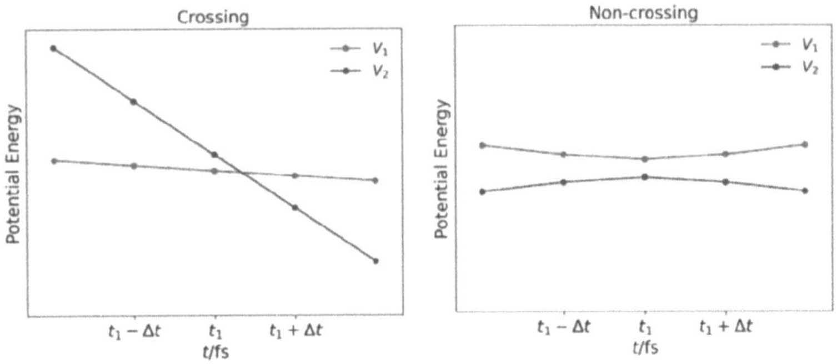

Figure 1.2 Attempting hopping point (t_1) at the minimum energy gap for the crossing and non-crossing cases.

In this study, the equations of motion for the classical nuclei have been integrated with a velocity-Verlet algorithm [60]. The time step used (Δt) was 1.00 fs. The cell size was decided from isobaric-isothermal ensemble simulation to evaluate short-range and long-rage interactions. Note that the variation of cell size is ignored at the short-time scale ab initio dynamics simulation. The initial coordinates were picked from the structures at 50 ns T-REMD simulation. The initial velocities have been generated based on the Maxwell–Boltzmann velocity distribution. The potential energies and the gradients for S_0 and T_1 states were obtained by the PME-ONIOM method at the same calculation level as mentioned in Section 1.2.4. A global switching algorithm [32, 61, 62] has been embraced to switch trajectories between potential energy surfaces as mentioned in Section 1.2.7. The S_0-T_1 energy gaps ΔE were compared for switching at every time steps. When the diabatic potential energies are crossing or have local minimum (<0.25 eV) with checking the back time and forward time as shown in Fig. 1.2, the switching of the trajectories was attempted. Equations (1.7) and (1.8) were used to get the transition probability for the crossing and non-crossing cases, respectively. The total number of trajectories is 55.

1.3 Results and Discussion

The equilibration, the electronic structure calculations of QM layer and PME-ONIOM MD simulations were carried out.

Table 1.1 The temperature replica exchange molecular dynamics (T-REMD) simulation details for the systems (**a**), (**b**), (**c**), (**2b**), (**3b**) and (**4b**)

System	N_{DMS}	N_{rep}	t_{tot} (ns)	τ_t (ps)	τ_p (ps)	τ_{rep} (ps)
(**a**)	512	8	100	0.1	2	1
(**b**)	512	8	100	0.1	2	1
(**c**)	512	8	100	0.1	2	1
(**2b**)	256	8	100	0.1	2	1
(**3b**)	768	8	100	0.1	2	1
(**4b**)	1728	8	100	0.1	2	1

1.3.1 Temperature Replica Exchange Molecular Dynamics Simulations

The T-REMD simulations for the isobaric-isothermal ensemble under the position restraints condition of the solute molecules were carried out for 100 ns for the systems (**a**) MIN/phenol/DMSO, (**b**) TS/phenol/DMSO and (**c**) acridone/CO_2/phenol/DMSO. The parameters with respect to the T-REMD are written in Table 1.1. The number of replicas is $N_{rep} = 8$. The eight simulations, i.e., simulation No. 0 to No. 7, have each reference temperature, which is 280.00, 286.66, 293.42, 300.33, 307.34, 314.50, 321.79 and 329.22 K, respectively. For systems (**a**), (**b**) and (**c**), the number of DMSO (N_{DMS}) is equal to 512. To monitor the effect of the difference of N_{DMS}, systems (**2b**), (**3b**) and (**4b**) have been set up. The molecule of QM layer is TS of dioxetanone and 256, 768 and 1728 of DMSO with 1 phenol are considered as MM layer in the systems (**2b**), (**3b**) and (**4b**), respectively.

Table 1.2 shows the average temperature for 100 ns simulations against the simulation No. 0 to No. 7 for the systems (**a**), (**b**), (**c**), (**2b**), (**3b**) and (**4b**). All of simulations were equilibrated to every reference temperature appropriately. The replica mixing parameter $m(T, t)$ for the system (**b**) in order to monitor the convergence of the T-REMD is shown in Fig. 1.3. The convergence was reached at 5 ns. The parameter $m(T) = 1 - \sqrt{N_{rep}}/N_{rep} \simeq 0.646$, which indicates all replicas equally stay at the temperature T.

Table 1.3 shows the significant geometry of the systems (**a**), (**b**) and (**c**) after 50 ns T-REMD simulation at the No. 3. The numbering of

Table 1.2 The average temperature (K) for 100 ns T-REMD simulation

Simulation number	(a)	(b)	(c)	(2b)	(3b)	(4b)
No. 0	280.49	280.57	280.50	280.70	280.58	280.55
No. 1	287.58	287.71	287.59	287.95	287.68	287.61
No. 2	294.34	294.47	294.36	294.80	294.41	294.34
No. 3	301.17	301.25	301.19	301.60	301.21	301.13
No. 4	308.12	308.23	308.13	308.47	308.16	308.06
No. 5	315.19	315.27	315.21	315.56	315.22	315.14
No. 6	322.42	322.49	322.41	322.77	322.45	322.36
No. 7	330.07	330.17	330.06	330.48	330.09	329.92

Note: No. 0–No. 7 represent each simulation number.

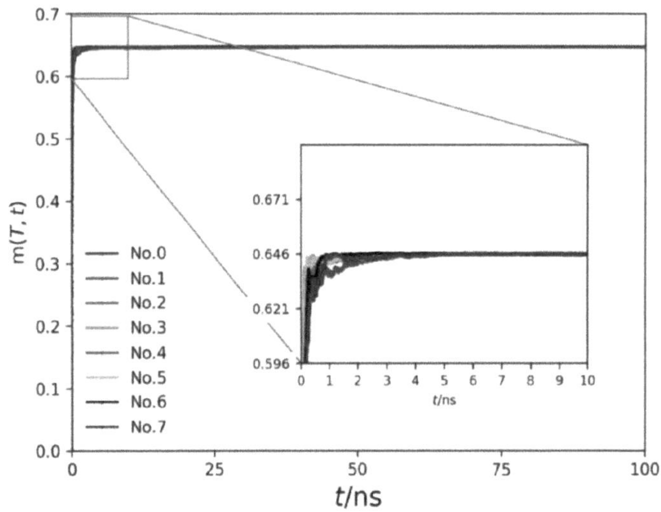

Figure 1.3 The replica mixing parameter $m(T, t)$ for the simulation No. 0 to No. 7 (the inset shows the $m(T, t)$ from 0 to 10 ns).

the atoms is shown in Fig. 1.4. The bond lengths between O3 and O5 for the systems (**a**) and (**b**) are r_{O3O5} =1.496 Å and $r_{\mathrm{O3O5}} = 2.071$ Å, while the C2–C4 bond lengths are $r_{\mathrm{C2C4}} = 1.560$ Å and $r_{\mathrm{C2C4}} = 1.498$ Å for the systems (**a**) and (**b**), respectively. It is suggested that the biradical region passes through the dissociation process to acridone with carbon dioxide as we expected from previous studies [16, 63]. The O3–O5 and C2–C4 bond lengths of the system (**c**) are 3.666

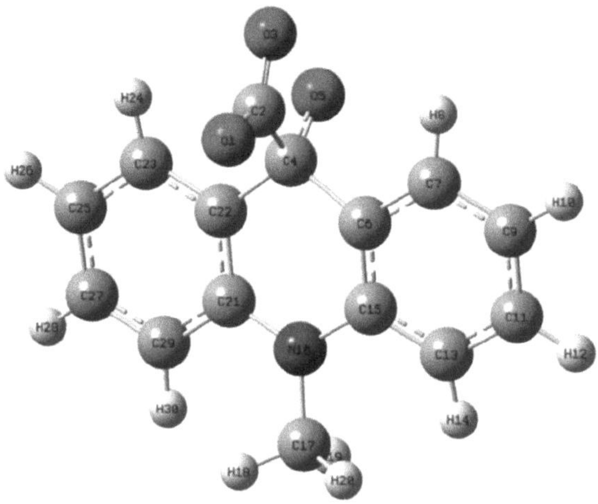

Figure 1.4 The numbering of the dioxetane atoms; the structure is at 50 ns T-REMD simulation of the simulation no. 3 of QM layer in system (b).

Å and 3.979 Å, which indeed shows that the system has acridone and CO_2. The dihedral angle for the system (**c**) is $d_{O3C2C4O5} = 20.9^\circ$, while O3, O5, C2 and C4 atoms are on a plane in systems (**a**) and (**b**). It is predicted that the out-of-plane motion of O3 from the O3–C2–C4–O5 plane happens during the dissociation. The bond angle $a_{O5C4N16} = 148.2^\circ$, 126.1° and 176.7° of the systems (**a**), (**b**) and (**c**), respectively, indicates the remaining O5 with C4 and N16 becomes linear due to removing the O1–C2–O3. For PME-ONIOM calculations, the cubical periodic box length (l) was obtained (see Table 1.4) from the volume under isobaric-isothermal ensemble for all systems. The lengths were obtained by Gaussview 6 from the results of NPT ensembles. The variation of volume has been already stable from 25 ns in T-REMD simulations. The length at 50 ns of T-REMD has been used for PME-ONIOM molecular dynamics simulations mentioned later.

1.3.2 Electronic Structure Calculation with Electrostatic Embedding

Figure 1.5 shows the geometry of the system (**b**) in the periodic box at 50 ns of the T-REMD simulation. The dioxetanone, phenol

Table 1.3 The key geometric parameters of the systems (**a**), (**b**) and (**c**) after 50 ns T-REMD simulation at the simulation No. 3

Geometry	r_{O3O5}	r_{C2C4}	$d_{O3C2C4O5}$	$a_{O5C4N16}$
(**a**)	1.496	1.560	2.3	148.2
(**b**)	2.071	1.498	1.0	126.1
(**c**)	3.666	3.979	20.9	176.7
$^1(\sigma,\sigma*)$-Min[a]	1.577	1.507		
$^1(\sigma,\sigma*)$-TS[a]	2.455	1.506		

Bond length *r* is given in angstrom (Å), and dihedral angle *d* and bond angle *a* are given in degrees.
[a]Ref. [14].

Table 1.4 The cubical periodic box length (l/Å) at 25.00, 50.00, 75.00 and 100.00 ns T-REMD simulation time (t) of the simulation No. 3 for each system

t (ns)	(**a**)	(**b**)	(**c**)	(**2b**)	(**3b**)	(**4b**)
25.00	43.6	43.9	43.4	35.2	50.6	63.2
50.00	45.0	43.8	45.0	35.6	49.1	64.5
75.00	44.6	43.4	45.5	35.5	50.0	65.5
100.00	45.1	44.0	45.0	36.6	49.4	65.1

and DMSO are described by ball and bond, tube and wireframe, respectively. The dioxetanone as QM parts and the phenol and DMSO as MM parts were computed. Table 1.5 shows the relative energy for QM layer of the systems (**a**), (**b**) and (**c**) included electrostatic embedding (EE) effects and excluded EE. The potential energy of the ground state of the system (**a**) was set as 0.00 eV. The relative energies of Table 1.5 are visualized in Fig. 1.6. There are no significant differences in the S_0-T_1 energy gaps ΔE in an EE effect. On the other hand, the energies of system (**b**) with EE and without EE are negative and positive. While the dioxetanone molecule of the system (**b**) without EE behaves as a "*Transition state*," the Coulomb interactions from the particle charges in MM layer make the potential energies stabilize by the EE effect. The S_0-T_1 energy gap ΔE of system (**c**), which is related to the emission energy, is

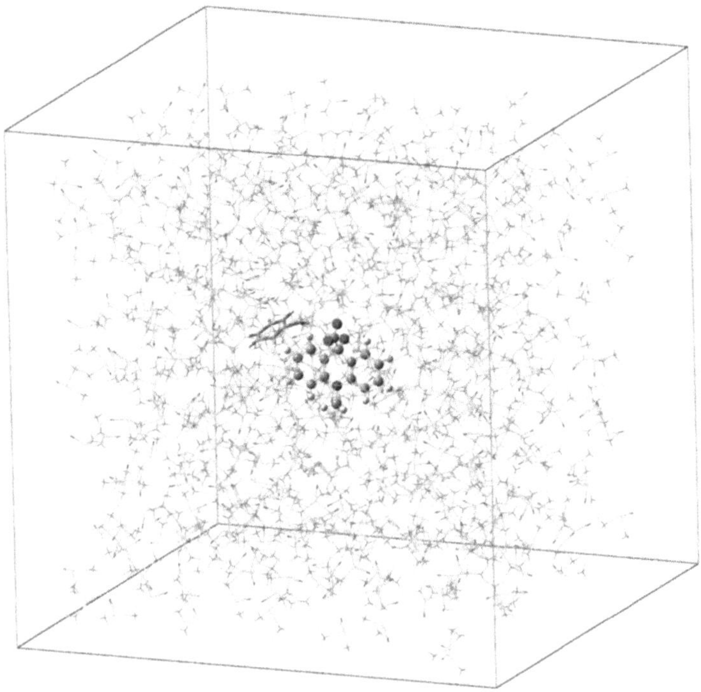

Figure 1.5 The geometry of the system (**b**) after 50 ns T-REMD simulation; the dioxetanone, phenol and DMSO are described by ball and bond, tube and wireframe, respectively. The color lines represent the 3D periodic boundary box.

2.58 eV at with-EE. The ΔE is a reliable value compared with the experimental fluorescence (FL) peak [5].

For the system (**c**), the potential energies of four lowest singlet states and four lowest triplet states by the density-fitting complete active space second-order perturbation theory (CASPT2) [64] with aug-cc-pVDZ have been calculated. The complete active space has been set as six electrons and six orbitals. Furthermore, the transition dipole moment between S_0 and S_j states (M_j) has been calculated with the following equation:

$$M_j = \sqrt{\sum_{\alpha=x,y,z} \left| \langle S_j | \hat{\mu}_\alpha | S_0 \rangle \right|^2} \tag{1.21}$$

Table 1.5 The relative energies with/without electron embedding (EE) for QM layer by DFT/B3LYP level at the 50 ns T-REMD structures and the energy gap $\Delta E = U(T_1)\text{–}U(S_0)$ with experimentally fluorescence (FL) peak of the 10-metyl-9(10 H)-acridone at pH 7[a]

	DFT/B3LYP						
	With EE			Without EE			FL peak[a]
QM energy	(a)	(b)	(c)	(a)	(b)	(c)	
$U(T_1)$	3.02	−0.93	−1.07	3.67	0.86	−0.40	
$U(S_0)$	0.00	−0.51	−3.65	0.53	1.07	−3.11	
ΔE	3.02	−0.41	2.58	3.14	−0.22	2.71	2.88 (430 nm)

Note: The potential energy is given in electron volt (eV).
[a]Ref. [5].

Table 1.6 The potential energies (E) and the transition dipole moments (M_j) for the system (**c**) by CASPT2 (6e,6o)/aug-cc-pVDZ level

Electronic state	E (eV)	M_j (a.u.)
T_4	5.75	
T_3	4.40	
T_2	3.87	
T_1	3.27	
S_3	5.30	0.67
S_2	4.78	0.43
S_1	3.60	1.22
S_0	0.00	

where $\hat{\mu}_\alpha$ represents the electric transition dipole moment for the Cartesian direction $\alpha \in \{x, y, z\}$. Table 1.6 shows the potential energies (E) and the transition dipole moments (M_j). The energy gaps between T_1-S_0 and S_1-S_0 are 3.27 and 3.60 eV, respectively. An ISC from T_1 to S_1 states might happen due to small energy differences (0.33 eV). However, 3.60 eV being the S_1-S_0 energy gap is higher than the experimental FL peak (2.88 eV). Note that the transition dipole moment between S_0 and S_1 states ($M_{S1} = 1.22$ a.u.) is feasible value to occur the photoemission from S_1 state. In this study, we are focusing on the dissociation dynamics of dioxetanone to acridone.

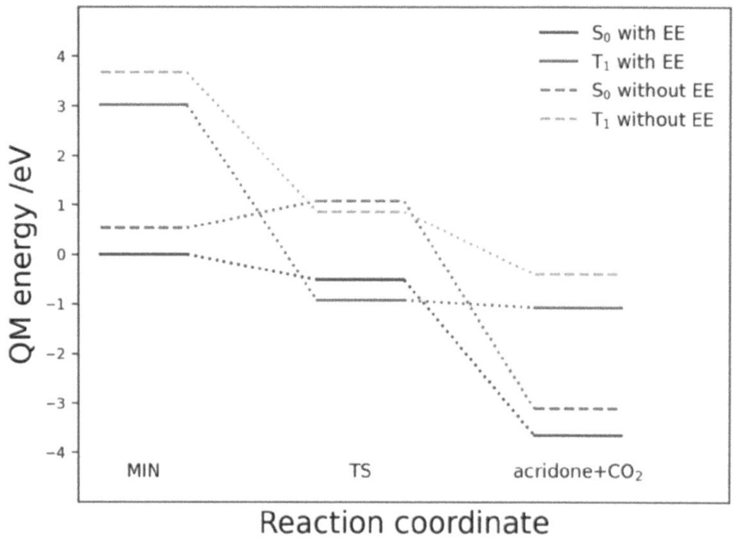

Figure 1.6 The relative energy for QM part of the systems (**a**, left), (**b**, center) and (**c**, right) with EE (solid) and without EE (dashed). The linear dotted lines connect each system.

The T_1-S_0 energy gap ΔE by B3LYP level is compared between the difference of the number of DMSO (N_{DMS}) in systems (**b**), (**2b**), (**3b**) and (**4b**) in Table 1.7. N'_{DMS} represents the number of DMSO in 1 Å^3 cube ($N'_{DMS} = N_{DMS}/l^3$) under 1 bar pressure and 301 K conditions. There is not much difference of N'_{DMS} between systems (**b**), (**2b**), (**3b**) and (**4b**); thus, the density of DMSO under 1 bar and room temperature conditions is expressed as a certain regularity by the NPT ensemble in T-REMD simulations. The ΔE obtained

Table 1.7 The T_1-S_0 energy gaps by B3LYP/aug-cc-pVDZ level under the condition of the different number of DMSO

System	N_{DMS}	l (Å)	N'_{DMS}(Å^{-3})	ΔE (eV)
(**b**)	512	43.8	6.1×10^{-3}	-0.41
(**2b**)	256	35.6	5.7×10^{-3}	-0.34
(**3b**)	768	49.1	6.5×10^{-3}	-0.52
(**4b**)	1728	64.5	6.4×10^{-3}	-0.45

by ONIOM-EE calculations of system (**b**) compared with the others is within the range of $\pm$ 0.11 eV. It has been considered that the dioxetanone has been affected by the appropriate solvate effects of DMSO.

1.3.3 On-the-Fly PME-ONIOM Molecular Dynamics

By using the structure at 50 ns T-REMD of the simulation No. 3 of the system (**b**) as the initial coordinate, the dynamics simulations were performed. The SOC values are required to carry out the global switching algorithm. In this study, the constant SOC $\left(V^{\mathrm{SOC}}\right)$, which is 50.0 cm^{-1}, was used due to reducing the computational cost. While the computational time to advance one step is roughly 20 min, it takes 4 h to get the SOC values. The value 50.0 cm^{-1} is chosen based on the preliminary 16 trajectories (19 points) and is shown in Table 1.8 with the key geometries. At all 19 points, the strengths of SOC are between 44.5 and 62.0 cm^{-1}.

All 55 trajectories start from the S_0 state. The dynamics simulations show time-evolved two-dimensional geometric distributions in terms of O3–O5 and C2–C4 bond lengths in Figs. 1.7 and 1.8. The distributions were described by Gaussian kernel density estimation (kde) of SciPy library of Python. Figure 1.7 represents the density on the lowest singlet state. The trajectories started from the coordinates of system (**b**) with velocities based on the Maxwell distribution at 300 K. The coordinates are located in the biradical region and the PESs in the adjacent area as shown in Fig. 1.6. At the time-evolution from 20 to 40 fs, the O3–O5 bond is shrinking or stretching without moving the C2–C4 bond. At 60 fs, the half of density returns to stable area of dioxetanone around 1.2 Å $<$ O3–O5 bond $<$ 2.0 Å and 1.2 Å $<$ C2–C4 bond $<$ 2.0 Å, while the rest is still trapped in the biradical region of 2.0 Å $<$ O3–O5 bond $<$ 3.2 Å and 1.2 Å $<$ C2–C4 bond $<$ 2.0 Å. Along the integration of the equations of motion from 60 fs to 80, to 100 and to 120 fs, the O3–O5 oscillation between the stable area and the biradical region is observed. A small quantity of trajectories dissociates to acridone and CO_2 on the S_0 state at 120 fs. This phenomenon takes place without chemiluminescence dissociation. On the other hand, Fig. 1.7 shows

Table 1.8 The strength of SOC (V^{SOC}/cm^{-1}) from 16 preliminary trajectories (19 point) with key geometric parameters

Trajectory number	r_{O3O5}	r_{C2C4}	$d_{O3C2C4O5}$	$a_{O5C4N16}$	V^{SOC}
No. 1	2.089	1.525	-3.4	125.7	60.7
No. 1	2.066	1.526	-4.6	125.9	60.4
No. 1	2.044	1.511	-5.4	125.7	60.6
No. 1	1.886	1.530	-10.9	129.8	58.8
No. 5	2.052	1.518	-3.2	126.8	60.1
No. 7	2.027	1.518	-0.4	128.8	46.7
No. 11	2.103	1.512	3.1	125.9	61.4
No. 17	2.061	1.529	1.2	128.1	49.1
No. 21	2.054	1.548	1.4	129.5	49.5
No. 22	2.054	1.517	-0.6	127.8	48.2
No. 26	1.940	1.479	-1.8	123.7	45.4
No. 28	2.092	1.472	1.5	126.2	59.3
No. 29	1.929	1.509	6.8	125.3	44.6
No. 30	2.068	1.486	2.5	128.0	50.1
No. 35	2.042	1.490	-0.5	125.3	48.9
No. 36	2.093	1.500	5.3	126.7	60.0
No. 38	2.090	1.501	1.6	127.5	61.8
No. 40	2.070	1.501	1.8	126.5	60.3
No. 46	2.037	1.529	-1.8	126.6	61.9

Note: Bond length r is given in angstrom (Å), and dihedral angle d and bond angle a are given in degrees.

the distribution of the trajectories leading to intersystem crossing (ISC) to the T_1 state. The ISC already happens at 20 fs in the biradical region. In the propagation, the density on the triplet state progresses the dissociation direction in a straightforward manner. The C2–C4 cleavage has been described from the O3–O5 biradical region not from a stable area in Fig. 1.8. All trajectories on the T_1 state eventually enter the acridone + CO_2 area (O3–O5 and C2–C4 are bigger than 2.8 and 2.0 Å) at 120 fs. Out of 55 trajectories, 15 have been switched to the T_1 state, and the remaining 40 trajectories stay on the S_0 state within 120 fs dynamics. The state population ratio could be changed for longer dynamics because some of trajectories are still trapped in the biradical region on the S_0 state at 120 fs illustrated in Fig. 1.7 and the switching area leading to the ISC

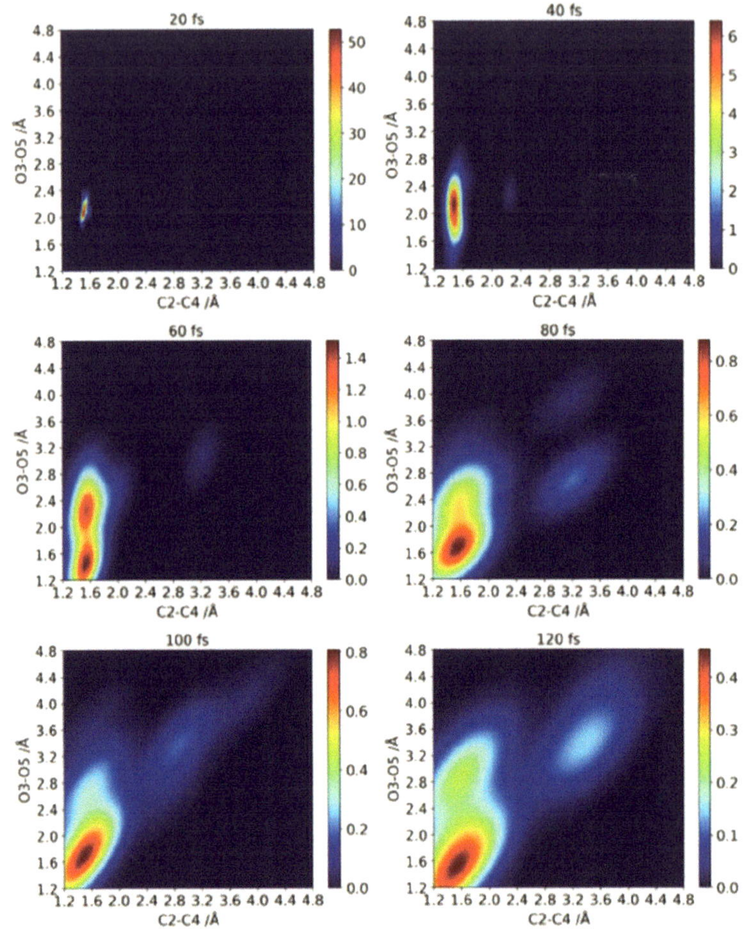

Figure 1.7 Two-dimensional geometric distributions in terms of O3–O5 and C2–C4 bond on S_0.

processes is located locally in the biradical region as shown in Fig. 1.9.

Figures 1.10 and 1.11 show the distribution in the O3–C2–C4–O5 dihedral angle and O3–O5 bond on S_0 and T_1 from 20 fs to 120 fs. As shown in Fig. 1.10, during periodic movements between the stable area and the biradical region of O3–O5 bond, the dihedral angle is twisted from 0 to 30° on the S_0 state.

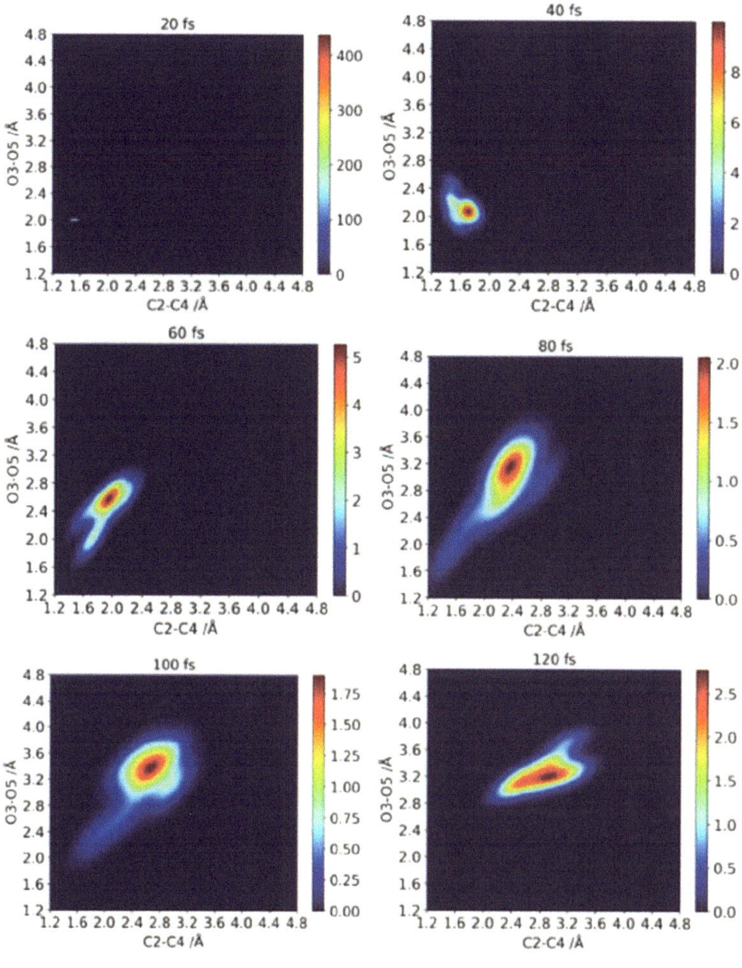

Figure 1.8 Two-dimensional geometric distributions in terms of O3–O5 and C2–C4 bond on T_1.

The normal mode frequency analysis for QM layer of system (**b**) at 50 ns T-REMD simulation's structure of No. 3 has been executed by Molpro 2021.2 under density-fitting second-order Møller–Plesset perturbation theory (MP2) [65] with aug-cc-pVDZ level. The electron embedding has been employed at the calculation. Seven imaginary modes have been obtained. Two imaginary modes

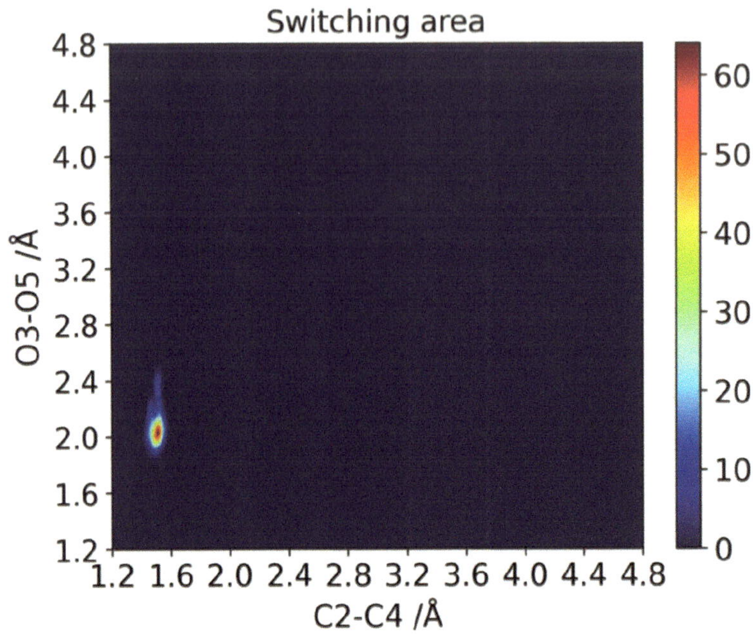

Figure 1.9 Surface switching area in C2–C4 bond and O3–O5 bond directions.

(ν_5 and ν_7) with the direction vector are shown in Fig. 1.12. The ν_5 and ν_7 correspond to the internal rotational mode of O1–C2–O3 and O3–O5 stretching motion, respectively. It has been estimated that the molecule descends along PESs slope from the starting saddle point by the combination of ν_5 and ν_7 modes despite of the initial velocities along Maxwell distribution at 300 K. At the same time, the twisting motion is not strongly correlated to intersystem crossing to the T_1 state as shown in Fig. 1.11.

Two-dimensional geometric distributions for O5-C4-N16 bond angle and O3–O5 bond on the S_0 and T_1 states are shown in Figs. 1.13 and 1.14. In case of the dynamics on S_0, the distribution has been moved along the ν_7 mode. When the O3–O5 bond is stretching, O5-C4-N16 is variated to an acute angle direction. On the other hand, when the molecule is located to a stable state from the biradical region, the bond angle becomes more obtuse. After the dissociation to acridone, the remaining O5 atom on acridone becomes planar for

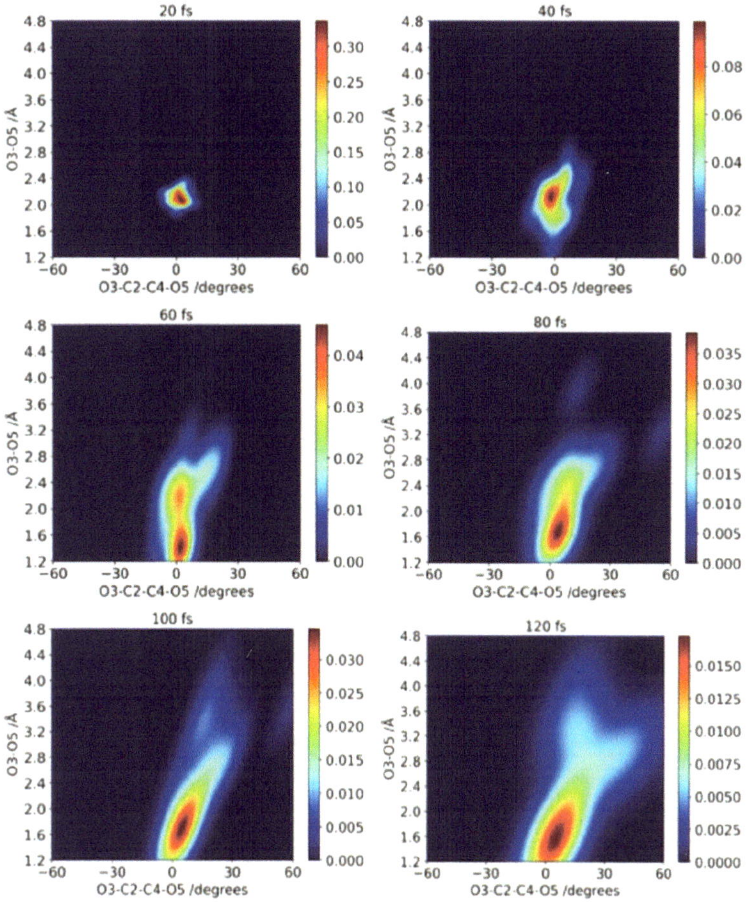

Figure 1.10 Two-dimensional geometric distributions in terms of O3–C2–C4–O5 dihedral angle and O3–O5 bond on S_0.

both of the cases on the S_0 and T_1 states as shown in Figs. 1.13 and 1.14.

1.4 Summary

In order to study the photodissociation mechanisms of dioxetanone in DMSO solvate, isothermal-isobaric, ab initio calculation of

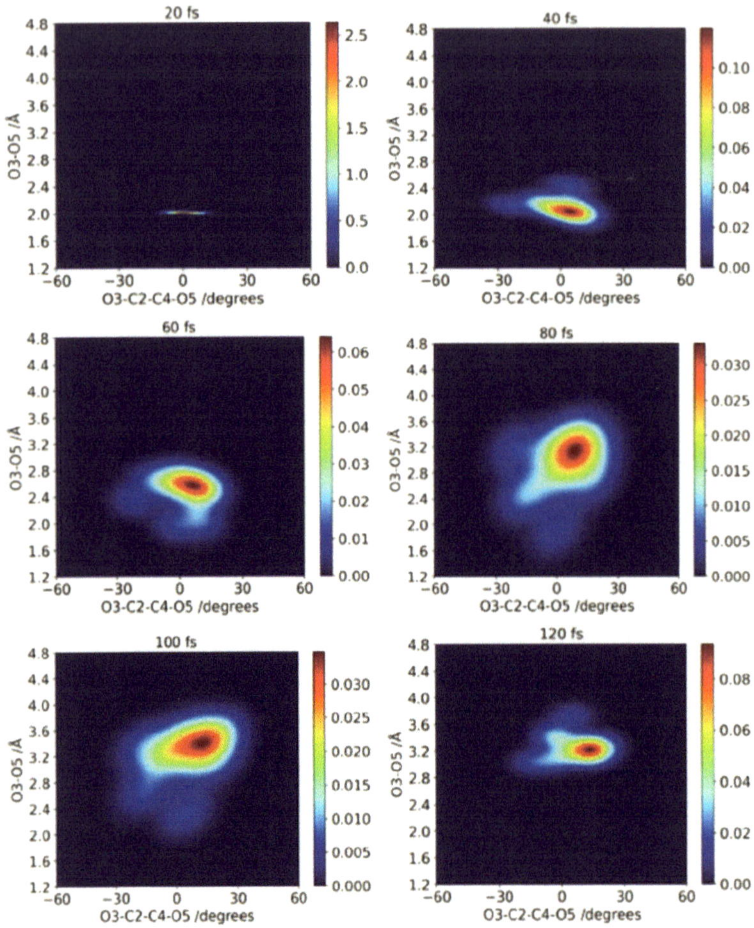

Figure 1.11 Two-dimensional geometric distributions in terms of O3–C2–C4–O5 dihedral angle and O3–O5 bond on T_1.

QM layer and PME-ONIOM molecular dynamics simulations were carried out.

The geometry optimization of the stable state (MIN) and transition state (TS) of dioxetanone, stable state of acridone and CO_2 were carried out by density functional theory (DFT) at the B3LYP/aug-cc-pVDZ level by means of Gaussian 16. The optimized structure of DMSO was also determined by MP2/aug-cc-pVTZ

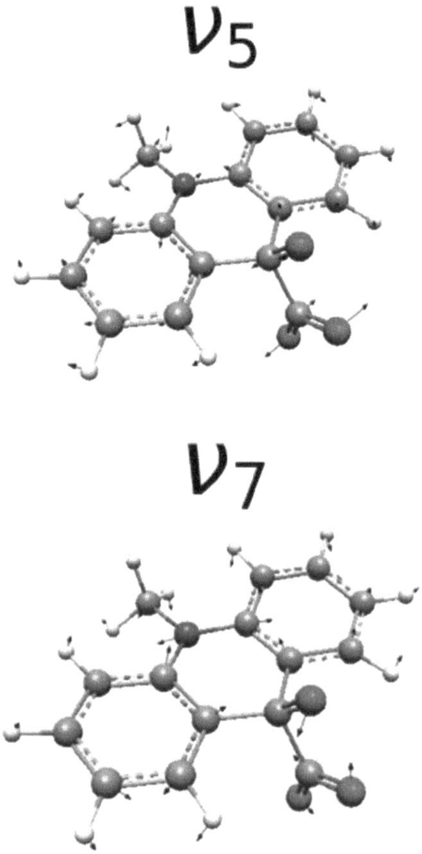

Figure 1.12 Two imaginary vibrational modes (ν_5 and ν_7) of the system (b) at 50 ns T-REMD simulation of the simulation No. 3.

level. The electrostatic potential (ESP) charges for all of them were obtained with the Merz–Singh–Kollman analysis. Systems (**a**) 1 MIN/1 phenol/512 DMSO, (**b**) 1 TS/1 phenol/512 DMSO and (**c**) 1 acridone/1 CO_2/1 phenol/512 DMSO were generated by Packmol. By using the ESP charge and General AMBER Force Field 2 (GAFF2), the topology files were generated by LEaP program of AMBER 18. After converting the topology files to the Gromacs format, the isothermal-isobaric ensembles were employed based on velocity rescaling thermostat and Parrinello–Rahman barostat with eight temperature replica exchange molecular dynamics (T-REMD)

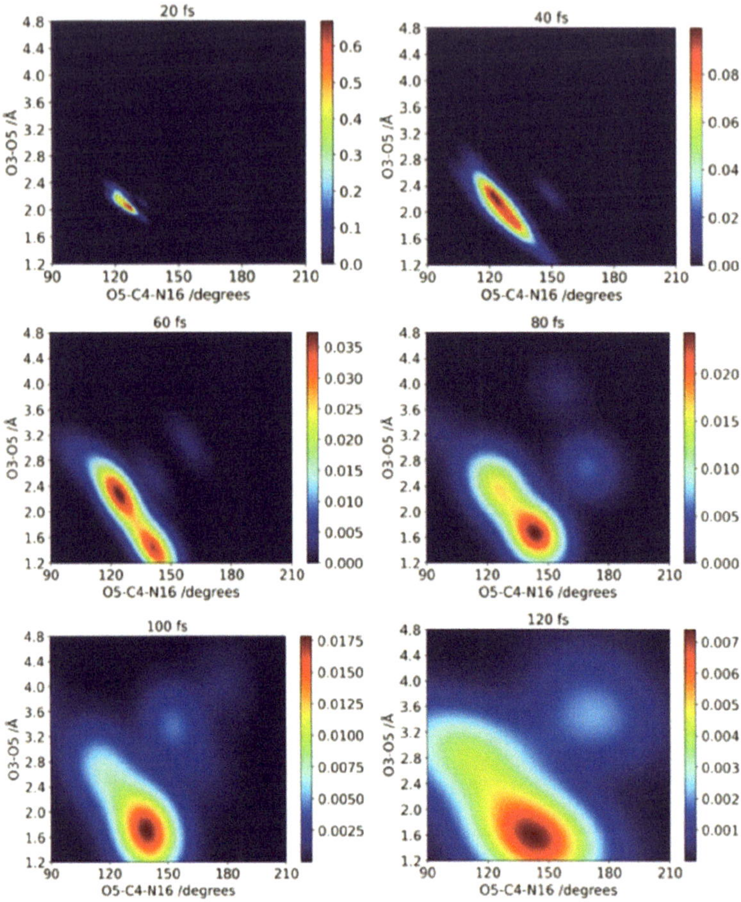

Figure 1.13 Two-dimensional geometric distributions in terms of O5–C4–N16 bond angle and O3–O5 bond on S_0.

simulations under 1 bar pressure condition for 100 ns. At 5 ns, T-REMD reached the convergence.

After the equilibration, ab initio calculations combined with electrostatic embedding (EE) method were employed. The potential energy gaps between the S_0 and T_1 states of DFT/aug-cc-pVDZ level by Molpro 2021.1 and 2021.2 were evaluated appropriately by comparing with an experimental fluorescence peak.

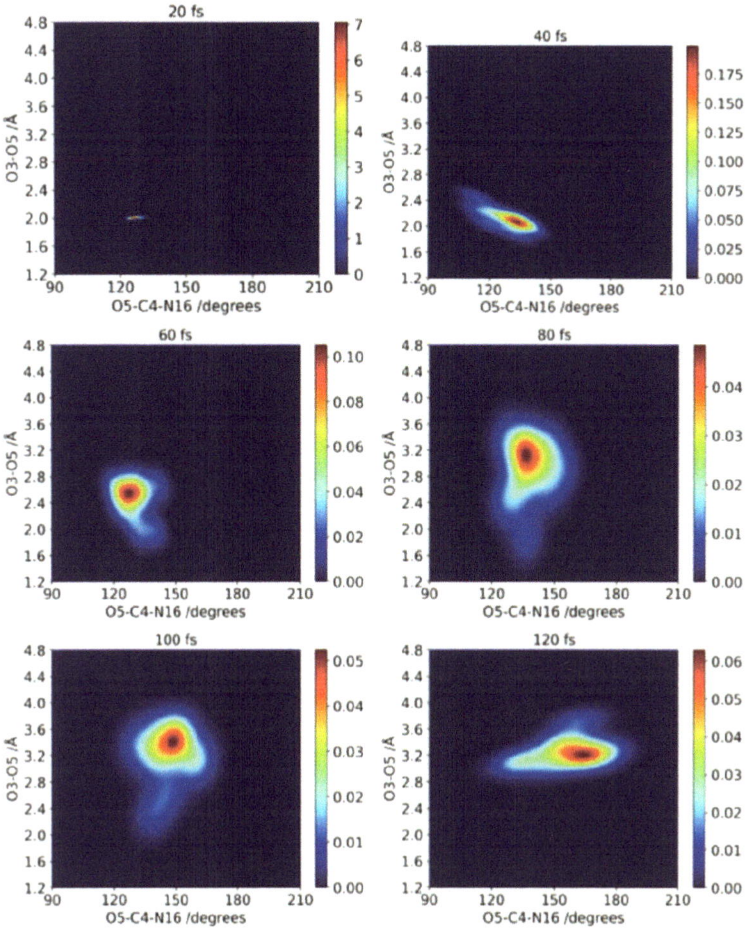

Figure 1.14 Two-dimensional geometric distributions in terms of O5–C4–N16 bond angle and O3–O5 bond on T_1.

On-the-fly molecular dynamics simulations by using the PME-ONIOM method, which can consider long-range Coulomb interaction and short-range VDW interaction, were carried out for system (**b**) equilibrated 1 bar pressure and 301 K at 50 ns T-REMD. The strength of the spin–orbit coupling (SOC) was calculated by the multireference configuration interaction (MRCI) method with the full Breit–Pauli Hamiltonian for the reference wave function of two lowest singlet and two lowest triplet states-averaged multireference self-consistent field program involved the complete active space of 18 electrons and 14 orbitals (4SA-CASSCF(18e,14o)). The SOC values for 19 points at preliminary trajectories were between 44.5 and 62.0 cm^{-1}. The intersystem crossing (ISC) probability was estimated by the Zhu–Nakamura formula with a global switching algorithm.

The dynamics on the S_0 state basically were moved along the O3–O5 bond stretching and O3–C2–C4–O5 twisting motion, which were correlated with two imaginary vibrational modes (ν_5 and ν_7). The intersystem crossing to the T_1 state observed 15 out of 55 trajectories under a constant SOC condition ($V^{SOC} = 50.0$ cm^{-1}). The ISC area was localized around the O3–O5 biradical region. All trajectories on the T_1 state were dissociated to 10-methyl-9(10 H)-acridone. From potential energies by complete active space second-order perturbation theory (CASPT2 (6e,6o)) and the dynamics results of current study, the photo emission of the mixed state (the from T_1 and S_1 states) of 10-methyl-9(10 H)-acridone in DMSO could be expected. Further understanding of the chemiluminescence dissociation reaction mechanisms of dioxetanone of acridinium ester has been provided by pursuing the propagated trajectories expressed by dioxetanone with phenol in DMSO solvate in a periodic boundary condition.

Acknowledgments

This work is supported by the Ministry of Education, Culture, Sports, Science and Technology (MEXT), Japan under grant (No. 17H06105, No. 18H01271, No. 18K05044 and No. 21K07574) of Grant-in-Aid for Scientific Research (S, B and C). The computer simulations in this work were taken with on-house computer system constructed by Concurrent Systems ltd., Japan.

References

1. Martelo, L., Periyasami, G., Fedorov, A. A., Baleizão, C., and Berberan-Santos, M. N. (2019). Chemiluminescence of naphthalene analogues of luminol in solution and micellar media. *Dye. Pigment.* **168**, 341–346.

2. Hananya, N., and Shabat, D. A. (2017). Glowing trajectory between bio- and chemiluminescence: from luciferin-based probes to triggerable dioxetanes. *Angew. Chemie - Int. Ed.* **56**, 16454–16463.
3. Hananya, N., Press, O., Das, A., Scomparin, A., Satchi-Fainaro, R., Sagi, I., and Shabat, D. (2019). Persistent chemiluminescent glow of phenoxy-dioxetane luminophore enables unique CRET-based detection of proteases. *Chem. - A Eur. J.* **25**, 14679–14687.
4. Green, O., Eilon, T., Hananya, N., Gutkin, S., Bauer, C. R., and Shabat, D. (2017). Opening a gateway for chemiluminescence cell imaging: distinctive methodology for design of bright chemiluminescent dioxetane probes. *ACS Cent. Sci.* **3**, 349–358.
5. Nakazono, M., Nanbu, S., Akita, T., and Hamase, K. (2020). Chemiluminescence of methoxycarbonylphenyl 10-methyl-10λ^4-2,7-disubstituted acridine-9-carboxylate derivatives. *J. Photochem. Photobiol. A Chem.* **403**, 112851.
6. White, E. H., Zafiriou, O., Kagi, H. H., and Hill, J. H. M. (1964). Chemilunimescence of luminol: the chemical reaction. *J. Am. Chem. Soc.* **86**, 940–941.
7. White, E. H., McCapra, F., Field, G. F., and McElroy, W. D. (1961). The structure and synthesis of firefly luciferin. *J. Am. Chem. Soc.* **83**, 2402–2403.
8. Seliger, H. H., McElroy, W. D., White, E. H., and Field, G. F. (1961). Stereospecificity and firefly bioluminescence, a comparison of natural and synthetic luciferins. *Proc. Natl. Acad. Sci. U. S. A.* **47**, 1129–1134.
9. Faulkner, K., and Fridovich, I. (1993). Luminol and lucigenin as detectors for $O_2^{\bullet -}$. *Free Radic. Biol. Med.* **15**, 447–451.
10. Zhang, Z., Lin, Y., Li, Z., Dong, G., Gao, Y., Ma, S., Li, J., Du, L., and Li, M. (2021). Bright chemiluminescent dioxetane probes for the detection of gaseous transmitter H_2S. *Bioorganic Med. Chem. Lett.* **46**, 128148.
11. Adam, W., and Baader, W. J. (1985). Effects of methylation on the thermal stability and chemiluminescence properties of 1,2-dioxetanes. *J. Am. Chem. Soc.* **107**, 410–416.
12. Yue, L., Yu, L., Xu, C., Zhu, C., and Liu, Y. (2020). Quantum yields of singlet and triplet chemiexcitation of dimethyl 1,2-dioxetane: ab initio nonadiabatic molecular dynamic simulations. *Phys. Chem. Chem. Phys.* **22**, 11440–11451.
13. De Vico, L., Liu, Y. J., Krogh, J. W., and Lindh, R. (2007). Chemiluminescence of 1,2-dioxetane. Reaction mechanism uncovered. *J. Phys. Chem. A* **111**, 8013–8019.

14. Liu, F., Liu, Y., De Vico, L., and Lindh, R. (2009). Theoretical study of the chemiluminescent decomposition of dioxetanone. *J. Am. Chem. Soc.* **131**, 6181–6188.
15. Navizet, I., Liu, Y. J., Ferré, N., Roca-Sanjuán, D., and Lindh, R. (2011). The chemistry of bioluminescence: an analysis of chemical functionalities. *ChemPhysChem* **12**, 3064–3076.
16. Francés-Monerris, A., Fdez. Galván, I., Lindh, R., and Roca-Sanjuán, D. (2017). Triplet versus singlet chemiexcitation mechanism in dioxetanone: a CASSCF/CASPT2 study. *Theor. Chem. Acc.* **136**, 1–8.
17. Nelson, N. C., Cheikh, A. B., Matsuda, E., and Becker, M. M. (1996). Simultaneous detection of multiple nucleic acid targets in a homogeneous format. *Biochemistry* **35**, 8429–8438.
18. McCapra, F., Richardson, D. G., and Chang, Y. C. (1965). Chemiluminescence involving peroxide decompositions. *Photochem. Photobiol.* **4**, 1111–1121.
19. McCapra, F. (1970). The chemiluminescence of organic compounds. *Pure Appl. Chem.* **24**, 611–629.
20. McCapra, F. (1976). Chemical mechanisms in bioluminescence. *Acc. Chem. Res.* **9**, 201–208.
21. White, E. H., Roswell, D. F., Dupont, A. C., and Wilson, A. A. (1987). Chemiluminescence involving acidic and ambident ion light emitters: the chemiluminescence of the 9-acridinepercarboxylate anion. *J. Am. Chem. Soc.* **109**, 5189–5196.
22. Dodeigne, C., Thunus, L., and Lejeune, R. (2000). Chemiluminescence as a diagnostic tool: a review. *Talanta* **51**, 415–439.
23. Pieńkos, M., Zadykowicz, B. (2020). Computational insights on the mechanism of the chemiluminescence reaction of new group of chemiluminogens-10-methyl-9-thiophenoxycarbonylacridinium cations. *Int. J. Mol. Sci.* **21**, 1–16.
24. Rak, J., Skurski, P., and Błażejowski, J. (1999). Toward an understanding of the chemiluminescence accompanying the reaction of 9-carboxy-10-methylacridinium phenyl ester with hydrogen peroxide. *J. Org. Chem.* **64**, 3002–3008.
25. Krzymiński, K., Ozóg, A., Malecha, P., Roshal, A. D., Wróblewska, A., Zadykowicz, B., and Błażejowski, J. (2011). Chemiluminogenic features of 10-methyl-9-(phenoxycarbonyl)acridinium trifluoromethanesulfonates alkyl substituted at the benzene ring in aqueous media. *J. Org. Chem.* **76**, 1072–1085.

26. Boużyk, A., Jóźwiak, L., Kolendo, A. Y., and Błażejowski, J. (2003). Theoretical interpretation of electronic absorption and emission transitions in 9-acridinones. *Spectrochim. Acta - Part A: Mol. Biomol. Spectrosc.* **59**, 543–558.

27. Nakazono, M., Oshikawa, Y., Nakamura, M., Kubota, H., and Nanbu, S. (2017). Strongly chemiluminescent acridinium esters under neutral conditions: synthesis, properties, determination, and theoretical study. *J. Org. Chem.* **82**, 2450–2461.

28. Nakazono, M., and Nanbu, S. (2018). Enhancement effect on the chemiluminescence of acridinium esters under neutral conditions. *Luminescence* **33**, 345–348.

29. Murakami, T., Nakazono, M., Kondorskiy, A., Ishida, T., and Nanbu, S. (2012). Photochemical dynamics of indolylmaleimide derivatives. *Phys. Chem. Chem. Phys.* **14**, 11546–11555.

30. Murakami, T., Ohta, A., Suzuki, T., Ikeda, K., Danielache, S. O., and Nanbu, S. (2015). Theoretical study of ultraviolet induced photodissociation dynamics of sulfuric acid. *Chem. Phys.* **452**, 17–24.

31. Ohta, A., Kobayashi, O., Danielache, S. O., and Nanbu, S. (2015). Nonadiabatic ab initio molecular dynamics of photoisomerization reaction between 1,3-cyclohexadiene and 1,3,5-cis-hexatriene. *Chem. Phys.* **459**, 45–53.

32. Yu, L., Xu, C., Lei, Y., Zhu, C., and Wen, Z. (2014). Trajectory-based nonadiabatic molecular dynamics without calculating nonadiabatic coupling in the avoided crossing case: trans–cis photoisomerization in azobenzene. *Phys. Chem. Chem. Phys.* **16**, 25883–25895.

33. Zhu, C., and Nakamura, H. (1992). The two-state linear curve crossing problems revisited. II. Analytical approximations for the stokes constant and scattering matrix: the Landau–Zener case. *J. Chem. Phys.* **97**, 8497–8514.

34. Zhu, C., and Nakamura, H. (1993). The two-state linear curve crossing problems revisited. III. Analytical approximations for stokes constant and scattering matrix: nonadiabatic tunneling case. *J. Chem. Phys.* **98**, 6208–6222.

35. Nakamura, H. (2012). *Nonadiabatic Transition: Concepts, Basic Theories and Applications*, 2nd ed, World Scientific Press, Singapore.

36. Dapprich, S., Komáromi, I., Byun, K. S., Morokuma, K., and Frisch, M. J. (1999). A new ONIOM implementation in Gaussian98. Part I. The calculation of energies, gradients, vibrational frequencies and electric field derivatives. *J. Mol. Struct. (Theocehem)* **461–462**, 1–21.

37. Frisch, M. J. *et. al.* (2009). Gaussian09, Gaussian Inc., Wallingford CT.
38. Vreven, T., Byun, K. S., Komáromi, I., Dapprich, S., Montgomery, J. A., Morokuma, K., and Frisch, M. J. (2006). Combining quantum mechanics methods with molecular mechanics methods in ONIOM. *J. Chem. Theory Comput.* **2**, 815–826.
39. Kobayashi, O., and Nanbu, S. (2015). Application of particle-mesh Ewald summation to ONIOM theory. *Chem. Phys.* **461**, 47–57.
40. Ohta, A., Kobayashi, O., Danielache, S. O., and Nanbu, S. (2017). Nonadiabatic ab initio molecular dynamics with PME-ONIOM scheme of photoisomerization reaction between 1,3-cyclohexadiene and 1,3,5-cis-hexatriene in solution phase. *Chem. Phys.* **485–486**, 45–59.
41. Lindahl, E., Abraham, M., Hess, B., and van der Spoel, D. (2021). GROMACS 2020.6.
42. Frisch, M. J. *et. al.* (2016). Gaussian 16, Gaussian Inc., Wallingford CT.
43. Martínez, L., Andrade, R., Birgin, E. G., and Martínez, J. M. (2009). Software news and update packmol: a package for building initial configurations for molecular dynamics simulations. *J. Comput. Chem.* **30**, 2157–2164.
44. Case, D. A. *et. al.* (2018). Amber 2018, University of California, San Francisco.
45. Darden, T., York, D., and Pedersen, L. (1993). Particle mesh Ewald: an $N \cdot \log(N)$ method for Ewald sums in large systems. *J. Chem. Phys.* **98**, 10089–10092.
46. Bussi, G., Donadio, D., and Parrinello, M. (2007). Canonical sampling through velocity rescaling. *J. Chem. Phys.* **126**, 014101.
47. Parrinello, M., and Rahman, A. (1981). Polymorphic transitions in single crystals: a new molecular dynamics method. *J. Appl. Phys.* **52**, 7182–7190.
48. Lockhart, C., and Klimov, D. K. (2012). Molecular interactions of Alzheimer's biomarker FDDNP with Aβ peptide. *Biophys. J.* **103**, 2341–2351.
49. Nayar, D., and Chakravarty, C. (2015). Free energy landscapes of alanine oligopeptides in rigid-body and hybrid water models. *J. Phys. Chem. B* **119**, 11106–11120.
50. Han, M., and Hansmann, U. H. E. (2011). Replica exchange molecular dynamics of the thermodynamics of fibril growth of Alzheimers Aβ_{42} peptide. *J. Chem. Phys.* **135**, 065101.

51. Polly, R., Werner, H. J., Manby, F. R., and Knowles, P. J. (2004). Fast Hartree–Fock theory using local density fitting approximations. *Mol. Phys.* **102**, 2311–2321.
52. Werner, H.-J., and Knowles, P. J. et al. (2021). Molpro, version 2021.1 and 2021.2, a package of ab initio programs, 2021.
53. Knowles, P. J., and Werner, H.-J. (1992). Internally contracted multiconfiguration-reference configuration interaction calculations for excited states. *Theor. Chim. Acta* **84**, 95–103.
54. Merchán, M., Serrano-Andrés, L., Robb, M. A., and Blancafort, L. (2005). Triplet-state formation along the ultrafast decay of excited singlet cytosine. *J. Am. Chem. Soc.* **127**, 1820–1825.
55. Werner, H.-J., and Knowles, P. J. (1985). A second order multiconfiguration SCF procedure with optimum convergence. *J. Chem. Phys.* **82**, 5053–5063.
56. Knowles, P. J., and Werner, H.-J. (1985). An efficient second order MCSCF method for long configuration expansions. *Chem. Phys. Lett.* **115**, 259–267.
57. Zhu, C. (1996). Unified semiclassical theory for the two-state system: analytical solutions for scattering matrices. *J. Chem. Phys.* **105**, 4159–4172.
58. Nikitin, E. E., Umanskii, S. Y. (1984). *Theory of Slow Atomic Collisions*, Springer, Berlin.
59. Tully, J. C. (1971). Trajectory surface hopping approach to nonadiabatic molecular collisions: the reaction of H^+ with D_2. *J. Chem. Phys.* **55**, 562–572.
60. Swope, W. C., Andersen, H. C., Berns, P. H., and Willson, K. R. A. (1982). Computer simulation method for the calculation of equilibrium constants for the formation of physical clusters of molecules: application to small water clusters. *J. Chem. Phys.* **76**, 637–649.
61. Xu, C., Yu, L., Zhu, C., Yu, J., and Cao, Z. (2016). Intersystem crossing-branched excited-state intramolecular proton transfer for o-nitrophenol: an ab initio on-the-fly nonadiabatic molecular dynamic simulation. *Sci. Rep.* **6**, 1–12.
62. Xu, C., Gu, F. L., Zhu, C. (2018). Ultrafast intersystem crossing for nitrophenols: ab initio nonadiabatic molecular dynamics simulation. *Phys. Chem. Chem. Phys.* **20**(8), 5606–5616.
63. Liu, J., and Miller, W. H. (2009). A simple model for the treatment of imaginary frequencies in chemical reaction rates and molecular liquids. *J. Chem. Phys.* **131**, 074113.

64. Győrffy, W., Shiozaki, T., Knizia, G., Werner, H. J. (2013). Analytical energy gradients for second-order multireference perturbation theory using density fitting. *J. Chem. Phys.* **138**, 53–60.

65. Werner, H. J., Manby, F. R., Knowles, P. J. (2003). Fast linear scaling second-order Møller-Plesset perturbation theory (MP2) using local and density fitting approximations. *J. Chem. Phys.* **118**, S8149–S8160.

Chapter 2

On-the-Fly Excited-State Molecular Dynamics Study Based on Spin-Flip Time-Dependent Density Functional Theory Approach: Photo-Branching Reaction of Stilbene and Stilbene Derivatives

Tetsuya Taketsugu,[a,b] Takuro Tsutsumi,[a] Yu Harabuchi,[a,b] and Takao Tsuneda[a,c]

[a] *Department of Chemistry, Faculty of Science, Hokkaido University, Sapporo 060-0810, Japan*
[b] *Institute for Chemical Reaction Design and Discovery, Hokkaido University, Sapporo 001-0021, Japan*
[c] *Graduate School of Science, Technology, and Innovation, Kobe University, Kobe 657-8501, Japan*
take@sci.hokudai.ac.jp

Excited-state branching reactions of photo-excited *cis*-stilbene (*cis*-SB), α-methyl-*cis*- stilbene (*cis*-mSB), and 1,1′-dimethyl-*cis*-stilbene (*cis*-dmSB) were examined by on-the-fly excited-state molecular dynamics simulations based on the spin-flip time-dependent density functional theory (SF-TDDFT). The basic concept of TDDFT and SF-

Time-Dependent Density Functional Theory: Nonadiabatic Molecular Dynamics
Edited by Chaoyuan Zhu

ISBN 978-981-4968-42-3 (Hardcover), 978-1-003-31921-4 (eBook)
www.jennystanford.com

TDDFT was described briefly, and then the applications to stilbene and stilbene derivatives were reported. The branching mechanism into the *trans*-form and 4a,4b-dihydrophenanthrene (DHP) form was elucidated by the static reaction pathways and dynamical trajectories on the $S_1(\pi\pi^*)$ potential energy surface. The substituent effects on excited-state dynamics were also discussed.

2.1 Introduction

In the past 60 years, laser techniques and spectroscopic experiments have revealed many aspects of the mechanisms and dynamics of photoreactions. In particular, transient absorption and emission spectra have clarified details of the dynamics of the excited molecules and have improved our understanding of the elementary reaction processes in the electronically excited state.

Nowadays, it is recognized that the overall picture of photoreactions cannot be drawn from experiments alone, and quantum chemical computations play an important role in elucidating the mechanisms of photoreactions [1–3]. Using ab initio wavefunction theory or density functional theory (DFT), we can characterize the vertical excitation energy and also explore the excited-state potential energy surfaces (PESs) to reveal the reaction pathway going through multiple excited states.

On-the-fly molecular dynamics (MD) is a classical trajectory method where atomic positions and velocities are propagated by solving Newton's equations of motion using energy gradients from quantum chemical computations [4–9]. The target of the on-the-fly MD approach has been extended to photochemical reactions with nonradiative transitions [10–12]. We have developed an on-the-fly MD code for excited-state dynamics for 20 years and have applied it to photochemical reactions in gas-phase [13–15] and in solution [16–18], dissociative recombination reactions [19–22], and photodissociation considering both nonadiabatic coupling and spin-orbit coupling terms [23], with the surface hopping scheme using Tully's surface hopping scheme [24]. In the Tully method, in addition to the time-evolution of nuclear degrees of freedom, the electronic wavefunction also evolves according to the time-dependent

Schrödinger equation using nonadiabatic coupling terms; the hopping probability is evaluated from the electronic wavefunctions and nonadiabatic coupling terms, which invokes the surface hopping by comparing with a uniform random number. In another approach, a global switching probability evaluated from Zhu–Nakamura formulae for nonadiabatic transition [25–27] has been used in a scheme of on-the-fly excited-state MD simulation [28].

In excited-state dynamics simulations, an electronic structure method should be chosen much more carefully than in ground-state ones because the quality of excited-state PESs sensitively depends on the level of the method. Actually, the electronic structure method is known to require well-balanced nondynamical and dynamical correlation effects to give quantitative excited-state PESs. In ab initio wavefunction theory calculations, multi-configuration theories have been employed to describe the nonadiabatic regions such as conical intersection (CI) regions of excited-state PESs [29–34]. Based on this type of theory, an automated search method for locating minimum energy conical intersections in multiple PESs has recently been developed [35–37]. A representative multiconfigurational approach is a state-averaged complete active space self-consistent field (SA-CASSCF) method, which has been reported to qualitatively describe the PESs of ground and excited electronic states. In the SA-CASSCF method, however, dynamic correlation effects are insufficient to quantitatively evaluate the energetics. Multireference (MR) configurational interaction or MR perturbation theories [38, 39] contain both nondynamical and dynamical correlation effects in a balanced manner, and, therefore, they can quantitatively provide the excited-state PESs. Since the MR calculations, however, require very high computational costs with the order of $O(N^6)$, where N is the number of electrons, they cannot be employed for the excited-state dynamics simulations of large systems. On the other hand, the accurate results of the MR calculations have revealed that SA-CASSCF cannot give even a qualitative picture of excited-state dynamics [40–42]. The trade-off between the accuracy and computational cost has been the most critical issue in excited-state dynamics simulations of ab initio wavefunction theories.

Time-dependent DFT (TDDFT) has been gradually accepted as an alternative method for performing excited-state dynamics simulations. In TDDFT, both nondynamical and dynamical correlation effects are included in the exchange-correlation functional of DFT. This makes it feasible to calculate excitation energies with low computational cost. In the application to excited-state dynamics, TDDFT, however, has two limitations: (1) TDDFT provides a discontinuous PES at the crossing point between the ground singlet state (S_0) and the first excited singlet state (S_1) because these states alternate with each other after passing through the crossing point; (2) Since TDDFT is a single-configuration theory, it includes no double-excitation electronic configuration effects. Spin-flip (SF)-TDDFT overcomes these limitations. SF-TDDFT is a sophisticated method for complementing double-excitation configurations in TDDFT by considering triplet (spin-flip) excitations from an open-shell triplet configuration [43–46]. As mentioned later, SF-TDDFT has been reported to improve TDDFT excitation energies when appropriate functionals being used. Despite its high availability, it is, so far, not usual to use SF-TDDFT in excited-state dynamics simulations probably due to the unpublicized situation.

In this chapter, we first introduce SF-TDDFT approach, and then review some applications of on-the-fly excited-state MD simulations based on SF-TDDFT calculations for photoisomerization of $\pi\pi^*$-excited *cis*-stilbene (*cis*-SB) [47, 48], 1,1′-dimethyl-*cis*-stilbene (*cis*-dmSB) [48, 49], and α-methyl-*cis*-stilbene (*cis*-mSB) [50]. Stilbene is one of the basic typical molecules in photo-chemistry, showing *cis*-*trans* photo-isomerization. The calculated results are compared with available experimental data, and the reaction pathways and reaction dynamics are discussed for the respective photo-isomerization.

2.2 Spin-Flip Time-Dependent Density Functional Theory Approach for Excited-State Dynamics Simulation

TDDFT in the linear-response formulation [51] is the most widely used theory for excited-state calculations of molecules with more than 60% of total calculations as of 2021, though it had been hardly used in the 20th century. Ab initio MR theories such as

MR configuration interaction and MR perturbation theories had occupied the excited-state calculations in the latest 20th and earliest 21st centuries because the high-level electron correlation of MR theories is essentially needed to investigate the excited states of molecules. Since excited states generally have very different electronic states from the ground state, they essentially require highly correlated theories involving many-body electron correlations. By definition, electron correlation is the difference between the exact and Hartree–Fock (HF) energies [52]. Electron correlation mainly consists of two types [53]: the first is "dynamical correlation" emerging from the correlation hole (correlation cusp [54]) of wavefunctions, which is generated by the repulsion of short-range opposite-spin electron pairs, and the second is "nondynamical correlation" attributed to kinetic energy loss by mixing near-degenerate electronic states. Quantitative discussions of excited states need to include both these two electron correlations in a balanced manner. Since MR theories can incorporate this well-balanced electron correlation, they usually provide excitation energies that quantitatively reproduce experimental ones. However, MR theories have some serious problems: high computational order (more than $O(N^5)$ for the number of electrons N), immense amounts of computational time, and poor convergence and arbitrariness in selecting electron configurations in CASSCF calculations. The high computational order, in particular, is the most critical problem in the calculations of large molecules. Linear-response TDDFT, therefore, has become the main theory in the excited-state calculations of large systems.

Due to its low computational order ($O(N^3)$ or less), TDDFT is accepted as the best alternative tool for investigating the excited states of large systems. So far this century, TDDFT has been applied to the studies on the photochemical reactions of large systems: e.g., the mechanism elucidation of photochemical reactions [55, 56]. For the excited-state calculations, TDDFT usually indicates the time-dependent Kohn–Sham (TDKS) theory, which is given in the following formulation based on the linear response of density [51]:

$$\omega\begin{pmatrix}\mathbf{X}\\-\mathbf{Y}\end{pmatrix}=\begin{pmatrix}\mathbf{A} & \mathbf{B}\\ \mathbf{B}^* & \mathbf{A}^*\end{pmatrix}\begin{pmatrix}\mathbf{X}\\ \mathbf{Y}\end{pmatrix}, \qquad (2.1)$$

where **A** and **B** matrices have the elements of

$$A_{ia\sigma,jb\tau} = \delta_{\sigma\tau}\delta_{ij}\delta_{ab}\left(\epsilon_{a\sigma} - \epsilon_{i\sigma}\right) + K_{ia\sigma,jb\tau}, \tag{2.2}$$

where $\epsilon_{i\sigma}$ is the i-th σ-spin orbital energy, and

$$B_{ia\sigma,jb\tau} = K_{ia\sigma,jb\tau}, \tag{2.3}$$

respectively, where $K_{ia\sigma,jb\tau}$ is provided as

$$K_{ia\sigma,jb\tau} = (ia\sigma \mid jb\tau) + \iint d^3\mathbf{r}_1 d^3\mathbf{r}_2 \varphi^*_{i\sigma}(\mathbf{r}_1)\,\varphi_{a\sigma}(\mathbf{r}_1)\, f_{xc}(\mathbf{r}_1, \mathbf{r}_2)\,\varphi_{j\tau}(\mathbf{r}_2)\,\varphi^*_{b\tau}(\mathbf{r}_2). \tag{2.4}$$

Note that the spins of the orbitals (σ, τ) are explicitly given in Eqs. (2.2) and (2.3) for purposes of accuracy. In Eq. (2.4), the first term of the right-hand side, which is called the Hartree integral, is provided as:

$$(ia\sigma \mid jb\tau) = \iint d^3\mathbf{r}_1 d^3\mathbf{r}_2 \varphi^*_{i\sigma}(\mathbf{r}_1)\,\varphi_{a\sigma}(\mathbf{r}_1)\, r_{12}^{-1}\varphi_{j\tau}(\mathbf{r}_2)\,\varphi^*_{b\tau}(\mathbf{r}_2). \tag{2.5}$$

The exchange-correlation integral kernel, f_{xc}, in Eq. (2.4) is assumed to have the local form,

$$f_{xc}(\mathbf{r}_1, \mathbf{r}_2) = \frac{\delta^2 E_{xc}}{\delta\rho^2(\mathbf{r}_1)}\delta(\mathbf{r}_1 - \mathbf{r}_2). \tag{2.6}$$

However, the TDKS method has a clear limit to the accuracy in excited-state calculations because it incorporates only single excitations [57]. Note that MR studies on excited states have revealed that double-excitation configurations are frequently mixed in highly excited states and appear even around equilibrium geometries in, e.g., the ground states of ozone molecules [58] and the lowest excited state of benzene molecules [59]. Photochemical reaction simulations including a wide range of geometries, therefore, demand the theory to incorporate double-excitation effects. Since MR theories including double-excitation effects need too much computational time to calculate the excited states of large systems, as mentioned above, the target systems of photochemical reaction simulations had been limited for many years.

SF-TDDFT recently attracts attention as a method for overcoming this TDKS problem by easily incorporating double-excitation effects in the TDKS method [45, 60]. The SF-TDDFT generates singlet ground and excited states incorporating double-excitation configurations by the SF excitations (i.e., excitations accompanying one α to β-spin alternation) of a triplet reference configuration. Consequently, this method only requires the supplement of the exchange term into the TDKS matrix through the exchange-correlation integral kernel,

$$K^{\mathrm{SF}}_{ia\sigma,\,jb\tau} = \iint d^3\mathbf{r}_1 d^3\mathbf{r}_2 \varphi^*_{i\sigma}(\mathbf{r}_1)\,\varphi_{a\sigma}(\mathbf{r}_1)\, f_{xc}(\mathbf{r}_1, \mathbf{r}_2)\,\varphi_{j\tau}(\mathbf{r}_2)\,\varphi^*_{b\tau}(\mathbf{r}_2)\,. \tag{2.7}$$

This is called the SF-TDKS method. Note that this $\boldsymbol{K}^{\mathrm{SF}}$ term only provides a nonzero value for the HF part of the exchange-integral kernel. Therefore, the SF-TDKS method using GGA functionals with no HF exchange integral yields the same results as the usual TDKS method in its form, though these results are generally different due to the discrepancy in the orbitals of the reference configurations. In contrast, hybrid and long-range corrected GGA functionals provide clear differences between TDKS and SF-TDKS excitation energies because they contain HF exchange integral parts. In the latter case, the SF-TDKS method can be a powerful tool for quantitatively investigating excited-state simulations, in which double-excitation effects generally play a significant role.

However, in the SF-TDKS methods, there are missing electronic configurations, which are not generated from the spin-flip (α spin $\rightarrow$ β spin) transition. Figure 2.1 shows five types of electronic configurations related to SF-TDKS calculations [43], i.e., the reference triplet state, the singlet ground state, the singlet excited state of HOMO-LUMO double excitation, the open-shell singlet and triplet states of HOMO-LUMO single excitation, and spin-mixed states of singlet, triplet, and quintet. As shown in this figure, spin-mixed states appear due to missing configurations denoted by blue slashes, which cause the spin-contamination in spin-flip excited states. To avoid the spin-contamination, the spin-symmetric configurations should be generated by using a spin-symmetric triplet reference configuration or by supplementing spin-complement configurations [61, 62].

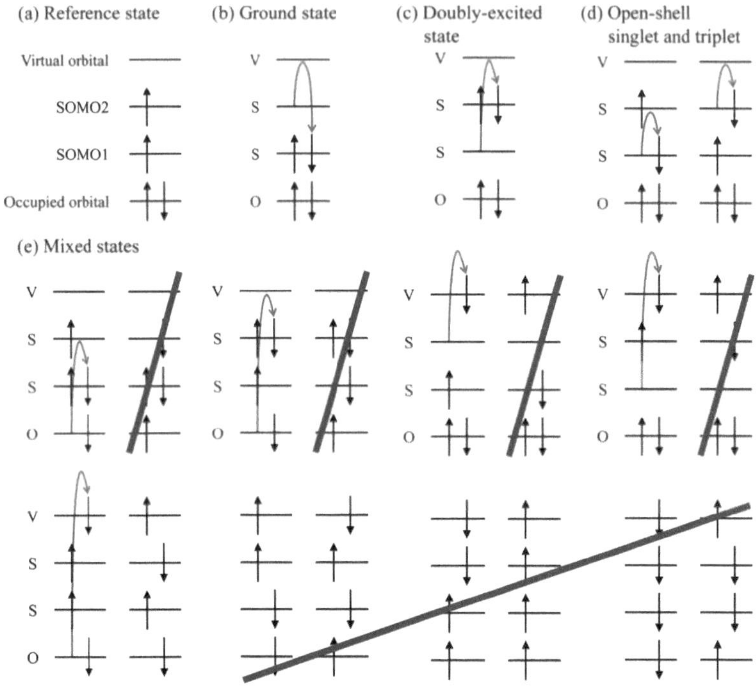

Figure 2.1 (a) The reference triplet state and (b–e) the electronic states generated by the one-electron spin-flip transition in SF-TDKS calculations, with their missing configurations depicted by blue slashes: (b) the singlet ground state, (c) the doubly excited state, (d) the open-shell singlet and triplet states, and (e) the mixed states of singlet, triplet, and quintet.

In the SF-TDDFT method, the expectation values of the squared total spin operator, $\langle S^2 \rangle$, for the singlet, triplet, and mixed states are normally 0, 2, and 1, respectively [61]. However, the previous study showed that, when the energy levels of some of these singlet, triplet, and mixed states get close to each other, their $\langle S^2 \rangle$ values are affected due to spin-contamination [47], and it becomes difficult to distinguish the respective states by $\langle S^2 \rangle$. Likewise, it is not easy to distinguish the target state by the energies and gradient vector due to the nature of the avoided crossing. In the on-the-fly excited state MD simulation on SB [47], a state tracking method was introduced newly to distinguish the target state by monitoring orbital coefficients and configuration interaction coefficients along the trajectory. After passing through a region of energetic proximity, if a different state was tracked, one needs to investigate the dynamics by rewinding the simulation time and reselecting the state. This procedure is somewhat cumbersome, and so, the T^{SF}-index method was proposed to extract only the singlet states by eliminating the triplet state [37]. In the on-the-fly excited-state MD simulation based on SF-TDDFT computations for dmSB [49] and mSB [50], this T^{SF}-index method was utilized.

Another problem, which has been reported as the most serious, in actual SF-TDKS calculations is that this method gives very poor results even for valence excitation energies when using most hybrid GGA functionals like B3LYP. The best ever hybrid functional suited to the SF-TDKS method is BHHLYP functional. In this functional, the Becke 1988 exchange functional [63] is combined with the HF exchange integral in a one-to-one ratio for the exchange part, has been used to avoid the underestimation of SF-TDKS excitation energies. Note, however, that even when using the BHHLYP functional, the calculated SF-TDKS excitation energies are considerably higher than the TDKS ones for excitations with negligible double excitations. It has very recently been reported that the long-range correction provides accurate SF-TDKS excitation energies with keeping the accuracy for single excitations with no significant double excitations. The SF-TDKS method using LC-BLYP functional gives very accurate excitation energies comparable to ab initio MR results for long-chain systems such as polyenes and oligoacenes [64] and reproduces TDKS results for nanographenes, in which double-excitation effects are essentially not included [65]. Since BHHLYP and LC-BLYP functionals are known to provide similar exchange energies for small molecules, this indicates that exchange interactions should be included in a balanced manner to give accurate SF-TDKS excitation energies.

Taking account of these problems, we can make the SF-TDKS method the best ever tool for performing photochemical reactions. To show the availability of the SF-TDKS method in photochemical reaction simulations, we introduce our studies on the photoisomerization of *cis*-stilbene and stilbene derivatives involving the potential energy surfaces and the ab initio excited-state molecular dynamics simulations in the following sections.

2.3 Applications to Photoreaction of *cis*-SB, *cis*-DMSB, and *cis*-MSB in $\pi\pi^*$ Excitation

2.3.1 Photoreaction of Stilbene

Stilbene (SB) is a basic molecule showing *cis-trans* photoisomerization. The photoisomerization mechanism of SB has been studied experimentally by many researchers [66–82]. Upon $\pi\pi^*$ excitation, *cis*-SB can transform to *trans*-SB within 1.0 ps, indicating the significance of the nonadiabatic transition from S_1 to S_0 state through a conical intersection (CI) region. Steady-state [72] and femtosecond time-resolved [73] fluorescence studies revealed two decay components with lifetimes of 0.23 and 1.2 ps. The initial dynamics of $\pi\pi^*$-excited *cis*-SB was investigated by femtosecond Raman spectroscopy, elucidating the origin of a temporal change of the vibrational spectrum as the twisting motion of the central C=C bond [79, 81]. In the photoreaction of *cis*-SB, in addition to *trans*-SB, 4,4-dihydrophenanthrene (DHP) is observed as a sub-product of the reaction. The branching ratio was reported as *cis*:*trans*:DHP = 55:35:10 [71, 75, 82]. Figure 2.2 shows a schematic picture of *cis*-form, *trans*-form, and DHP-form of SB and SB derivatives ($\mathbf{R}^1$ and $\mathbf{R}^2$ are substituents). In this chapter, we introduce application dynamics studies for normal SB ($\mathbf{R}^1 = \mathbf{R}^2 =$ H), dmSB ($\mathbf{R}^1 = \mathbf{R}^2 = CH_3$), and mSB ($\mathbf{R}^1 =$ H and $\mathbf{R}^2 = CH_3$). Interestingly, a recent combined experimental and theoretical study elucidated that tetraphenylethene ($\mathbf{R}^1 = \mathbf{R}^2 = C_6H_5$), known as aggregation-induced emission (AIE) molecule, shows a structural transformation by photoexcitation like SB in isolated conditions, which is important to understand the AIE property [83].

Many theoretical studies have also been reported to elucidate photo-isomerization mechanism of SB [66, 79, 84–94]. Theoretical studies indicated that $\pi\pi^*$-excited *cis*-SB evolves on the $S_1(\pi\pi^*)$ PES with the twisting motion of the central C=C bond [87, 90, 93, 94]. An accessible minimum energy conical intersection (MECI) point of S_0 and S_1 states (denoted as S_1/S_0-MECI) was located near the minimum on the S_1-PES, which corresponds to the C=C bond twisting structure (denoted as *twist*) [90, 92–94]. The *cis-trans* photoisomerization occurs through a "*hula-twist*" motion where the

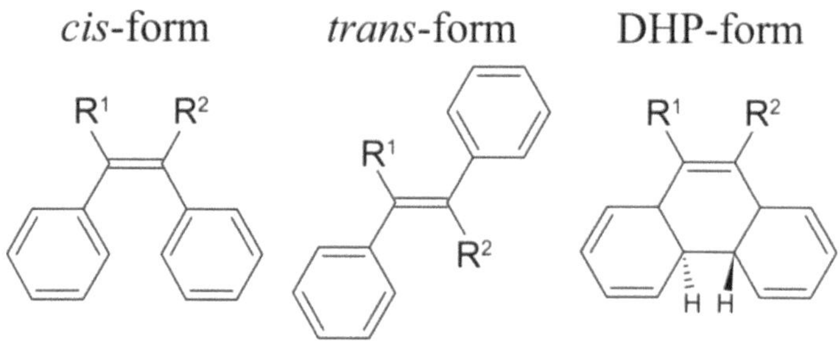

Figure 2.2 A schematic picture of *cis*-form, *trans*-form, and DHP-form of SB and SB derivatives (**R**^{1} and **R**^{2} are substituents).

C–H group of the central C=C bond rotates out of the plane while the other atoms reorient to keep coplanar [67]. Minezawa and Gordon employed the SF-TDDFT method to investigate reaction pathways in the relaxation process of $\pi\pi^*$-excited *cis*-SB [90]. Their study found geometries of minima and S_1/S_0-MECIs for *twist* and DHP regions on the S_1-PES, showing that the photocyclization to DHP-form is in competition with the photoisomerization to *trans*-form, which was later confirmed by the multireference method [89].

The $\pi\pi^*$-excitation energies of *cis*-SB significantly depend on the level of electron correlation included. Table 2.1 compares the calculated vertical excitation energies for *cis*-SB in the $S_1(\pi\pi^*)$ state with the experimental one [47]. The SA-CASSCF method overestimates the excitation energy largely probably due to the insufficient dynamic correlation, and then the MR perturbation theory significantly improves the energy. Note that the SA-CASSCF with the larger active space such as (14,14) does not improve the excitation energy. The table shows that SF-TDDFT using BHHLYP functional gives the most accurate excitation energy as usual. Though the TDDFT with PBE0 functional also gives reasonable excitation energy, it is not applicable to the nonadiabatic region, as described above. These results indicate that high-level electron correlation is needed to produce accurate excitation energies of this system.

For 1,1′-dimethylstilbene (dmSB), Berndt et al. reported transient absorption spectra in solution [95]. They noted a very interesting feature of the ultralong lifetime for the perpendicular

Table 2.1 Experimental and calculated vertical excitation energies of *cis*-SB in the $S_1(\pi\pi^*)$ state

Level	ΔE/eV
Experiment	4.6[a]
CASSCF(2,2)/6-31G(d,p)	6.07[b]
CASPT2(2,2)/cc-pVDZ	4.23[c]
XMCQDPT2(14,14)/cc-pVDZ	4.43[d]
TDDFT(PBE0)/6-311+G(2d,2p)	4.09[e]
SF-TDDFT(BHHLYP)/DH(d,p)	4.78[f]
SF-TDDFT(B3LYP)/6-31G(d)	3.86[c]
SF-TDDFT(BHHLYP)/6-31G(d)	4.96[c]
SF-TDDFT(PBE0)/6-31G(d)	4.02[c]
SF-TDDFT(BLYP)/6-31G(d)	6.64[c]
SF-TDDFT(BOP)/6-31G(d)	6.66[c]
SF-TDDFT(PBE)/6-31G(d)	6.66[c]

[a]Ref. [73]; [b]Ref. [92]; [c]Ref. [47]; [d]Ref. [89]; [e]Ref. [88]; [f]Ref. [90].

conformation (denoted as P*) in the $\pi\pi^*$ excited state. The lifetime of a transient band observed around 330 nm (P* band) is 19 ps in hexane and 2.9 ps in acetonitrile, which are much longer than the lifetime of the corresponding P* band of *cis*-SB [69]. Another transient band, associated with the *cis*-form (denoted as CIS*), was also observed around 600 nm (CIS* band), with a lifetime of 0.2 ps, which is shorter than the corresponding CIS* band lifetime of *cis*-SB (0.92 ps). The unusual longer lifetime of P* in dmSB could be attributed to the large difference in geometry between the minimum and the conical intersection around P*, but there had been no report which examines the relaxation process of dmSB in the S_1 state.

α-methyl-*cis*-stilbene (*cis*-mSB) should have a character between those of *cis*-SB and *cis*-dmSB because of the asymmetric substituents for $\mathbf{R}^1$ and $\mathbf{R}^2$. However, there have been only a few experimental reports on this molecule [96, 97], and it is interesting to examine the reaction pathways and on-the-fly trajectories in the $S_1(\pi\pi^*)$ state. The comparison of the reaction mechanism and dynamics with those of *cis*-SB and *cis*-dmSB should provide significant physical insight into the photoreaction processes.

In the following, we introduce theoretical studies to elucidate the relaxation mechanism and dynamics of $\pi\pi^*$ excited *cis*-SB, *cis*-

dmSB, and *cis*-mSB for the *cis-trans* photoisomerization and the photocyclization to DHP-form based on the SF-TDDFT computations [47–50]. The substituent effect for the relaxation mechanism is discussed by comparing geometries, reaction pathways, and on-the-fly trajectories on the excited state PES for SB, dmSB, and mSB. Finally, the origin of the unusual lifetime of P* in dmSB is discussed.

2.3.2 Geometries and Reaction Pathways on the $\pi\pi^*$ Excited State of SB, dmSB, and mSB

In our previous theoretical studies on SB [47], dmSB [49], and mSB [50], we chose SF-TDDFT(BHHLYP)/6-31G(d) level of theory to investigate the reaction mechanism and the dynamics of the $\pi\pi^*$-excited *cis*-forms. Minima and TS geometries in the S_0 and $S_1(\pi\pi^*)$ states, as well as S_1/S_0-MECI geometries, were optimized, followed by the normal-mode analyses to verify geometrical characters (minimum or saddle point) for the respective stationary points. The steepest descent pathways, which are defined in mass-weighted Cartesian coordinate, from the Franck-Condon (FC) structure of *cis*-forms in the $\pi\pi^*$ state (meta intrinsic reaction coordinate: meta-IRC) and the IRC pathways connecting the *cis* (or DHP) minimum and the *twist* minimum on the S_1-PES were also determined. Geometry optimizations and IRC calculations were performed by employing the GRRM14 program [98] with GAMESS [99].

Geometries of the *cis* minimum in the S_0 state (denoted by $(S_0)_{cis-\mathrm{min}}$), several minima in the $S_1(\pi\pi^*)$ state, and S_1/S_0-MECIs are shown for (a) SB, (b) dmSB, and (c) mSB in Fig. 2.3. In the S_1 state, dmSB has *cis*, DHP, and *twist* minima (denoted by $(S_1)_{cis-\mathrm{min}}$, $(S_1)_{\mathrm{DHP-min}}$, and $(S_1)_{twist-\mathrm{min}}$), but there is no *cis* minimum for SB and mSB. Note that the existence of $(S_1)_{cis-\mathrm{min}}$ is very sensitive to the computational level because $(S_1)_{cis-\mathrm{min}}$ is located in a very flat region on the S_1-PES. As for DHP-form, there is one minimum on the S_1-PES in all cases. In the *twist* region, SB and dmSB have only one minimum, whereas mSB has two different minima. For MECIs, SB and dmSB have one MECI in the DHP region ($(S_1/S_0)_{\mathrm{DHP}}$) and two in the *twist* region ($(S_1/S_0)_{twist1}$ and $(S_1/S_0)_{twist2}$); however, the number of MECIs for mSB is twice of those for SB and dmSB: $(S_1/S_0)_{\mathrm{DHP1}}$, $(S_1/S_0)_{\mathrm{DHP2}}$, $(S_1/S_0)_{twist1}$, $(S_1/S_0)_{twist2}$, $(S_1/S_0)_{twist3}$, and

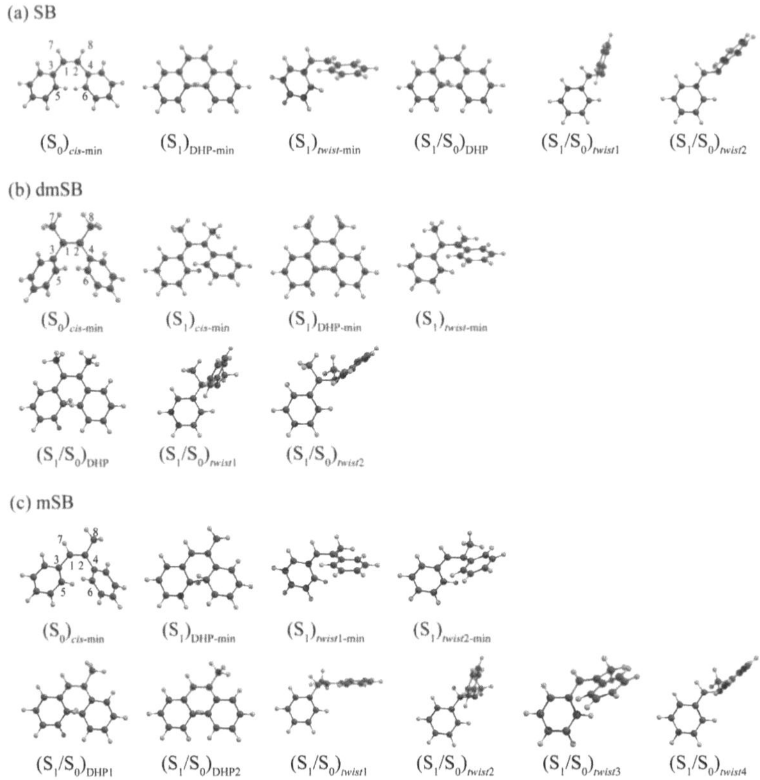

Figure 2.3 Geometries of minima in the S_0 and S_1 states and S_1/S_0-MECI for (a) SB, (b) dmSB, and (c) mSB. Definitions of atom numbering are given in $(S_0)_{cis-\text{min}}$.

$(S_1/S_0)_{twist4}$. These tendencies reflect the asymmetric substituents of $\mathbf{R}_1$ (H) and $\mathbf{R}_2$ (CH_3) for mSB. Focusing on the molecular geometries, in SB, $(S_1)_{twist-\text{min}}$, $(S_1/S_0)_{twist1}$, and $(S_1/S_0)_{twist2}$ are all characterized as a *twisted pyramidal* structure. On the other hand, in dmSB, $(S_1)_{twist-\text{min}}$ is intermediate between pyramidal and planar (the dihedral angle$d_{\text{C3C1C2C7}} = 153.3°$), while $(S_1/S_0)_{twist1}$ and $(S_1/S_0)_{twist2}$ are pyramidal ($d_{\text{C3C1C2C7}} = 114.0°$ and $116.3°$, respectively) because of the steric hindrance between phenyl and methyl groups. This difference should be related to the unusual long lifetime of P* in the *twist* region of dmSB.

Figure 2.4 shows significant geometrical structures (minimum, TS, and MECI) and reaction pathways (IRC and meta-IRC) projected on a two-dimensional configurational space spanned by r_{C5C6} (which characterizes the transformation to the DHP) and d_{C3C1C2C4} (which characterizes the transformation to the *twist* form), for (a) SB, (b) dmSB, and (c) mSB. The very interesting finding is that the meta-IRC pathway starting from $(\mathrm{S}_1)_{\mathrm{FC}}$ $(= (\mathrm{S}_0)_{cis-\mathrm{min}})$ leads to not $(\mathrm{S}_1)_{twist-\mathrm{min}}$ but $(\mathrm{S}_1)_{\mathrm{DHP-min}}$ in all cases, even though the previous experimental studies on SB showed that the $\pi\pi^*$-excited *cis*-SB prefers to lead to the twisted structure [71, 75, 82]. Actually, in SB, the meta-IRC initially goes toward the *twist* region (only d_{C3C1C2C4} increases), but once it reaches the IRC pathway between $(\mathrm{S}_1)_{\mathrm{DHP-min}}$ and $(\mathrm{S}_1)_{twist-\mathrm{min}}$, the meta-IRC changes its direction suddenly toward $(\mathrm{S}_1)_{\mathrm{DHP-min}}$. This sharply curved region corresponds to an energetically flat region on the S_1-PES, indicating that the photoreaction of *cis*-SB should have branching products of the twisted structure and the DHP structure. Therefore, the consideration of dynamic effect is essential for discussing the photoreaction mechanisms involved in the branching ratio of the products on the S_1-PES. To understand the geometrical changes along the meta-IRC, variations of significant internal coordinates, r_{C5C6}, d_{C3C1C2C4}, d_{X7C1C2C4}, and d_{X8C2C1C3} (X = H or C), are also shown for (a) SB, (b) dmSB, and (c) mSB in Fig. 2.5. In the initial stage (s = 0 ~3.1 bohr amu$^{1/2}$) of SB, d_{C3C1C2C4} and d_{H7C1C2C4} $(= d_{\mathrm{H8C2C1C3}})$ linearly change: two hydrogen atoms connected in the central C=C part move from a position where the central ethylenic (HC=CH) part is planar ($s = 0$) to that where two $\mathrm{C_6H_5}$–CH parts are planar (s = 3.1). Simultaneously, r_{C5C6} continues to decrease and then forms the C5–C6 bond, leading to the DHP-form.

It is interesting to compare the reaction pathways for dmSB (Fig. 2.4b) with those for SB (Fig. 2.4a). In dmSB, there is the minimum of *cis*-form in the S_1 state, and three minima, $(\mathrm{S}_1)_{\mathrm{DHP-min}}$, $(\mathrm{S}_1)_{cis-\mathrm{min}}$, and $(\mathrm{S}_1)_{twist-\mathrm{min}}$, are separated by two TSs, $(\mathrm{S}_1)_{TS1}$ and $(\mathrm{S}_1)_{TS2}$. Along the meta-IRC from the $(\mathrm{S}_1)_{\mathrm{FC}}$, the molecule approaches $(\mathrm{S}_1)_{\mathrm{TS1}}$ and then enters the $(\mathrm{S}_1)_{\mathrm{DHP-min}}$ region. As shown in Fig. 2.4b, the meta-IRC direction is toward $(\mathrm{S}_1)_{\mathrm{DHP-min}}$ from the beginning, and thus, photocyclization to the DHP-form is enhanced compared with *cis*-SB. Since $(\mathrm{S}_1)_{\mathrm{DHP-min}}$ is very similar to $(\mathrm{S}_1/\mathrm{S}_0)_{\mathrm{DHP}}$ in dmSB, the

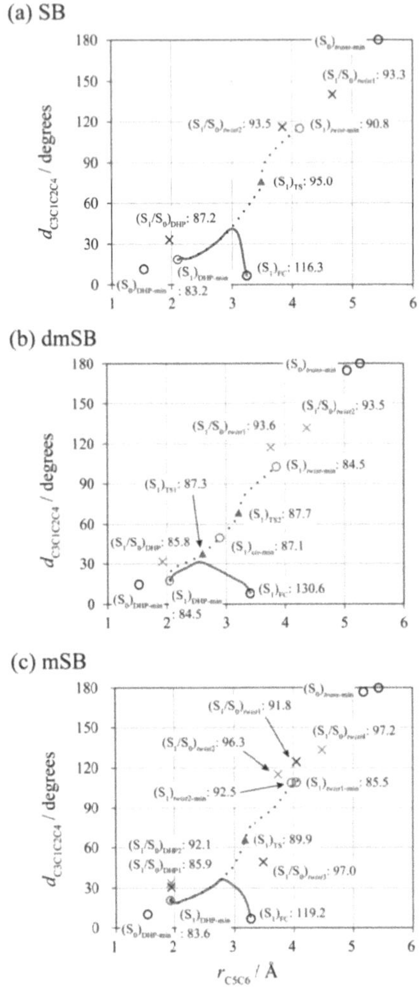

Figure 2.4 Significant stationary points and MECIs, as well as the reaction pathways on the S_1-PES, projected on a two-dimensional configurational space determined by r_{C5C6} and $d_{C3C1C2C4}$, for (a) SB, (b) dmSB, and (c) mSB. Minima in the S_0 state are denoted by black circles, and minima and TS geometries in the S_1 state are denoted by red circles and red triangles. S_1/S_0-MECIs are also denoted by cross marks. The meta-IRC from $(S_1)_{FC}$ is denoted by a red solid line, and the IRC pathway is denoted by a blue dotted line. The potential energy relative to $(S_0)_{cis-\min}$ is also given.

molecule entering the $(S_1)_{DHP-min}$ region is expected to reach the $(S_1/S_0)_{DHP}$ region easily, resulting in the nonradiative transition to the ground state. Additionally, the energy barriers at $(S_1)_{TS1}$ and $(S_1)_{TS2}$ are not so high, which suggests that the molecule near the $(S_1)_{cis-min}$ region can easily enter both the DHP and the *twist* regions. By comparing Figs. 2.5a and 2.5b, we can find that the rate of changes in $d_{C3C1C2C4}$, $d_{X7C1C2C4}$, and $d_{X8C2C1C3}$ are different, although the following trend is consistent: r_{C5C6} continues to decrease while $d_{C3C1C2C4}$ first increase and then decrease. Former is caused by a difference in mass of moving fragments, i.e., CH_3 (dmSB) and H (SB) [49]. In mSB, since both CH_3 and H parts of the central C=C bond can become moving fragments, the reaction pathway has a mid-character between SB and dmSB, as shown in Fig. 2.4c. Figure 2.5c shows that variations of $d_{H7C1C2C4}$ and $d_{C8C2C1C3}$ are different from each other: the variation of $d_{H7C1C2C4}$ is similar to SB, while that of $d_{C8C2C1C3}$ is similar to dmSB. The migration of the H atom, which is faster than the CH_3 fragment, affects the reaction pathway and probably the branching ratio [50].

2.3.3 Excited-State MD Simulations on Photo-Branching Reactions for SB, dmSB, and mSB

Photoexcitation of *cis*-SB invokes an excited-state branching reaction into the DHP-form and the *trans*-form. Experimentally, the branching ratio and time constants were reported, but the detailed reaction mechanism is not clear only from the experimental data. The static analyses of excited-state PES; i.e., the location of minima, TSs, and MECIs, and determination of meta-IRC and IRC pathways, can provide the insight into the reaction mechanism as described in the previous section; however, to discuss the branching ratio and the time constant, the dynamics simulations are required. As discussed above, in SB, dynamic correlation effects are very important to determine the relative energies of the excited state and the ground state, so the SA-CASSCF approach is not enough as the electronic structure method. The ab initio multireference theory, such as CASPT2 and MR-SDCI, is time-consuming to perform the on-the-fly MD simulation for such a medium-size molecule, and also, the standard TDDFT is not recommended to describe

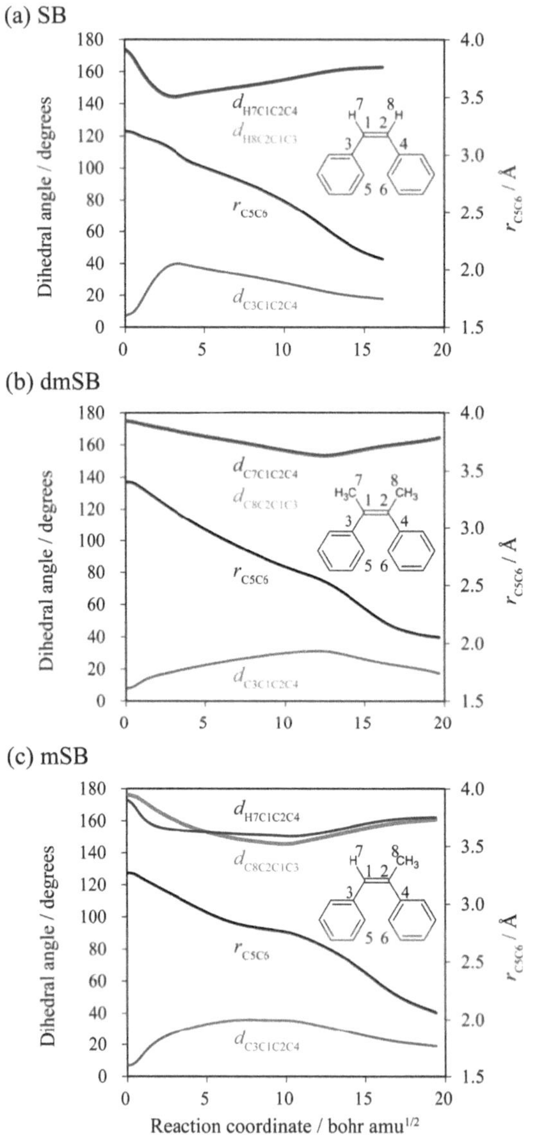

Figure 2.5 Variations of significant internal coordinates, r_{C5C6}, $d_{C3C1C2C4}$, $d_{X7C1C2C4}$, and $d_{X8C2C1C3}$ (X = H or C), along the meta-IRC pathway from $(S_1)_{FC}$ on the S_1-PES for (a) SB, (b) dmSB, and (c) mSB.

nonadiabatic coupling regions. Thus, SF-TDDFT is a good alternative for describing the photoreaction processes of SB.

In a previous section, we summarized theoretical results of geometries and reaction pathways on the S_1-PES of SB, dmSB, and mSB at the SF-TDDFT level. In this section, we describe the results of on-the-fly excited-state MD simulation at the SF-TDDFT level with BHHLYP functional and 6-31G(d) basis sets for *cis*-SB [47], *cis*-dmSB [49], and *cis*-mSB [50]. The MD simulations were started upon the excitation to the $S_1(\pi\pi^*)$ state with initial conditions determined by the normal mode sampling for the ground state equilibrium geometry of *cis*-form: the atomic coordinates and velocities were generated randomly by adding the energy of kT to each normal mode under the Boltzmann distribution at 300 K. To avoid the mixed singlet and triplet states in the SF-TDDFT calculations around the CI region along the trajectory, the T^{SF}-index method [37] was employed for dmSB and mSB. The time step was set to 0.2 fs, and 50, 33, and 40 trajectories were calculated for SB, dmSB, and mSB, respectively, over the simulation time (1.5 ps for SB and 1 ps for dmSB and mSB). The on-the-fly MD simulations were carried out by using our general on-the-fly MD code, SPPR [100].

As discussed in Section 2.3.2, there are two different S_1/S_0-decay channels in the $S_1(\pi\pi^*)$ state, i.e., DHP and *twist* CI regions. As shown in Fig. 2.4, these two regions are separated from each other in the configuration space, and once the molecule drops off to the ground state via DHP-CI, the trajectory will branch into DHP and *cis*-form, while once the molecule drops off via *twist*-CI, the trajectory will branch into *trans* and *cis*-form. Therefore, to examine the branching ratio of *cis*, *trans*, and DHP, one needs to do the on-the-fly excited-state MD simulation with the nonadiabatic transition to the ground state. In our previous studies [47–50], however, each trajectory was terminated when reaching the CI region, where the energy difference between the S_0 and S_1 states became less than 0.2 eV. This was because the nonadiabatic coupling term at the SF-TDDFT level was not available. Thus, the previous studies focused on the branching dynamics toward DHP- and *twist*-regions in the $S_1(\pi\pi^*)$ state. It should be noted that the terminal points of the respective trajectories are not exactly the same as the S_1/S_0-MECI points because the S_1/S_0-crossing regions are distributed around the S_1/S_0-MECI points in configuration space.

We first examine the branching ratio for DHP and *twist* regions of trajectories in the $S_1(\pi\pi^*)$ state. In SB, 13 trajectories (13/50) reach the DHP-CI region, while 35 trajectories (35/50) reach the *twist*-CI region; and the remain two trajectories (2/50) first proceed to the *twist* region and then enter the DHP region [47]. Henceforth, the two trajectories that proceed to the *twist* region and then arrive at the DHP region are excluded to simplify the discussion for SB. The dominant product is expected to be *trans*-SB with a minor product of DHP, even though it is opposite to the expectation from the static reaction pathway. Indeed, the branching ratio from this theoretical simulation (*twist*:DHP = 35:13) is in quantitive agreement with the experimental ratio (*trans*:DHP = 35:10) [71, 75, 82]. This result indicates that the photo-branching mechanism in the $\pi\pi^*$ state of SB is controlled by dynamics. In dmSB, 19 trajectories (19/33) directly reach the DHP-CI region, nine trajectories (9/33) enter the DHP-region and then move to the *twist*-region, four trajectories (4/33) enter the *twist*-region and stay there, and one trajectory (1/33) enter the *twist*-region and then move to the DHP-region. According to the branching ratio (*twist*:DHP = 13:20), dmSB prefers photo-cyclization to *cis-trans* isomerization, which is opposite to the tendency of SB. In mSB, 11 trajectories (11/40) reach the DHP-CI region, while 29 trajectories (29/40) reach the *twist*-CI region, indicating the branching ratio of *twist*:DHP = 29:11. This result is similar to the SB case.

The difference in the branching ratio for SB/mSB and dmSB can be understood from the difference in the reaction pathways in Figs. 2.4 and 2.5. In SB, the $\pi\pi^*$-excited *cis*-SB starts moving towards *twist*-region because the meta-IRC points in the direction of *twist*-region, and during descending on the slope of the PES, the torsional motion about the central C=C bond receives kinetic energy; then, majority of trajectories enter the *twist* region first due to the law of inertia. Note that *cis*-mSB also behaves in the same reaction process as SB because it has an H atom in the $\mathbf{R}^1$ position. This mechanism results in a large branching ratio of *twist*-region for *cis*-SB and *cis*-mSB. On the other hand, in *cis*-dmSB, most trajectories start moving towards the TS between $(S_1)_{DHP-min}$ and $(S_1)_{cis-min}$ and then enter the DHP-region smoothly.

Next, we discuss the time constant obtained from dynamics simulations, with an assumption that the trajectories reaching S_1/S_0-CI regions hop to the ground state. Figure 2.6 shows the calculated S_1-population decay of *cis*-SB, *cis*-dmSB, and *cis*-mSB in the $\pi\pi^*$ state where the population decays in the DHP region, in the *twist* region, and in both of them are denoted by a red, blue, and black line, respectively. In *cis*-SB (Fig. 2.6a), the total population of the $\pi\pi^*$ state decreases to 0.42 at 1.0 ps, reflecting that 28/48 trajectories reached the S_1/S_0-CI regions before 1.0 ps. The population decays in each region also support that the lifetime in the DHP region (13) is relatively longer than that in the *twist* region (35). The femtosecond time-resolved fluorescence spectra for the photoreaction process of *cis*-SB proposed a two-step mechanism with time constants of 0.2 ps and 1.2 ps [73]. The observed fluorescence wavelength (420 nm) [73] well coincides with the S_0-S_1 energy difference in the *cis* region on the S_1-PES, while that in other regions along the IRC is too small. Hence, the decay components in the experimental spectra with the time constants of 0.2 ps and 1.2 ps could be attributed to the molecules, after descending the meta-IRC, entering the *twist* region and the *cis*/DHP region [47]. This suggestion is consistent with the dynamical behaviors of *cis*-SB: most molecules (35/48) enter the *twist* region first due to the initial motion after $\pi\pi^*$ excitation, while others (13/48) enter the DHP region with staying around the *cis* region in the S_1 state [47].

In *cis*-dmSB (Fig. 2.6b), the molecule at the DHP-region has a short lifetime while the molecule at the *twist*-region has a very long lifetime, which is just opposite to *cis*-SB. In the transient absorption spectra of *cis*-dmSB [95], the lifetime of the CIS* band at 600 nm and the P* band at 330 nm in hexane were reported to be 0.2 ps and 19 ps, respectively. According to the results of on-the-fly excited-state MD trajectories, the population of the CIS* state should increase initially and gradually decrease through a movement to the DHP region or the *twist* region. Indeed, the calculated S_1-population for all trajectories decreases during the initial 0.2 ps. This time span is in very good agreement with the experimental lifetime of the CIS*

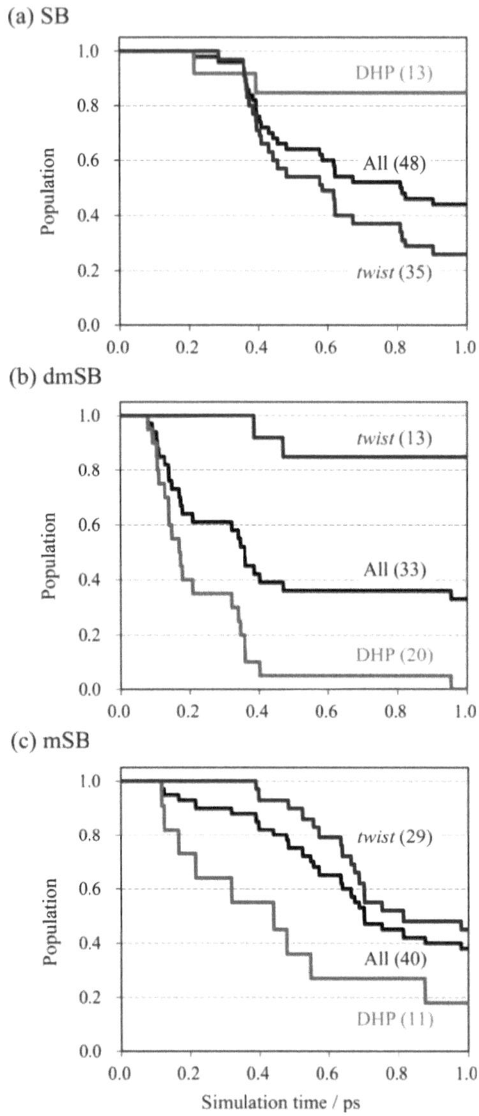

Figure 2.6 The decay of S_1-population as a function of time estimated from on-the-fly MD simulations for (a) SB, (b) dmSB, and (c) mSB. The S_1-population for all the trajectories is plotted in black, while those for the trajectories staying at the DHP region and at the *twist* region are plotted in red and in blue, respectively.

band ($\tau = 0.2$ ps) [95]. On the other hand, most trajectories entering the *twist* region stay around $(S_1)_{twist-\min}$ for a long time without reaching the S_1/S_0-CI region. Thus, the molecule should stay at the *twist* region in the S_1 state with a long lifetime, which corresponds to the long lifetime of P* band observed in the experiment [95].

The *cis*-mSB (Fig. 2.6c) is expected to be the middle of *cis*-SB and *cis*-dmSB. Indeed, the trajectories in the DHP region reach the CI more quickly than in the *twist* region, like *cis*-dmSB, but the trajectories in the *twist* region also reach the CI smoothly. It means that *cis*-mSB has no long-lifetime component in the S_1 state, unlike *cis*-SB and *cis*-dmSB. This feature can be explained by the initial direction of the meta-IRC, oriented between the *twist* (*cis*-SB) and DHP (*cis*-dmSB) regions. In the *twist* region, the decay rate for mSB is slightly longer than that for SB because of the larger energy difference of S_1-minima and S_1/S_0-MECIs. The present on-the-fly excited-state MD simulations show the possibility to tune the lifetime of the excited species by introducing light (H) and relatively heavy (CH_3) fragments as substituents.

Finally, we would discuss the transient absorption spectra for dmSB. In the experimental spectra for dmSB in Fig. 2.7a [95], the P* signal increases to $\sim$0.8 till 0.2 ps, with showing a shoulder feature at 0.2 $\sim$ 0.3 ps, and increases again to 1.0 within 0.6 ps; then, the P* signal gradually decreases. The experimental spectra also showed that the CIS* signal decays with a lifetime of 0.2 ps, with a distinct shoulder feature at 0.2 $\sim$ 0.3 ps [95]. To understand the shoulder feature in the profiles of CIS* and P* bands, the S_1-population for *cis*/DHP- and *twist*-region estimated from the trajectories were plotted as a function of time. In Fig. 2.6b, the S_1-population decay was discussed based on the trajectories reaching the S_1/S_0-CI regions; for a more detailed discussion, we define the *cis*/DHP-region and *twist*-region by using the dihedral angle, $d_{C3C1C2C4}$, at $(S_1)_{TS2}$: the structures with $d_{C3C1C2C4} \leq 68.7°$ are regarded as those in the *cis*/DHP-region (a red line in Fig. 2.7b), while the structures with $d_{C3C1C2C4} \leq 68.7°$ are regarded as those in the *twist*-region (a blue line in Fig. 2.7b) [49]. In on-the-fly excited-state MD simulations, all trajectories initially enter the *cis*-region, corresponding to the appearance of the transient absorption band at 600 nm (CIS* band). The S_1-population of *cis*/DHP-region decreases with a lifetime of

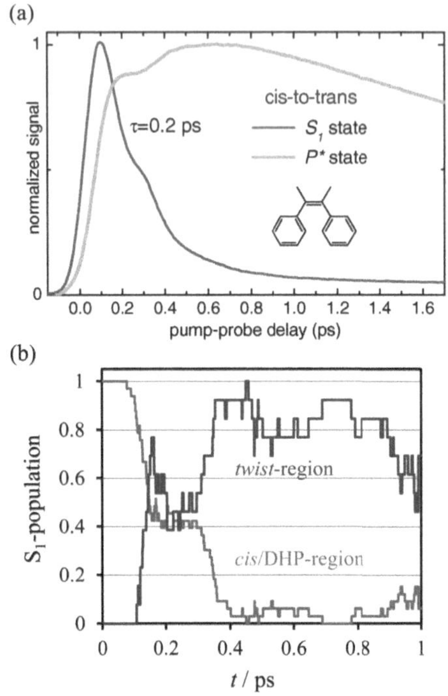

Figure 2.7 The transient absorption spectra for dmSB: (a) Experimental spectrum. Decay of the CIS* band (a red line) and development of the P* band (a cyan line) for *cis*-dmSB (in acetonitrile). Reprinted from Ref. [95], Copyright (2017), with permission from Elsevier. (b) Theoretical spectrum generated from on-the-fly excited-state MD simulations [49]. The rate of S_1-population in the *cis*/DHP-region ($d_{C3C1C2C4} \leq 68.7°$) is plotted in red, while that in *twist*-region ($d_{C3C1C2C4} \leq 68.7°$) is plotted in blue. The plot for *twist*-region is normalized so that the maximum value is equal to 1.

ca. 0.3 ps since some trajectories move out to the *twist*-region or reach the S_1/S_0-CI regions within the DHP-region, which is in good agreement with the experimental lifetime, 0.2 ps, of the CIS* band. The S_1-population of *twist*-region increases during 0.6 ps due to the transfer of trajectories from the *cis*/DHP-region, which is in good agreement with the finite rise time in the P* band. In Fig. 2.7b, there are shoulders in the plots of both the *cis*/DHP- and *twist*-region during 0.2 ~0.3 ps. These shoulders are related to trajectories that once move in the *twist*-region and then go back to the *cis*/DHP-region. This feature can explain the shoulder in the experimental time profiles of the CIS* and P* bands in Fig. 2.7a [95].

2.4 Concluding Remarks

In this chapter, we introduce our recent application studies of on-the-fly excited-state MD simulation based on the SF-TDDFT method for photoreaction of *cis*-SB, *cis*-dmSB, and *cis*-mSB. The on-the-fly MD method without potential functions is a powerful approach to examine the mechanism and dynamics of the photoreaction elementary process because it is difficult to prepare excited-state potential energy surfaces in an empirical function form. It is also required to treat nonadiabatic transition in dynamics, which becomes significant in the energetically crossing regions of multiple electronic states. To make the on-the-fly MD result reliable, the choice of an appropriate electronic structure method is important. Ab initio multiconfigurational methods such as SA-CASSCF can describe the bond dissociation and bond formation processes correctly, but the accuracy of the calculated potential energy surfaces is only qualitative. Ab initio multireference theory can describe the potential energy surfaces accurately, but it is usually very time-consuming. The spin-flip time-dependent density functional theory (SF-TDDFT) is the best alternative for describing the excited-state reaction process including nonadiabatic regions.

In Section 2.2, SF-TDDFT is briefly introduced as a powerful tool for performing excited-state calculations of large systems and on-the-fly MD photoreaction simulations. SF-TDDFT generates singlet ground and excited states containing double-excitation configurations by spin-flip excitations of a reference triplet state. As a result, SF-TDDFT can easily incorporate double-excitation effects in the usual single-excitation TDDFT. This method is a very powerful tool for high-speed on-the-fly excited-state MD simulations due to its much less computational order compared to ab initio multireference theories. However, SF-TDDFT has three major problems: spin-contamination, missing electronic configurations, and poor excitation energies when using conventional GGA functionals. Since these problems, however, are improved or solved by using appropriate functionals and a correction method, SF-TDDFT is the best ever tool for on-the-fly excited-state MD simulations.

The applications of on-the-fly excited-state MD simulations using this SF-TDDFT to photobranching reactions of *cis*-SB, *cis*-dmSB, and *cis*-mSB are reviewed in Section2.3. It was shown that the meta-IRC from the FC structure of *cis*-SB in the $\pi\pi^*$ state reached the S_1-minimum of DHP via a very flat region of the S_1-PES. On-the-fly MD simulations, however, showed that more trajectories reached the *twist*-CI region, compared with the DHP-CI region, and the branching ratio was in good agreement with the experiment. The discrepancy between the meta-IRC and on-the-fly MD simulations can be understood from geometrical features of the reaction pathway on the excited-state PES, which was concluded that the $\pi\pi^*$-excited *cis*-SB proceeding to the very flat *cis*-region in the S_1-PES propagates primarily toward the *twist*-region due to dynamic effects, with partial branching to the DHP-region. For *cis*-dmSB, the meta-IRC from the FC structure is oriented toward the DHP region from the beginning, in contrast to SB. The minima and MECI geometries suggested that the molecule entering the DHP region could easily decay to the ground state, but the molecule entering the *twist* region will stay there for a long time because of the large difference in geometry between S_1-minimum and S_1/S_0-MECI, which is consistent with the experimental observation of the long lifetime of the perpendicular P* structure. The on-the-fly excited-state MD simulations showed that the trajectories entering the *cis*/DHP-region reached the S_1/S_0-CI region with $\sim$0.2 ps, while, in the *twist*-region, very few trajectories reached S_1/S_0-CI even after 1 ps. In the *cis*-mSB, the meta-IRC pathway showed intermediate characteristics between SB and dmSB as a result of the introduction of an inequivalent substituent at the central C=C bond. On-the-fly MD simulations showed a similar trend to *cis*-SB in terms of preferring to enter the *twist*-region rather than the DHP-region, but unlike *cis*-SB and *cis*-dmSB, the S_1/S_0-MECI region was reached smoothly in both DHP- and *twist*-regions. These results suggest the possibility of controlling the dynamics of the excited states by introducing inequivalent substituents into the central C=C bond.

Acknowledgments

We are sincerely grateful to Ms. Rina Yamamoto, Prof. Satoshi Maeda, Dr. Kristopher Keipert, Dr. Federico Zahariev, Prof. Mark Gordon, Prof. Satoshi Takeuchi, and Dr. Tahei Tahara for collaborated works on the target photoreaction dynamics.

References

1. Domcke, W., and Stock, G. (1997). Theory of ultrafast nonadiabatic excited-state processes and their spectroscopic detection in real time. *Adv. Chem. Phys.*, **100**, 1–169.
2. Maeda, S., Taketsugu, T., Ohno, K., and Morokuma, K. (2015). From roaming atoms to hopping surfaces: mapping out global reaction routes in photochemistry. *J. Am. Chem. Soc.*, **137**, 3433–3445.
3. Sobolewski, A. L., Domcke, W., Dedonder-Lardeux, C., and Jouvet, C. (2002). Excited-state hydrogen detachment and hydrogen transfer driven by repulsive $^1\pi\sigma^*$ states: a new paradigm for nonradiative decay in aromatic biomolecules. *Phys. Chem. Chem. Phys.*, **4**, 1093–1100.
4. Chen, W., Hase, W. L., and Schlegel, H. B. (1994). Ab initio classical trajectory study of $H_2CO \rightarrow H_2 + CO$ dissociation. *Chem. Phys. Lett.*, **228**, 436–442.
5. Gordon, M. S., Chaban, G., and Taketsugu, T. (1996). Interfacing electronic structure theory with dynamics. *J. Phys. Chem.*, **100**, 11512–11525.
6. Sun, L., Song, K., and Hase, W. L. (2002). A S_N2 reaction that avoids its deep potential energy minimum. *Science*, **296**, 875–878.
7. Taketsugu, T., and Gordon, M. S. (1995). Dynamic reaction path study of $SiH_4 + H^-$ .fwdarw. SiH_5^- and the Berry pseudorotation mechanism. *J. Phys. Chem.*, **99**, 14597–14604.
8. Taketsugu, T., and Gordon, M. S. (1995). Dynamic reaction path analysis based on an intrinsic reaction coordinate. *J. Chem. Phys.*, **103**, 10042–10049.
9. Taketsugu, T., and Gordon, M. S. (1995). Dynamic reaction coordinate analysis: an application to $SiH_4 + H^-$ .fwdarw. SiH_5^-. *J. Phys. Chem.*, **99**, 8462–8471.

10. Ben-Nun, M., Quenneville, J., and Martínez, T. J. (2000). Ab initio multiple spawning: photochemistry from first principles quantum molecular dynamics. *J. Phys. Chem. A*, **104**, 5161–5175.
11. Kaledin, A. L., and Morokuma, K. (2000). An ab initio direct-trajectory study of the photodissociation of ClOOCl. *J. Chem. Phys.*, **113**, 5750–5762.
12. Vreven, T., Bernardi, F., Garavelli, M., Olivucci, M., Robb, M. A., and Schlegel, H. B. (1997). *Ab initio* photoisomerization dynamics of a simple retinal chromophore model. *J. Am. Chem. Soc.*, **119**, 12687–12688.
13. Nakayama, A., Harabuchi, Y., Yamazaki, S., and Taketsugu, T. (2013). Photophysics of cytosine tautomers: new insights into the nonradiative decay mechanisms from MS-CASPT2 potential energy calculations and excited-state molecular dynamics simulations. *Phys. Chem. Chem. Phys.*, **15**, 12322.
14. Nakayama, A., Yamazaki, S., and Taketsugu, T. (2014). Quantum chemical investigations on the nonradiative deactivation pathways of cytosine derivatives. *J. Phys. Chem. A*, **118**, 9429–9437.
15. Ootani, Y., Satoh, K., Nakayama, A., Noro, T., and Taketsugu, T. (2009). Ab initio molecular dynamics simulation of photoisomerization in azobenzene in the $n\pi*$ state. *J. Chem. Phys.*, **131**, 194306.
16. Kina, D., Arora, P., Nakayama, A., Noro, T., Gordon, M. S., and Taketsugu, T. (2009). Ab initio QM/MM excited-state molecular dynamics study of coumarin 151 in water solution. *Int. J. Quantum Chem.*, **109**, 2308–2318.
17. Kina, D., Nakayama, A., Noro, T., Taketsugu, T., and Gordon, M. S. (2008). *Ab initio* QM/MM molecular dynamics study on the excited-state hydrogen transfer of 7-azaindole in water solution. *J. Phys. Chem. A*, **112**, 9675–9683.
18. Nakayama, A., Arai, G., Yamazaki, S., and Taketsugu, T. (2013). Solvent effects on the ultrafast nonradiative deactivation mechanisms of thymine in aqueous solution: excited-state QM/MM molecular dynamics simulations. *J. Chem. Phys.*, **139**, 214304.
19. Kayanuma, M., Taketsugu, T., and Ishii, K. (2006). Ab initio surface hopping simulation on dissociative recombination of H_3O^+. *Chem. Phys. Lett.*, **418**, 511–518.
20. Kayanuma, M., Taketsugu, T., and Ishii, K. (2008). *An ab* initio molecular dynamics study on the dissociative recombination reaction of $HD_2O^+ + e^-$. *Theor. Chem. Acc.*, **120**, 191–198.

21. Taketsugu, T., and Kobayashi, Y. (2019). Ab initio surface hopping molecular dynamics on the dissociative recombination of CH_3^+. *Comput. Theor. Chem.*, **1150**, 1–9.
22. Taketsugu, T., Tajima, A., Ishii, K., and Hirano, T. (2004). *Ab initio* direct trajectory simulation with nonadiabatic transitions of the dissociative recombination reaction $HCNH^+ + e^- \rightarrow HNC/HCN + H$. *Astrophys. J.*, **608**, 323–329.
23. Kamiya, M., and Taketsugu, T. (2019). Ab initio surface hopping excited-state molecular dynamics approach on the basis of spin–orbit coupled states: an application to the A-band photodissociation of CH_3I. *J. Comput. Chem.*, **40**, 456–463.
24. Tully, J. C. (1990). Molecular dynamics with electronic transitions. *J. Chem. Phys.*, **93**, 1061–1071.
25. Zhu, C. (1996). Unified semiclassical theory for the two-state system: analytical solutions for scattering matrices. *J. Chem. Phys.*, **105**, 4159–4172.
26. Zhu, C., and Nakamura, H. (1992). The two-state linear curve crossing problems revisited. II. Analytical approximations for the Stokes constant and scattering matrix: the Landau–Zener case. *J. Chem. Phys.*, **97**, 8497–8514.
27. Zhu, C., and Nakamura, H. (1993). The two-state linear curve crossing problems revisited. III. Analytical approximations for Stokes constant and scattering matrix: nonadiabatic tunneling case. *J. Chem. Phys.*, **98**, 6208–6222.
28. Nanbu, S., Ishida, T., and Nakamura, H. (2010). Future perspectives of nonadiabatic chemical dynamics. *Chem. Sci.*, **1**, 663.
29. Bernardi, F., Olivucci, M., and Robb, M. A. (1996). Potential energy surface crossings in organic photochemistry. *Chem. Soc. Rev.*, **25**, 321.
30. Domcke, W., and Yarkony, D. R. (2012). Role of conical intersections in molecular spectroscopy and photoinduced chemical dynamics. *Annu. Rev. Phys. Chem.*, **63**, 325–352.
31. Levine, B. G., Ko, C., Quenneville, J., and Martínez, T. J. (2006). Conical intersections and double excitations in time-dependent density functional theory. *Mol. Phys.*, **104**, 1039–1051.
32. Levine, B. G., and Martínez, T. J. (2007). Isomerization through conical intersections. *Annu. Rev. Phys. Chem.*, **58**, 613–634.
33. Mori, T., and Kato, S. (2009). Dynamic electron correlation effect on conical intersections in photochemical ring-opening reaction of cyclohexadiene: MS-CASPT2 study. *Chem. Phys. Lett.*, **476**, 97–100.

34. Yarkony, D. R. (1998). Conical intersections: diabolical and often misunderstood. *Acc. Chem. Res.*, **31**, 511–518.
35. Harabuchi, Y., Maeda, S., Taketsugu, T., Minezawa, N., and Morokuma, K. (2013). Automated search for minimum energy conical intersection geometries between the lowest two singlet states S_0/S_1-MECIs by the spin-flip TDDFT method. *J. Chem. Theory Comput.*, **9**, 4116–4123.
36. Harabuchi, Y., Taketsugu, T., and Maeda, S. (2015). Exploration of minimum energy conical intersection structures of small polycyclic aromatic hydrocarbons: toward an understanding of the size dependence of fluorescence quantum yields. *Phys. Chem. Chem. Phys.*, **17**, 22561–22565.
37. Maeda, S., Harabuchi, Y., Taketsugu, T., and Morokuma, K. (2014). Systematic exploration of minimum energy conical intersection structures near the Franck–Condon region. *J. Phys. Chem. A*, **118**, 12050–12058.
38. Nakano, H. (1993). Quasidegenerate perturbation theory with multiconfigurational self-consistent-field reference functions. *J. Chem. Phys.*, **99**, 7983–7992.
39. Nakano, H. (1993). MCSCF reference quasidegenerate perturbation theory with Epstein—Nesbet partitioning. *Chem. Phys. Lett.*, **207**, 372–378.
40. Harabuchi, Y., Ishii, M., Nakayama, A., Noro, T., and Taketsugu, T. (2013). A multireference perturbation study of the NN stretching frequency of *trans*-azobenzene in $n\pi^*$ excitation and an implication for the photoisomerization mechanism. *J. Chem. Phys.*, **138**, 064305.
41. Yamazaki, S., and Taketsugu, T. (2012). Nonradiative deactivation mechanisms of uracil, thymine, and 5-fluorouracil: a comparative ab initio study. *J. Phys. Chem. A*, **116**, 491–503.
42. Yu, X., Yamazaki, S., and Taketsugu, T. (2011). Concerted or stepwise mechanism? CASPT2 and LC-TDDFT study of the excited-state double proton transfer in the 7-azaindole dimer. *J. Chem. Theory Comput.*, **7**, 1006–1015.
43. Bernard, Y. A., Shao, Y., and Krylov, A. I. (2012). General formulation of spin-flip time-dependent density functional theory using non-collinear kernels: theory, implementation, and benchmarks. *J. Chem. Phys.*, **136**, 204103.
44. Minezawa, N., and Gordon, M. S. (2009). Optimizing conical intersections by spin–flip density functional theory: application to ethylene. *J. Phys. Chem. A*, **113**, 12749–12753.

45. Shao, Y., Head-Gordon, M., and Krylov, A. I. (2003). The spin–flip approach within time-dependent density functional theory: theory and applications to diradicals. *J. Chem. Phys.*, **118**, 4807–4818.

46. Wang, F., and Ziegler, T. (2004). Time-dependent density functional theory based on a noncollinear formulation of the exchange-correlation potential. *J. Chem. Phys.*, **121**, 12191.

47. Harabuchi, Y., Keipert, K., Zahariev, F., Taketsugu, T., and Gordon, M. S. (2014). Dynamics simulations with spin-flip time-dependent density functional theory: photoisomerization and photocyclization mechanisms of *cis*-stilbene in $\pi\pi^*$ states. *J. Phys. Chem. A*, **118**, 11987–11998.

48. Taketsugu, T., and Harabuchi, Y. (2018). *Ab initio* molecular dynamics study on photoisomerization reactions: applications to azobenzene and stilbene, in *Frontiers of Quantum Chemistry* (Wójcik, M. J., Nakatsuji, H., Kirtman, B., and Ozaki, Y., eds), Springer Singapore, Singapore, pp. 431–453.

49. Harabuchi, Y., Yamamoto, R., Maeda, S., Takeuchi, S., Tahara, T., and Taketsugu, T. (2016). *Ab initio* molecular dynamics study of the photoreaction of 1,1′-dimethylstilbene upon $S_0 \rightarrow S_1$ excitation. *J. Phys. Chem. A*, **120**, 8804–8812.

50. Tsutsumi, T., Harabuchi, Y., Yamamoto, R., Maeda, S., and Taketsugu, T. (2018). On-the-fly molecular dynamics study of the excited-state branching reaction of α-methyl-*cis*-stilbene. *Chem. Phys.*, **515**, 564–571.

51. Casida, M. E. (1996). Time-dependent density functional response theory of molecular systems: theory, computational methods, and functionals, in *Recent Developments and Applications of Modern Density Functional Theory* (Seminario, J. M., ed), Elsevier, Amsterdam, pp. 391–439.

52. Löwdin, P.-O. (1955). Quantum theory of many-particle systems. III. Extension of the Hartree–Fock scheme to include degenerate systems and correlation effects. *Phys. Rev.*, **97**, 1509–1520.

53. Sinanoglu, O. (1964). Many-electron theory of atoms, molecules and their interactions. *Adv. Chem. Phys.*, **6**, 315–412.

54. Kato, T. (1957). On the eigenfunctions of many-particle systems in quantum mechanics. *Commun. Pure Appl. Math.*, **10**, 151–177.

55. Nakatsuka, Y., Tsuneda, T., Sato, T., and Hirao, K. (2011). Theoretical investigations on the photoinduced phase transition mechanism of tetrathiafulvalene-p-chloranil. *J. Chem. Theory Comput.*, **7**, 2233–2239.

56. Suzuki, S., Tsuneda, T., and Hirao, K. (2012). A theoretical investigation on photocatalytic oxidation on the TiO_2 surface. *J. Chem. Phys.*, **136**, 024706.

57. Maitra, N. T., Burke, K., and Woodward, C. (2002). Memory in time-dependent density functional theory. *Phys. Rev. Lett.*, **89**, 023002.

58. Tsuneda, T., Nakano, H., and Hirao, K. (1995). Study of low-lying electronic states of ozone by multireference Mo/ller–Plesset perturbation method. *J. Chem. Phys.*, **103**, 6520–6528.

59. Hashimoto, T., Nakano, H., and Hirao, K. (1996). Theoretical study of the valence $\pi \rightarrow \pi^*$ excited states of polyacenes: benzene and naphthalene. *J. Chem. Phys.*, **104**, 6244–6258.

60. Krylov, A. I. (2001). Size-consistent wave functions for bond-breaking: the equation-of-motion spin-flip model. *Chem. Phys. Lett.*, **338**, 375–384.

61. Li, Z., Liu, W., Zhang, Y., and Suo, B. (2011). Spin-adapted open-shell time-dependent density functional theory. II. Theory and pilot application. *J. Chem. Phys.*, **134**, 134101.

62. Sears, J. S., Sherrill, C. D., and Krylov, A. I. (2003). A spin-complete version of the spin-flip approach to bond breaking: what is the impact of obtaining spin eigenfunctions? *J. Chem. Phys.*, **118**, 9084–9094.

63. Becke, A. D. (1988). Density-functional exchange-energy approximation with correct asymptotic behavior. *Phys. Rev. A*, **38**, 3098–3100.

64. Tsuneda, T., Singh, R. K., and Nakata, A. (2016). Relationship between orbital energy gaps and excitation energies for long-chain systems. *J. Comput. Chem.*, **37**, 1451–1462.

65. Tsuneda, T., Singh, R. K., and Nakata, A. (2017). On low-lying excited states of extended nanographenes. *J. Comput. Chem.*, **38**, 2020–2029.

66. Dobryakov, A. L., Ioffe, I., Granovsky, A. A., Ernsting, N. P., and Kovalenko, S. A. (2012). Femtosecond Raman spectra of *cis*-stilbene and *trans*-stilbene with isotopomers in solution. *J. Chem. Phys.*, **137**, 244505.

67. Fuß, W., Kosmidis, C., Schmid, W. E., and Trushin, S. A. (2004). The photochemical *cis–trans* isomerization of free stilbene molecules follows a hula-twist pathway. *Angew. Chemie Int. Ed.*, **43**, 4178–4182.

68. Ishii, K., Takeuchi, S., and Tahara, T. (2004). A 40-fs time-resolved absorption study on *cis*-stilbene in solution: observation of wavepacket motion on the reactive excited state. *Chem. Phys. Lett.*, **398**, 400–406.

69. Kovalenko, S. A., Dobryakov, A. L., Ioffe, I., and Ernsting, N. P. (2010). Evidence for the phantom state in photoinduced *cis–trans* isomerization of stilbene. *Chem. Phys. Lett.*, **493**, 255–258.

70. Moore, W. M., Morgan, D. D., and Stermitz, F. R. (1963). The photochemical conversion of stilbene to phenanthrene. the nature of the intermediate. *J. Am. Chem. Soc.*, **85**, 829–830.

71. Muszkat, K. A., and Fischer, E. (1967). Structure, spectra, photochemistry, and thermal reactions of the 4a,4b-dihydrophenanthrenes. *J. Chem. Soc. B Phys. Org.*, 662.

72. Nakamura, T., Takeuchi, S., Suzuki, N., and Tahara, T. (2008). Revised steady-state fluorescence spectrum and nature of the reactive S_1 state of *cis*-stilbene in solution. *Chem. Phys. Lett.*, **465**, 212–215.

73. Nakamura, T., Takeuchi, S., Taketsugu, T., and Tahara, T. (2012). Femtosecond fluorescence study of the reaction pathways and nature of the reactive S_1 state of *cis*-stilbene. *Phys. Chem. Chem. Phys.*, **14**, 6225.

74. Pedersen, S., Bañares, L., and Zewail, A. H. (1992). Femtosecond vibrational transition-state dynamics in a chemical reaction. *J. Chem. Phys.*, **97**, 8801–8804.

75. Petek, H., Yoshihara, K., Fujiwara, Y., Lin, Z., Penn, J. H., and Frederick, J. H. (1990). Is the nonradiative decay of S_1 *cis*-stilbene due to the dihydrophenanthrene isomerization channel? Suggestive evidence from photophysical measurements on 1,2-diphenylcycloalkenes. *J. Phys. Chem.*, **94**, 7539–7543.

76. Repinec, S. T., Sension, R. J., Szarka, A. Z., and Hochstrasser, R. M. (1991). Femtosecond laser studies of the *cis*-stilbene photoisomerization reactions: the *cis*-stilbene to dihydrophenanthrene reaction. *J. Phys. Chem.*, **95**, 10380–10385.

77. Rodier, J.-M., Ci, X., and Myers, A. B. (1991). Resonance Raman spectra of 4a,4b-dihydrophenanthrene, the photocyclization product of *cis*-stilbene. *Chem. Phys. Lett.*, **183**, 55–62.

78. Sajadi, M., Dobryakov, A. L., Garbin, E., Ernsting, N. P., and Kovalenko, S. A. (2010). Time-resolved fluorescence spectra of *cis*-stilbene in hexane and acetonitrile. *Chem. Phys. Lett.*, **489**, 44–47.

79. Takeuchi, S., Ruhman, S., Tsuneda, T., Chiba, M., Taketsugu, T., and Tahara, T. (2008). Spectroscopic tracking of structural evolution in ultrafast stilbene photoisomerization. *Science*, **322**, 1073–1077.

80. Todd, D. C., Jean, J. M., Rosenthal, S. J., Ruggiero, A. J., Yang, D., and Fleming, G. R. (1990). Fluorescence upconversion study of *cis*-stilbene isomerization. *J. Chem. Phys.*, **93**, 8658–8668.

81. Weigel, A., and Ernsting, N. P. (2010). Excited stilbene: intramolecular vibrational redistribution and solvation studied by femtosecond stimulated raman spectroscopy. *J. Phys. Chem. B*, **114**, 7879–7893.

82. Wismonski-Knittel, T., Fischer, G., and Fischer, E. (1974). Temperature dependence of photoisomerization. Part VIII. Excited-state behaviour of 1-naphthyl-2-phenyl- and 1,2-dinaphthyl-ethylenes and their photocyclisation products, and properties of the latter. *J. Chem. Soc. Perkin Trans. 2*, **15**, 1930.

83. Kokado, K., Machida, T., Iwasa, T., Taketsugu, T., and Sada, K. (2018). Twist of C=C bond plays a crucial role in the quenching of AIE-active tetraphenylethene derivatives in solution. *J. Phys. Chem. C*, **122**, 245–251.

84. Berweger, C. D., van Gunsteren, W. F., and Müller-Plathe, F. (1998). Molecular dynamics simulation with an ab initio potential energy function and finite element interpolation: the photoisomerization of *cis*-stilbene in solution. *J. Chem. Phys.*, **108**, 8773–8781.

85. Dou, Y., and Allen, R. E. (2003). Another important coordinate in the photoisomerization of *cis*-stilbene. *Chem. Phys. Lett.*, **378**, 323–329.

86. Dou, Y., and Allen, R. E. (2003). Detailed dynamics of a complex photochemical reaction: *cis–trans* photoisomerization of stilbene. *J. Chem. Phys.*, **119**, 10658–10666.

87. Dou, Y., and Allen, R. E. (2004). Dynamics of the photocyclization of *cis*-stilbene to dihydrophenanthrene. *J. Mod. Opt.*, **51**, 2485–2491.

88. Improta, R., and Santoro, F. (2005). Excited-state behavior of *trans* and *cis* isomers of stilbene and stiff stilbene: a TD-DFT study. *J. Phys. Chem. A*, **109**, 10058–10067.

89. Ioffe, I. N., and Granovsky, A. A. (2013). Photoisomerization of stilbene: the detailed XMCQDPT2 treatment. *J. Chem. Theory Comput.*, **9**, 4973–4990.

90. Minezawa, N., and Gordon, M. S. (2011). Photoisomerization of stilbene: a spin-flip density functional theory approach. *J. Phys. Chem. A*, **115**, 7901–7911.

91. Molina, V., Merchán, M., and Roos, B. O. (1999). A theoretical study of the electronic spectrum of *cis*-stilbene. *Spectrochim. Acta Part A Mol. Biomol. Spectrosc.*, **55**, 433–446.

92. Quenneville, J., and Martínez, T. J. (2003). Ab initio study of *cis–trans* photoisomerization in stilbene and ethylene. *J. Phys. Chem. A*, **107**, 829–837.

93. Amatatsu, Y. (1999). Ab initio study on the electronic structures of stilbene at the conical intersection. *Chem. Phys. Lett.*, **314**, 364–368.

94. Bearpark, M. J., Bernardi, F., Clifford, S., Olivucci, M., Robb, M. A., and Vreven, T. (1997). Cooperating rings in *cis*-stilbene lead to an S_0/S_1 conical intersection. *J. Phys. Chem. A*, **101**, 3841–3847.

95. Berndt, F., Dobryakov, A. L., Quick, M., Mahrwald, R., Ernsting, N. P., Lenoir, D., and Kovalenko, S. A. (2012). Long-lived perpendicular conformation in the photoisomerization path of 1,1′-dimethylstilbene and 1,1′-diethylstilbene. *Chem. Phys. Lett.*, **544**, 39–42.

96. Brandt, P., Hedberg, C., and Andersson, P. G. (2003). New mechanistic insights into the iridium–phosphanooxazoline-catalyzed hydrogenation of unfunctionalized olefins: a DFT and kinetic study. *Chem. - A Eur. J.*, **9**, 339–347.

97. Serreli, V., Lee, C.-F., Kay, E. R., and Leigh, D. A. (2007). A molecular information ratchet. *Nature*, **445**, 523–527.

98. Maeda, S., Harabuchi, Y., Osada, Y., Taketsugu, T., Morokuma, K., and Ohno K. GRRM14, https://iqce.jp/GRRM/index_e.shtml (accessed May 6, 2022).

99. Schmidt, M. W., Baldridge, K. K., Boatz, J. A., Elbert, S. T., Gordon, M. S., Jensen, J. H., Koseki, S., Matsunaga, N., Nguyen, K. A., Su, S., Windus, T. L., Dupuis, M., and Montgomery, J. A. (1993). General atomic and molecular electronic structure system. *J. Comput. Chem.*, **14**, 1347–1363.

100. Harabuchi, Y., Okai, M., Yamamoto, R., Tsutsumi, T., Ono, Y., and Taketsugu, T. SPPR (a development version), 2016, Hokkaido Univ., Sapporo.

Chapter 3

Nonadiabatic Dynamics Simulations on the Excited States of Carbon-Related Materials with Time-Dependent Density Functional Theory

Shunwei Chen,[a,b] Naeem Ullah,[b] and Ruiqin Zhang[b]

[a]*School of Materials Science and Engineering,*
Qilu University of Technology (Shandong Academy of Sciences), Jinan, China
[b]*Department of Physics, City University of Hong Kong, Kowloon, Hong Kong SAR, China*
swchen5-c@my.cityu.edu.hk, nullah3-c@my.cityu.edu.hk, aprqz@cityu.edu.hk

Studying the excited-state decay process is important for materials science since the fate of the excited states governs the materials' efficiencies for various applications, including photoluminescence and photocatalysis. However, the high computational cost hinders the application of high-accuracy theoretical methods for investigating research systems containing a substantial number of atoms. In this chapter, we review the application of time-dependent density functional theory (TDDFT) in excited-state dynamic simulations of graphene-related materials and graphitic carbon nitride to investigate their optoelectronic properties, that is, the excited state decays of a nanographene epoxide and heptazine-water clusters, by properly constructing structural models and cooperating with

Time-Dependent Density Functional Theory: Nonadiabatic Molecular Dynamics
Edited by Chaoyuan Zhu

ISBN 978-981-4968-42-3 (Hardcover), 978-1-003-31921-4 (eBook)
www.jennystanford.com

complete active space self-consistent field (CASSCF) calculations. Taking the above examples, we show the significance of conical intersections involved in the excited-state decay processes of these two materials, which in the first case limit the photoluminescence and in the second case relate to the photocatalytic water splitting reaction. This work may serve as a reference to study the excited state decays of various materials.

3.1 Introduction

3.1.1 Graphene-Based Luminescent Nanomaterials

As well known, carbon atoms have a robust ability to form different hybridization states (that is, sp, sp^2, and sp^3), and thus are capable of creating numerous carbon structures, ranging from small molecules to structurally large and complicated three-dimensional (3D) topologies (Fig. 3.1) [1]. The various carbon nanoallotropes typically include zero-dimensional (0D) materials such as fullerenes, carbon dots (CDs), and graphene quantum dots (GQDs), 1D structures like carbon nanotubes (CNTs) and graphene nanoribbons (GNRs), and 2D structures in particular including graphene [2]. Even until now, the carbon material family is still constantly growing, providing numerous members for various potential applications in diverse areas. Graphene, the most important material found in recent years, has rapidly been the most attractive carbon material since 2004 of its discovery and attracted unprecedented interest in various fields [3]. Lots of experiments and theoretical studies have been performed to fully explore its intrinsic potential for various purposes by utilizing its superior mechanical, electronic, and thermal properties [4]. As it is a zero-bandgap material, its application in many fields has been seriously restricted. Therefore, developing graphene-related materials that possess specific bandgaps has become a hot research direction [5]. A very early computational work had already demonstrated the band-gap tunability of graphene fragments [6]. Among the experimental approaches, reducing its lateral dimension has been a practical way for the effectiveness of bandgap tuning [7, 8]. To date, many kinds of graphene-based low-

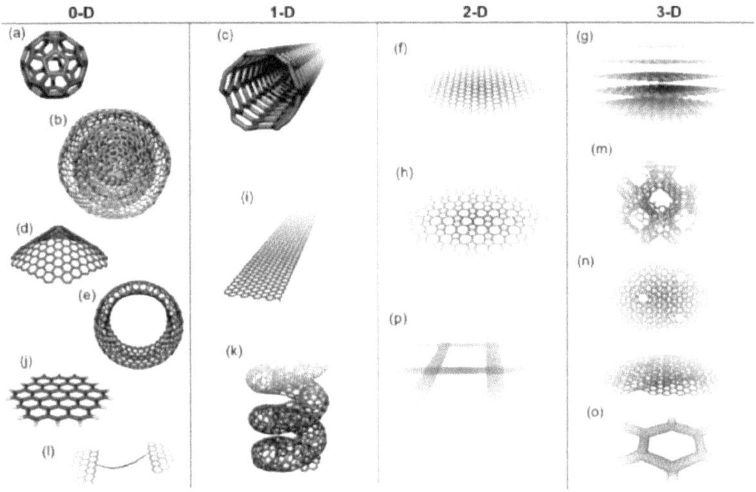

Figure 3.1 Typical structures of versatile graphene-related materials with diverse dimensions. Adapted from Ref. [1].

dimensional materials have been produced, typically including GQDs and GNRs which show extraordinary semiconducting properties [9–11]. Among the properties, photoluminescence has been an important research aspect [11]. To date, light emission from these nanomaterials can be facilely tuned to cover the full visible light region, and emission colors from the deep ultraviolet to far-infrared are also reachable [12]. Experimental methods like engineering the shape/size, heteroatom doping, and chemical functionalization are commonly utilized strategies to tune their luminescence [13–15]. Highly-tech areas including quantum communication and quantum computing become potential application fields of these materials by acting as single-photon emitters [16]. These nanomaterials are receiving more and more research attention from different disciplines like materials, biology, chemistry, and physics to exploit their great potential.

However, the large heterogeneity in the compositions of the synthesized samples seriously restricts the further development of graphene science [12, 17, 18]. The various surface chemical groups contained by the samples are known to critically influence the light emission, but the detailed effects of an individual functional group on the optical properties are difficult to be confirmed, making the luminescence tuning by chemical modification uncertain. Among various chemical groups, the epoxy group (–COC–) is a commonly existing species in the synthesized samples and is usually believed to enhance the photoluminescence intensity upon binding on the graphene surface [19, 20]. However, this viewpoint has been challenged by many other experiments [21, 22]. Our recent work, on one hand, reveals that surface functional groups, including epoxy and hydroxyl groups, may activate the otherwise forbidden lowest optical transitions, and, on the other hand, may localize the excited electron and hole to their proximities, indicting a role in facilitating the luminescence [20, 23, 24]. Since the mechanism of epoxy groups in assisting the non-radiative excited state decays is largely unclear, and given that the above theoretical work only performed static calculations, which ignores the dynamic nature of the excited states, thus we further performed excited state dynamics simulations to investigate the possible excited-state decay pathways of epoxy group modified graphene nanostructures [25]. Such work may not only provide an in-detail understanding of epoxy group-mediated nonradiative excited-state decays but also show the importance of chemical species in facilitating both radiative and nonradiative decays competitively determining the material's luminescent efficiencies. Thus, it may provide useful guidance and reference to relevant research.

3.1.2 Graphitic Carbon Nitride Photocatalyst

Developing visible light-enabled photocatalytic materials have gained tremendous research attention due to the relatively unlimited nature of the sunlight resource. Among the catalysts, graphitic carbon nitride (g-C_3N_4), which is polymeric π-conjugated material with the framework formed by carbon and nitrogen atoms, has been extensively studied as a future photocatalyst for hydrogen production (see structures in Fig. 3.2) [26–30]. This material has a suitable electronic bandgap of 2.7 eV, can be easily synthesized under environmentally friendly conditions, and is relatively physicochemical stable. In 2006, Goettmann et al. firstly used g-C_3N_4 as a heterogeneous catalyst [31]. In 2009, Wang et al.

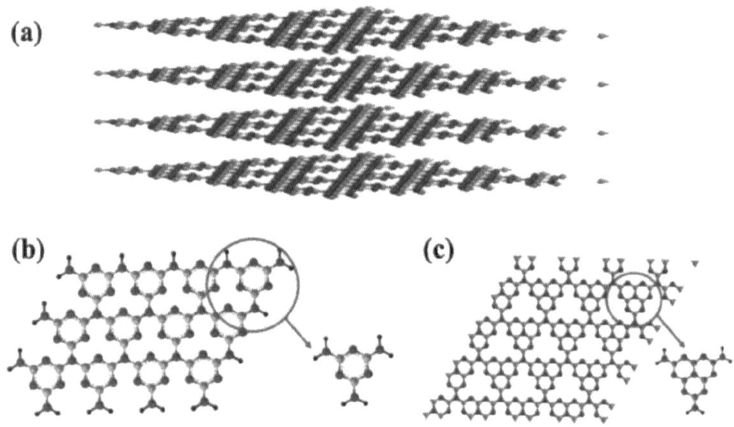

Figure 3.2 Structures of graphitic carbon nitride (g-C_3N_4). (a) Stacked g-C_3N_4 layers (b) S-triazine-based g-C_3N_4 monolayer (c) Tri-s-triazine-based g-C_3N_4 monolayer. In this work, we have studied the case of tri-s-triazine-based g-C_3N_4 since it is reported to be the most likely structure synthesized in experiments. Adapted from Ref. [30].

reported its promise for metal-free photocatalytic applications [32]. Different from most other photocatalysts, the novel g-C_3N_4 materials may form novel hybrid photocatalysts of superior tailorable properties, and importantly the compositions, thickness, sizes, morphologies, and pore structures of the materials can be facilely tailored. Thus, the development of g-C_3N_4-based photocatalysts with multipurpose functions through appropriate modifications is a hot research direction in many fields. In very recent years, many innovative structures of g-C_3N_4-based photocatalysis materials, for example, nanorods, polymeric chains, nanosheets, and complex 3D hierarchical topologies, have been constructed in experiments [30].

Although a tremendous number of researchers have devoted themselves to the field of polymeric g-C_3N_4 for photocatalytic applications, the insufficient fundamental understandings of the excited-state dynamics of g-C_3N_4 importantly hinder the further development, especially the way to largely enhance the photocatalysis efficiency is lacking [33–37]. Importantly, recent theoretical studies on the heptazine-H_2O complex proposed that heptazine molecule is likely to bond hydrogen atoms broken from the surrounding

H_2O molecule via an electron/proton transfer mechanism [38]. And an electron-driven proton transfer reaction was revealed for the bulk pyridine-H_2O system [39]. These studies have provided basic molecular-level understandings of photocatalytic water-splitting reactions. Inspired by the fundamental framework for possible molecular-level water-splitting reactions, we herein investigated the excited-state dynamics of heptazine-water clusters to reveal the predicted mechanisms [40]. As reported by experiments, the small heptazine-based oligomers of g-C_3N_4 surprisingly show even higher hydrogen evolution efficiencies than the bulk g-C_3N_4 materials. Thus, though this work uses a notably small material model (that is, a heptazine molecule), the results are still believed to provide an informative understanding of the photocatalytic water splitting mechanism for g-C_3N_4 materials.

3.1.3 Applications of Excited-State Dynamics Simulations

The investigation of the nonradiative processes in molecules involving conical intersections is one of the biggest successes in the field of theoretical chemistry [41]. In recent years, the importance of conical intersections influencing the fate of excited states has been extended in materials science, since the nonradiative recombination enabled by conical intersections is found to greatly limit the emission efficiencies of the materials for optoelectronic applications [42]. However, the large computation cost for studying bulk materials seriously hinders the application of current theoretical methods into this research area. Among the choices of theoretical methods, TDDFT has become a compromised but a suitable choice that may balance the calculation efficiency and accuracy to satisfy the need [43, 44]. There have been many examples in nanomaterials utilizing TDDFT for studying the photochemical and photophysical phenomena, highlighting the importance of potential energy surface crossing between the ground state and the excited state in nanoscience [45]. The trajectory surface hopping (TSH) method base on TDDFT has been successfully applied in many research systems, for example, studying the photoexcitation promoted proton transfer and atmospheric photochemical reactions [43, 46–48]. Of course, concerning nonadiabatic excited-state dynamics, the most significant problems expected for TDDFT include the failure to correctly depict the conical intersections between the excited state and the ground state. Thus, the use of TDDFT for excited-state dynamics must be cautious, and to ensure the reliability of the results, further confirmation with a higher-level method is optimal.

3.2 Ground-State Structures and Absorption

Covalently attached epoxy groups on graphene surfaces widely exist in experimentally synthesized samples, and recent studies have reported the probability of electron-driven migration of epoxy oxygen on graphene surfaces [49–51]. According to the experiments, we constructed a graphene nanosheet epoxide to carry out our study (Fig. 3.3) [25]. The ground-state structures were fully optimized by density functional theory (DFT) with the long-range and dispersion corrected ωB97XD functional together with the 6-31G(d,p) standard basis set as implemented in the Gaussian 09 software [52, 53]. This nanostructure has a limited number of atoms due to the high computational cost. However, it shall be an appropriate representative of the bulk materials containing epoxy groups. This nanographene epoxide is predicted to be stable as its formation

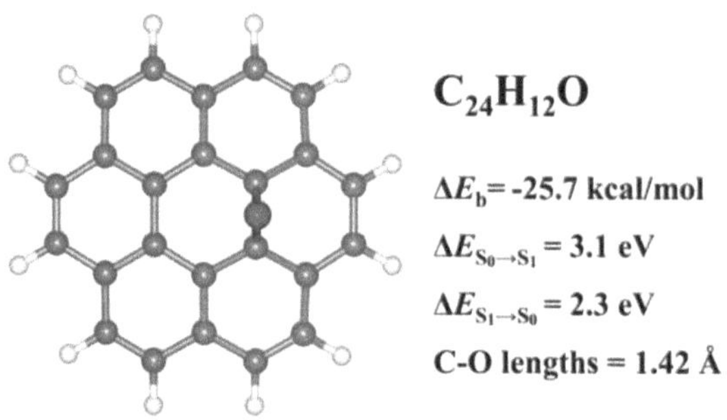

Figure 3.3 Optimized ground-state structure of the nanographene epoxide model. ΔE_b, the binding energy, representing the energy difference of $C_{24}H_{12}O$ with individual $C_{24}H_{12}$ and a triplet oxygen atom. $\Delta E_{S0 \to S1}$ and $\Delta E_{S1 \to S0}$ are the transition energies between the corresponding states. Both the C–O bonds are predicted to be ~1.42 Å in length. Adapted from Ref. [25].

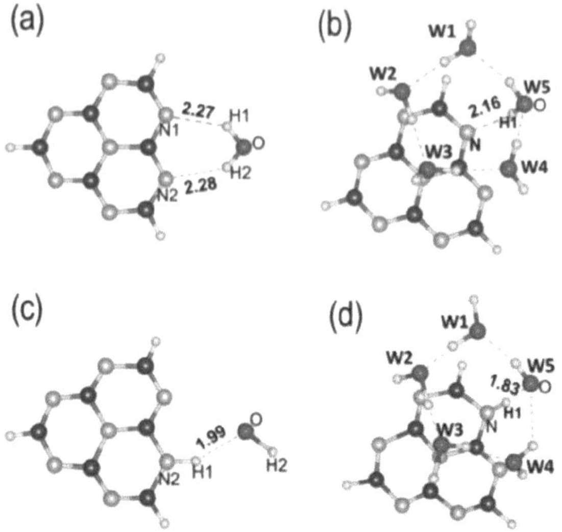

Figure 3.4 Models for heptazine-water systems: (a) Heptazine-w1 and (b) Heptazine-w5 (The structures have been relaxed at the ground state). (c, d) The corresponding structures close to the conical intersection found in TDDFT excited-state dynamic simulations. The representative distances have been shown along with the structures (in Å). Adapted from Ref. [40].

from the pristine graphene nanosheet and one oxygen atom is exothermic (25.7 kcal/mol). Structurally, the lengths of the C—O bonds are predicted to be ~1.42 Å. The excited-state properties of the studied systems were simulated by TDDFT at the ωB97XD/6-31G(d,p) level of theory. After optimizing its structure at the ground (S_0) state and the lowest excited singlet (S_1) state, the S_0-S_1 and S_1-S_0 transitions were calculated to be ~3.1 and ~2.3 eV, respectively. The initial geometries for starting the simulations were chosen within the absorption energy region of 2.5 $\pm$ 0.3 eV, corresponding to the S_0-S_1 transition. As a result, 30 simulations were individually performed.

The studied heptazine-H_2O clusters are shown in Fig. 3.4 [40]. In the first cluster (namely, heptazine-w1), the heptazine and H_2O molecules are coplanar due to the formation of hydrogen bonds. As for the heptazine-5w cluster, each two nearby H_2O molecules form a hydrogen bond, resulting in a loop formed by O and H atoms. Due to

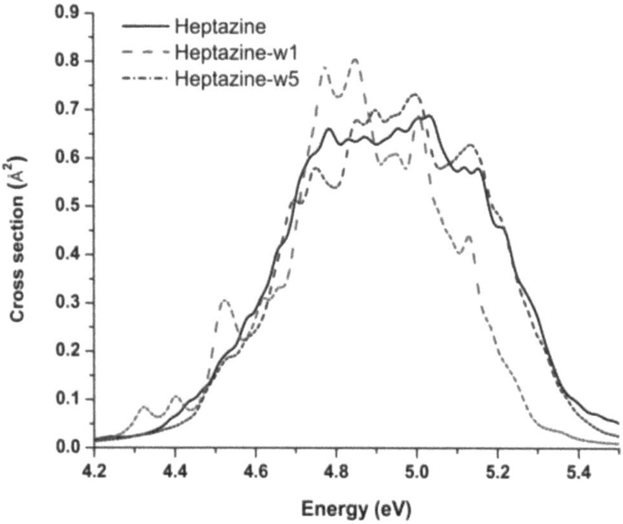

Figure 3.5 Absorption cross section of individual heptazine, heptazine-w1, and heptazine-w5, simulated for the lowest 10 excitations of each complex by considering 200 points for each excited state. Adapted from Ref. [40].

the O—H$\cdots\pi$ bonding interaction, five water molecules preferably stay above the plane of heptazine.

Figure 3.5 shows the absorption spectra of individual heptazine and the studied heptazine-H_2O clusters, obtained by simulating the first 10 excitations for each structure. For the individual heptazine molecule (in the gas phase), it has an absorption maximum locating at ~246 nm, while the strongest absorption region of heptazine-w1 is ~4.81 eV (corresponding to ~257 nm). Increasing water molecules (heptazine-w5) slightly blue-shifts the absorption maximum to ~252 nm. Two different cases starting the dynamics were considered: The first one is to start all the simulations for the three systems from the S_1 state, while the other one is from the S_6 state only for heptazine-w1. For the first case, geometries corresponding to the spectral window of 3.0 $\pm$ 0.3 eV were used as the initial geometries of the simulations in all three systems, but for the second case, 4.9 $\pm$ 0.2 eV energy interval was chosen as the criterion to choose the initial geometries. This leads to 30, 39, and 26 geometries for heptazine, heptazine-w1, and heptazine-w5 systems starting from S_1 state, respectively, and 30 trajectories for the heptazine-w1 system starting from the S_6 state.

3.3 Nonadiabatic Excited-State Simulations

The excited-state dynamic simulations were carried out by Gaussian 09 interfaced with the Newton-X code [48, 54, 55]. A time step of 0.5 fs was chosen for integrating the classical equations, while a time step of 0.025 fs for the quantum equations by interpolating quantities between two adjacent classical steps. Each trajectory ran for a maximum of 300 fs unless the S_1-S_0 energy gap becoming smaller than 0.20 eV; such a small S_1-S_0 energy difference is treated as an indication of conical intersections, akin to the previous studies. Figure 3.6 summarizes the simulation results for the nanographene epoxide. A total number of 30 trajectories were individually simulated, with the initial structures choosing by the Wigner distribution method within a specific excitation energy window corresponding to the interested excited states. In theory, a much larger number of trajectories is preferential to achieve statistical results, but we have made a compromise to the high computational cost. Instead, we focus on revealing the mechanisms other than quantitative analysis. After populating the electron from the ground state to the S_1 state, the simulations started to simulate the dynamic decays. The non-adiabatic nature between two adjacent states was interpreted by the fewest switches surface hopping method with decoherence corrections taking into consideration. All the initial structures show emission energy of ~3.0 eV at the first step, then a quick decrease in the S_1-S_0 energy (to ~2.3 eV) was found. This energy corresponds to the energy at the S_1 state minimum. Finally, we found a substantial part (33.3%) of the trajectories stopped after the S_1-S_0 state energy being smaller than 0.2 eV, the energy threshold for indicating the presence of possible conical intersections. Among all trajectories finally reaching a conical intersection, the fastest decay happens within 37 fs, and then, the left ones successively decayed within 300 fs (300 fs is the setup of the longest simulation time). The average decay time is computed to be 167.0 fs. For the other 20 trajectories with no conical

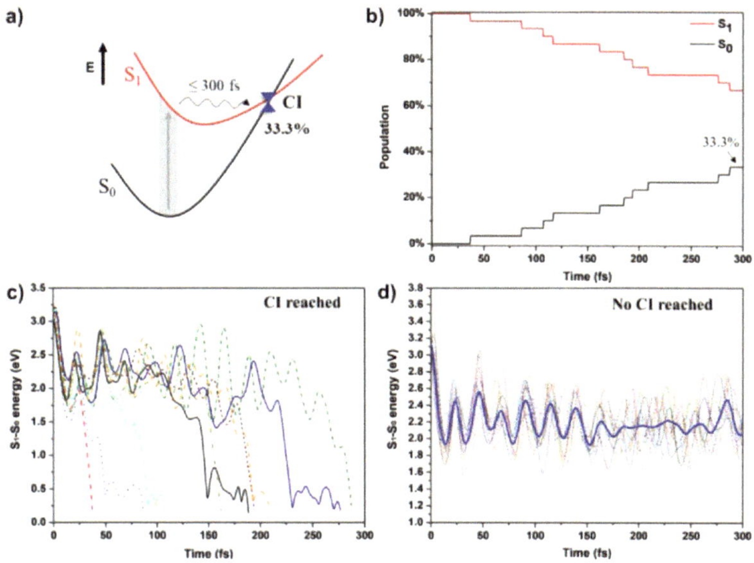

Figure 3.6 (a) Summary for the excited-state dynamic simulations of the nanographene epoxide system. (b) The population of S_0 and S_1 state during the 300 fs duration time for the 30 trajectories. Once the S_1-S_0 energy difference becomes smaller than 0.2 eV, the simulation stops and is tentatively taken as surface crossing at conical intersections. (c) S_1-S_0 energy evolution of the 10 trajectories encountering conical intersections. The blue line in bold represents the average value over all trajectories. Adapted from Ref. [25].

intersection found, the S_1-S_0 energy differences after a fast decay deviate around ~2.3 eV till the simulation end. As an indication, since the average decay time (167.0 fs) is relatively much faster than the typical photoluminescence lifetimes of graphene-related materials, such fast nonradiative decays of the excited states might greatly compete with the radiative decays, consequently decreasing the light emission efficiencies [56, 57].

It was found that the accessing of conical intersections accompanies the breakage of one ether bond. The average CO distance of the broken C–O bonds of the 10 systems is ~2.35 Å, while the remaining C–O bond is ~1.42 Å. Taking one trajectory as a representative (Fig. 3.7), it takes 188.5 fs to reach the conical intersection, at which the S_1-S_0 state energy difference was found to be 0.14 eV. The

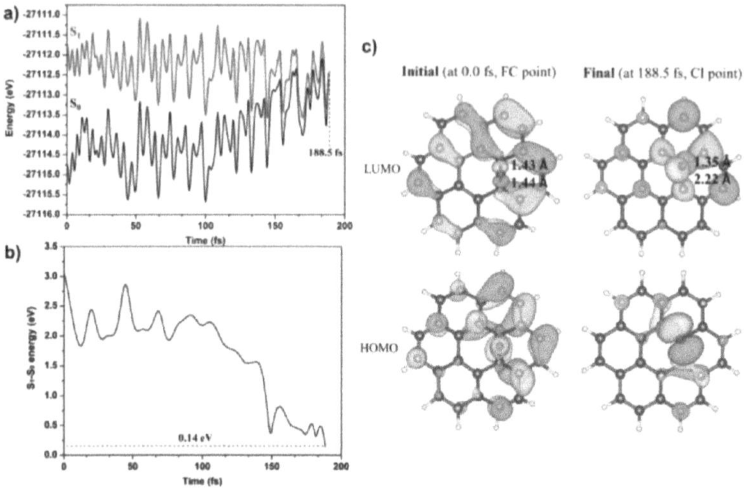

Figure 3.7 Simulation results for an example trajectory. (a) Changes of total electronic energy of the S_1 and S_0 state during the simulation. At the conical intersection, the S_1-S_0 energy difference is 0.14 eV. (c) HOMO and LUMO spatial distribution at the initial and final molecular structure of the simulation. The values show the distance between the nearby carbon and oxygen atoms in Å. Adapted from Ref. [25].

initial molecular structure features C–O bond lengths of 1.43 and 1.44 Å. However, the parameters at the conical intersection are 1.35 and 2.22 eV, respectively, that is, one of the C–O bonds is broken. Apparently, C–O bond breaking, the most prominent structural change, is responsible for the formation of the conical intersection. Such a phenomenon was also found in oxidized silicon clusters [42]. Further inspection of the frontier molecular orbitals reveals the dominant role of oxygen in redistributing the highest occupied and lowest unoccupied molecular orbitals (HOMO and LUMO), further confirming the capability of the epoxy group to induce non-radiative excited state decays. In addition, we also performed a rigid scan for O–C–C angles to profile the energy curves of the S_1 and S_0 state, as shown in Fig. 3.8, which importantly indicates the presence of conical intersections due to C–O bond breaking, in accordance with the TDDFT simulation results.

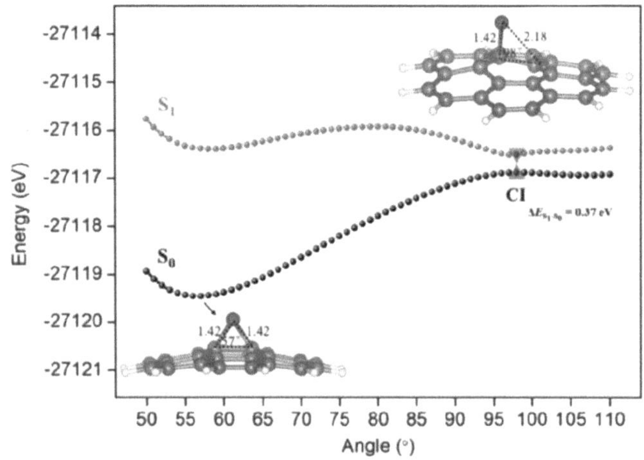

Figure 3.8 Rigid scan for O–C–C angles of the epoxy moiety. The curves show evolutions of the S_0 and S_1 potential energy surfaces as a function of the angles. The initial and final structures during the scanning with the key structural parameters are shown. At the point that the S_0 and S_1 potential energy surfaces are closest in energy, the geometry shows similar features to the conical intersection found by the dynamic simulations. Adapted from Ref. [25].

For the g-C_3N_4 systems, the simulations first start with geometries corresponding to the spectral region of 3.0 $\pm$ 0.3 eV, that is, the first excited singlet (S_1) state. Each trajectory ran for a maximum of 300 fs unless the S_1-S_0 energy gap becoming smaller than 0.15 eV, which is a sign of the presence of possible conical intersections. According to simulation results (Fig. 3.9), ~23% of the trajectories under the participation of water molecules decayed in energy from the S_1 states to S_1-S_0 energy difference being smaller than 0.15 eV, which is predicted to be a sign for reaching conical intersections. Whereas, the rest of the trajectories fluctuate in energies that are far from being a crossing between two energy surfaces. Further examination reveals that electron-driven proton transfer from H_2O to heptazine is responsible for the formation of the conical intersections, where the hydrogen atom was finally found to bond with nitrogen atoms of the heptazine molecule. By contrast, the individual heptazine has no trajectories decayed to approach the

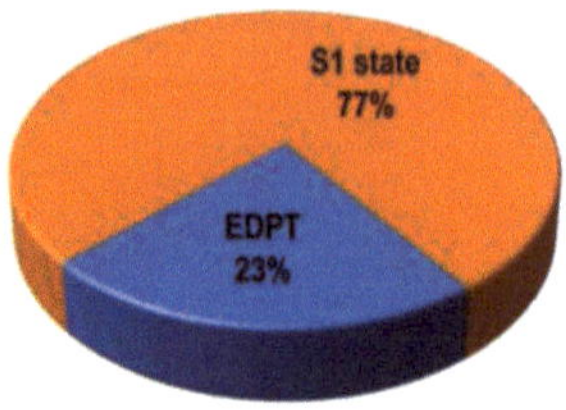

Figure 3.9 Summary of the excited state dynamic simulations for the heptazine-water systems. During the simulations, 23% of the trajectories stopped after approaching a conical intersection, wherein the electron-driven proton transfer is the main decay path. In comparison, the left (77%) trajectories fluctuate in energy far from the set energy threshold for conical intersections. Adapted from Ref. [40].

ground state, that is, the S_1-S_0 energy difference is still relatively too large to be seen as a surface crossing.

A representative trajectory was chosen to in-detail show the proton-transfer physical pictures. Figure 3.10 shows the distances between the mainly involved atoms (N, H, and O atoms, connected

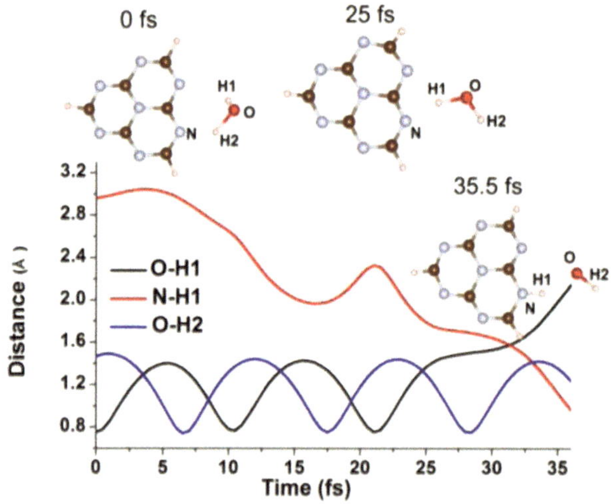

Figure 3.10 Bond length involution during the simulation of a representative trajectory. Proton transfer from H_2O to heptazine molecule is observed and is the dominating structural change of the system. Adapted from Ref. [40].

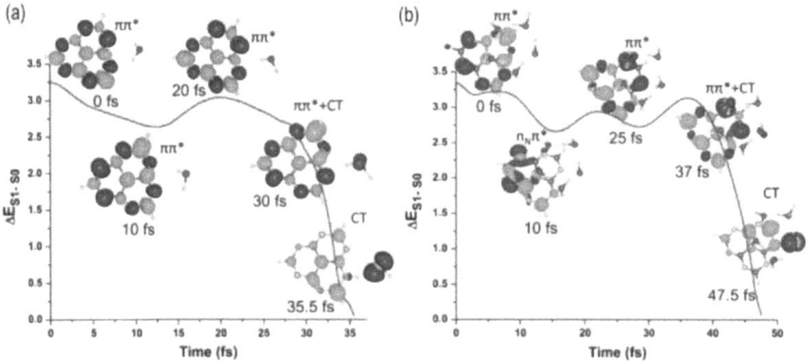

Figure 3.11 Evolution of the S_1-S_0 state energy difference of representative trajectories for heptazine-w1 and heptazine-w5. The isosurfaces show charge density differences at given times. Green nodes show charge gaining, while the blue ones indicate a loss. Adapted from Ref. [40].

via hydrogen bonding), which clearly shows O–H bond breaking and N–H bond formation. Such phenomenon is shared by all trajectories involving electron-driven proton transfer. To reveal the underlying driven mechanism, the electronic structures of the complexes at several points of the simulation were inspected. Figure 3.11 displays the charge density changes involved during the deexcitation process. Initially (at 0 fs), the $\pi\pi^*$ state is formed upon photoexcitation. Then, the structural relaxation leads to a charge transfer state (at 30 fs), due to the charge transfer from H_2O to heptazine. Likely, the strong force due to the electronic charge separation drove the proton transfer from H_2O to the N atom of the heptazine molecule. Finally, the simulation encountered a conical intersection, at which the S_1-S_0 energy gap is 0.09 eV, producing an OH radical and a heptazine radical. In the representative trajectory for the heptazine-w5 system, the simulation shows basically the same physical picture. During the decay process, the hydrogen network formed by H_2O molecules rearranges their positions and eventually transfers one H atom to heptazine forming a conical intersection with the S_1-S_0 energy gap being 0.12 eV.

We noted that the electronic transition from the S_0 to S_1 state is optically forbidden since this transition features zero oscillator strength. Thus, though the dynamic simulations revealed

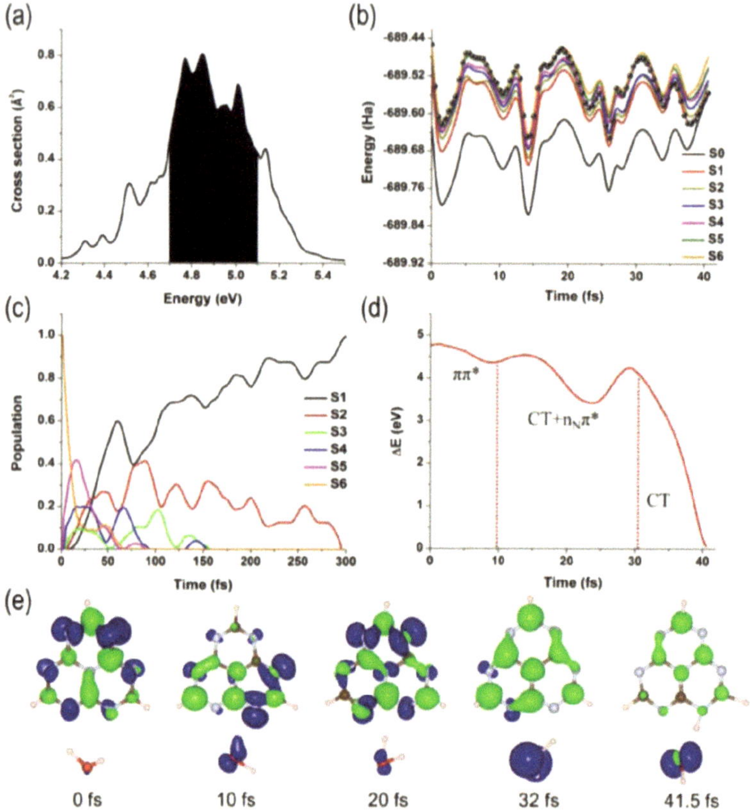

Figure 3.12 Results for a representative trajectory of the heptazine-w1 system that encounters a conical intersection at the end of the simulation. (a) Absorption cross section simulated for the lowest 10 excitations of the heptazine-w1 complex by considering 200 points for each state. The shadowed area indicates the spectral energy window for choosing the initial conditions. (b) Potential energy surfaces of S_0-S_6 states as a function of simulation time for the representative trajectory starting from an optically bright S_6 state. The block dots indicate which state is the system stays at a given time. (c) The average population of all trajectories for the heptazine-water systems. (d) Evolution of the energy gap between the instantaneous state and ground state for the representative trajectory. (e) Charge density difference for the representative trajectory at several key points. Green nodes show charge gaining, while the blue ones indicate a loss. Adapted from Ref. [40].

the nonradiative decay channel from the S_1 state, this phenomenon may happen with low possibility directly from the photoexcitation-induced S_1 state. Instead, the excited state may first decay from higher states and finally reach the S_1 state. As such, we have further carried out simulations from a higher excited state, i.e., the S_6 state, which is about 1.68 eV higher in energy than the S_1 state. The results are shown in Fig. 3.12. During the simulation, hydrogen detachment was also found to happen from H_2O to heptazine. As shown in Fig 3.12b for a representative trajectory, a fast observed conical intersection was encountered at 41 fs after starting the simulation. According to our results, the S_6 states in all 30 trajectories rapidly decayed to the lowest excited state. This agrees with Kasha's rule, stating that photochemistry and photophysical processes primarily happen at the S_1 state [58]. Further inspection revealed that the decay process after photoexcitation to the bright S_6 state shares a common mechanism for forming conical intersections to the case of the S_1 state, that is, the conical intersection is induced by electron-driven proton transfer. Thus, it can be concluded that conical intersections are an important intermediate state for photocatalytic water splitting by g-C_3N_4 materials. Importantly, the finding of the OH radical supports the experimental observations [32, 33, 39].

3.4 Confirmation by Higher-Level Theoretical Method—Complete Active Space Self-Consistent Field

It is noted that the 10 conical intersections found for nanographene epoxide feature notably different structural parameters, as shown in Fig. 3.13, which is primarily ascribed to the limitation of TDDFT in accurately treating the region close to conical intersections [41]. Therefore, CASSCF calculations have been performed to further ascertain the TDDFT results, in particular, the conical intersection [59]. By optimizing the structures at the CASSCF(10,10)/6-21G level of theory, the optimized geometry at the conical intersection features an S_1-S_0 state energy gap of 0.002 eV and the same molecular characteristics with the TDDFT results. In addition,

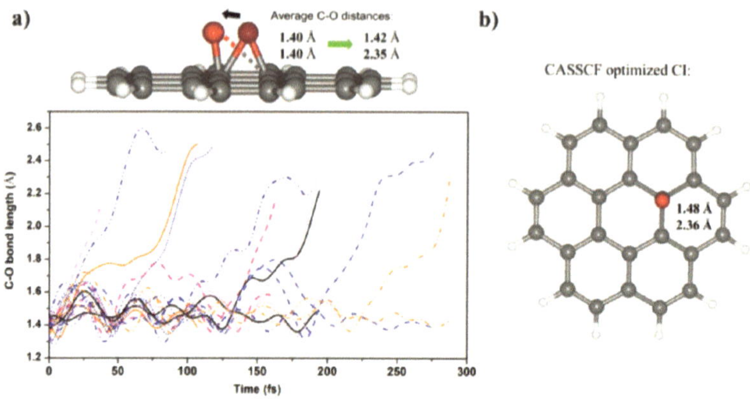

Figure 3.13 (a) Schematic of the structural changes of the nanographene epoxide during the simulation approaching the conical intersection. The upper panel shows detailed changes of the C–O bond between the initial and conical intersection geometry of the representative trajectory. (b) Conical intersection geometry obtained by CASSCF optimization. Adapted from Ref. [25].

during the simulations for the heptazine-water systems, an energy gap of 0.15 eV has been considered as a crossing threshold of two potential energy surfaces, this is, tentatively treated as a conical intersection after the gap being smaller than 0.15 eV. Since TDDFT has the typical drawback to deal with the conical intersections and may result in artificial charge transfer [41]. For confirmation, CASSCF calculations have been performed to check the TDDFT results for the heptazine-water system. First, using the conical intersection geometry of the simulations as the initial input for CASSCF single-point calculation, the S_1-S_0 gap was predicted to be 0.37 eV, which is 0.25 eV higher than the TDDFT result. Further, CASSCF optimization was carried out, with a good agreement on the S_1-S_0 energy difference obtained (0.09 vs. 0.12 eV). In addition, the conical intersection structural features for the heptazine-w1 system from these two methods are basically in good agreement, as shown in Fig 3.14. All these satisfying agreements in the TDDFT and CASSCF results prove the feasibility of studying excited-state dynamics of various materials by FSSH based TDDFT method coupled with a higher level of theory.

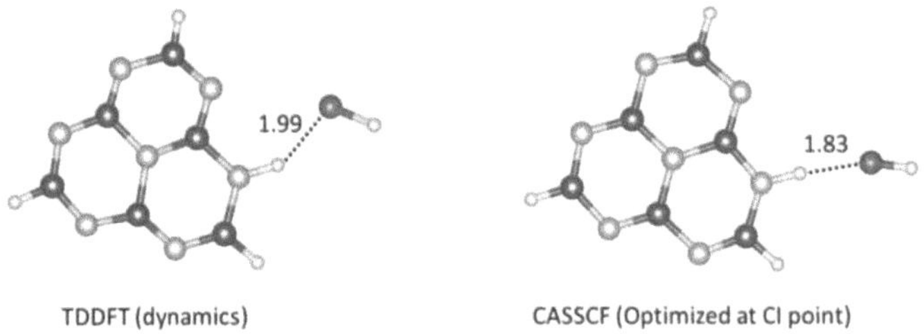

Figure 3.14 Structural comparison of the conical intersection geometries of the heptazine-w1 system obtained by TDDFT and CASSCF. The values show the O···H distances in Å. CASSCF optimization found basically the same conical intersection to the TDDFT excited-state dynamic simulation. Adapted from Ref. [40].

3.5 Summary

In summary, this work utilizes TDDFT for excited-state dynamics simulations to study the photo-relaxation of graphitic carbon-related nanostructures (nanographene epoxide and heptazine-water clusters) to gain insights into their corresponding properties, that is, photoluminescence and photocatalysis.

In detail, the deexcitation of a nanographene epoxide upon photoexcitation from the lowest excited S_1 state has been studied by nonadiabatic dynamics simulations. We found a fast nonradiative decay pathway via conical intersection between the S_1 and the S_0 state, which happens within 300 fs for 33.3% of the studied trajectories. The conical intersections are induced by breaking one C–O bond at the epoxy moiety. We have shown that conical intersections are very likely to be a key factor influencing the excited state decays in graphene-related materials. This fast nonradiative decay is predicted to likely compete with relatively slow radiative decays. As an indication, such competition between radiative and nonradiative excited-state decays is also possible to be induced by other chemical species, thus governing the light emission efficiencies of related materials. For the heptazine-water complexes, we show that photorelaxation via conical intersections facilitated by electron-driven proton transfer is an important decay pathway for photoexcited heptazine–H_2O clusters. During the excited-state decay process, the splitting of one water molecule happens rapidly within tens of femtoseconds, leading to the formation of an OH radical and a heptazine radial. These results contribute a valuable physical picture for the photochemical water splitting mechanism and H_2 evolution and provide implications on how to improve the photocatalytic H_2 producing efficiencies of g-C_3N_4 materials.

There are now increasing interests in understanding the nonradiative recombination in materials science. Though TDDFT has been widely shown to be less accurate to predict excited state properties, in particular the nonradiative decays, TDDFT coupled with higher accuracy theory method might be is a compromised, practical way to investigate the excited state decays of various materials, especially those with a large number of atoms needed for modeling.

Acknowledgments

Grants from the Innovation Committee of Shenzhen Municipality (JCYJ20170818104105891), and the Research Grants Council of Hong Kong SAR (11305618, 11334716, and 11304415), the National Natural Science Foundation of China (21703190, 11874081, and 12004209), and the Colleges and Universities Twenty Terms Foundation of Jinan City (2019GXRC034) are sincerely acknowledged.

References

1. Terrones, M., Botello-Méndez, A. R., Campos-Delgado, J., López-Urías, F., Vega-Cantú, Y. I., Rodríguez-Macías, F. J., Elías, A. L., Muñoz-Sandoval, E., Cano-Márquez, A. G., Charlier, J.-C., Terrones, H. (2010). Graphene and graphite nanoribbons: morphology, properties, synthesis, defects and applications. *Nano Today* **5**(4), 351–372.
2. Tasis, D., Tagmatarchis, N., Bianco, A., Prato, M. (2006). Chemistry of carbon nanotubes. *Chem. Rev.* **106**(3), 1105–1136.
3. Novoselov, K. S., Geim, A. K., Morozov, S. V., Jiang, D., Zhang, Y., Dubonos, S. V., Grigorieva, I. V., Firsov, A. A. (2004). Electric field effect in atomically thin carbon films. *Science* **306**(5696), 666–669.

4. Bonaccorso, F., Colombo, L., Yu, G., Stoller, M., Tozzini, V., Ferrari, A. C., Ruoff, R. S., Pellegrini, V. (2015). Graphene, related two-dimensional crystals, and hybrid systems for energy conversion and storage. *Science* **347**(6217), 1246501.
5. Hao, Y., Wang, L., Liu, Y., Chen, H., Wang, X., Tan, C., Nie, S., Suk, J. W., Jiang, T., Liang, T., Xiao, J., Ye, W., Dean, C. R., Yakobson, B. I., McCarty, K. F., Kim, P., Hone, J., Colombo, L., Ruoff, R. S. (2016). Oxygen-activated growth and bandgap tunability of large single-crystal bilayer graphene. *Nat. Nanotechnol.* **11**(5), 426–431.
6. Zhang, R. Q., Bertran, E., Lee, S. T. (1998). Size dependence of energy gaps in small carbon clusters: the origin of broadband luminescence. *Diamond Relat. Mater.* **7**(11), 1663–1668.
7. Bacon, M., Bradley, S. J., Nann, T. (2014). Graphene quantum dots. *Part. Part. Syst. Charact.* **31**(4), 415–428.
8. Zheng, X. T., Ananthanarayanan, A., Luo, K. Q., Chen, P. (2015). Glowing graphene quantum dots and carbon dots: properties, syntheses, and biological applications. *Small* **11**(14), 1620–1636.
9. Shen, J., Zhu, Y., Yang, X., Li, C. (2012). Graphene quantum dots: emergent nanolights for bioimaging, sensors, catalysis and photovoltaic devices. *Chem. Comm.* **48**(31), 3686–3699.
10. Cao, L., Meziani, M. J., Sahu, S., Sun, Y.-P. (2013). Photoluminescence properties of graphene versus other carbon nanomaterials. *Acc. Chem. Res.* **46**(1), 171–180.
11. Eda, G., Lin, Y.-Y., Mattevi, C., Yamaguchi, H., Chen, H.-A., Chen, I.-S., Chen, C.-W., Chhowalla, M. (2010). Blue photoluminescence from chemically derived graphene oxide. *Adv. Mater.* **22**(4), 505–509.
12. Zhu, S., Song, Y., Wang, J., Wan, H., Zhang, Y., Ning, Y., Yang, B. (2017). Photoluminescence mechanism in graphene quantum dots: quantum confinement effect and surface/edge state. *Nano Today* **13**, 10–14.
13. Li, L., Wu, G., Yang, G., Peng, J., Zhao, J., Zhu, J.-J. (2013). Focusing on luminescent graphene quantum dots: current status and future perspectives. *Nanoscale* **5**(10), 4015–4039.
14. Johari, P., Shenoy, V. B. (2011). Modulating optical properties of graphene oxide: role of prominent functional groups. *ACS Nano* **5**(9), 7640–7647.
15. Cao, L., Meziani, M. J., Sahu, S., Sun, Y.-P. (2012). Photoluminescence properties of graphene versus other carbon nanomaterials. *Acc. Chem. Res.* **46**(1), 171–180.

16. Zhao, S., Lavie, J., Rondin, L., Orcin-Chaix, L., Diederichs, C., Roussignol, P., Chassagneux, Y., Voisin, C., Müllen, K., Narita, A., Campidelli, S., Lauret, J.-S. (2018). Single photon emission from graphene quantum dots at room temperature. *Nat. Commun.* **9**(1), 3470.
17. Dave, S. H., Gong, C., Robertson, A. W., Warner, J. H., Grossman, J. C. (2016). Chemistry and structure of graphene oxide via direct imaging. *ACS Nano* **10**(8), 7515–7522.
18. Jiao, L., Wang, X., Diankov, G., Wang, H., Dai, H. (2010). Facile synthesis of high-quality graphene nanoribbons. *Nat. Nanotechnol.* **5**(5), 321–325.
19. Naumov, A., Grote, F., Overgaard, M., Roth, A., Halbig, C. E., Nørgaard, K., Guldi, D. M., Eigler, S. (2016). Graphene oxide: a one- versus two-component material. *J. Am. Chem. Soc.* **138**(36), 11445–11448.
20. Chen, S., Ullah, N., Wang, T., Zhang, R.-Q. (2018). Tuning optical properties of graphene quantum dots by selective oxidation: a theoretical perspective. *J. Mater. Chem. C* **6**, 6875–6883.
21. Chien, C.-T., Li, S.-S., Lai, W.-J., Yeh, Y.-C., Chen, H.-A., Chen, I. S., Chen, L.-C., Chen, K.-H., Nemoto, T., Isoda, S., Chen, M., Fujita, T., Eda, G., Yamaguchi, H., Chhowalla, M., Chen, C.-W. (2012). Tunable photoluminescence from graphene oxide. *Angew. Chem. Int. Ed.* **51**(27), 6662–6666.
22. Loh, K. P., Bao, Q., Eda, G., Chhowalla, M. (2010). Graphene oxide as a chemically tunable platform for optical applications. *Nat. Chem.* **2**(12), 1015–1024.
23. Chen, S., Zhao, Y., Ullah, N., Wan, Q., Zhang, R. (2019). Revealing the trap emission in graphene-based nanostructures. *Carbon* **150**, 439–445.
24. Chen, S., Ullah, N., Zhang, R.-Q. (2018). Exciton self-trapping in sp^2 carbon nanostructures induced by edge ether groups. *J. Phys. Chem. Lett* **9**(17), 4857–4864.
25. Chen, S., Ullah, N., Zhao, Y., Zhang, R. (2019). Nonradiative excited-state decay via conical intersection in graphene nanostructures. *ChemPhysChem* **20**(21), 2754–2758.
26. Ullah, N., Chen, S., Zhang, R. (2019). Mechanism of the charge separation improvement in carbon-nanodot sensitized g-C_3N_4. *Appl. Surf. Sci.* **487**, 151–158.
27. Ullah, N., Chen, S., Zhang, R. (2018). Mechanism of charge separation and frontier orbital structure in graphitic carbon nitride and graphene quantum dots. *ChemPhysChem* **19**(19), 2534–2539.
28. Huang, C., Wen, Y., Ma, J., Dong, D., Shen, Y., Liu, S., Ma, H., Zhang, Y. (2021). Unraveling fundamental active units in carbon nitride for photocatalytic oxidation reactions. *Nat. Commun.* **12**(1), 320.

29. Inagaki, M., Tsumura, T., Kinumoto, T., Toyoda, M. (2019). Graphitic carbon nitrides (g-C_3N_4) with comparative discussion to carbon materials. *Carbon* **141**, 580–607.

30. Wen, J., Xie, J., Chen, X., Li, X. (2017). A review on g-C_3N_4-based photocatalysts. *Appl. Surf. Sci.* **391**, 72–123.

31. Goettmann, F., Fischer, A., Antonietti, M., Thomas, A. (2006). Metal-free catalysis of sustainable Friedel–Crafts reactions: direct activation of benzene by carbon nitrides to avoid the use of metal chlorides and halogenated compounds. *Chem. Comm.* **43**, 4530–4532.

32. Wang, X., Maeda, K., Thomas, A., Takanabe, K., Xin, G., Carlsson, J. M., Domen, K., Antonietti, M. (2009). A metal-free polymeric photocatalyst for hydrogen production from water under visible light. *Nat. Mater.* **8**(1), 76–80.

33. Ye, S., Wang, R., Wu, M.-Z., Yuan, Y.-P. (2015). A review on g-C_3N_4 for photocatalytic water splitting and CO_2 reduction. *Appl. Surf. Sci.* **358**, 15–27.

34. Mishra, A., Mehta, A., Basu, S., Shetti, N. P., Reddy, K. R., Aminabhavi, T. M. (2019). Graphitic carbon nitride (g–C_3N_4)–based metal-free photocatalysts for water splitting: a review. *Carbon* **149**, 693–721.

35. Wang, S., Zhang, J., Li, B., Sun, H., Wang, S. (2021). Engineered graphitic carbon nitride-based photocatalysts for visible-light-driven water splitting: a review. *Energy Fuels* **35**(8), 6504–6526.

36. Chen, S., Ullah, N., Zhang, R. (2020). Engineering the excited state of graphitic carbon nitride nanostructures by covalently bonding with graphene quantum dots. *Theor. Chem. Acc.* **139**(2), 20.

37. Xiong, W., Chen, S., Huang, M., Wang, Z., Lu, Z., Zhang, R. Q. (2018). Crystal-face tailored graphitic carbon nitride films for high-performance photoelectrochemical cells. *ChemSusChem* **11**(15), 2497–2501.

38. Ehrmaier, J., Karsili, T. N. V., Sobolewski, A. L., Domcke, W. (2017). Mechanism of photocatalytic water splitting with graphitic carbon nitride: photochemistry of the heptazine–water complex. *J. Phys. Chem. A* **121**(25), 4754–4764.

39. Pang, X., Jiang, C., Xie, W., Domcke, W. (2019). Photoinduced electron-driven proton transfer from water to an N-heterocyclic chromophore: nonadiabatic dynamics studies for pyridine–water clusters. *Phys. Chem. Chem. Phys.* **21**(26), 14073–14079.

40. Ullah, N., Chen, S., Zhao, Y., Zhang, R. (2019). Photoinduced water–heptazine electron-driven proton transfer: perspective for water splitting with g-C_3N_4. *J. Phys. Chem. Lett* **10**(15), 4310–4316.

41. Crespo-Otero, R., Barbatti, M. (2018). Recent advances and perspectives on nonadiabatic mixed quantum–classical dynamics. *Chem. Rev.* **118**(15), 7026–7068.
42. Shu, Y., Fales, B. S., Levine, B. G. (2015). Defect-induced conical intersections promote nonradiative recombination. *Nano Lett.* **15**(9), 6247–6253.
43. Barbatti, M. (2011). Nonadiabatic dynamics with trajectory surface hopping method. *Wiley Interdiscip. Rev. Comput. Mol. Sci.* **1**(4), 620–633.
44. Adamo, C., Jacquemin, D. (2013). The calculations of excited-state properties with time-dependent density functional theory. *Chem. Soc. Rev.* **42**(3), 845–856.
45. Zhao, M., Yang, F., Xue, Y., Xiao, D., Guo, Y. (2014). A time-dependent DFT study of the absorption and fluorescence properties of graphene quantum dots. *ChemPhysChem* **15**(5), 950–957.
46. Yue, L., Liu, Y., Zhu, C. (2018). Performance of TDDFT with and without spin-flip in trajectory surface hopping dynamics: cis–trans azobenzene photoisomerization. *Phys. Chem. Chem. Phys.* **20**(37), 24123–24139.
47. Fazzi, D., Barbatti, M., Thiel, W. (2015). Modeling ultrafast exciton deactivation in oligothiophenes via nonadiabatic dynamics. *Phys. Chem. Chem. Phys.* **17**(12), 7787–7799.
48. Barbatti, M. (2014). Photorelaxation induced by water–chromophore electron transfer. *J. Am. Chem. Soc.* **136**(29), 10246–10249.
49. Radovic, L. R., Silva-Tapia, A. B., Vallejos-Burgos, F. (2011). Oxygen migration on the graphene surface. 1. Origin of epoxide groups. *Carbon* **49**(13), 4218–4225.
50. Yan, J.-A., Chou, M. Y. (2010). Oxidation functional groups on graphene: structural and electronic properties. *Phys. Rev. B* **82**(12), 125403.
51. Loh, K. P., Bao, Q., Eda, G., Chhowalla, M. (2010). Graphene oxide as a chemically tunable platform for optical applications. *Nat. Chem.* **2**(12), 1015.
52. Krishnan, R., Binkley, J. S., Seeger, R., Pople, J. A. (1980). Self-consistent molecular orbital methods. XX. A basis set for correlated wave functions. *J. Chem. Phys.* **72**(1), 650–654.
53. Hehre, W. J., Stewart, R. F., Pople, J. A. (1969). Self-consistent molecular-orbital methods. I. Use of Gaussian expansions of Slater-type atomic orbitals. *J. Chem. Phys.* **51**(6), 2657–2664.
54. Frisch, M., Trucks, G., Schlegel, H. B., Scuseria, G. E., Robb, M. A., Cheeseman, J. R., Scalmani, G., Barone, V., Mennucci, B., Petersson,

G., Nakatsuji, H. (2009). Gaussian 09, Revision A. 02. Gaussian Inc., Wallingford, CT.

55. Barbatti, M., Ruckenbauer, M., Plasser, F., Pittner, J., Granucci, G., Persico, M., Lischka, H. (2014). Newton-X: a surface-hopping program for nonadiabatic molecular dynamics. *Wiley Interdiscip. Rev. Comput. Mol. Sci.* **4**(1), 26–33.
56. Deng, X., Sun, J., Yang, S., Shen, H., Zhou, W., Lu, J., Ding, G., Wang, Z. (2015). The emission wavelength dependent photoluminescence lifetime of the N-doped graphene quantum dots. *Appl. Phys. Lett.* **107**(24), 241905.
57. Wang, L., Zhu, S.-J., Wang, H.-Y., Qu, S.-N., Zhang, Y.-L., Zhang, J.-H., Chen, Q.-D., Xu, H.-L., Han, W., Yang, B., Sun, H.-B. (2014). Common origin of green luminescence in carbon nanodots and graphene quantum dots. *ACS Nano* **8**(3), 2541–2547.
58. Valeur, B., Berberan-Santos, M. N. (2012). *Molecular Fluorescence: Principles and Applications*, John Wiley & Sons.
59. Roos, B. O., Taylor, P. R., Sigbahn, P. E. M. (1980). A complete active space SCF method (CASSCF) using a density matrix formulated super-CI approach. *Chem. Phys.* **48**(2), 157–173.

Chapter 4

Mixed-Reference Spin-Flip Time-Dependent Density Functional Theory (MRSF-TDDFT) as a Method of Choice for Nonadiabatic Molecular Dynamics

Seunghoon Lee,[a] Woojin Park,[b] Hiroya Nakata,[c] Michael Filatov,[b] and Cheol Ho Choi[b]

[a] *Division of Chemistry and Chemical Engineering, California Institute of Technology, Pasadena, California 91125, USA*

[b] *Department of Chemistry, Kyungpook National University, Daegu 41566, South Korea*

[c] *R&D Center Kagoshima, Kyocera, 1-4 Kokubu Yamashita-cho, Kirishimashi, Kagoshima 899-4312, Japan*

cchoi@knu.ac.kr

The recently developed MRSF-TDDFT method is capable of alleviating the major limitations of the DFT, standard linear-response TDDFT as well as the original SF-TDDFT methods, thus providing a promising protocol for computing a wide range of molecular systems, from weakly correlated to strongly correlated. Hence, MRSF-TDDFT can be a method of choice for running the non-adiabatic

Time-Dependent Density Functional Theory: Nonadiabatic Molecular Dynamics
Edited by Chaoyuan Zhu

ISBN 978-981-4968-42-3 (Hardcover), 978-1-003-31921-4 (eBook)
www.jennystanford.com

molecular dynamics simulations (NAMD). In this contribution, the main aspects of the new methodology are documented and the advantages of using it are highlighted.

4.1 Introduction

There is growing interest in studying the non-adiabatic dynamics of electronically excited states, which is implicated in a wide range of phenomena, from the fundamental processes, such as photosynthesis and vision, to technologically important processes of light-to-energy conversion [1] and directed molecular motion [2]. To understand the underlying photochemistry and photo-physics, non-adiabatic molecular dynamics (NAMD) simulations have become standard tools in the repertoire of researchers. Hence, the accurate and efficient description of the excited electronic states has become more important than ever.

In order for NAMD to be successful, it is necessary to perform theoretical simulations using quantum chemical methodologies that: (i) provide for a balanced and accurate account of the dynamic correlation as well as the multi-reference (MR) characteristics (including important double excitations) of the electronic states, (ii) are capable of correctly describing the conical intersections including ground and excited electronic states as shown in Fig. 4.1, and (iii) are efficient enough to perform statistical dynamical sampling for several picoseconds and beyond. Although there is a multitude of quantum chemical computational methods oriented at the excited states, the most accurate methods, such as the MR coupled cluster (MRCC), the MR perturbation theory (MRPT) and the equations-of-motion coupled cluster (EOMCC)methods, are much too expensive to be applied to systems of realistic size. Therefore, relatively economical methods, such as the complete active space self-consistent field (CASSCF) method, the algebraic-diagrammatic construction (ADC)method and the linear-response time-dependent density functional theory (LR-TDDFT) method have become the workhorses of the current NAMD simulations. However, the simplified methodologies have a number of drawbacks, e.g., the CASSCF neglects the dynamic electron correlation, the ADC(2) and

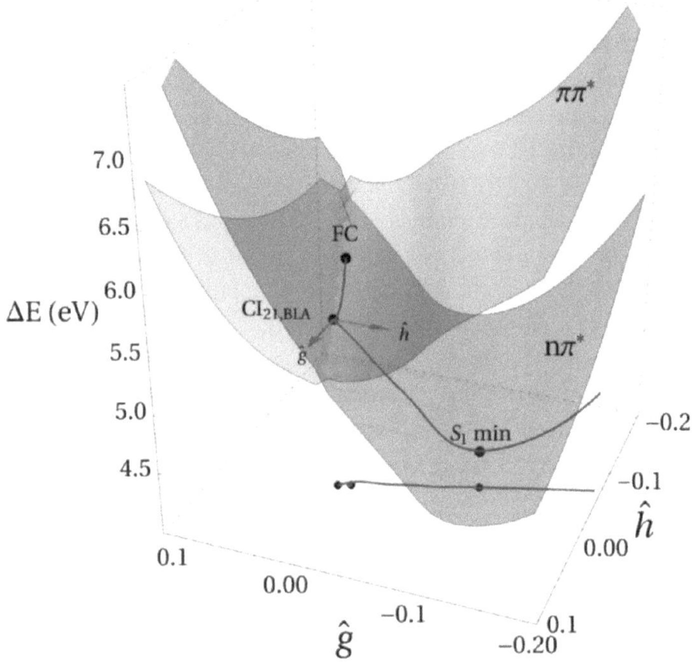

Figure 4.1 A schematic diagram of conical intersection. Reprinted with permission from Ref. [3]. Copyright (2021) American Chemical Society.

the TDDFT methods cannot describe the proper dimensionality of the S_1/S_0 conical intersection seam [4], which limit the reliability of their predictions considerably [3].

There are a number of well-known failures of the popular LR-TDDFT method, e.g., when describing the energy of long-range charge transfer excitations [5–9], excited states with substantial double excitation character [10–13], excited states of molecules undergoing bond breaking [13–15], and real and avoided crossings between the ground and excited states of molecules [16–19]. Some of these drawbacks, the incorrect description of the S_1/S_0 conical intersections and the poor description of multi-reference electronic states, can be corrected to some extent by the use of the spin-flip (SF)-TDDFT method [20, 21], which employs an open-shell high-spin, e.g., triplet, reference state instead of the closed-shell reference of LR-TDDFT. However, the conventional formulation of SF-TDDFT selects only one component, e.g., $|\alpha\alpha\rangle$, of the degenerate triplet state as a reference, which leads to considerable spin contamination of the resulting excited electronic states; perhaps, with the exception of a few low-lying states [22]. The spin contamination of SF-TDDFT is different from that of the unrestricted Hatree–Fock (UHF) wavefunction. In SF-TDDFT based on the restricted open-shell HF (ROHF) formalism, the main source of the spin contamination is the spin incompleteness of the response configurations. Therefore, a key to solving this problem is to expand the response space of SF-TDDFT such that it includes the missing configurations; thus, making the response space complete.

Several approaches have been developed to tackle spin-contamination of the SF configuration interaction with single excitations (CIS), [19, 22–24] which is a wavefunction version of SF-TDDFT. However, unlike SF-CIS, a considerable challenge for TDDFT is to go beyond the adiabatic approximation and to account for more than single excitations [25, 26]. Because of this difficulty, only a few methods have been developed to address the spin-contamination problem of SF-TDDFT. A possible way of increasing the response space is to use higher excitation operators [13]. Beyond the time-dependent Kohn–Sham (TD–KS) formalism, on the other hand, tensor equation-of-motion (TEOM) approaches have achieved considerable progress, yielding a series of spin-adapted (SA)-SF-DFT methodologies [26–29]. These approaches can produce correct spin eigenstates by applying tensor operators to a tensor reference state. However, the matrix elements of the TEOM are evaluated using the Wigner–Eckart theorem, which is not satisfied by the approximate density functionals. Thus, the SA-SF-DFT formalism [29] requires an *a posteriori* DFT correction to the SA-SF-CIS equations [29]. Due to the complexity of TEOM, the analytic energy gradient for SA-SF-DFT has yet to be derived.

Rather than using high excitation operators [13] or using tensor operators with a tensor reference [26], we have recently shown a way of expanding the response space by combining more than one reference state into a single *mixed* reference [30, 31]. Because LR-TDDFT based on the TD–KS equation is derived from a single Slater determinant [32, 33], this type of mixed-reference approach has never been considered because of violation of the *idempotency* condition. However, we have shown that the condition is preserved if one can build a *hypothetical* single reference from not only the $|\alpha\alpha\rangle$ ($M_S = +1$) but also the $|\beta\beta\rangle$ ($M_S = -1$) components of an ROHF triplet. To achieve this, a transformation to spinor-like mixed spin functions was proposed, which combines the two components into a *hypothetical* single reference state, finally yielding an *idempotent* mixed-reference reduced density matrix (MR-RDM). The TD–KS equations for this particular MR-RDM double the number of response configurations generated in TDDFT. The resulting mixed-reference SF-TDDFT (MRSF-TDDFT) has several advantages over the conventional collinear formulation of SF-TDDFT. The spin contamination of the response states of SF-TDDFT is nearly perfectly removed, thereby considerably simplifying identification of the excited states as singlets and triplets; especially, in "black-box"-type applications, such as automatic geometry optimization, reaction path following, or molecular dynamics simulations.

In recent years, a series of benchmark studies have been performed to reveal the full potential of the newly developed MRSF-TDDFT formalism. In the following, the basic formulation of MRSF-TDDFT is given, followed by the results of various benchmark calculations illustrating the performance of MRSF-TDDFT.

4.2 Mixed-Reference Spin-Flip Time-Dependent Density Functional Theory (MRSF-TDDFT)

4.2.1 Eliminating Spin-Contamination of SF-TDDFT

Detailed formulation of the MRSF-TDDFT can be found elsewhere [30, 31] and here only a brief overview is given. The derivation of the MRSF-TDDFT method [30] is based on the density-matrix formulation of the time-dependent Kohn–Sham methodology [32]. In MRSF-TDDFT, the zeroth-order MR-RDM, $\rho_0^{\mathrm{MR}}(x, x')$, is defined to satisfy the idempotence relation,

$$\rho_0^{\mathrm{MR}}(x, x') = \int \rho_0^{\mathrm{MR}}(x, x'')\rho_0^{\mathrm{MR}}(x'', x')dx'', \qquad (4.1)$$

with its diagonal part (the density), $\rho_0^{\text{MR}}(x) = \rho_0^{\text{MR}}(x, x)$, obtained as an equiensemble of the $M_S = +1$ and $M_S = -1$ components of a triplet (reference) state,

$$\rho_0^{\text{MR}}(x) = \frac{1}{2}\left\{\rho_0^{M_S=+1}(x) + \rho_0^{M_S=-1}(x)\right\}. \tag{4.2}$$

According to the ensemble DFT [34], the spatial parts of the molecular orbitals (MOs) optimized by using the MR density in Eq. (4.2) are identical to those optimized by ROHF formalism with the open-shell high-spin triplet reference. Therefore, even though the MR density of Eq. (4.2) requires two configurations, no additional multi-configurational orbital optimization is necessary, greatly simplifying overall computational procedures while taking the full advantage of multi-configurational features. As a result, the spatial parts of MOs of SF-TDDFT are identical to those of MRSF-TDDFT. It was shown that the formal idempotency of the MR-RDM, which is the central property of single-reference linear response theory, can be recovered by rotating spin function of the MOs in the singly occupied MO space (the "O" space). The rotated spin functions become mixed spin functions, labeled in Eq. (4.3) as s_1 and s_2.

$$s_1 \equiv \frac{(1+i)\alpha + (1-i)\beta}{2}, \quad s_2 \equiv \frac{(1-i)\alpha + (1+i)\beta}{2}. \tag{4.3}$$

Here, i denotes the imaginary unit. Within the Tamm–Dancoff approximation [35–37], the use of MR-RDM in the linear-response formalism yields two completely decoupled linear-response equations for the singlet and the triplet excited states, respectively [30],

$$\left(\mathbf{A}_{\text{S}}^{(0)} + \mathbf{A}_{\text{S}}'\right)\mathbf{X}_{\text{S}}^{I} = \Omega_{\text{S}}^{I}\mathbf{X}_{\text{S}}^{I}, \tag{4.4}$$

$$\left(\mathbf{A}_{\text{T}}^{(0)} + \mathbf{A}_{\text{T}}'\right)\mathbf{X}_{\text{T}}^{I} = \Omega_{\text{T}}^{I}\mathbf{X}_{\text{T}}^{I}, \tag{4.5}$$

where S and T in subscript denote singlet and triplet states, $\mathbf{A}_{\text{S}}^{(0)}$ and $\mathbf{A}_{\text{T}}^{(0)}$ are the orbital Hessian matrices derived within the linear response formalism, and $\mathbf{A}_{\text{S}}'$ and $\mathbf{A}_{\text{T}}'$ are coupling matrices between the configurations originating from different components, $M_S = +1$ and $M_S = -1$, of the mixed reference state [30, 31]. $\mathbf{X}_{\text{S}}^{I}$, $\mathbf{X}_{\text{T}}^{I}$ and Ω_{S}^{I}, Ω_{T}^{I} are the amplitude vectors and excitation energies of the Ith excited state with respect to the reference state, respectively. Because the singlet and triplet response states are generated by different response equations, there is no mixing between the two different groups of response configurations. This is a great advantage of MRSF-TDDFT over SF-TDDFT, since only one group of response states is produced by the latter theory, requiring further identification of the response states as singlets or triplets.

The advantage of having a two-component reference state is shown in Fig. 4.2, where the $M_S = +1$ and -1 triplet components as well as their corresponding response configurations are shown with the black and red arrows, respectively. The response configurations shown by the red arrows represent the configurations missing in the conventional SF-TDDFT, the account of which largely eliminates the spin contamination of the response states in MRSF-TDDFT. The O $\rightarrow$ O type configurations shown in Fig. 4.2 by the blue arrows originate from both $M_S = +1$ and $M_S = -1$ components of the MR state. They cover the **G**, **D**, **L** and **R** configurations defined in Fig. 4.2, which are important for proper description of the open-shell singlet and triplet states of diradicals. As these configurations occur in both the SF and MRSF methods, it might be expected that both methods should be equally able to describe diradicals and to eliminate the spin-contamination. However, as the contributions of **L** and **R** to the SF response states depend on the particular spin-flip transitions, a mismatch between their contributions into the final response state may become a main source of spin contamination. For example, in the SF method, the $\alpha \rightarrow \beta$ O1 $\rightarrow$ O1 spin-flip transition yields the **L** configuration, while the $\alpha \rightarrow \beta$ O2 $\rightarrow$ O2 spin-flip transition yields the **R** configuration. As the two configurations occur due to two distinct spin-flip transitions, their contributions into the final SF response state may become unequal, which may yield substantial spin-contamination. The problem of the mismatch of the contributions of the **L** and **R** configurations was solved in MRSF by externally contracting the same configurations originating from the different spin-flip orbital transitions from the two components, $M_S = +1$ and $M_S = -1$, of the MR state [30].

It is noteworthy that not all of the electronic configurations shown in Fig. 4.2 can be recovered using the MR-RDM of Eq. (4.2). Thus, four out of six type IV configurations (i.e., those shown by the gray arrows in Fig. 4.2) remain unencountered for. Typically, these configurations represent high-lying excited states

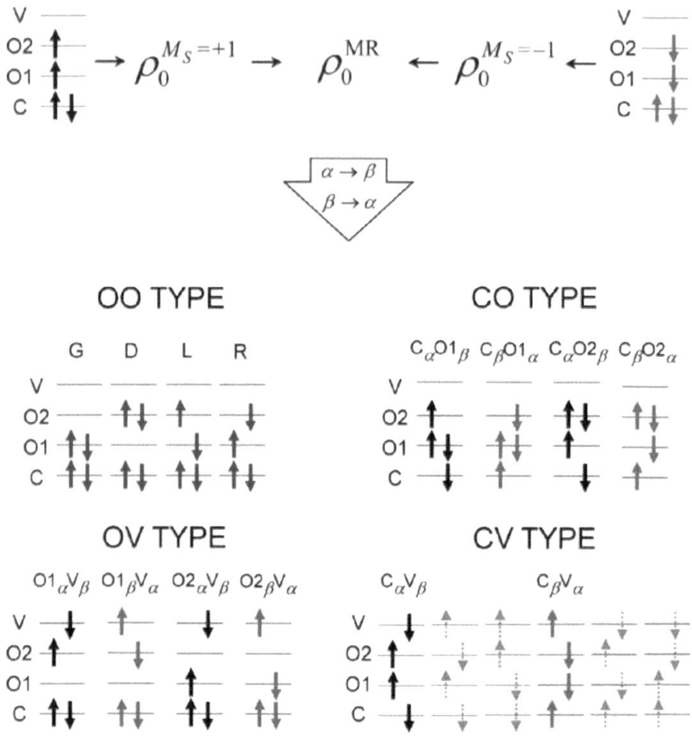

Figure 4.2 Upper panel shows the two references of MRSF-TDDFT denoted by black and red arrows. The zeroth-order MR-RDM which combines $M_S = +1$ and -1 RDMs is used in MRSF-TDDFT, while only the $M_S = +1$ RDM are used in SF-TDDFT. In the lower panel, the electronic configurations that can be generated by spin-flip linear responses from the MR-RDM are given by blue, black, and red arrows. The blue ones are generated from both references, which require a symmetrization procedure to eliminate the OO-type spin contamination. The black and red ones are generated from Ms = +1 and –1 references, respectively. By contrast, those of SF-TDDFT are only the blue and black ones. Configurations that cannot be obtained even in the MRSF-TDDFT are denoted by gray dashed arrows.

and make insignificant contributions to the low-lying states of organic molecules [30]. Thus, the effect of the missing configurations on the spin contamination is expected to be small.

Dimension of the response space of SF-TDDFT is $n_{occ}^{\alpha} n_{vir}^{\beta}$, where n_{occ}^{α} and n_{vir}^{β} are the number of occupied α and virtual β MOs of the

$M_S = +1$ reference. While, in case of MRSF-TDDFT, the dimensions of the singlet and triplet response spaces are $n^{\alpha}_{occ}n^{\beta}_{vir} - 1$ and $n^{\alpha}_{occ}n^{\beta}_{vir} - 3$, respectively. Thus, the total dimension is $2n^{\alpha}_{occ}n^{\beta}_{vir} - 4$, as the response space of MRSF-TDDFT originate from two reference states. The –4 contribution eliminates the redundant configurations shown by the blue arrows in Fig. 4.2.

4.2.2 Combining Response States from Individual References

Although MR-RDM of Eq. (4.2) employs two reference states, their respective response configurations are generated independently within the linear response space. Instead of resorting to higher non-linear formulations, an *a posteriori* coupling was introduced [30, 31] as:

$$(\mathbf{A}'_{\mathrm{S}})_{pq,rs} = c_H \left\langle \Psi^{M_S=+1}_{p\alpha q\beta} \middle| \hat{H} \middle| \Psi^{M_S=-1}_{r\beta s\alpha} \right\rangle, \quad (\mathbf{A}'_{\mathrm{T}})_{pq,rs} = -(\mathbf{A}'_{\mathrm{S}})_{pq,rs}, \tag{4.6}$$

where $\Psi^{M_S=+1}_{p\alpha q\beta}$ denotes configuration generated by spin-flip transition from alpha p MO to beta q MO of $M_S = +1$ reference and $\Psi^{M_S=-1}_{r\beta s\alpha}$ originates from beta r MO to alpha s MO of $M_S = -1$ reference.

4.3 Performance Analysis of MRSF-TDDFT

4.3.1 Doubly Excited Configurations

As shown in Fig. 4.2, a great advantage of MRSF-TDDFT over conventional TDDFT is its ability to include the doubly excited configurations in its response space. The doubly excited configurations are obtained by spin-flip transitions from the high-spin triplet reference state without requiring genuine double excitations. The advantage of having the doubly excited configurations can be illustrated by the avoided crossings between the potential energy surfaces (PES) of the $1^1\Sigma_g^+$ and the $2^1\Sigma_g^+$ states as well as the $2^1\Sigma_g^+$ and the $3^1\Sigma_g^+$ states of H_2. Although H_2 is the simplest many-electron molecule, the accurate description of its ground and excited

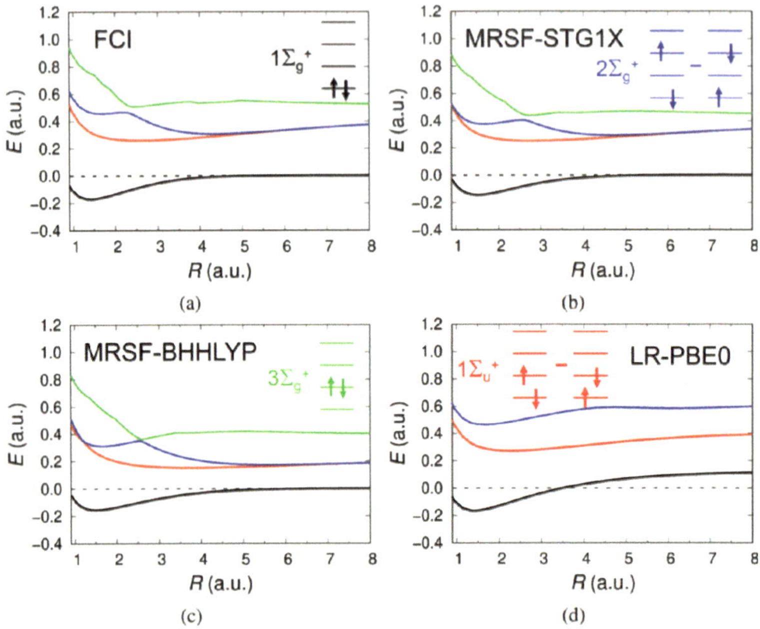

Figure 4.3 Potential energy surfaces for $^1H_2$ dissociation calculated with (a) FCI, (b) MRSF-STG1X, (c) MRSF-BHHLYP, and (d) LR-PBE0, methods. The all-electron cc-pVTZ [40] basis set is employed in all the calculations. The black, red, blue, and green curves correspond to the S_0, S_1, S_2, and S_3 states, respectively. The leading configurations in the states at the Franck-Condon geometry (R = 1.40 a.u.) are shown as insets with the corresponding color. After passage through the avoided crossing region, the characteristics of the intersecting states are swapped. Reprinted with permission from Ref. [39]. Copyright (2021) American Chemical Society.

states, which undergo avoided crossings along the bond dissociation coordinate, is not as a straightforward task as it may seem [13, 38]. The corresponding PESs are inspected along the whole dissociation path of the molecule and compared with the values obtained by the full configuration interaction (FCI) method; see Fig. 4.3 [39]. Near the equilibrium bond length ($\sim$ 1.4 a.u.), the ground electronic state of H_2 is dominated by a single Slater determinant with the doubly occupied $1\sigma_g$ MO. The $1^1\Sigma_u^+$ and $2^1\Sigma_g^+$ states are dominated by the open-shell singlet configurations $\frac{1}{\sqrt{2}}(|1\sigma_g 1\bar{\sigma}_u\rangle - |1\bar{\sigma}_g 1\sigma_u\rangle)$ and

$\frac{1}{\sqrt{2}}(|1\sigma_g 2\bar{\sigma}_g\rangle - |1\bar{\sigma}_g 2\sigma_g\rangle)$, respectively. The $3\Sigma_g^+$ state is a doubly excited state with the $1\sigma_u$ orbital doubly occupied, see insets in Fig. 4.3.

As the H–H bond is stretched, the doubly excited $|1\sigma_u 1\bar{\sigma}_u\rangle$ configuration slides down in energy and beyond *ca.* 2.5 a.u. it becomes strongly mixed with the $\frac{1}{\sqrt{2}}(|1\sigma_g 2\bar{\sigma}_g\rangle - |1\bar{\sigma}_g 2\sigma_g\rangle)$ configuration of $2^1\Sigma_g^+$. As a result, the $2^1\Sigma_g^+$ and $3\Sigma_g^+$ states become strongly mixed and undergo an avoided crossing at $R \sim 2.5$ a.u. At the same time, the doubly excited configuration starts making greater contribution to the ground electronic state, which leads to flattening the dissociation curve at the correct dissociation limit shown by the dashed line.

Compared to the full configuration interaction (FCI) curves, the MRSF method used in connection with three density functionals, BHHLYP [41], STG1X [42], and HFLYP [42] correctly reproduces the main features of the dissociation curves of the four lowest singlet electronic states of H_2, demonstrating the importance of doubly excited configurations, see Fig. 4.3. For comparison, the dissociation curves obtained by the LR-TDDFT method with the PBE0 functional for the lowest singlet states are shown in the lower right panel of Fig. 4.3. As LR-TDDFT misses the doubly excited configurations among its response states, only three singlet states result in the displayed region of energies and the shape of the $1^1\Sigma_g^+$ and $2^1\Sigma_g^+$ dissociation curves is incorrect.

4.3.2 Nonadiabatic Coupling Matrix Elements (NACME)

When describing non-adiabatic transitions between adiabatic electronic states, e.g., in NAMD simulations, the probability of the transition depends critically on the first-order non-adiabatic coupling matrix element (NACME) defined below.

$$\tau_{IJ}^{x} = \langle \Psi_I | \frac{\partial}{\partial x} | \Psi_J \rangle. \tag{4.7}$$

The NACME between the adiabatic states $\langle \Psi_I |$ and $| \Psi_J \rangle$ (also known as the derivative coupling vector, DCV), (together with the energy gradient difference vector, GDV) defines the branching plane of conical intersections (CIs) [43–47] as well as the rate of the non-adiabatic population transfer at and near conical intersections and avoided crossings between the states I and J. As only a handful of quantum chemical computational methods are capable of properly describing the topology of CIs [4, 16, 49, 50, 56], the computational studies of NACME (or DCV) in the scientific literature are relatively rare. This is caused not only by a limited repertoire of computational methods enabling its proper computation, but also by the complexity of the analytic formulation of its evaluation.

Within the finite difference approximation, NACME in Eq. (4.7) is obtained from the overlap integrals (IO) between the (*non-orthogonal*) states I and J. A series of fast and accurate specialized algorithms for the OI evaluation on the basis of truncated Leibniz formula was proposed by some of us [52] and this series was successfully utilized for obtaining the time-derivative non-adiabatic coupling of TDDFT and SF-TDDFT methods for the NAMD simulations. The truncated Leibniz formula of order-one (TLF(1)) and order-two (TLF(2)) methods require computational effort that scales on the order of $O(n^3)$, thus achieving a considerable speedup of the computations. Quadratic scaling ($O(n^2)$) can also be achieved with the zeroth-order (TLF(0)), which was especially useful in recent NAMD simulations [3]. With these developments, the computational burden of obtaining OIs is effectively eliminated.

The same TLF series has been recently applied to the computation of NACME in connection with the MRSF method [39]. Using FCI, MRSF and LR, the nonzero NACME elements between the $1^1\Sigma_g^+$ and $2^1\Sigma_g^+$ states were calculated along the dissociation path of H_2, see Fig. 4.4. MRSF-STG1X and MRSF-HFLYP calculations reproduce the reference FCI curves of the $\tau_{1^1\Sigma_g^+/2^1\Sigma_g^+}$ coupling elements rather well. The BHHLYP functional yields a reasonable description of the $\tau_{1^1\Sigma_g^+/2^1\Sigma_g^+}$ coupling element.

By contrast to MRSF, the LR-TDDFT methodology fails completely to reproduce the NACME values, as shown in Fig. 4.4. The dashed curve in the upper panel of Fig. 4.4 shows little similarity to the reference FCI curve. This comparison displays superiority of the MRSF method for describing the non-adiabatic couplings between the electronic states.

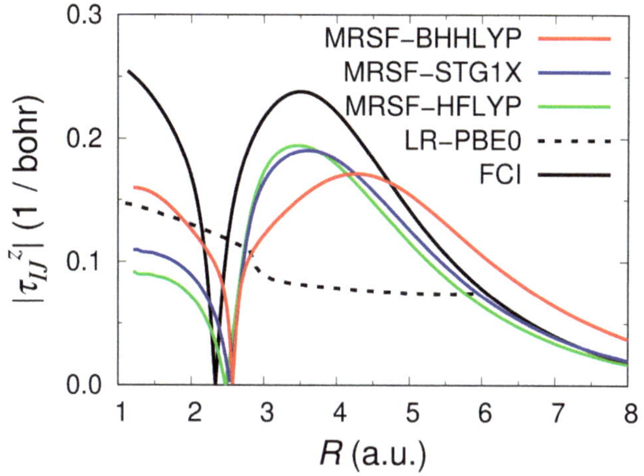

Figure 4.4 Absolute value of z-component of NACME obtained with MRSF and FCI methods for H_2 between the $1^1\Sigma_g^+$ and $2^1\Sigma_g^+$ states. NACME data for LR-PBE0 were adopted from the previous work [51]. Reprinted with permission from Ref. [39]. Copyright (2021) American Chemical Society.

4.3.3 Conical Intersections between S_1 and S_0 States ($CI_{1/0}$)

As stated in the introduction, the LR-TDDFT [53–55], fails to yield the correct dimensionality ($3N - 8$) of the S_1/S_0 CI seam and predicts a linear crossing (i.e., $3N - 7$ dimensional seam) instead [16, 48, 57], see Fig. 4.5. The problem is rooted in the absence of coupling between the S_0 state (the reference state in the LR formalism) and the excited states (the response states) [16, 48, 57]. This failure of LR-TDDFT can be corrected by using MRSF [30] without the spin-contamination pitfalls of the SF-TDDFT formalism [20, 21, 58, 59].

The accuracy of the MRSF method was tested in the calculation of S_1/S_0 CIs occurring in several organic molecules, see Fig. 4.6 and Table 4.1. This set of molecules was previously used to benchmark several other methods including the conventional SF-TDDFT [49]. In the absence of the experimental data on geometries and relative energies of CIs (which are, probably, impossible to obtain), the results of geometry optimization by the multi-reference configura-

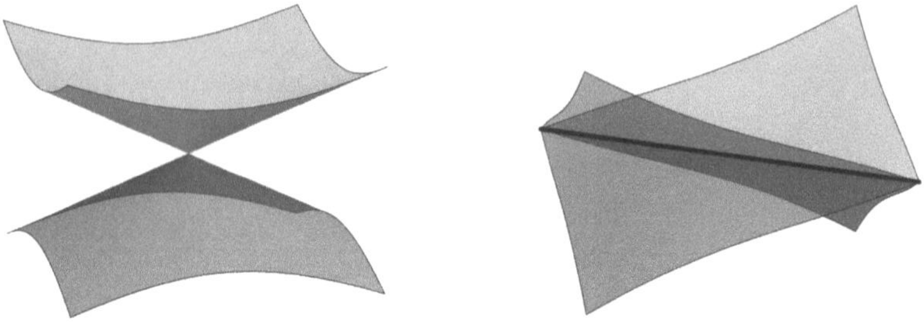

Figure 4.5 Examples of conical intersection (left panel) and linear intersection (right panel) of two potential energy surfaces. Reprinted with permission from Ref. [60]. Copyright (2019) American Chemical Society.

tion interaction with singles and doubles (MRCISD) utilizing the 6-31+G** basis set are used as the reference data [49].

The CIs in Fig. 4.6 and Table 4.1 originate from two types of transitions, the $\pi \rightarrow \pi^*$ and $n \rightarrow \pi^*$ transitions between the frontier orbitals [49, 61]. The CIs due to the $\pi \rightarrow \pi^*$ transitions originate in all the molecules shown in Fig. 4.6 with the exception of ketene. In the latter molecule, as well as the ethylidene CI of ethylene and the methylimine CI of methyliminium, the CIs originate from the $n \rightarrow \pi^*$ transitions. The former CIs can be further distinguished by the geometric distortion occurring at the CI geometry and classified as the twisted–pyramidalized (tw-pyr) or twisted–BLA (tw-BLA; BLA: bond length alternation) CIs [49, 62]. The S_0 and S_1 states at the geometry of these CIs have distinct electronic characteristics, which are associated with the diradical (Dir) or charge transfer (CT) nature of the electronic state [48, 62]. Typically, tw-pyr CIs occur in molecules where the isomerizing π-bond dissociates by homolytic mechanism and the tw-BLA CIs occur in the case of heterolytic π-bond dissociation [62]. The $n \rightarrow \pi^*$ CIs occur due to crossing between a closed-shell lone pair electronic configuration and an open-shell singlet configuration due to the one-electron transition [49].

Table 4.1 collects the relative energies ΔE (in eV) calculated with respect to the S_0 equilibrium conformation. Vertical excitation energies (VEEs) at the Franck-Condon (FC) geometry are also shown in Table 4.1. For molecules which exist in trans- and cis-conformations, both FC geometries are shown. The available

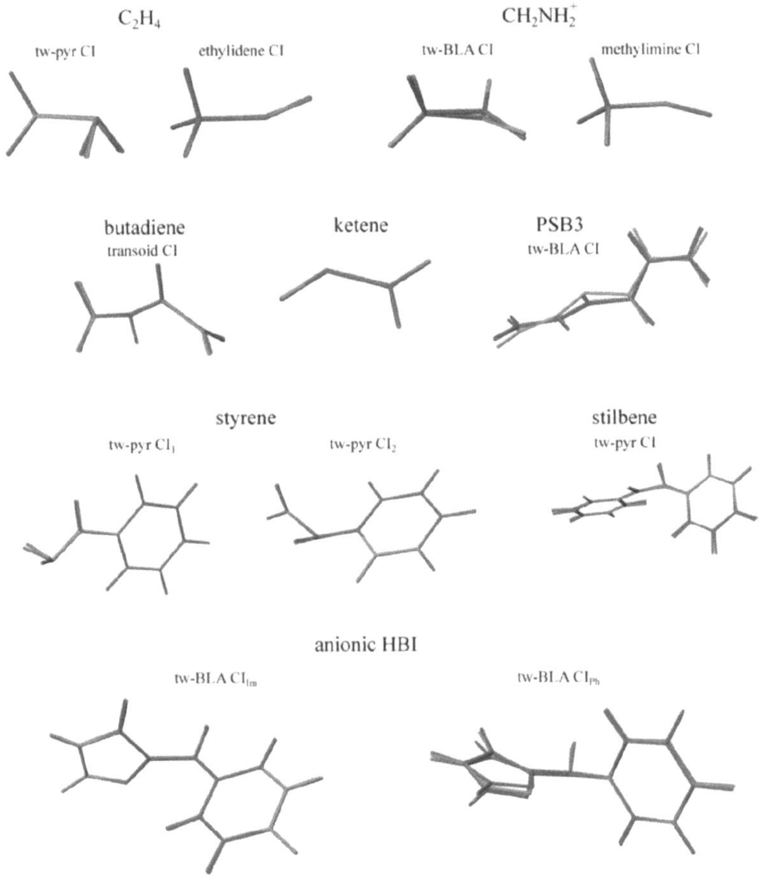

Figure 4.6 Superimposed geometries of the S_1/S_0 CIs optimized using the MRCISD/6-31+G** method (gray; geometries from Ref. [49]), SF-TD-DFT/6-31+G** method (green; from Ref. [49], and MRSF-TD-DFT/6-31+G** method (red; this work). Reprinted with permission from Ref. [60]. Copyright (2019) American Chemical Society.

experimental VEEs or their best theoretical estimates (TBE) are shown in the second column of Table 4.1.

In comparison with the MRCISD relative energies, both methods, SF and MRSF, show similar accuracy. The mean absolute deviation (MAD) of the MRSF energies reported in Table 4.1 from the MRCISD energies is 0.41 eV. The geometries obtained by MRSF also agree

Table 4.1 Relative energies ΔE (in eV) with respect to the S_0 equilibrium conformations and root-mean-square deviations (RMSD, in Å) of the geometries optimized with the DFT methods with respect to the MRCISD optimized geometries

geometry	Lit.[a]	MRCISD[b]	MRSF[c]	
			ΔE	RMSD[e]
		ethylene		
FC	7.80 [63]	8.11	7.21	na[f]
tw-pyr CI		4.79	4.94	0.066
eth CI		4.69	4.38	0.022
		methyliminium		
FC		9.52	9.28	na
tw-BLA CI		3.82	4.25	0.099
meth CI		5.63	5.43	0.016
		butadiene		
trans-FC	6.18 [63]	6.89	6.04	na
cis-FC		6.68	6.40	na
$\Delta E^{S_0}_{\text{cis-trans}}$ [g]		0.07	0.06	
transoid CI		4.89	5.23	0.043
		ketene		
FC	3.84 [64]	4.01	4.00	na
CI		2.43	2.81	0.016
		penta-2,4-dieniminium, PSB3		
trans-FC		4.47	4.40	na
cis-FC	4.20 [65]	4.24	4.45	na
$\Delta E^{S_0}_{\text{cis-trans}}$		0.16	0.0	
tw-BLA CI		2.38	2.83	0.069
		styrene		
FC	4.88 [66]	5.76	5.51	na
tw-pyr CI_1		4.68	4.13	0.097
tw-pyr CI_2		5.28	4.53	0.025
		stilbene		
trans-FC	4.1 [67]	5.35	4.35	na
cis-FC	4.6 [67]	5.74	4.68	na
$\Delta E^{S_0}_{\text{cis-trans}}$		0.13	0.29	
tw-pyr CI		4.50	3.98	0.093

(*Continued*)

Table 4.1 (*Continued*)

geometry	Lit.[a]	MRCISD[b]	MRSF[c]	
			ΔE	RMSD[e]
		HBI anion		
FC	3.06 [68]	3.88	3.25	na
tw-BLA CI_{Im}		2.87	2.91	0.050
tw-BLA CI_{Ph}		3.20	3.30	0.205

Source: Reprinted with permission from Ref. [60]. Copyright (2019) American Chemical Society.
a) Theoretical or experimental estimate of ΔE from the literature.
b) MRCISD/6-31+G** results from Ref. [49].
c) MRSF-TD-DFT/6-31+G** results obtained in this work.
d) SF-TD-DFT/6-31+G** results from Ref. [49].
e) Root-mean-square deviation from the MRCISD geometry.
f) MRCISD Franck-Condon geometry not available in Ref. [49].
g) Energy difference between the S_0 equilibrium *cis*- and *trans*-conformations.

rather well with the MRCISD geometries; the RMSD is 0.067 Å. Unlike SF-TDDFT, the MRSF method eliminates spin-contamination as well as ambiguity with identification of the state while preserving the correct $3N - 8$ dimensionality of the intersection space.

Visual inspection of the superimposed geometries of the CIs in Fig. 4.6 obtained with the MRCISD (gray color), SF (green), and MRSF (red) shows that in most cases the two TDDFT methods agree with one another. There are only two cases where the SF and MRSF geometries display a noticeable deviation from each other. In the case of the tw-pyr CI_{Ph} of the HBI anion, MRSF displays somewhat greater torsion of the imidazole ring about the methine bridge, see Fig. 4.6. However, this does not lead to a noticeable alteration of the energetic position of tw-pyr CI_{Ph} relative to the other CI, tw-pyr CI_{Im}, or the FC point of HBI, see Table 4.1.

A more considerable deviation of the SF-TDDFT geometry from both the MRSF-TDDFT and the MRCISD geometries occurs in the tw-BLA CI of the PSB3 cation. The SF-TDDFT geometry features pronounced pyramidalization of the allylic carbon atom adjacent to the central double bond of PSB3. As was analyzed in [49], the excessive pyramidalization of this atom as predicted by SF-TDDFT is caused by an overestimation of the relative stability of the ionic electronic configuration resulting from heterolytic breaking of the

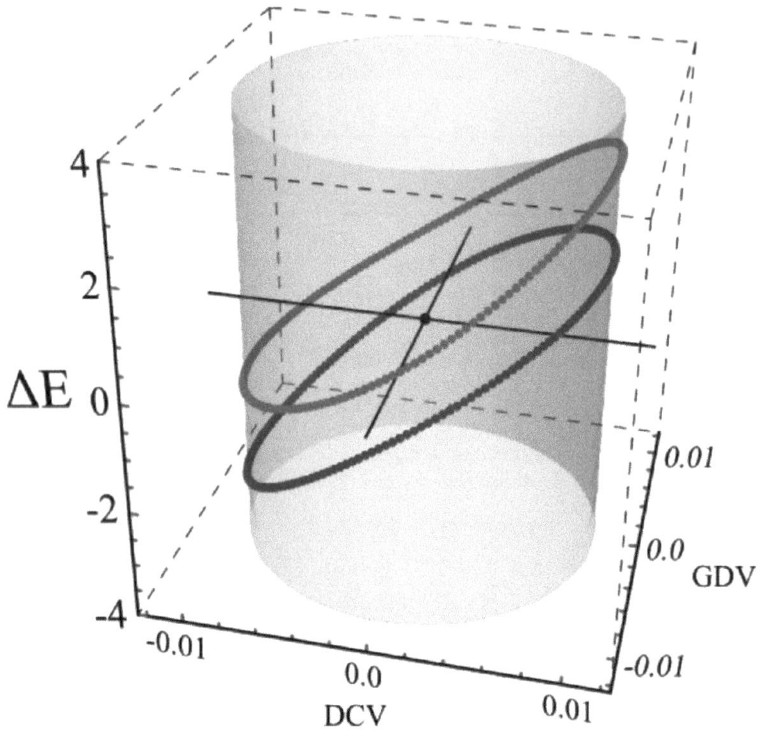

Figure 4.7 The relative energies ΔE (kcal/mol; with respect to the MECI energy) of the S_1 (red) and S_0 (blue) states round the loops obtained with the MRSF-TDDFT. The GDV and DCV vectors are measured in Å. Reprinted with permission from Ref. [60]. Copyright (2019) American Chemical Society.

central π-bond of PSB3. It is gratifying that MRSF-TDDFT formalism corrects this deficiency of SF-TDDFT simultaneously eliminating noticeable spin-contamination, $\langle S^2 \rangle = 0.579$ of SF-TDDFT.

The ability of the MRSF method to produce the correct double cone topology of CIs is illustrated in Fig. 4.7, where the S_1–S_0 energy difference as well as the S_1 and S_0 energies are shown round a loop in the branching plane around the tw-BLA MECI. MRSF correctly predicts the non-zero energy differences round the whole loop; hence, the correct double cone topology is reproduced. All the loops reported there confirm the double cone topology of the CIs optimized with MRSF-TD-DFT.

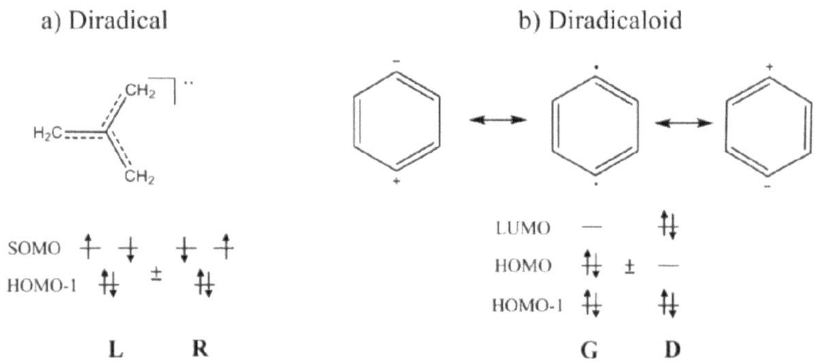

Figure 4.8 Major Electronic configurations of (a) diradical and (b) diradicaloid molecule. (a) Two electronic configurations **L** and **R** can be combined as **L**−**R** giving **OSS** or as **L**+**R** yielding the low-spin triplet component. (b) Ionic and covalent resonance forms that contribute to a diradicaloid. The **G** and **D** configurations yield two singlet states, **G**−**D** and **G**+**D**. The former singlet is lower in energy than the latter. Reprinted with permission from [69]. Copyright (2021) American Chemical Society.

Correcting the shortcomings of SF-TDDFT, the MRSF-TDDFT method predicts the geometries of the CIs studied here and their relative energies with respect to the S_0 equilibrium conformations in a good agreement with the reference MRCISD calculations. The results obtained in this work suggest that MRSF-TDDFT is a reliable method that can be confidently used to study the ground and excited PESs of molecules and run on-the-fly NAMD simulations of photochemical reactions.

4.3.4 Diradicals and Singet/Triplet Gap

Diradicals are molecules with two unpaired electrons occupying two (nearly) degenerate molecular orbitals [70–76].

Due to their multi-configurational nature [77] diradicals and diradicaloids represent a challenge for quantum chemical computational methodologies [78]. Especially challenging is the accurate description of their singlet-triplet (ST) energy gaps [79, 80], the knowledge of which is important for designing new compounds for optoelectronic [81] and photovoltaic [82] applications. The single-reference methods of wavefunction *ab initio* theory [83], as well as the conventional KS-DFT methods [84] fail to accurately describe the ST gaps.

As the MRSF method includes the doubly excited configurations—which are important for describing the multiconfigurational states—in its response space, it is an efficient and accurate method for diradicals. For a set of diradical and diradicaloid molecules, comprising o-, m-, p-benzynes (By), m-xylene (Xy), α,2-, α,3-, α,4-didehydrotoluenes (DHT), trimethylenemethane (TMM), 1,8-naphthoquinone (1,8-NQ), and 2,7-naphthoquinone (2,7-NQ) [90], the *adiabatic* ST gaps $\Delta E_{ST} = E_S - E_T$ were computed using MRSF [69] and are reported in Table 4.2, where they are compared with the experiment and the other theoretical methods. With respect to the experimental ST gaps, MRSF yields sufficiently small MAD of 0.14 eV, which is comparable with the MAD (0.09

Table 4.2 Adiabatic singlet-triplet gaps (ΔE_{ST}, in eV) and mean absolute errors (MAE) (in eV) as compared to experiments

	MRSF-BHHLYP	EOM-SF-CCSD(dT)[a]	Other	*Exp*
^{3}m-Xy	0.46	0.45	0.48[h]	0.42[g]
^{3}2,7-NQ	0.68	-	0.48[g]	0.55[g]
^{3}TMM ($^1A_1$)	0.91	0.85	0.92[i]	0.79[e]
^{3}TMM ($^1B_2$)	0.73	0.70	0.79[i]	-
$^3\alpha$,2-DHT	0.21	0.29	0.28[c]	-
$^3\alpha$,4-DHT	0.15	0.25	0.29[c]	-
$^1\alpha$,3-DHT	−0.01	−0.06	−0.09[c]	−0.22 to −0.13[h]
^{1}o-By	−1.86	−1.62	−1.55[b]	−1.66[d]
^{1}m-By	−0.96	−0.89	−0.69[b]	−0.91[d]
^{1}1,8-NQ	−0.13	-	−0.11[f]	−0.03[f]
^{1}p-By	−0.16	−0.17	−0.16[b]	−0.16[d]
MAE	0.14	0.09		

Note: The negative values of the gap indicate the singlet state is more stable than the triplet state. The superscript in front of the compound name refers to the multiplicity of the ground state.
Source: Reprinted with permission from [69]. Copyright (2021) American Chemical Society.
a) ref [59].
b) CASSCF/ANO result from Ref. [85].
c) DDCI2(2,2)/6-31G** result from Ref. [86].
d) Ref. [83].
e) Ref. [80].
f) CASPT2/aug-cc-pVDZ result from Ref. [87].
g) CASPT2/aug-cc-pVDZ result from Ref. [88].
h) CASPT2/6-31G* result from Ref. [89].
i) EOM-SF-CCSD/cc-pVQZ result from Ref. [59].

eV) of the EOM-SF-CCSD(dT) method. This demonstrates that MRSF-TDDFT can be a practical and accurate method for the study of diradicals and ST Gaps [69].

4.3.5 Jahn–Teller Distortion

Accurate study of the Jahn–Teller distortion (JTD) requires a theoretical method capable of describing multi-configurational electronic structure. To demonstrate the ability of MRSF-TDDFT to describe JTD, the trimethylenemethane (TMM) diradical is studied [69]. TMM diradical has a triplet ground state, $^3A_2'$ in the D_{3h} symmetry. At the D_{3h} symmetric geometry, the two singly occupied MOs (SOMOs), $1a_2$ and $2b_1$ in C_{2v}, become degenerate and the lowest singlet state of TMM is a doubly degenerate $^1E''$ state, where the two degenerate components transform according the B and the A irreducible representations of the C_2 group (subgroup of D_{3h}). The ^1B state is represented by an open shell singlet (OSS) electronic configuration comprising the two degenerate orbitals and the ^1A state is an out-of-phase linear combination of the **G** and **D** configurations.

When the D_{3h} symmetry is imposed, the degeneracy between the ^1B and ^1A states is maintained by both methods, MRSF and SF, and the equilibrium geometries predicted by both methods are similar, with the C–C bond length of 1.401 Å. As the degeneracy is lifted by distorting the symmetry away from D_{3h}, which according to the Jahn–Teller theorem occurs along two directions in the space of the internal molecular coordinates [91], the manifold of the D_{3h} geometries qualifies as a (symmetry imposed) CI and will be labeled as ^1CI in the following.

In [69], the minimum energy paths connecting stationary points on the singlet as well as the triplet potential energy surfaces of TMM were optimized using the MRSF and SF-TDDFT methods. For reference, the high-level XMS-CASPT2 calculations have been performed along the paths. MRSF as well as SF yield similar C–C bond lengths at the triplet equilibrium geometry, R(C–C) = 1.400 Å, and the spin-contamination in the SF calculation is minimal. In the singlet state, the degeneracy of the ^1B and ^1A states at the ^1CI geometry is lifted by geometric distortions transforming according

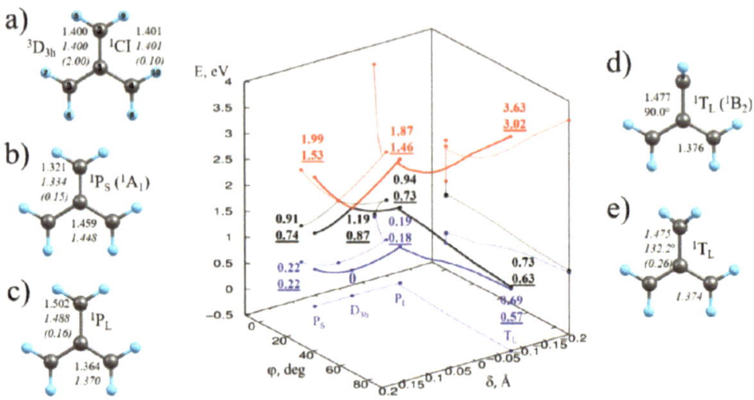

Figure 4.9 The minimum energy paths connecting stationary points on the lowest energy triplet (blue) and two lowest energy singlet (black and red) potential energy surfaces of TMM optimized using MRSF- and SF-BHHLYP for (a) triplet minimum ($^3D_{3h}$) as well as conical intersection point (^{1}CI) between the second and first singlet excited states, (b) singlet planar minimum with one short bond (^1P_S), (c) singlet planar structure with one long bond (^1P_L), (d) MRSF and (e) SF singlet twisted minimum with one long bond (^1T_L). The atom numbering is shown in (a). The numbers written with normal font indicate the distances (in Å) or dihedral angles (in°) obtained with MRSF, while those in italic are from SF with the values in parentheses showing its spin contamination. Zero energy corresponds to the energy of the optimized $^3D_{3h}$ structure. The numbers with normal and underlined fonts indicate the MRSF and XMS-CASPT2 single points energies (in eV) with respect to the D_{3h} structure. The point with D_{3h} symmetry is marked as 'D_{3h}'. The rest of the points have C_{2v} symmetry. The [δ, φ] coordinates of the P_S, D_{3h}, P_L, T_L points are [0.07, 0], [0, 0], [−0.10, 0], [−0.07, 90] respectively. Reprinted with permission from Ref. [69]. Copyright (2021) American Chemical Society.

the B and A irreps of the C_2 group. Thus, the A-symmetric distortion, which lengthens two out of three C–C bonds and shortens the remaining C–C bond, stabilizes the ^{1}A state and the B-symmetric distortion, which corresponds to torsion of one of the carbene groups along the respective C–C bond, stabilizes the ^{1}B state. This is illustrated in Fig. 4.9, where cross sections of the A and B potential energy surfaces are shown along the two distortion modes, the difference between the C–C bond lengths (δ) and the CH_2 torsion angle (φ).

Displacement along the δ distortion coordinate maintains the C_{2v} symmetry. Starting in the ^{1}CI (D_{3h}-symmetric) geometry motion in the $+\delta$ direction leads to the ^1P_S local minimum (with one short and two long C–C bonds, see Fig. 4.9b) and in the $-\delta$ direction to the ^1P_L local minimum (see Fig. 4.9c). The ^1P_S geometry corresponds to the $^1A_1$ state in the C_{2v} symmetry and the ^1P_L minimum to the $^1B_2$ state. As seen in Fig. 4.9, the MRSF and SF optimized geometries of the C_{2v} minima start to differ by *ca.* 0.1–0.2 Å, even though the spin-contamination of the SF states remains sufficiently low (0.15–0.16) at these geometries.

Torsion about a C–C bond described by the dihedral angle φ reduces the symmetry to C_2. The planar C_{2v}-symmetric ^1P_L local minimum is unstable against this symmetry reduction (it has an imaginary frequency of $i256$ cm^{-1} corresponding to the torsion) and rotation of one of the terminal carbene groups leads to the global singlet minimum that has C_{2v} symmetry with $\varphi = 90.0^\circ$. This minimum corresponds to the $^1B_2$ state in the C_{2v} symmetry. By contrast to MRSF, SF predicts that the 90° twisted geometry is not a minimum but a transition state connecting two equivalent minima with φ $\pm 132.2^\circ$ as shown in Fig. 4.9e. The existence of the two minima predicted by the SF formalism can be associated with the spin-contamination of the SF singlet state. Indeed, at $\varphi = 90^\circ$ SF yields the expectation value of the total spin operator $\langle \hat{S}^2 \rangle =$ 0.91, which indicates a very strong mixture between a true singlet state ($\langle \hat{S}^2 \rangle = 0$) and a triplet state ($\langle \hat{S}^2 \rangle = 2$). As discussed in the theoretical background, this mixing originates from the mismatched weights of the **L** and **R** configurations. As the triplet state has a planar geometry, i.e., $\varphi = 0^\circ$, it can be expected that a torsion away from the 90° value lowers the energy of the SF singlet state due to mixing with the triplet (i.e., the spin-contamination).

As reported in Table 4.2, the $^1B_2$ state of TMM lies 0.73 eV above the $^3A_2'$ minimum. This value is in a good agreement with the adiabatic ST gap obtained with EOM-SF-CCSD(dT)/cc-pVQZ [59]. The experimentally estimated adiabatic ST gap is 0.79 eV, where, however, the singlet state was identified as the $^1A_1$ state. The ST gap between the $^3A_2'$ and the $^1A_1$ minima obtained with MRSF is 0.91 eV. This value is still in a fair agreement with the experiment and in a good agreement with the gap reported at the EOM-SF-CCSD(dT)/cc-pVQZ level [59]. Experimentally, the diradical species of TMM was characterized by NIPES (Negative Ion PhotoElectron Spectroscopy), where the photodetachment signal of an electron from the $TMM^{\cdot -}$ anion-radical was measured. The neutral TMM diradical formed was assumed to be planar, since $TMM^{\cdot -}$ has a planar geometry, according to the B3LYP/aug-cc-pVDZ theoretical calculation [92].

The single point energies (underscored) obtained using XMS-CASPT2 in connection with the MRSF optimized geometries are also presented in Fig. 4.9. Although the singlet state energies from XMS-CASPT2 are generally lower by 0.01∼0.61 eV depending on the geometry, this method also finds $^1B_2$ to be the global minimum. Overall, MRSF-TDDFT can successfully describe the complicated electronic structure along the JTD coordinate of TMM, in agreement with high-level multiconfigurational wavefunction calculations.

4.4 Nonadiabatic Molecular Dynamics (NAMD)

Currently, two most popular computational methods used in connection with NAMD are CASSCF and TDDFT. Here, we present a comparative study of the accuracy of MRSF-TDDFT by applying this method to photodynamics of thymine, which is of high interest in view of the potential solar UV damage inflicted on DNA.

Figure 4.10 shows three possible mechanisms of the internal conversion of photoexcited thymine proposed in the literature. First, the S_1-trapping model involves a sequential $\pi\pi^* \rightarrow n\pi^* \rightarrow S_0$ internal conversion, where the first step ($\pi\pi^* \rightarrow n\pi^*$) occurs on an ultrafast timescale [93] and the second step ($n\pi^* \rightarrow S_0$) lasts for several picoseconds. Second, the S_2-trapping model was proposed on the basis of the *ab initio* multiple spawning (AIMS) [94] as well as the trajectory surface hopping (TSH) simulations based on the CASSCF PES [95, 96]. In this model, the short time constant was explained by relaxation of the S_2 state away from the FC region towards a minimum on the S_2 surface where the molecule remains trapped for a long time (*ca.* 5 ps) [95–97]. Finally, the S_2&S_1-trapping model was proposed based on quantum nuclear dynamics simulations [98], where it was shown that a substantial

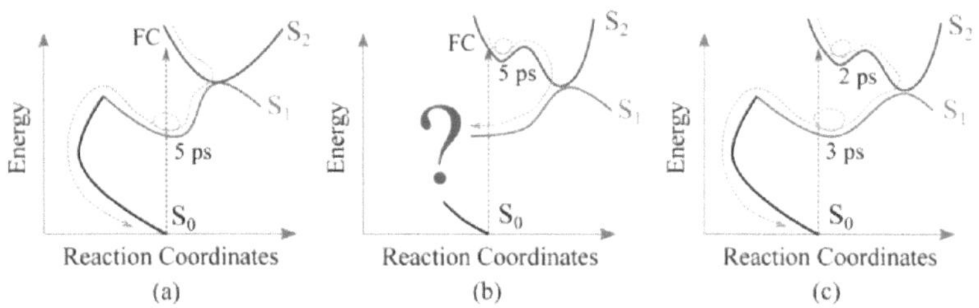

Figure 4.10 Previously proposed excited thymine decay pathways: (a) S_1-trapping model, (b) S_2-trapping model, and (c) S_1&S_2-trapping model.

redistribution of the population occurs between the S_2 and S_1 states during the first *ca.* 50 fs.

As can be judged from theoretical simulations presented in the literature, none of the previous theoretical NAMD simulations was able to reproduce both the ultrafast and the intermediate timescales of the thymine dynamics in quantitative agreement with the experimental measurements. Furthermore, the results of theoretical simulations display a marked dependence on the electronic structure method used in NAMD.

The results of recent TSH NAMD simulations with MRSF/BH&HLYP [3] are shown in Fig. 4.11, which displays the time evolution of the populations of the S_2, S_1, and S_0 adiabatic states. Starting in the optically bright S_2 state, the population rapidly (within the first *ca.* 50 fs) drops below 0.3 due to the $S_2 \rightarrow S_1$ ($\pi\pi^* \rightarrow n_O\pi^*$) population transfer. This result is in a good agreement with the experimental observations reported by [93] ($\tau_{\pi\pi^*\rightarrow n\pi^*} = 60 \pm 30$ fs). At longer times, $t \gtrsim 100$ fs, almost all of the S_2 population ends up in the S_1 state, where it slowly decays to the S_0 state. The exponential decay parameter obtained by fitting the S_1 population by a mono-exponential function is 6.1$\pm$0.035 ps, which is in a quantitative agreement with the experimental estimates of 5–7 ps.

At short simulation times, $t \lesssim 50$ fs, the pyrimidine ring remains planar and the major geometric changes involve elongation of the $C_4{=}O_8$ and $C_5{=}C_6$ double bonds and shortening of the $C_4{-}C_5$ single bond (see Fig. 4.11 for the atom numbering). This leads to bond length alternation (BLA) in the pyrimidine ring. The so-defined

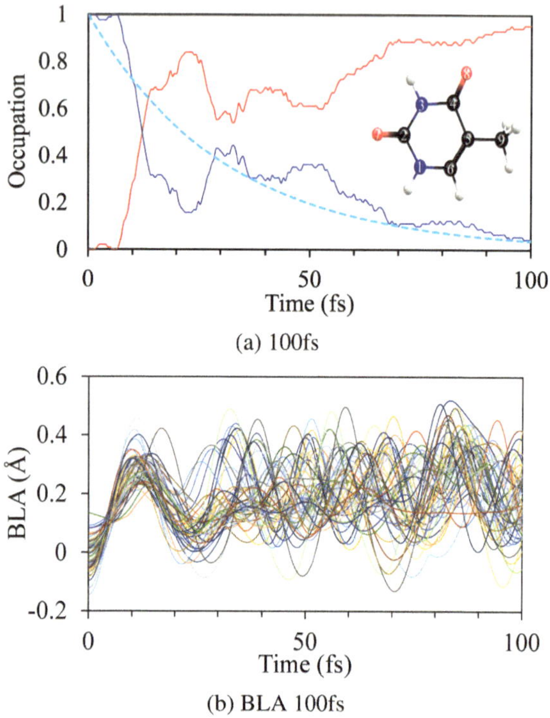

Figure 4.11 Time evolution of the adiabatic S_0, S_1, and S_2 populations for (a) the first 100 fs.Panel (b) shows the time evolution of bond length alternation (BLA) along the trajectories within the first 100 fs of the simulation. The BLA coordinate is defined here as the difference between the average increments of the lengths of the double bonds and the decrease of the single bond, $\text{BLA} = \frac{1}{2}(\Delta R_{C_4=O_8} + \Delta R_{C_5=C_6}) - \Delta R_{C_4-C_5}$, where ΔR's are displacements with respect to the S_0 equilibrium geometry. Reprinted with permission from [3]. Copyright (2021) American Chemical Society.

BLA characterizes the magnitude of the geometric distortion away from the FC region and its time evolution, see Fig. 4.11b, displays a pronounced synchronicity among all the trajectories during the first *ca.* 20–30 fs. This suggests that a concerted displacement along the BLA coordinate causes an ultrafast transfer of the S_2 population to S_1; see Fig. 4.11a. The synchronicity is rapidly lost, becoming incoherent after *ca.* 50 fs.

The ultrafast decay of the S_2 population at the early propagation times occurs through an S_2/S_1 CI, which was located from the snapshots of the NAMD trajectories at which the $S_2 \rightarrow S_1$ surface hops occur. The located S_2/S_1 CI features planar pyrimidine ring geometry and can be reached from the FC region by the BLA distortion alone. Using the MRSF/BH&HLYP optimized geometries of the FC point, $CI_{21,BLA}$, and $S_{1,min}$, all of which are planar, a minimum energy path (MEP) was constructed on the S_2 and S_1 PESs by the nudged elastic band (NEB) method [99, 100]. In Figs. 4.12a,b, the S_2 and S_1 PES profiles along the MEP are shown; the MEP is characterized by the BLA distortion. The MRSF results are compared with the PES profiles obtained along the same MEP by other theoretical methods.

Inferred from the MRSF results, the S_2 and S_1 PESs undergo a crossing at a BLA value of 0.14 Å and then the S_1 PES rapidly descends to the $S_{1,min}$ geometry at BLA = 0.215 Å. The MRSF curves (solid curves) are in good qualitative agreement with the EOM-CCSD/6-31G* method (dashed curves). As the S_2 ($\pi\pi^*$) and S_1 ($n\pi^*$) states have different symmetry, the intersection is allowed and the EOM-CC formalism does not produce any artifacts [50]. The NEVPT2 and the XMS-CASPT2 methods are multi-reference methodologies, which do not suffer from possible artifacts inherent in single-reference methods, such as EOM-CC, ADC(2), or LR-TDDFT. Both methods include the dynamic electron correlation and the S_2 and S_1 potential energy profiles produced by these methods along the BLA path in Fig. 4.12b show very similar shapes to the MRSF profiles.

In Fig. 4.12b, the curves obtained using CASSCF (solid lines) are compared with the NEVPT2 (dotted lines), and the XMS-CASPT2 (dashed lines) curves. Distinctively, the CASSCF method, which neglects the dynamic electron correlation, does not predict crossing between the S_2 and S_1 states along the BLA path. Instead, it predicts a shallow minimum on the S_2 PES, which is separated by more than 2 eV from the minimum on the S_1 PES. In the CASSCF description, $S_2 \rightarrow S_1$ population transfer is only possible due to a pronounced puckering distortion of the pyrimidine ring ($CI_{21,\phi}$) [94–96], which results in a slow $\pi\pi^* \rightarrow n\pi^*$ relaxation time and supports the S_2-trapping model. Therefore, the lack of the dynamic electron

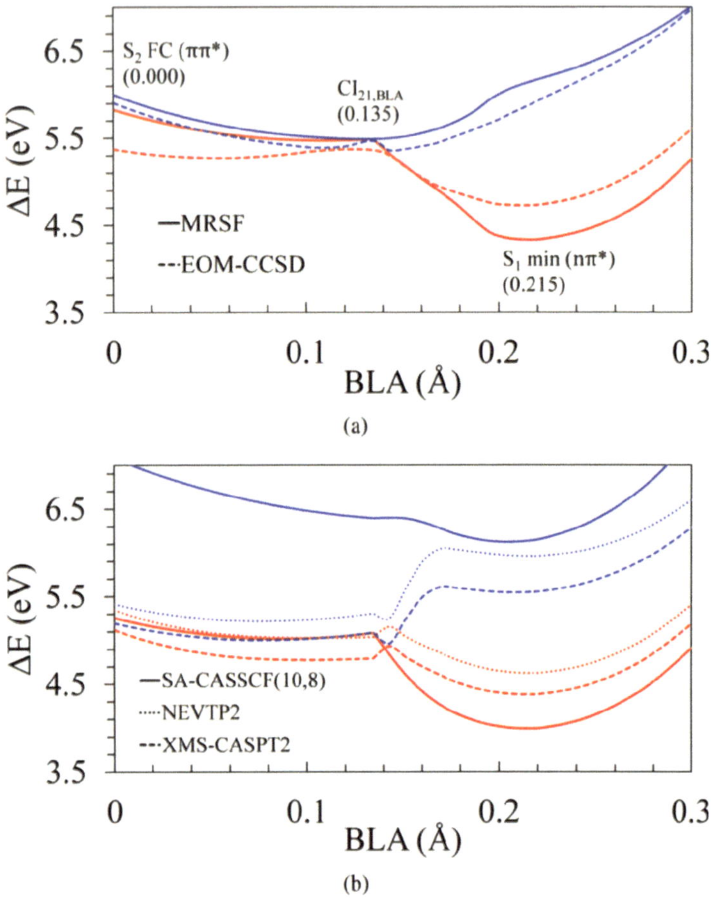

Figure 4.12 MEPs on the S_2 and S_1 PESs optimized using the NEB method in connection with MRSF and connecting the FC region, the $CI_{21,BLA}$, and the $S_{1,min}$ geometries; the respective BLA values are given parenthetically. For all other electronic structure methods, the MRSF MEP geometries are utilized employing a 6-31G* basis set with Cs symmetry restriction. Panel (a) compares the S_2 (blue) and S_1 (red) PES profiles along the MRSF MEPs (solid lines) with the EOM-CCSD curves (dashed lines). Panel (b) shows the S_2 and S_1 PES profiles obtained with the SA-CASSCF(10,8) (solid lines), the eXtended Multi-State Complete Active Space second-order Perturbation Theory (XMS-CASPT2, dashed lines), and the n-electron valence state perturbation theory (NEVPT2, dotted lines). Reprinted with permission from Ref. [3]. Copyright (2021) American Chemical Society.

correlation in CASSCF destabilizes the $\pi\pi^*$ (S_2) state relative to the $n\pi^*$ (S_1) state and leads to the absence of the S_2/S_1 crossing along the BLA path. The proper account of the dynamic correlation by MRSF corrects this deficiency and recovers an ultrafast S_2/S_1 internal conversion measured experimentally [93], thus resolving the long-standing controversy.

4.5 Conclusions

As MRSF obtains all the singlet (as well as triplet) states from spin-flip orbital transitions generated for an ensemble of two components ($M_S = +1$ and -1) of the triplet reference state without requiring for expensive multi-configurational orbital optimization, it can properly and economically introduce the multi-reference characteristics into the response states, both ground and excited states alike. In particular, important doubly excited configurations are naturally introduced into its response space. MRSF provides for an *explicit* (through both singly and some important doubly excited configurations) as well as an *implicit* (through the XC functional) account of electron correlation, thus allowing for a balanced description of the *dynamic* and the *static* electron correlation effects. These advantages let MRSF to describe open shell singlet states of diradicals, bond dissociation in the ground and excited states, excited states with double excitation characteristics, and, most importantly, coupling between the ground and excited electronic states necessary for obtaining the correct topology of conical intersections. The inability to describe these situations has been the major drawback of the conventional TDDFT formalism. The particular advantage of MRSF before the usual SF-TDDFT is that it eliminates the notorious spin-contamination of the response electronic states. The superiority of MRSF over ground state Kohn–Sham DFT as well as other response methodologies in a number of specific situations has been documented in this chapter, highlighting that MRSF is a practical and accurate method for general use as well as for NAMD simulations.

Acknowledgements

This work was supported by the Samsung Science and Technology Foundations grant SSTF-BA1701-12 (to C.H.C.) for fundamental theory developments and the NRF grants 2019H1D3A2A02102948 (to M.F.) and 2020R1A2C2008246 and 2020R1A5A1019141 (to C.H.C.) funded by the Ministry of Science and ICT for applications of the developed methodologies.

References

1. Hedley, G. J., Ruseckas, A., and Samuel, I. D. (2017). Light harvesting for organic photovoltaics, *Chem. Rev.* **117**, 2, pp. 796–837.
2. Roke, D., Stuckhardt, C., Danowski, W., Wezenberg, S. J., and Feringa, B. L. (2018). Light-gated rotation in a molecular motor functionalized with a dithienylethene switch, *Angew. Chem.* **130**, 33, pp. 10675–10679.
3. Park, W., Lee, S., Huix-Rotllant, M., Filatov, M., and Choi, C. H. (2021). Impact of the dynamic electron correlation on the unusually long excited-state lifetime of thymine, *J. Phys. Chem. Lett.* **12**, pp. 4339–4346.
4. Tuna, D., Lefrancois, D., Wolanski, Ł., Gozem, S., Schapiro, I., Andruniów, T., Dreuw, A., and Olivucci, M. (2015). Assessment of approximate coupled-cluster and algebraic-diagrammatic-construction methods for ground-and excited-state reaction paths and the conical-intersection seam of a retinal-chromophore model, *J. Chem. Theor. Comput.* **11**, 12, pp. 5758–5781.
5. Dreuw, A., Weisman, J. L., and Head-Gordon, M. (2003). Long-range charge-transfer excited states in time-dependent density functional theory require non-local exchange, *J. Chem. Phys.* **119**, 6, pp. 2943–2946.
6. Dreuw, A., and Head-Gordon, M. (2004). Failure of time-dependent density functional theory for long-range charge-transfer excited states: The zincbacteriochlorin- bacteriochlorin and bacteriochlorophyll-spheroidene complexes, *J. Am. Chem. Soc.* **126**, 12, pp. 4007–4016.
7. Dev, P., Agrawal, S., and English, N. J. (2012). Determining the appropriate exchange-correlation functional for time-dependent density functional theory studies of charge-transfer excitations in organic dyes, *J. Chem. Phys.* **136**, 22, p. 224301.

8. Maitra, N. T. (2005). Undoing static correlation: Long-range charge transfer in time-dependent density-functional theory, *J. Chem. Phys.* **122**, 23, p. 234104.
9. Baerends, E., Gritsenko, O., and Van Meer, R. (2013). The Kohn–Sham gap, the fundamental gap and the optical gap: The physical meaning of occupied and virtual Kohn–Sham orbital energies, *Phys. Chem. Chem. Phys.* **15**, 39, pp. 16408–16425.
10. Cave, R. J., Zhang, F., Maitra, N. T., and Burke, K. (2004). A dressed TDDFT treatment of the 21Ag states of butadiene and hexatriene, *Chem. Phys. Lett.* **389**, 1-3, pp. 39–42.
11. Neugebauer, J., Baerends, E. J., and Nooijen, M. (2004). Vibronic coupling and double excitations in linear response time-dependent density functional calculations: Dipole-allowed states of n 2, *J. Chem. Phys.* **121**, 13, pp. 6155–6166.
12. Maitra, N. T., Zhang, F., Cave, R. J., and Burke, K. (2004). Double excitations within time-dependent density functional theory linear response, *J. Chem. Phys.* **120**, 13, pp. 5932–5937.
13. Casida, M., and Huix-Rotllant, M. (2016). Many-body perturbation theory (MBPT) and time-dependent density-functional theory (TD-DFT): MBPT insights about what is missing in, and corrections to, the TD-DFT adiabatic approximation, in N. Ferré, M. Filatov and M. Huix-Rotllant (eds.), *Density-Functional Methods for Excited States*, Topics in Current Chemistry, Vol. 368 (Springer, Heidelberg), pp. 1–60.
14. Aryasetiawan, F., Gunnarsson, O., and Rubio, A. (2002). Excitation energies from time-dependent density-functional formalism for small systems, *EPL (Europhys. Lett.)* **57**, 5, p. 683.
15. Filatov, M. (2016). *Density-Functional Methods for Excited States* (Springer).
16. Levine, B. G., Ko, C., Quenneville, J., and Martínez, T. J. (2006). Conical intersections and double excitations in time-dependent density functional theory, *Mol. Phys.* **104**, 5-7, pp. 1039–1051.
17. Huix-Rotllant, M., Filatov, M., Gozem, S., Schapiro, I., Olivucci, M., and Ferré, N. (2013). Assessment of density functional theory for describing the correlation effects on the ground and excited state potential energy surfaces of a retinal chromophore model, *J. Chem. Theor. Comput.* **9**, 9, pp. 3917–3932.
18. Sears, J. S., Sherrill, C. D., and Krylov, A. I. (2003). A spin-complete version of the spin-flip approach to bond breaking: What is the impact of obtaining spin eigenfunctions? *J. Chem. Phys.* **118**, pp. 9084–9094.

19. Casanova, D., and Head-Gordon, M. (2008). The spin-flip extended single excitation configuration interaction method, *J. Chem. Phys.* **129**, p. 064104.
20. Shao, Y., Head-Gordon, M., and Krylov, A. I. (2003). The spin-flip approach within time-dependent density functional theory: Theory and applications to diradicals, *J. Chem. Phys.* **118**, pp. 4807–4818.
21. Wang, F., and Ziegler, T. (2004). Time-dependent density functional theory based on a noncollinear formulation of the exchange-correlation potential, *J. Chem. Phys.* **121**, pp. 12191–12196.
22. Sears, J. S., Sherrill, C. D., and Krylov, A. I. (2003). A spin-complete version of the spin-flip approach to bond breaking: What is the impact of obtaining spin eigenfunctions? *J. Chem. Phys.* **118**, pp. 9084–9094.
23. Tsuchimochi, T. (2015). Spin-flip configuration interaction singles with exact spin-projection: Theory and applications to strongly correlated systems, *J. Chem. Phys.* **143**, p. 144114.
24. Mato, J., and Gordon, M. S. (2018). A general spin-complete spin-flip configuration interaction method, *Phys. Chem. Chem. Phys.* **20**, pp. 2615–2626.
25. Li, Z., Liu, W., Zhang, Y., and Suo, B. (2011). Spin-adapted open-shell time-dependent density functional theory. ii. theory and pilot application, *J. Chem. Phys.* **134**, p. 134101.
26. Li, Z., and Liu, W. (2011). Spin-adapted open-shell time-dependent density functional theory. iii. an even better and simpler formulation, *J. Chem. Phys.* **135**, p. 194106.
27. Li, Z., and Liu, W. (2010). Spin-adapted open-shell random phase approximation and time-dependent density functional theory. i. theory, *J. Chem. Phys.* **133**, p. 064106.
28. Li, Z., Liu, W., Zhang, Y., and Suo, B. (2011). Spin-adapted open-shell time-dependent density functional theory. ii. theory and pilot application, *J. Chem. Phys.* **134**, p. 134101.
29. Zhang, X., and Herbert, J. M. (2015). Spin-flip, tensor equation-of-motion configuration interaction with a density-functional correction: A spin-complete method for exploring excited-state potential energy surfaces, *J. Chem. Phys.* **143**, p. 234107.
30. Lee, S., Filatov, M., Lee, S., and Choi, C. H. (2018). Eliminating spin-contamination of spin-flip time dependent density functional theory within linear response formalism by the use of zeroth-order mixed-reference (MR) reduced density matrix, *J. Chem. Phys.* **149**, 10, p. 104101.

31. Lee, S., Kim, E. E., Nakata, H., Lee, S., and Choi, C. H. (2019). Efficient implementations of analytic energy gradient for mixed-reference spin-flip time-dependent density functional theory (MRSF-TDDFT), *J. Chem. Phys.* **150**, 18, p. 184111.

32. McWeeny, R. (1992). *Methods of Molecular Quantum Mechanics* (Academic press).

33. de Wergifosse, M., and Grimme, S. (2018). Nonlinear-response properties in a simplified time-dependent density functional theory (STD-DFT) framework: Evaluation of the first hyperpolarizability, *J. Chem. Phys.* **149**, 2, p. 024108.

34. Filatov, M. (2015). Spin-restricted ensemble-referenced Kohn-Sham method: Basic principles and application to strongly correlated ground and excited states of molecules, *WIREs Comput. Mol. Sci.* **5**, pp. 146–167.

35. Tamm, I. E. (1945). Relativistic interaction of elementary particles, *J. Phys. (USSR)* **9**, pp. 449–460.

36. Dancoff, S. M. (1950). Non-adiabatic meson theory of nuclear forces, *Phys. Rev.* **78**, pp. 382–385.

37. Hirata, S., and Head-Gordon, M. (1999). Time-dependent density functional theory within the Tamm–Dancoff approximation, *Chem. Phys. Lett.* **314**, 3, pp. 291–299.

38. Filatov, M., Huix-Rotllant, M., and Burghardt, I. (2015). Ensemble density functional theory method correctly describes bond dissociation, excited state electron transfer, and double excitations, *J. Chem. Phys.* **142**, p. 184104.

39. Lee, S., Horbatenko, Y., Filatov, M., and Choi, C. H. (2021). Fast and accurate computation of nonadiabatic coupling matrix elements using the truncated Leibniz formula and mixed-reference spin-flip time-dependent density functional theory, *J. Phys. Chem. Lett.* **12**, pp. 4722–4728.

40. Dunning, T. H. (1989). Gaussian basis sets for use in correlated molecular calculations. i. the atoms boron through neon and hydrogen, *J. Chem. Phys.* **90**, 2, pp. 1007–1023.

41. Becke, A. D. (1993). A new mixing of Hartree–Fock and local density-functional theories, *J. Chem. Phys.* **98**, 2, pp. 1372–1377.

42. Horbatenko, Y., Lee, S., Filatov, M., and Choi, C. H. (2019). Performance analysis and optimization of mixed-reference spin-flip time-dependent density functional theory (MRSF-TDDFT) for vertical excitation energies and singlet–triplet energy gaps, *J. Phys. Chem. A* **123**, 37, pp. 7991–8000.

43. Teller, E. (1937). The crossing of potential surfaces. *J. Phys. Chem.* **41**, 1, pp. 109–116.
44. Herzberg, G., and Longuet-Higgins, H. (1963). Intersection of potential energy surfaces in polyatomic molecules, *Discuss. Faraday Soc.* **35**, pp. 77–82.
45. Longuet-Higgins, H. C. (1975). The intersection of potential energy surfaces in polyatomic molecules, *Proc. R. Soc. Lond.* **344**, 1637, pp. 147–156.
46. Atchity, G. J., Xantheas, S. S., and Ruedenberg, K. (1991). Potential energy surfaces near intersections, *J. Chem. Phys.* **95**, pp. 1862–1876.
47. Manaa, M. R., and Yarkony, D. R. (1993). On the intersection of two potential energy surfaces of the same symmetry. systematic characterization using a Lagrange multiplier constrained procedure, *J. Chem. Phys.* **99**, pp. 5251–5256.
48. Gozem, S., Melaccio, F., Valentini, A., Filatov, M., Huix-Rotllant, M., Ferré, N., Frutos, L. M., Angeli, C., Krylov, A. I., Granovsky, A. A. *et al.* (2014). Shape of multireference, equation-of-motion coupled-cluster, and density functional theory potential energy surfaces at a conical intersection, *J. Chem. Theor. Comput.* **10**, 8, pp. 3074–3084.
49. Nikiforov, A., Gamez, J. A., Thiel, W., Huix-Rotllant, M., and Filatov, M. (2014). Assessment of approximate computational methods for conical intersections and branching plane vectors in organic molecules, *J. Chem. Phys.* **141**, 12, p. 124122.
50. Kjønstad, E. F., Myhre, R. H., Martínez, T. J., and Koch, H. (2017). Crossing conditions in coupled cluster theory, *J. Chem. Phys.* **147**, p. 164105.
51. Send, R., and Furche, F. (2010). First-order nonadiabatic couplings from time-dependent hybrid density functional response theory: Consistent formalism, implementation, and performance, *J. Chem. Phys.* **132**, p. 044107.
52. Lee, S., Kim, E., Lee, S., and Choi, C. H. (2019). Fast overlap evaluations for nonadiabatic molecular dynamics simulations: Applications to SF-TDDFT and TDDFT, *J. Chem.* **15**, 2, pp. 882–891.
53. Casida, M. E., Jamorski, C., Bohr, F., Guan, J. G., and Salahub, D. R. (1996). Optical properties from density-functional theory, in Karna, S. P., and Yeates, A. T. (ed.), *Nonlinear Optical Materials: Theory and Modeling, ACS Symposium Series*, Vol. 628 (Am. Chem. Soc., Div. Comp. Chem.), pp. 145–163, Symposium on Nonlinear Optical Materials - Theory and Modeling, at the 208th National Meeting of the American-Chemical-Society, Washington, DC, August 21–25, 1994.

54. Marques, M. A. L., and Gross, E. K. U. (2003). Time-dependent density functional theory, in C. Fiolhais, F. Nogueira and M. A. L. Marques (eds.), *A Primer in Density-Functional Theory, Lecture Notes in Physics*, Vol. 620 (Springer, Berlin), pp. 144–184.
55. Casida, M. E., and Huix-Rotllant, M. (2012). Progress in time-dependent density-functional theory, *Annu. Rev. Phys. Chem.* **63**, pp. 287–323.
56. Gozem, S., Melaccio, F., Valentini, A., Filatov, M., Huix-Rotllant, M., Ferré, N., Frutos, L. M., Angeli, C., Krylov, A. I., Granovsky, A. A., Lindh, R., and Olivucci, M. (2014). On the shape of multireference, EOM-CC, and DFT potential energy surfaces at a conical intersection, *J. Chem. Theory Comput.* **10**, pp. 3074–3084.
57. Huix-Rotllant, M., Nikiforov, A., Thiel, W., and Filatov, M. (2016). *Density-Functional Methods for Excited States* (Springer).
58. Rinkevicius, Z., Vahtras, O., and Ågren, H. (2010). Spin-flip time dependent density functional theory applied to excited states with single, double, or mixed electron excitation character, *J. Chem. Phys.* **133**, p. 114104.
59. Bernard, Y. A., Shao, Y., and Krylov, A. I. (2012). General formulation of spin-flip time-dependent density functional theory using non-collinear kernels: Theory, implementation, and benchmarks, *J. Chem. Phys.* **136**, p. 204103.
60. Lee, S., Shostak, S., Filatov, M., and Choi, C. H. (2019). Conical intersections in organic molecules: Benchmarking mixed-reference spin–flip time-dependent DFT (MRSF-TD-DFT) vs spin–flip TD-DFT, *J. Phys. Chem. A* **123**, 30, pp. 6455–6462.
61. Filatov, M. (2013). Assessment of density functional methods for obtaining geometries at conical intersections in organic molecules, *J. Chem. Theory Comput.* **9**, pp. 4526–4541.
62. Filatov, M., and Olivucci, M. (2014). Designing conical intersections for light-driven single molecule rotary motors: From precessional to axial motion, *J. Org. Chem.* **79**, pp. 3587–3600.
63. Schreiber, M., Silva-Junior, M. R., Sauer, S. P. A., and Thiel, W. (2008). Benchmarks for electronically excited states: Caspt2, cc2, ccsd, and cc3, *J. Chem. Phys.* **128**, p. 134110.
64. Chiang, S.-Y., Bahou, M., Wu, Y.-J., and Lee, Y.-P. (2002). Experimental and theoretical studies on Rydberg states of CH2CO in the region 120-220 nm, *J. Chem. Phys.* **117**, pp. 4306–4316.
65. Valsson, O., Angeli, C., and Filippi, C. (2012). Excitation energies of retinal chromophores: Critical role of the structural model, *Phys. Chem. Chem. Phys.* **14**, pp. 11015–11020.

66. Leopold, D., Hemley, R., Vaida, V., and Roebber, J. (1981). Direct absorption spectra of higher excited states of jet-cooled monosubstituted benzenes: Phenylacetylene, styrene, benzaldehyde, and acetophenone, *J. Chem. Phys.* **75**, 10, pp. 4758–4769.

67. Rice, J. K., and Baronavski, A. P. (1992). Ultrafast studies of solvent effects in the isomerization of cis-stilbene, *J. Phys. Chem.* **96**, pp. 3359–3366.

68. Filippi, C., Zaccheddu, M., and Buda, F. (2009). Absorption spectrum of the green fluorescent protein chromophore: A difficult case for ab initio methods? *J. Chem. Theor. Comp.* **5**, 8, pp. 2074–2087.

69. Horbatenko, Y., Sadiq, S., Lee, S., Filatov, M., and Choi, C. H. (2021). Mixed-reference spin-flip time-dependent density functional theory (MRSF-TDDFT) as a simple yet accurate method for diradicals and diradicaloids, *J. Chem. Theory Comput.* **17**, 2, pp. 848–859.

70. Borden, W. T. (2015). *Diradicals—A Fifty Year Fascination*, Vol. 1209, chap. 11, pp. 251–303.

71. Abe, M. (2013). Diradicals, *Chem. Rev.* **113**, 9, pp. 7011–7088.

72. Peng, D., Hu, X., Devarajan, D., Ess, D. H., Johnson, E. R., and Yang, W. (2012). Variational fractional-spin density-functional theory for diradicals, *J. Chem. Phys.* **137**, 11, p. 114112.

73. Yang, Y., Peng, D., Davidson, E. R., and Yang, W. (2015). Singlet–triplet energy gaps for diradicals from particle–particle random phase approximation, *J. Phys. Chem. A* **119**, 20, pp. 4923–4932.

74. Rajca, A. (1994). Organic diradicals and polyradicals: From spin coupling to magnetism? *Chem. Rev.* **94**, 4, pp. 871–893.

75. Gryn'ova, G., Coote, M. L., and Corminboeuf, C. (2015). Theory and practice of uncommon molecular electronic configurations, *WIREs Comput. Mol. Sci.* **5**, 6, pp. 440–459.

76. Abe, M., Ye, J., and Mishima, M. (2012). The chemistry of localized singlet 1, 3-diradicals (biradicals): From putative intermediates to persistent species and unusual molecules with a π-single bonded character, *Chem. Soc. Rev.* **41**, 10, pp. 3808–3820.

77. Wang, E. B., Parish, C. A., and Lischka, H. (2008). An extended multireference study of the electronic states of para-benzyne, *J. Chem. Phys.* **129**, 4, p. 044306.

78. Nakano, M., Fukuda, K., Ito, S., Matsui, H., Nagami, T., Takamuku, S., Kitagawa, Y., and Champagne, B. (2017). Diradical and ionic characters

of open-shell singlet molecular systems, *J. Phys. Chem. A* **121**, 4, pp. 861–873.

79. Ferré, N., Guihéry, N., and Malrieu, J.-P. (2015). Spin decontamination of broken-symmetry density functional theory calculations: deeper insight and new formulations, *Phys. Chem. Chem. Phys.* **17**, pp. 14375–14382.
80. Zimmerman, P. M. (2017). Singlet–triplet gaps through incremental full configuration interaction, *J. Phys. Chem. A* **121**, 24, pp. 4712–4720.
81. Milián-Medina, B., and Gierschner, J. (2012). Computational design of low singlet–triplet gap all-organic molecules for OLED application, *Org. Electron.* **13**, 6, pp. 985–991.
82. Ito, S., and Nakano, M. (2015). Theoretical molecular design of heteroacenes for singlet fission: Tuning the diradical character by modifying π-conjugation length and aromaticity, *J. Phys. Chem. C* **119**, 1, pp. 148–157.
83. Wenthold, P. G., Squires, R. R., and Lineberger, W. (1998). Ultraviolet photoelectron spectroscopy of the o-, m-, and p-benzyne negative ions. electron affinities and singlet- triplet splittings for o-, m-, and p-benzyne, *J. Am. Chem. Soc.* **120**, 21, pp. 5279–5290.
84. Squires, R. R., and Cramer, C. J. (1998). Electronic interactions in aryne biradicals. ab initio calculations of the structures, thermochemical properties, and singlet- triplet splittings of the didehydronaphthalenes, *J. Phys. Chem. A* **102**, 45, pp. 9072–9081.
85. Cramer, C. J., Nash, J. J., and Squires, R. R. (1997). A reinvestigation of singlet benzyne thermochemistry predicted by CASPT2, coupled-cluster and density functional calculations, *Chem. Phys. Lett.* **277**, 4, pp. 311–320.
86. Cabrero, J., Ben-Amor, N., and Caballol, R. (1999). Singlet- triplet gap in α-n-dehydrotoluene and related biradicals: An ab initio configuration interaction study, *J. Phys. Chem. A* **103**, 31, pp. 6220–6224.
87. Yang, Z., Hrovat, D. A., Hou, G.-L., Borden, W. T., and Wang, X.-B. (2019). Negative ion photoelectron spectroscopy confirms the prediction of a singlet ground state for the 1, 8-naphthoquinone diradical, *J. Phys. Chem. A* **123**, 14, pp. 3142–3148.
88. Yang, Z., Hrovat, D. A., Hou, G.-L., Borden, W. T., and Wang, X.-B. (2018). Negative ion photoelectron spectroscopy confirms the prediction of the relative energies of the low-lying electronic states of 2, 7-naphthoquinone, *J. Phys. Chem. A* **122**, 21, pp. 4838–4844.

89. Hrovat, D. A., A. Murcko, M., M. Lahti, P., and Borden, W. T. (1998). The effects of heteroatom substitution on the singlet-triplet energy differences in diradicals—ab initio calculations of δe_{st} in meta-benzoquinomethane and in 1,3-naphthoquinomethane, *J. Chem. Soc., Perkin Trans. 2* , pp. 1037–1044.

90. de Wergifosse, M., Bannwarth, C., and Grimme, S. (2019). A simplified spin-flip time-dependent density functional theory approach for the electronic excitation spectra of very large diradicals, *J. Phys. Chem. A* **123**, 27, pp. 5815–5825.

91. Jahn, H. A., Teller, E., and Donnan, F. G. (1937). Stability of polyatomic molecules in degenerate electronic states—I: Orbital degeneracy, *Proc. Roy. Soc. London. Ser. A* **161**, pp. 220–235.

92. Wenthold, P. G., Hu, J., Squires, R. R., and Lineberger, W. C. (1999). Photoelectron spectroscopy of the trimethylenemethane negative ion, *J. Am. Soc. Mass Spectrom.* **10**, 9, pp. 800–809.

93. Wolf, T. J. A., Myhre, R. H., Cryan, J. P., Coriani, S., Squibb, R. J., Battistoni, A., Berrah, N., Bostedt, C., Bucksbaum, P., Coslovich, G., Feifel, R., Gaffney, K. J., Grilj, J., Martinez, T. J., Miyabe, S., Moeller, S. P., Mucke, M., Natan, A., Obaid, R., Osipov, T., Plekan, O., Wang, S., Koch, H., and Gühr, M. (2017). Probing ultrafast $\pi\pi^*/n\pi^*$ internal conversion in organic chromophores via k-edge resonant absorption, *Nat. Commun.* **8**, p. 29.

94. Hudock, H. R., Levine, B. G., Thompson, A. L., Satzger, H., Townsend, D., Gador, N., Ullrich, S., Stolow, A., and Martinez, T. J. (2007). Ab initio molecular dynamics and time-resolved photoelectron spectroscopy of electronically excited uracil and thymine, *J. Phys. Chem. A* **111**, 34, pp. 8500–8508.

95. Szymczak, J. J., Barbatti, M., Soo Hoo, J. T., Adkins, J. A., Windus, T. L., Nachtigallová, D., and Lischka, H. (2009). Photodynamics simulations of thymine: Relaxation into the first excited singlet state, *J. Phys. Chem. A* **113**, 45, pp. 12686–12693.

96. Barbatti, M., Aquino, A. J., Szymczak, J. J., Nachtigallová, D., Hobza, P. and Lischka, H. (2010). Relaxation mechanisms of uv-photoexcited DNA and RNA nucleobases, *Proc. Natl. Acad. Sci. U. S. A* **107**, 50, pp. 21453–21458.

97. Asturiol, D., Lasorne, B., Worth, G. A., Robb, M. A., and Blancafort, L. (2010). Exploring the sloped-to-peaked s 2/s 1 seam of intersection of thymine with electronic structure and direct quantum dynamics calculations, *Phys. Chem. Chem. Phys.* **12**, 19, pp. 4949–4958.

98. Picconi, D., Barone, V., Lami, A., Santoro, F., and Improta, R. (2011). The interplay between $\pi\pi^*/n\pi^*$ excited states in gas-phase thymine: A quantum dynamical study, *ChemPhysChem* **12**, pp. 1957–1968.

99. Jónsson, H., Mills, G., and Jacobsen, K. W. (1998). *Nudged Elastic Band Method for Finding Minimum Energy Paths of Transitions*, chapter 16 (World Scientific, Singapore), pp. 385–404.

100. Kästner, J., Carr, J. M., Keal, T. W., Thiel, W., Wander, A., and Sherwood, P. (2009). DL-FIND: An open-source geometry optimizer for atomistic simulations, *J. Phys. Chem. A* **113**, 43, pp. 11856–11865.

Chapter 5

Conformationally Controlled Photochemistry Studied by Trajectory Surface Hopping

Enrico Tapavicza
Department of Chemistry and Biochemistry, California State University, Long Beach, 1250 Bellflower Boulevard, Long Beach, California 90840, USA
enrico.tapavicza@csulb.edu

5.1 Introduction

The influence of the conformation of flexible molecules on the photochemical reactivity has been an active field of research since the early work of Havinga [71], which was motivated by understanding vitamin D photochemistry. Later, Havinga formulated the concept of nonequilibrium excited state rotamers (NEER) on the basis of the reactivity of simple hexatriene [3, 69] molecules and vitamin D related model compounds [32]. The NEER principle states that the photochemical reactivity and photoproduct distribution of a flexible molecule is dictated by the conformational equilibrium of the molecule in the electronic ground state and not by the

Time-Dependent Density Functional Theory: Nonadiabatic Molecular Dynamics
Edited by Chaoyuan Zhu

ISBN 978-981-4968-42-3 (Hardcover), 978-1-003-31921-4 (eBook)
www.jennystanford.com

conformational equilibrium that would result from the potential energy surface of the excited singlet state. The rationale of this argument is that after photoexcitation of the ensemble of conformers, the lifetime of the reactive excited state is too short to allow for structural adaptation to the new equilibrium of rotamers in the excited state and thus the outcome of a fast photochemical reaction is a consequence of the conformational equilibrium at the time of photoexcitation.

This principle has been found to be useful to understand the photoreactivity in a variety of hexatriene systems, most prominently in vitamin D photochemistry, but also in the photochemical synthesis of ikarugamycin [153], and the photochemistry of alkyl-substituted hexatrienes [15, 77]. Besides hexatriene systems, conformationally controlled photochemistry has also been found involved in the reactivity of substituted ketones [57, 84, 86, 150], in photochemical reactions of bifunctional molecules [149], in the triplet reactivity of conjugated dienes [88], in the photoinduced ring-opening of substituted hexadienes [124], in the photoreactivity of alkyl azides [102], in intramolecular photocyclization of dienes [106], α-(aminoalkyl)styrenes [83] and diarylethenes [96], in the wavelength-dependent photochemistry of cyclocarbenes [163], and the photoreactivity of bilirubin [79, 92]. With the recent developments in photochemical switches, light-driven motors [38, 67], and photoactive proteins and peptides [144], strategies to control the conformationally dependent photochemistry has attracted new interest [37, 66, 96, 114]. However, despite its success in predicting the outcome of numerous photochemical reactions, the NEER principle has failed in several cases [11, 12, 21, 35, 41, 60]; this anti-NEER behavior is attributed to low energetic barriers for the rotational isomerization in the excited state, allowing for rapid interconversion of rotamers in the excited states [42]. Furthermore, admixtures between these two extreme cases occur, where the excited state relaxation affects the distribution of rotamers to some extent [45].

The development of accurate quantum chemistry methods to describe electronically excited states such as time-dependent density functional theory (TDDFT) [115], complete active space self-consistent field (CASSCF) [87] and its perturbative variant CASPT2 [7, 8], and coupled cluster (CC) methods [122] allows to support photochemical studies by theoretical calculations. In a large number of early theoretical studies, electronic structure calculations based on single geometries at crucial points of the potential energy surface (PES), or minimum energy pathways (MEP) along a reaction coordinate are used to support experimental results [39, 64]. These static calculations do not include dynamic effects arising from the relaxation from the Franck-Condon region and from the electron–nuclear coupling. These effects are, however, often crucial to accurately describe the outcome of photochemical reactions. Therefore, methods that are able to describe the dynamics of the system are more promising to yield accurate results. In principle the description of photochemical dynamics requires a full quantum description of the coupled nuclear-electronic motion according to the time-dependent Schrödinger equation. This goal is usually out of reach for large and medium-sized molecules. While a number of highly accurate methods, such as the multiconfigurational time-dependent Hartree [23] method, allow to carry out full-quantum dynamics simulations, these methods are limited to small systems due to their large computational cost. In addition, these methods often require precalculated potential energy surfaces [48], which is cumbersome for systems with a large number of degrees of freedom. Computationally more efficient are approximative semiclassical "on-the-fly" non-adiabatic ab initio molecular dynamics (AIMD) methods [46, 53, 134], which are based on a classical description of the nuclear motion, but treat electrons quantum mechanically. Semiclassical non-adiabatic methods can be divided into i) mean-field or Ehrenfest approaches [9, 85, 139], which are based on single trajectories computed on potential energy surfaces averaged over several electronic states, and ii) trajectory surface hopping (TSH) [46, 53, 134, 145, 161] methods, in which a large number of trajectories is used to describe the dynamics of the ensemble. Besides the purely semiclassical methods, the ab initio multiple spawning (AIMS) [5, 26] method incorporates the quantum nature of nuclear motion to some extent. These methods have allowed to study the dynamics of photochemical reaction mechanisms in a large number of systems. On one hand, non-adiabatic simulations have been shown to be useful to predict excited state lifetimes, product distributions, branching ratios, and wavelength dependent product quantum yields [137, 142]. On the other hand, the simulation of time-resolved spectra [90, 116, 156] and other observables has served as a means to verify time constants and branching ratios obtained from non-adiabatic simulations by direct comparison of their experimental counterparts. In contrast to mean-field methods, in which nuclear trajectories are propagated on an excited state potential energy surface that is obtained as a superposition of several electronic states, the TSH molecular dynamics method has been shown to be particularly useful for the description of branching of the nuclear trajectories. Due to its computational efficiency, TDDFT is often the method of choice for solving the underlying electronic structure problem for electronically excited states. In this article I will review the efforts of applying TSH methods based on TDDFT in investigating conformationally controlled photochemical reactions in my group. I will first lay out the theoretical methods to study conformationally controlled photochemical reactions and then show how to apply these methods on the example of some systems that have been studied in my group.

Theoretical steps include the generation of Boltzmann ensembles of structures, the computation of electronic absorption spectra of conformationally flexible molecules, the TDDFT trajectory surface hopping method (TDDFT-SH). Furthermore, I will outline how to compute wavelength dependent quantum yields using the TDDFT-SH method.

5.2 Theoretical Methods

5.2.1 Generating Boltzmann Ensembles

The accurate simulation of a photochemical experiment critically depends on a reliable description of the initial conditions of the system at the time the system is photoexcited. Typically the most important parameters of interest are the nuclear positions and their momenta, but in certain cases molecular orientation and impact energies are required [16]. Thus the first step is usually to generate an ensemble of structures and velocities for the TSH simulations.

Generating Boltzmann ensembles of structures of molecules for TSH is typically done by either generating a Wigner distribution [18, 31] of structures based on vibrational spectra and vibrational normal modes computed within the harmonic oscillator approximation, or based on semiclassical molecular dynamics (MD) simulations. The Wigner approach has the advantage that it takes into account the quantum nature of the nuclei. This is particularly important if the molecule is assumed to be in the lowest vibrational states at the time of photoexcitation, which is the case for reactions at low temperatures or for reactions of rigid molecules with high vibrational frequencies compared to the available thermal energy. In contrast, for conformationally flexible systems that possess low vibrational frequencies, one can assume a classical behavior of the nuclei and thus, a classical MD simulation should be able to accurately generate a Boltzmann ensemble of structures and velocities. Furthermore, a MD simulation gives access to the distribution of distinct conformations, which is not possible in the case of the Wigner distribution, because it is usually based on a single minimum energy structure.

Several approaches exist that intend to incorporate some of the advantages of classical sampling but still contain some of the quantum behavior of the nuclei. Here to mention are the approaches using ring-polymer dynamics [47], the generalized Langevin equations [40], which has been used to generate initial conditions in TSH [79], and mixed Franck-Condon methods combined with MD simulations, which have been used to simulate electronic spectra [162].

For conformationally flexible molecules with increasing number of rotational degrees of freedom, however, sampling the ensemble of rotational isomers becomes increasingly challenging. This is especially the case when energetic minima are separated by energetic barriers that are large compared to the thermal energy available by the system. Thus to describe conformationally controlled photochemistry, special care has to be taken to ensure proper sampling of the initial ensemble of structures. To achieve this goal a variety of enhanced sampling methods [61] are available, including umbrella sampling, meta-dynamics [100], or parallel tempering, also known as replica exchange molecular dynamics (REMD) [128]. While all these methods have been developed to accelerate sampling in systems with large energetic barriers, I will focus here on the description of REMD, which can easily be implemented in the context of classical MD and AIMD. REMD has been successfully applied to generate Boltzmann ensembles of structures for spectra simulations and initial conditions for TSH to study conformationally flexible systems [45, 50, 123, 133, 137, 142].

In REMD, several molecular dynamics trajectories at different temperatures are simulated in parallel. To ensure proper sampling of a canonical ensemble, the temperature in each simulation is controlled by a Nosé-Hoover thermostat [105]. After a certain number of MD steps, the probability that two structures from trajectories at temperature T_i and T_j will be switched is evaluated as

$$P(i \rightarrow j) = \min\left[1, \frac{\exp\left(-\frac{E_j}{k_B T_i} - \frac{E_i}{k_B T_j}\right)}{\exp\left(-\frac{E_i}{k_B T_i} - \frac{E_j}{k_B T_j}\right)}\right], \tag{5.1}$$

where k_B is the Boltzmann constant and E_i and E_j are the total energies of structures at T_i and T_j, respectively. If the switching probability $P(i \rightarrow j)$ is 1, the two structures at different temperatures are switched and the nuclear coordinates are propagated at the new temperature. In case the probability is below 1, a random number from the interval [0,1] is generated and the structures are only switched if the random number is smaller than the switching probability. After a switch between two structures, velocities are rescaled to be consistent with the new temperatures. An ensemble of structures generated in this fashion has been shown to lead to the thermodynamically correct distribution of structures. Trajectories at high temperatures enable the system to overcome energetic barriers, exploring parts of the conformational space that would not have been accessible at low temperature within a reasonable simulation time. Furthermore, the method retains the relative populations of energetic minima if the simulation is carried out for a long enough time interval.

5.2.2 Calculation of Absorption Spectra

The initial event of a photochemical reaction is light-absorption. Therefore, a critical component in describing photochemical reactions is an accurate assessment of the light-absorption properties of the system of interest. For instance, the prediction of wavelength dependent product quantum yields relies on an accurate prediction of electronic absorption spectra.

Neglecting vibrational effects, for monochromatic radiation of wavelength λ, the number of photons absorbed during a transition between two electronic states is proportional to the oscillator strength [74]. The calculation of excitation energies and oscillator strengths is routinely available from many different electronic structure methods. In wavefunction based-methods the oscillator strengths f are obtained from the transition dipole moment computed for the initial and final electronic states, whereas in TDDFT f is computed from the transition density between ground and excited state [62]. Recently, also machine learning models have been developed and applied to predict excitation energies and oscillator strengths [113, 138, 152, 158]. Oscillator strengths can be converted through a line function $\rho(\lambda)$ to the absorption cross section $\sigma(\lambda)$ or, equivalently, to the extinction coefficient $\epsilon(\lambda)$ [74]:

$$\epsilon(\lambda) = 2.303 \frac{4\pi^2 q_e}{3\hbar c} \rho(\lambda) N_A f, \tag{5.2}$$

where $\hbar$ is Planck's reduced constant, c denotes the speed of light, q_e is the electronic charge, and N_A is Avogadro's number. Typically, Lorentzian and Gaussian lineshapes are used for broadening in case of homogeneous and inhomogeneous broadening, respectively. For a more accurate description of the envelope of absorption spectra conformational sampling and the inclusion of vibronic effects might be necessary. In case of flexible molecules that exist in a large number of different conformations, a Boltzmann ensemble of structures obtained from molecular dynamics, as described in the previous section, can be used. In this case, excitation energies and oscillator strengths computed for a subset of snapshot structures from a MD or REMD trajectory are converted into absorption spectra according to Eq. 5.2 [58]. An average of spectra computed for a large number of snapshot structures yields the macroscopic

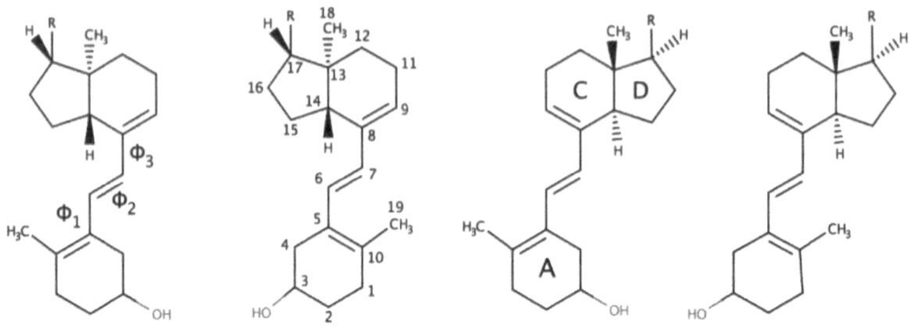

tEt cEt tEc cEc

Figure 5.1 Rotational isomers of truncated tachysterol. R indicates a methyl group. Dihedral angles Φ_1 and Φ_3 are defined by atoms C4-C5-C6-C7 and C6-C7-C8-C9, respectively. Reproduced with permission from C. Cisneros, T. Thompson, N. Baluyot, A. C. Smith, E. Tapavicza, *Phys. Chem. Chem. Phys.* 19 (8), 5763–5777 (2017) the PCCP Owner Societies.

absorption spectrum of the ensemble of structures. The procedure is exemplified for the flexible molecule tachysterol, which occurs in four major rotamers cEc, cEt, tEc, and tEt (Fig. 5.1). After generating an ensemble of structures of tachysterol using REMD [45], excitation energies and oscillator strengths for the electronic transition between the ground state (S_0) and the first excited state (S_1) have been computed using TDDFT and the second-order approximate coupled cluster singles and doubles (CC2) method [44]. Plotting CC2 excitation energies and oscillator strengths as a function of the dihedral angles Φ_1 and Φ_2 (Fig. 5.2), we see that both, absorption energy and intensity, are sensitive to the conformation of the molecule. Converting excitation energies and oscillator strengths to extinction coefficients using a Gaussian lineshape with a full width at half maximum (FWHM) of 0.1 eV, and averaging over all snapshot structures, yields the macroscopic absorption spectrum of the system (black, Fig. 5.3). The same procedure can be applied individually to the specific rotamers (magenta, blue, green, and red in Fig. 5.3), which allows to gauge the influence of conformation on the overall absorption spectrum.

However, in some cases the envelope of an absorption spectrum is caused by the quantum nature of the nuclear vibrations and not by the ensemble of conformations. This is particularly the case for

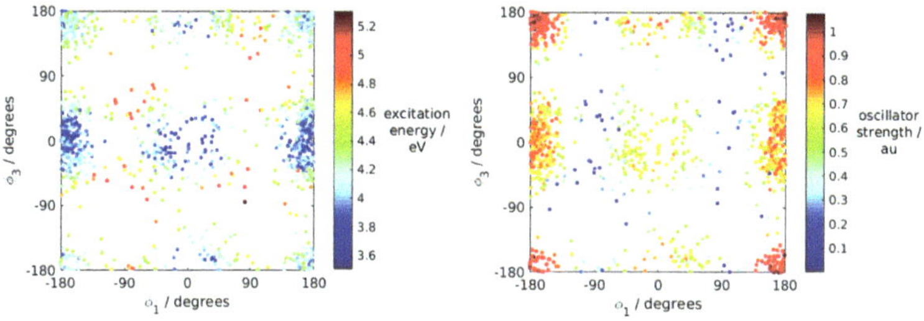

Figure 5.2 S_1 excitation energies (left) and oscillator strengths (right) as a function of dihedral angles Φ_1 and Φ_3, as are indicated in Fig. 5.1. Reproduced from C. Cisneros, T. Thompson, N. Baluyot, A. C. Smith, E. Tapavicza, *Phys. Chem. Chem. Phys.* 19 (8), 5763–5777 (2017) with permission from the PCCP Owner Societies.

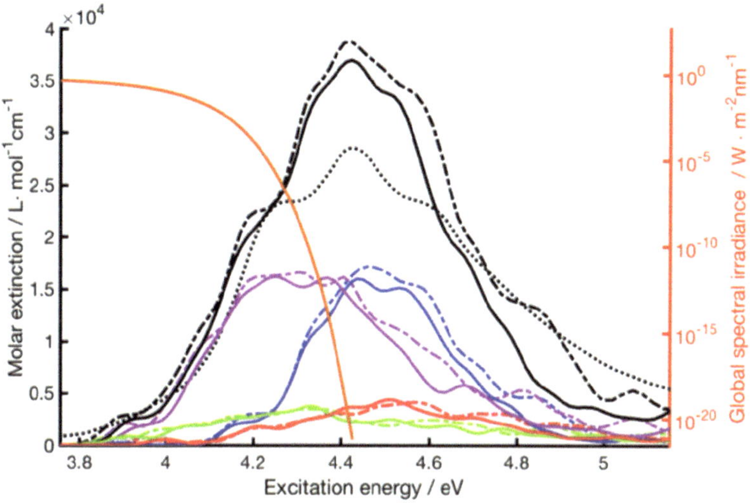

Figure 5.3 Tachysterol absorption spectrum for the different rotamers computed by TDPBE0 (solid) and CC2 (dashed). tEt: blue, tEc: magenta, cEt: red, cEc: green. Experimental spectrum (dotted) measured in ether [91]. TDPBE0 and CC2 absorption bands were shifted by to match the position of the peak maximum of the experimental spectrum. As example, the global irradiation energy density for a zenith angle of 30° is shown (brown) [68]. These conditions approximately correspond to the irradiation in Berlin on July 8 at noon [1]. Reproduced from Cisneros, C., Thompson, T., Baluyot, N., Smith, A. C., and Tapavicza, E. (2017), *Phys. Chem. Chem. Phys.* 19, pp. 5763–5777 with permission from the PCCP Owner Societies.

rigid molecules that possess high vibrational frequencies. In this case, it is necessary to apply a quantum mechanical treatment of the nuclear vibrations, which can be done using Franck–Condon methods [130]. The vibronic absorption cross section at frequency ω and temperature T is given by [132]

$$\sigma_{abs}(\omega, T) = \frac{4\pi^2\omega}{3c}|\mu_{if}|^2 \sum_{\mathbf{v}_i}\sum_{\mathbf{v}_f} P_{\mathbf{v}_i}(T)|\langle\theta_{\mathbf{v}_i}|\theta_{\mathbf{v}_f}\rangle|^2\delta(\Delta E_{if} + E_{\mathbf{v}_i} - E_{\mathbf{v}_f} - \omega), \tag{5.3}$$

where the electronic transition dipole moment μ_{if} is defined through the wavefunctions of initial (ground) and final (excited) electronic state, c is the speed of light and the Franck–Condon factors $|\langle\theta_{\mathbf{v}_i}|\theta_{\mathbf{v}_f}\rangle|^2$ are defined through initial and final vibrational wavefunctions θ_i and θ_f, respectively. ΔE_{if} denotes the adiabatic excitation energy, and $E_{\mathbf{v}_i}$ and $E_{\mathbf{v}_f}$ are the initial and final vibrational energies, respectively. $P_{\mathbf{v}_i}(T)$ denotes the occupation of initial vibrational state with quantum numbers $\mathbf{v}_i$ [130], dictated by the Boltzmann distribution. To compute the absorption spectrum according to Eq. 5.3, one needs to compute Franck–Condon factors between vibrational states of the initial and final electronic states. Two general approaches in order to do so exist: time-independent (TI) [29, 51, 78, 119, 125] and time-dependent (TD) [13, 28, 73, 80, 98, 103, 108, 126, 132, 159] approaches. In TI approaches Franck–Condon factors are computed directly in the frequency domain using analytical formulas for either the harmonic oscillator approximation [148] or for the anharmonic approximations [20]. TD approaches, in contrast, use a *time-correlation function* $G(t)$, whose Fourier transform yields the vibronic absorption spectrum, according to

$$\sigma(\omega) = \frac{4\pi^2\omega}{3c}|\mu_{if}|^2\frac{1}{2\pi}\int_{-\infty}^{\infty} dt \exp\left[it(\Delta E_{if} + E_{\mathbf{0}_i} - \omega)\right]G(t), \tag{5.4}$$

where $E_{\mathbf{0}_i}$ is the zero point vibrational energy in the ground state. Equation 5.4 is formulated for the special case at zero temperature, but expressions for finite temperature have been derived as well [25, 109]. While TI approaches are relatively simple to implement, the computational effort for their computation increases rapidly with the number of vibrational degrees of freedom N, as they usually require an N-fold loop, unless specific algorithms, such as the Beyer–Swinehart [27] and the related Rabinovich–Stein algorithm [125], are employed that lead to improved scaling with N. However, TD approaches exhibit better scaling with system size and are therefore suited for larger systems. Furthermore, TD approaches allow to easily incorporate Duschinsky mode [56] mixing, which expands the vibrational modes in the excited state $\mathbf{Q}_\mathrm{f}$ as a linear combination of vibrational modes in the ground state $\mathbf{Q}_\mathrm{i}$ (and reverse), according to

$$\mathbf{Q}_\mathrm{f} = \mathbf{J}\mathbf{Q}_\mathrm{i} + \mathbf{D}, \tag{5.5}$$

where, $\mathbf{J}$ is the Duschinsky rotation matrix and $\mathbf{D}$ is the geometric displacement matrix, accounting for the difference in ground and excited state structures.

The computation of the time-correlation (or generating) function can be done efficiently based on ground and excited state equilibrium structures and their vibrational frequencies and normal modes. On the basis of the latter, the time-correlation function reads

$$G(t) = 2^{\frac{N}{2}} \left(\frac{\det(\mathbf{S}^{-1}\Omega_\mathrm{i}\Omega_\mathrm{f})}{\det(\mathbf{L})\det(\mathbf{M})} \right)^{\frac{1}{2}} \exp\left[\mathbf{D}^\mathsf{T}(\Omega_\mathrm{f}\mathbf{B}\mathbf{J}\mathbf{M}^{-1}\mathbf{J}^\mathsf{T}\Omega_\mathrm{f}\mathbf{B} - \Omega_\mathrm{f}\mathbf{B})\mathbf{D}\right]. \tag{5.6}$$

Matrices Ω_i, Ω_f, $\mathbf{S}$, $\mathbf{B}$ are diagonal and depend on time t and vibrational frequencies in the ground (ω_k^i) and excited state (ω_k^f): $(\Omega_\mathrm{i})_{kk} = \omega_k^i$, $(\Omega_\mathrm{f})_{kk} = \omega_k^f$, $S_{kk} = \sinh(i\omega_k^f t)$, and $B_{kk} = \tanh(i\omega_k^f t/2)$. $^\mathsf{T}$ denotes the transpose of the matrices. Matrices $\mathbf{L}$ and $\mathbf{M}$ are given by $\mathbf{M} = \mathbf{J}^\mathsf{T}\Omega_\mathrm{f}\mathbf{B}\mathbf{J} + \Omega_\mathrm{i}$ and $\mathbf{L} = \mathbf{J}^\mathsf{T}\Omega_\mathrm{f}\mathbf{B}^{-1}\mathbf{J} + \Omega_\mathrm{i}$, respectively. Using Eq. 5.4 together with Eq. 5.6 allows to compute the absorption spectrum for the zero temperature case. Alternatively, the time-correlation function can be computed using wavepacket propagation methods [72, 73] or semiclassical methods based on trajectories [24].

For rigid molecules, the previously discussed method using snapshot structures from MD trajectories can lead to an erroneous modeling of the absorption spectrum, as shown in Fig. 5.4 for 7,7,8,8,-tetracyanoquinodimethane anion (TCNQ^-). As we can see from Fig. 5.4, the MD approach (red) is not able to reproduce

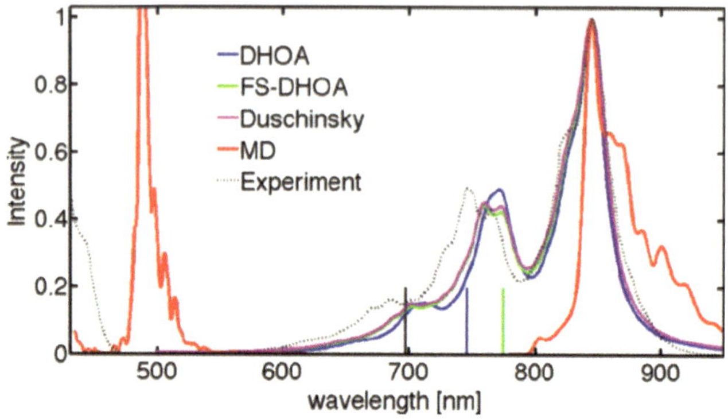

Figure 5.4 Comparison between the calculated and experimental (dashed) spectra. The vertical excitation energy is indicated by the black bar (1.778 eV). The original position of the spectrum obtained from the displaced harmonic oscillator approximation (DHOA) is indicated by the blue bar (1.664 eV). The DHOA spectrum was red-shifted by 106 nm from the adiabatic excitation energy to match the 0–0 transition of the experimental spectrum. The frequency-shifted DHOA (FS-DHOA) spectrum (green) and the spectrum calculated using Duschinsky rotation (magenta) were shifted by 71 nm from the original position of the corrected 0–0 transition (green bar, 1.602 eV). The peak of the spectrum obtained from MD simulations (red) was shifted by 135 nm. Reprinted with permission from E. Tapavicza, F. Furche, D. Sundholm, J. Chem. Theory Comput. 2016, 12, 5058–5066. Copyright 2016 American Chemical Society.

the experimentally measured absorption spectrum (black, dashed), because it misses the bands between 600 and 800 nm, which originate from vibronic effects. In contrast, all approaches based on the generating function approach (DHOA, FS-DHOA, and Duschinsky) are able to account for the bands in this spectral region in $TCNQ^-$.

5.2.3 Linear Response Time-Dependent Density Functional Surface Hopping

A mixed quantum-classical treatment of the coupled nuclear and electronic dynamics [131] of a photochemical reaction requires the solution of the time-dependent electronic Schrödinger equation

$$i\hbar\frac{\partial}{\partial t}\Psi(t|\mathbf{R}) = \hat{H}_{el}(t|\mathbf{R})\Psi(t|\mathbf{R}), \tag{5.7}$$

with $\Psi(t|\mathbf{R})$ being the time-dependent wavefunction, at nuclear coordinates $\mathbf{R}$. The electronic Hamiltonian is defined as

$$\hat{H}_e(t|\mathbf{R}) = \hat{V}_{NN}(\mathbf{R}) + \hat{T}_e + \hat{V}_{Ne}(\mathbf{R}) + \hat{V}_{ee} + \hat{V}_{ext}(t), \tag{5.8}$$

where $\hat{V}_{NN}$ denotes the nuclear-nuclear repulsion, $\hat{T}_e$ is the kinetic energy of the electrons, $\hat{V}_{Ne}$ denotes the attraction between nuclei and electrons, $\hat{V}_{ee}$ is the electron–electron repulsion, and $\hat{V}_{ext}$ is a time-dependent external potential, for example an oscillating laser field.

In quantum-classical dynamics, the nuclear motion evolves purely classically according to Newton's equation of motion

$$F_j = m_j \ddot{R}_j, \tag{5.9}$$

where F_j is the force acting on nuclear degree of freedom j with mass m_j, and $\ddot{R}_j$ denotes the acceleration. In Ehrenfest dynamics, the force is evaluated as the negative derivative of an energy expectation value $\langle E \rangle$, averaged over several electronic states. In TSH, in contrast, the force is obtained as the derivative of the energy of a *pure* adiabatic state m:

$$F_j = -\frac{\partial E_m(\mathbf{R})}{\partial R_j}. \tag{5.10}$$

For the purpose of deriving the working equations of TSH, we expand the time-dependent electronic wavefunction in terms of the adiabatic states according to the Born-Oppenheimer expansion [54]

$$\psi(t|\mathbf{R}) = \sum_k C_k(t)\Phi_k(\mathbf{R}). \tag{5.11}$$

Here, $\Phi_k(\mathbf{R})$ denote the adiabatic states. $C_k(t)$ denote the complex time-dependent expansions coefficients that define the density matrix of the system. In TSH based on linear response TDDFT (TDDFT-SH), however, the time-dependent wavefunction (Eq. 5.11) is never directly computed; instead, only a differential equation for the state coefficients is propagated in time:

$$i\dot{\mathbf{C}} = [\mathbf{H} - i\sigma + \mathbf{V}_{ext}]\mathbf{C}, \tag{5.12}$$

where the first-order derivative non-adiabatic coupling (NAC) [43, 54, 140, 141] is given by

$$\sigma_{kj} = \langle \Phi_k | \frac{\partial}{\partial t} \Phi_j \rangle. \tag{5.13}$$

The Hamiltonian matrix **H** contains the adiabatic eigenenergies $H_{nm} = E_{nm}\delta_{nm}$. Within linear response TDDFT [62], the ground state energy E_{00} is obtained from a ground state density functional theory (DFT) calculation, whereas for excited states E_{nn} are obtained by adding the excitation energies ω_n to the ground state energy.

Typically, a photochemical reaction is initiated by absorption of light, promoting the system to an electronically excited state. Therefore, in the beginning of a TDDFT-SH simulation the system is assumed to be in a pure adiabatic state and the population of the initial state m is unity ($C_m^* C_m = 1$). Excited state nuclear forces are then computed using linear response TDDFT [62] and used to propagate the nuclear degrees of freedom on an adiabatic Born-Oppenheimer (BO) surface. At each time step of the TSH simulation, the $3N$-dimensional Cartesian derivative coupling vector between the state of interest and neighboring electronic states is computed

$$\mathbf{d}_{kj}(\mathbf{R}) = \langle \Phi_k(\mathbf{R}) | \nabla_{\mathbf{R}} \Phi_j(\mathbf{R}) \rangle. \tag{5.14}$$

Using the chain rule, the NAC can be computed as the product of the nuclear velocities and the Cartesian NAC vector

$$\sigma_{kj} = \dot{\mathbf{R}} \cdot \mathbf{d}_{kj}\,. \tag{5.15}$$

Within localized basis sets the NAC vector can be efficiently calculated using the coupled perturbed Kohn-Sham equations [118]. From the NAC and the state amplitudes (Eq. 5.12), a hopping probability is computed at each time step

$$g_{nm} = \frac{2\Delta t}{|C_n|^2}\left[\Im(C_m^*(H_{nm} - V_{ext})C_n) - \Re(c_m \sigma_{mn} C_n)\right]. \tag{5.16}$$

The probability is compared to a random number from $\zeta \in [0, 1]$ and a surface hop from state m to n ($m \neq n$) is accepted if and only if ζ satisfies

$$\sum_{k<n} g_{mk} < \zeta < \sum_{n<k} g_{mk}. \tag{5.17}$$

After a surface hop, the nuclear forces of the new electronic state are used to propagate the nuclear coordinates. In case the new state is the electronic ground state, the forces are computed using DFT. Furthermore, changing the electronic state requires scaling of the nuclear momenta in order to conserve the total energy of the system.

According to Tully's recipe [145], scaling should be done parallel to the first order derivative NAC vector (Eq. 5.14). In cases, where the NAC vector is not available, isotropic velocity scaling has been applied [53, 134]. However, proper velocity scaling has been found to be crucial for an accurate description of the dynamics in some systems [17, 147].

After having generated the initial conditions of the system of interest (see Section 5.2.1), typically several hundreds of representatives of this ensemble are used to carry out the excited state non-adiabatic simulations. Each trajectory will follow a distinct relaxation pathway, allowing to observe several reaction channels.

The TDDFT-SH method has been implemented in several variations and applied to study a variety of systems [133, 134, 154, 155]. For molecular systems, its implementation with localized basis sets [14, 131] offers better efficiency compared to plane wave basis sets [134]. An important advantage of localized basis sets lies in the efficient application of hybrid exchange-correlation functionals, which are crucial in the description of excited states with charge-transfer character [55, 135].

Regarding the accuracy of TDDFT, one has to be aware of shortcomings regarding the following issues in excited state simulations: (i) charge-transfer excitations, (ii) Rydberg excitations, (iii) conical intersections, and (iv) double excitations. While charge-transfer and Rydberg excitations can be treated sufficiently accurate with hybrid exchange-correlation functionals, it is still advisable to carefully benchmark results by comparison with more accurate wavefunction based methods, such as the CC2 method. Double excitations can be described within TDDFT using a frequency dependent exchange-correlation kernel [94]. The erroneous description of conical intersections between ground and excited states within TDDFT [82, 136] arises from its single-reference character. It can cause problems in excited state simulations, due to the reduced dimension of the intersection space and due to discontinuous potential energy surfaces. While practical approaches have been used to handle these problems [131], recently, several TDDFT methods based on an ensemble description have been proposed to cure these deficiencies [59, 81].

5.2.4 Prediction of Product Quantum Yields

According to the IUPAC, the wavelength-dependent product quantum yield (PQY) of a photochemical reaction is defined [34] as the number of product molecules formed upon excitation at wavelength λ ($N^{\mathrm{P}}_{\mathrm{molecules}}(\lambda)$) divided by the total number of photons of wavelength λ absorbed ($N^{\mathrm{tot}}_{\mathrm{photons}}(\lambda)$) [142]:

$$\Phi(\lambda) = \frac{N^{\mathrm{P}}_{\mathrm{molecules}}(\lambda)}{N^{\mathrm{tot}}_{\mathrm{photons}}(\lambda)}. \tag{5.18}$$

Hence, to predict $\Phi(\lambda)$ from first principles it is necessary to know two contributions: a) the efficiency of light absorption, which is given by the absorption cross section $\sigma(\lambda)$ or the extinction coefficient (Eq. 5.2) and b) the probability that the specific product is formed once the molecule is excited. We will refer to latter probability to as the *branching ratio*. For conformationally flexible molecules, the macroscopic absorption cross section often strongly depends on the equilibrium distribution of conformers present in the ground state, which we have seen on the example of tachysterol (Fig. 5.3). The branching ratio of a photochemical reaction is governed by the dynamics induced by photoexcitation, which inherently depends on the topology and shape of the excited and ground state potential energy surfaces, non-adiabatic effects, and the temperature. Moreover, the course of a trajectory depends on the initial conformation of a molecule. Thus, one can suspect that different conformers will exhibit different photodynamics and potentially lead to different photoproducts. Consequently, in order to predict wavelength-dependent PQYs, it is necessary to account for the dependency of the molecular conformation on the absorption properties, as well as for the dependency of the molecular conformation on the photodynamical behavior. Since TSH is based on ensemble approach that allows trajectories to follow different reaction pathways, it is well-suited to account for both contributions.

In the definition of the PQY (Eq. 5.18), the number of product molecules $N^{\mathrm{P}}_{\mathrm{molecules}}(\lambda)$ is equivalent to the number of photons absorbed by reactants that successfully form product P ($N^{\mathrm{P}}_{\mathrm{photons}}(\lambda)$), assuming that that one photon is needed to produce one product molecule. This allows us to rewrite Eq. 5.18 as

$$\Phi(\lambda) = \frac{N^{\mathrm{P}}_{\mathrm{photons}}(\lambda)}{N^{\mathrm{tot}}_{\mathrm{photons}}(\lambda)}. \tag{5.19}$$

Considering monochromatic radiation of wavelength λ, the number of photons absorbed by a material is proportional to the oscillator strength, which is related through a line function to the absorption cross section $\sigma(\lambda)$ or, equivalently, to the extinction coefficient $\epsilon(\lambda)$ (Eq. 5.2). In the framework of TSH, $N^{\mathrm{P}}_{\mathrm{photons}}(\lambda)$ is proportional to the average absorption cross section of the initial structures of trajectories that successfully lead to a given product $\sigma_P(\lambda)$:

$$N^{\mathrm{P}}_{\mathrm{photons}}(\lambda) = k\frac{1}{N^{\mathrm{tot}}_{\mathrm{traj}}}\sum_{i}^{N^{\mathrm{P}}_{\mathrm{traj}}}\sigma_i(\lambda) = k\sigma_P(\lambda). \tag{5.20}$$

$N^{\mathrm{tot}}_{\mathrm{photons}}(\lambda)$ is proportional to the average absorption cross section of the initial structures of all trajectories $\sigma_{tot}(\lambda)$:

$$N^{\mathrm{tot}}_{\mathrm{photons}}(\lambda) = k\frac{1}{N^{\mathrm{tot}}_{\mathrm{traj}}}\sum_{i}^{N^{\mathrm{tot}}_{\mathrm{traj}}}\sigma_i(\lambda) = k\sigma_{tot}(\lambda). \tag{5.21}$$

Here, k is a constant relating the number of photons to the absorption cross section, and $N^{\mathrm{P}}_{\mathrm{traj}}$ and $N^{\mathrm{tot}}_{\mathrm{traj}}$ are the number of trajectories that successfully form products and the total number of trajectories, respectively. Hence, within TSH, the product quantum yield can be computed as

$$\Phi(\lambda) = \frac{\sum_i^{N^{\mathrm{P}}_{\mathrm{traj}}}\sigma_i(\lambda)}{\sum_i^{N^{\mathrm{tot}}_{\mathrm{traj}}}\sigma_i(\lambda)} = \frac{\sigma_P(\lambda)}{\sigma_{tot}(\lambda)}. \tag{5.22}$$

For the practical computation of the PQY using the outlined approach, trajectories need to be grouped according to their products formed. Then, for each group the average absorption spectrum $\sigma_P(\lambda)$ is computed by averaging the broadened spectra of each TSH initial structure. Analogously, the total spectrum $\sigma_{tot}(\lambda)$ is computed by averaging the spectra of all initial structures. Dividing the product spectrum by the total spectrum of all initial structure, according to Eq. 5.22, yields the PQY. This approach avoids binning of the initial structures according to their excitation energy.

In most TSH studies, the PQY is simply taken as the ratio of products forming trajectories [30, 120, 133, 154, 160], which we have termed the branching ratio. In case of a wavelength independent photochemistry, the branching ratio converges to the PQY. However, in the general case the absorption energy and efficiency of the initial structures has to be taken into account to obtain PQYs according to the IUPAC definition (Eq. 5.18). The above outlined strategy achieves this goal.

5.3 Applications

5.3.1 Photochemistry of Z-Hexatriene Derivatives

Substituted Z-hexatriene derivatives have been one of the first systems investigated that led to the hypothesis of the NEER principle [3, 69]. One of the major conclusions of these studies was that the "various conformers of the ground state upon absorption of UV light yield different excited species which do not interconvert during their short (singlet excited) lifetime" [3]. Thus each conformer is thought to produce a distinct photoproduct distribution. To investigate the validity of this hypothesis, we carried out TDDFT-SH simulations of 3Z-hexa-1,3,5-triene (HT) and several of its substituted derivatives (Fig. 5.5), [137] namely 2,5-dimethyl-HT (DMHT), 2-isopropyl-5-methyl-HT (IMHT), and 2,5,-diisopropyl-HT (DIHT), depicted in the first column of Fig. 5.7. To generate a Boltzmann ensemble of structures for these molecules we carried out REMD based on DFT using the PBE approximation to the exchange-correlation [110]. The ensemble of structures can be presented in the space of dihedral angles Φ_1 and Φ_3 (see Fig. 5.5 for definitions), similar to the

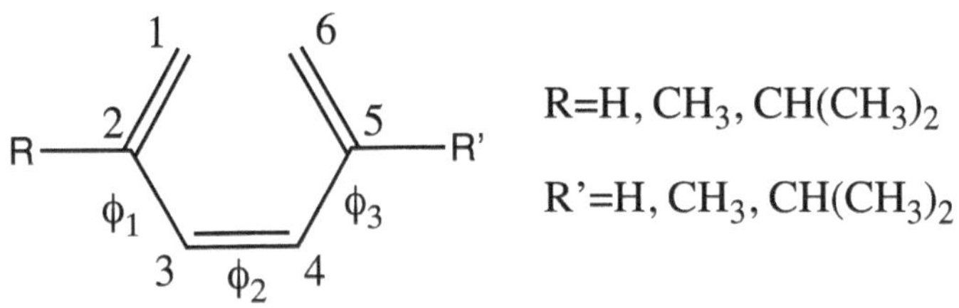

Figure 5.5 General structure of substituted 3Z-hexatriene derivatives. Dihedral angles Φ_1, Φ_2, and Φ_3 are defined by carbon atoms C1-C2-C3-C4, C2-C3-C4-C5, and C3-C4-C5-C6, respectively.

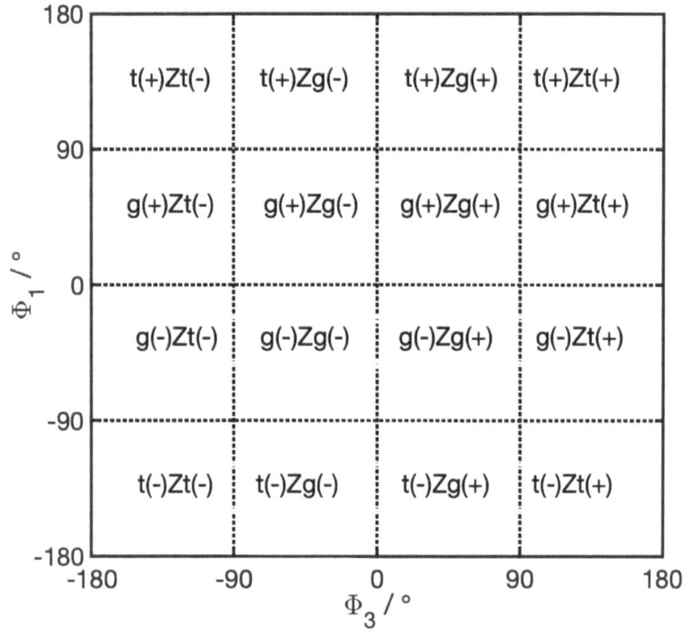

Figure 5.6 Nomenclature used to describe conformational isomers in 3Z-1,3,5-hexatriene derivatives. Dihedral angles Φ_1 and Φ_3 are defined in Fig. 5.5.

well known Ramachandran plot used in biochemistry to analyze dihedral angle conformations of peptides [2]. In the following, we will adopt the nomenclature of Sension et al. [123] for labeling the 3Z-hexatriene conformers: the structures in which both dihedral angles lie between -90° and 90° are labeled gZg (rather than cZc, which is also found in the literature [77, 133]) to highlight their helical structure. The more planar structures, are labeled tZt, tZg, and gZt. In addition we use the sign of the dihedral angle to highlight the specific location in the dihedral space (see Fig. 5.6), if necessary. Analyzing the density of structures from the 300 K REMD trajectory in the dihedral space (middle column of Fig. 5.7), we note that unsubstituted HT mainly adopts a tZt conformation, with the highest density located at the corners of the Φ_1/Φ_3-plot. To a lesser degree, also tZg and gZt conformations are present; in the Φ_1/Φ_3-plot they are located at the borders of the plot between -90° and 90°. Very

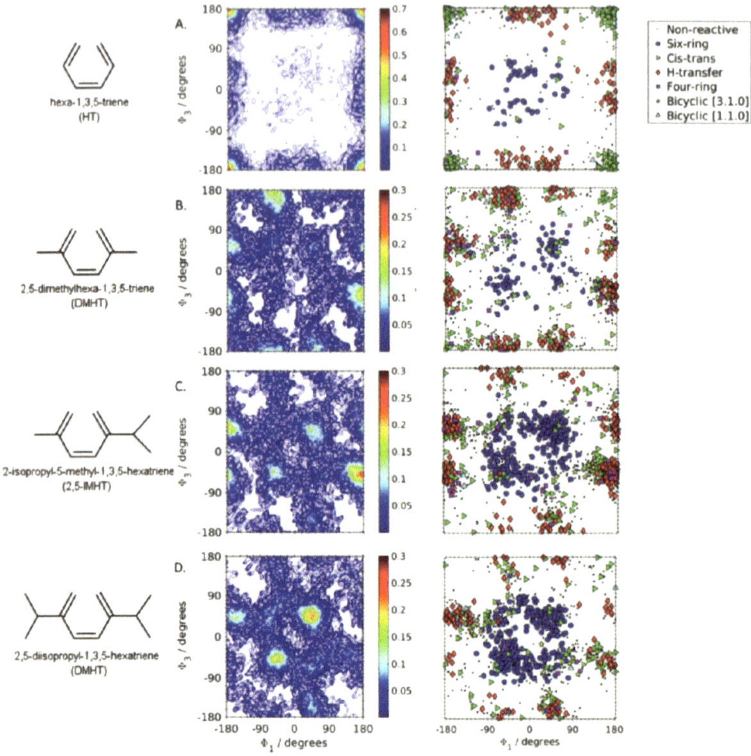

Figure 5.7 Relationship between ground state rotamers and photochemical product distribution: Left: Lewis drawing of hexatriene derivative. Middle: density distribution in the space of ϕ_1 and ϕ_3. Right: Photochemical product distribution as a function of ϕ_1 and ϕ_3. Reproduced from Tapavicza, E., Thompson, T., Redd, K., and Kim, D. (2018), *Phys. Chem. Chem. Phys.* 20, pp. 24807–24820 with permission from the PCCP Owner Societies.

few structures are present in the gZg conformation (center of the Φ_1/Φ_3-plot). In contrast, for the substituted HT derivatives, we note that with increasing size of the substituents, more density is found in the central gZg region. The highest amount of gZg structures can be found in the case DIHT, which almost shows no tZt conformations. The steric repulsion of the bulky isopropyl substituents forces the molecules into the gZg conformation.

To investigate the dependency of the photochemical reactivity on the dihedral angle conformation, we carried out TDDFT-SH simulations with each of the HT derivatives. Between 800–1400 starting structures and velocities were chosen randomly from the ensemble of structures generated by REMD (middle column of Fig. 5.7). In TDDFT-SH simulations, we used the hybrid PBE0 approximation to the exchange-correlation functional [111], which contains one fourth exact (Hartree–Fock like) exchange. To obtain accurate excitation energies and potential energies, usage of PBE0 in TDDFT (TDBPE0) is crucial. At time zero of the TDDFT-SH simulations, the molecules were prepared in the first excited singlet state (S_1), which possesses the highest oscillator strength among the 10 lowest excitations within TDPBE0. The nuclear coordinates are then propagated using the excited state nuclear forces of this state. An example trajectory of HT, undergoing a Z/E-isomerization of the central double bond, is shown in Fig. 5.8. This is the main reaction channel of HT, with a branching ratio of 22% (Fig. 5.9). Besides minor reaction channels forming products **3–5**, we also find a photoinduced [1,5]-hydrogen shift forming product **2** and an electrocyclic ring-opening forming cyclohexadiene (CHD), product **1**. Analyzing the trajectories of the substituted DMHT (Fig. 5.10), we find an increased amount of electrocyclic ring-closing reaction relative to unsubstituted HT (formation of product **7**). In contrast, the amount of Z/E-isomerization is reduced compared to unsubstituted HT. For the other substituted derivatives IMHT and DIHT, the trend continues exhibiting increased amount ring-closing reactions and decreased Z/E-isomerizations (Fig. 5.11).

To understand the relationship between initial conformation and the outcome of the simulation, we plotted the reaction products as a function the initial dihedral angle conformation (right column in Fig. 5.7). This analysis shows that CHD derivatives formed by electrocyclic ring-closing only originate from initial structures in the gZg conformation. This behavior can be explained by the fact that only in the gZg conformations the critical carbon atoms (C1 and C6, according to Fig. 5.5) that form the new σ-bond are close enough for this reaction to happen. Because of the short excited state lifetime, ranging from 115–150 fs, no major changes in dihedral angle conformation occur, which is confirmed by the example trajectory of DIHT in Fig. 5.12. From Fig. 5.7, we also infer that hydrogen-transfer is a typical reaction for tZg and gZt isomers. This behavior is also

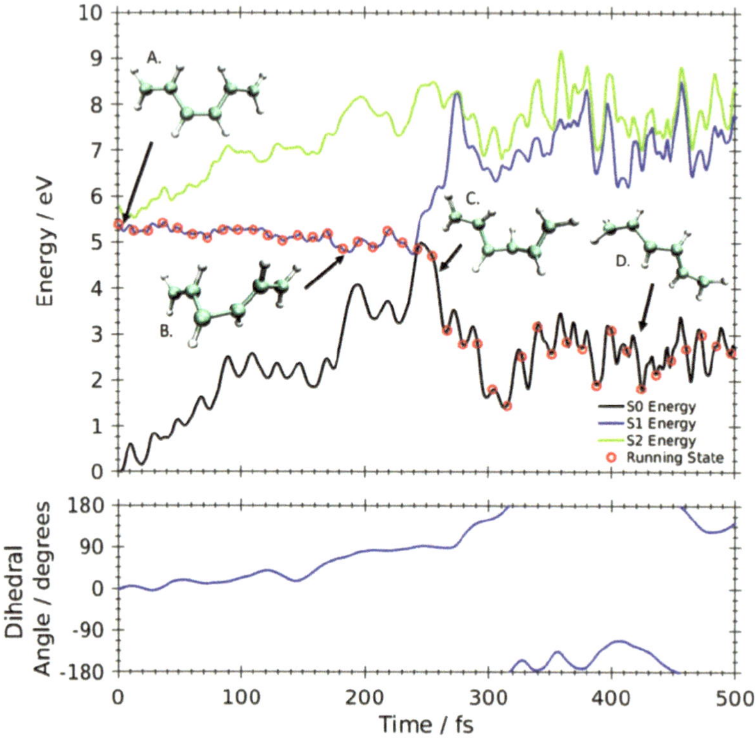

Figure 5.8 Upper panel: Potential energy surfaces and snapshot structures of an example trajectory that forms all-trans hexatriene via Z/E-isomerization in HT. Snapshot structures were taken at times 0 fs (A), 186 fs (B), 263 fs (C), and 429 fs (D). Lower panel: Time evolution of the dihedral angle (Φ_2) of the central double bond. Reproduced from Tapavicza, E., Thompson, T., Redd, K., and Kim, D. (2018), *Phys. Chem. Chem. Phys.* 20, pp. 24807–24820 with permission from the PCCP Owner Societies.

easily understood from structural considerations: only in these two conformations the critical hydrogen atom is located in close vicinity of one of the unsaturated carbon atoms C1 or C6, as can be seen from the second reaction channel from the left in Fig. 5.9 and from channel B in Fig. 5.10. Inspecting a trajectory of photoexcited tZg HT confirms this behavior (Fig. 5.13). Also here, the initial conformation is largely conserved during the excited state dynamics. As another major trend we observe that Z/E-isomerizations occur mainly from

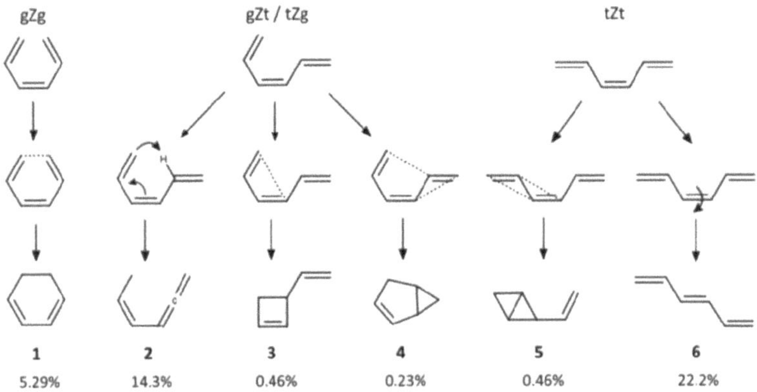

Figure 5.9 Reaction channels observed in TDDFT-SH simulations for unsubstituted hexatriene. 1-isopropyl-4-methyl-HT (1,4-IMHT) is not further discussed here. Branching rations for the different compounds are given in percent. Reproduced from Tapavicza, E., Thompson, T., Redd, K., and Kim, D. (2018), *Phys. Chem. Chem. Phys.* 20, pp. 24807–24820 with permission from the PCCP Owner Societies.

tZt conformers. This is not directly understandable from structural consideration. While the isomerization of the central double bond could occur more easily in the tZt conformer, it is also possible that in the tZt conformation there is no competition with the other two mentioned reaction channels (namely electrocyclic ring-closing and sigmatropic hydrogen shift) and consequently the probability of Z/E-isomerization increases. Comparing all derivatives, we can clearly see a positive, almost linear correlation between the amount of gZg conformers in the ground state and the amount CHD derivatives formed (Fig. 5.11). In contrast, the amount of Z/E-isomerization decreases linearly between the amount of gZg conformers in the ground state. In summary, we can say that in the case of the substituted Z-hexatriene derivatives, simulation show that it is indeed the composition of ground state rotamers at the time of photoexcitation that determines the outcome of the photochemical reaction, confirming the validity of the NEER principle. No major rearrangements of the dihedral angle conformation is observed during the short excited state lifetime of the presented molecules.

Another aspect that follows from this principle is the characteristic wavelength dependency caused by the fact that the different

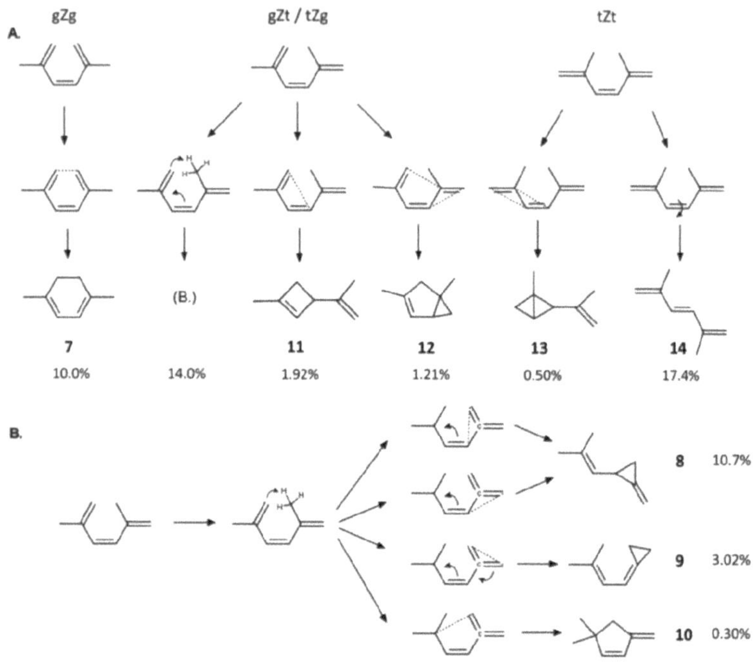

Figure 5.10 Reaction channels observed in TDDFT-SH simulations for 2,4-dimethyl-hexatriene. Branching rations for the different compounds are given in percent. Reproduced from Tapavicza, E., Thompson, T., Redd, K., and Kim, D. (2018), *Phys. Chem. Chem. Phys.* 20, pp. 24807–24820 with permission from the PCCP Owner Societies.

ground state rotamers exhibit distinct absorption properties. Inspecting the dihedral angle dependency of the S_1 excitation energies of DIHT (Fig. 5.14), we see that the gZg conformers in the center of the plot exhibit lower excitation energies compared to the remaining ensemble of conformers; all substituted HT derivatives discussed here follow this trend. As discussed above, these are the structures in which the distance between carbon C1 and C6 is close enough to form a covalent bond after photoexcitation. Excitation at low energies selectively excites those gZg conformers and, consequently, excitation at low energies increases the probability of inducing ring-closing reactions. In turn, selective excitation of tZt structures, which can be achieved by excitation at higher energies, will lead

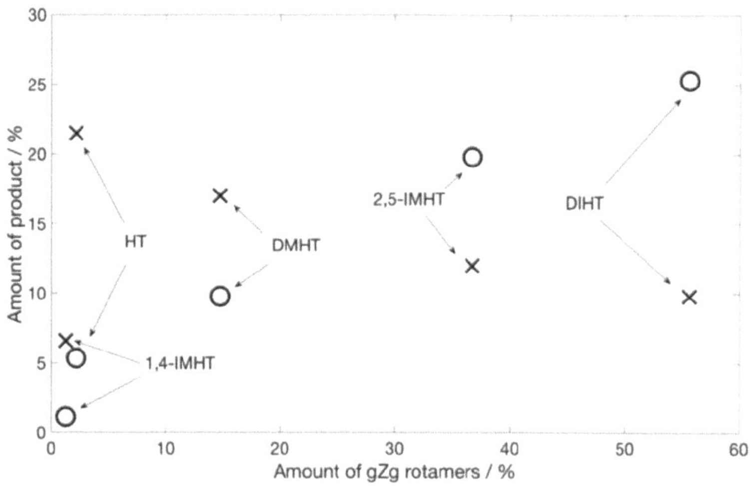

Figure 5.11 Branching ratio as a function of the percentage of gZg rotamers in the ground state. Reproduced from Tapavicza, E., Thompson, T., Redd, K., and Kim, D. (2018), *Phys. Chem. Chem. Phys.* 20, pp. 24807–24820 with permission from the PCCP Owner Societies.

to an increase in probability of inducing the Z/E-isomerization. Hence, excitation at higher energies leads predominantly to Z/E-isomerization. This pattern is characteristic for all 3Z-hexatriene derivatives studied here, and, as we will see in the next section, it also applies to the 3Z-hexatriene derivative previtamin D.

5.3.2 Vitamin D Photochemistry

Vitamin D is an important prohormone that is involved in a variety of biological processes, including calcium absorption from the blood stream [75], prevention of cancer [65] and autoimmune diseases [151]. It has also been found to have protective effects against acute respiratory infections [95]. The largest part of the world population obtain their major part of vitamin D through natural photoproduction in the epidermal skin cells, induced by ultraviolet (UVB) light. Natural vitamin D photosynthesis is a prominent paradigm for a conformationally controlled photochemical system in biology. The initial step of vitamin D photosynthesis is the photo-induced 6π-electrocyclic ring-opening of the central hexadiene unit

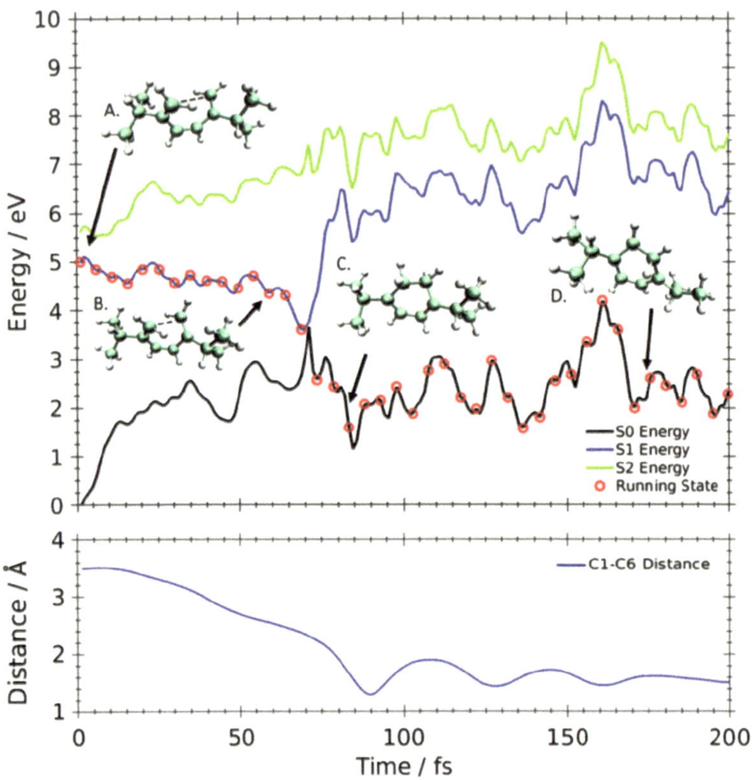

Figure 5.12 Upper panel: Potential energy surfaces and snapshot structures of an example trajectory of DIHT forming 1,4-diisopropyl-cyclohexadiene. Total simulation times of the snapshot structures are 0 fs (A), 60 fs (B), 83 fs (C), and 177 fs (D). Lower panel: C1–C6 distance as a function of time. Atom numbers are defined in Fig. 5.5. Reproduced from Tapavicza, E., Thompson, T., Redd, K., and Kim, D. (2018), *Phys. Chem. Chem. Phys.* 20, pp. 24807–24820 with permission from the PCCP Owner Societies.

in provitamin D (Pro), forming previtamin D (Pre), as indicated in the Fig. 5.15. A biexponential excited state decay has been determined with lifetimes ranging from a few hundreds of femtoseconds to a few picoseconds [6, 63, 99, 129]. According to the Woodward–Hoffmann rules [157], the ring-opening occurs in a conrotatory fashion. Its quantum yield has been determined to be between 20–65% [104]. While the optimal wavelength for Pro/Pre conversion

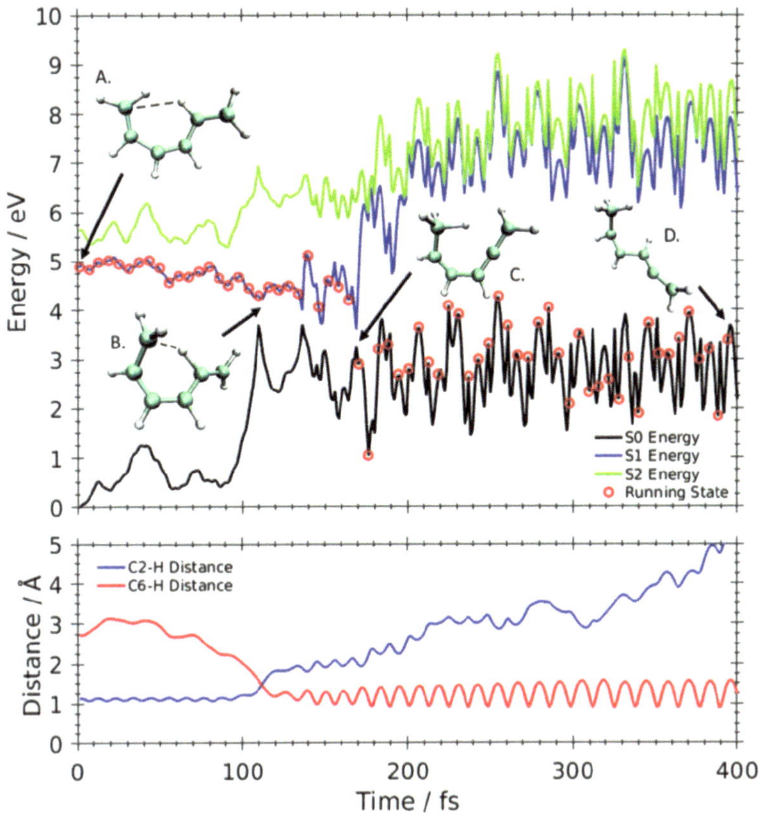

Figure 5.13 Upper panel: Potential energy surfaces and snapshot structures of an example trajectory that forms allene **2** via [1,5]-sigmatropic hydrogen shift in HT. Lower panel: Time evolution of the distance between carbon C2 and hydrogen (blue) and carbon C6 and hydrogen (red). Reproduced from Tapavicza, E., Thompson, T., Redd, K., and Kim, D. (2018), *Phys. Chem. Chem. Phys.* 20, pp. 24807–24820 with permission from the PCCP Owner Societies.

is around 290 nm, no pronounced wavelength dependency has been observed for this reaction [146]. After successful ring-opening, Pre is formed in the gZg conformation, which is thought to rapidly undergo rotational isomerization to tZg conformers [123]. Eventually, an equilibrium distribution of six different major conformers is formed [4, 133], each of which can further be divided into structures having

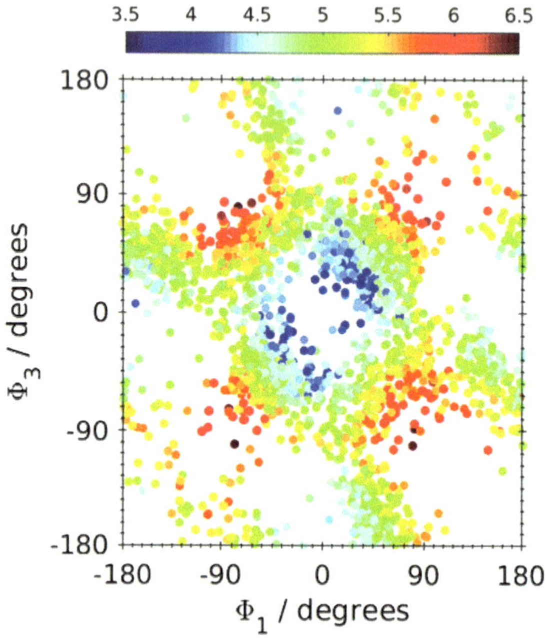

Figure 5.14 ADC(2)/COSMO $S_1 \leftarrow S_0$ excitation energies of DIHT in eV as a function of dihedral angles Φ_1 and Φ_3. Reproduced (adapted) from Tapavicza, E., Thompson, T., Redd, K., and Kim, D. (2018), *Phys. Chem. Chem. Phys.* 20, pp. 24807–24820 with permission from the PCCP Owner Societies.

the OH group in axial or equatorial position [22]. The time until an equilibrium distribution of conformers is adopted is about 100 ps and depends on the solvent viscosity [129]. The formed Pre can react thermally to vitamin D via a [1,7]-sigmatropic hydrogen shift (Fig. 5.15). However, this process only occurs if carbon C9 and C19 (see Fig. 5.15 for atom numbers) are in close vicinity to each other, which is only possible in the gZg conformation. Furthermore, Pre can also undergo several other photochemical reactions, which are highly wavelength dependent and conformationally controlled [133]: excitation at the red side of the spectrum predominantly leads to 6π-electrocyclic ring-closing, forming either Pro or lumisterol (Lumi), a stereoisomer of Pro (Fig. 5.15). Prolonged irradiation and excitation at the blue side of the spectrum predominantly leads to tachysterol (Tachy) formation. Besides these major photoproducts,

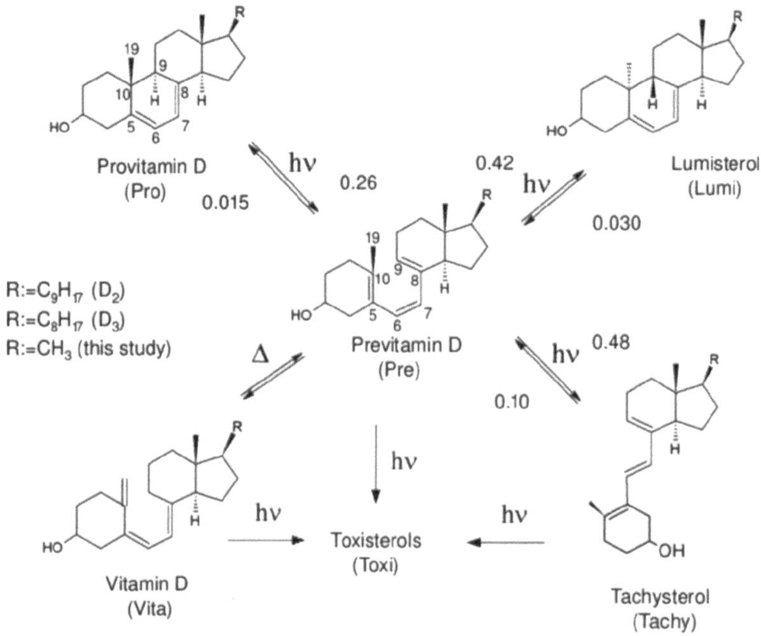

Figure 5.15 Photochemical and thermal reactions involved in the natural photosynthesis of vitamin D. Quantum yields measured at 253 nm are given on the arrows [70]. Reproduced from Cisneros, C., Thompson, T., Baluyot, N., Smith, A. C., and Tapavicza, E. (2017), *Phys. Chem. Chem. Phys.* 19, pp. 5763–5777 with permission from the PCCP Owner Societies.

a variety of so-called *toxisterols* have been determined [32, 33, 77]. Their biological role and further metabolism is largely unknown. In living organism, these photochemical and thermal processes occur in the phospholipid bilayer of epidermal skin cells. It has been found that both the presence of the membrane and the spectral composition of the irradiation source strongly affect quantum yields of the involved reactions [91]. For instance, Pre formation continuously increases with increasing radiation density if monochromatic light at 295 nm is used. In contrast, if a solar simulator is used, Pre formation only increases at low radiation density; at higher radiation density Pre formation plateaus [91]. Similar variations have also been found for Lumi and Tachy formation. Obviously, an intrinsic regulation occurs due to the

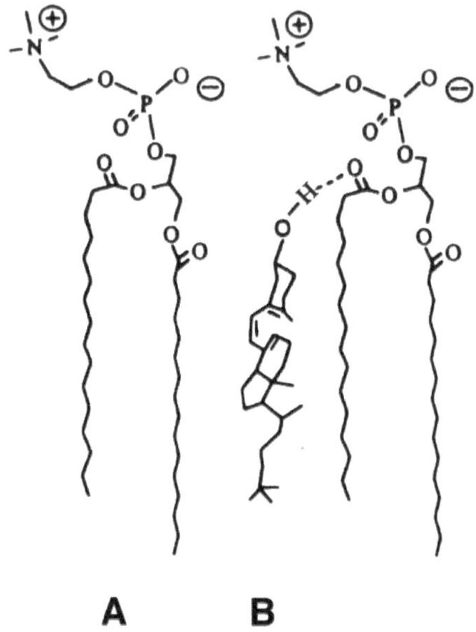

Figure 5.16 Proposed structural model for the localization of the gZg-previtamin D3 in a phospholipids membrane. Picture taken from Tian, X. Q., and Holick, M. F. (1999), J. Biol. Chem. 274, 7, 4174–4179, licensed under Creative Commons CC-BY.

spectral composition of the sun, preventing overproduction of vitamin D upon prolonged sun exposure. The same study [91] also found slight decrease in photochemical Pre formation in skin cells compared to tetrahydrofuran solution. In contrast, the reaction rate of the isomerization of Pre to vitamin D has been found to be enhanced by a factor of 10 in epidermal skin cells and in artificial dipalmitoylphosphatidylcholine (DPPC) liposomes compared to in organic solutions [76, 143]. It has been hypothesized that this increase in vitamin D formation is due to the "cholesterol-like" interactions between Pre and the phospholipid molecules that trap the molecule in the gZg conformation (Fig. 5.16), facilitating the [1,7]-sigmatropic hydrogen transfer reaction. According to theoretical studies [97], both g(+)Zg(+) and g(-)Zg(-) can convert to vitamin D. In constrained ground state AIMD simulations we could confirm this reaction to happen in the g(-)Zg(-) conformation at a distance of about 2.56 Å between carbons C9 and C19 (Fig. 5.17).

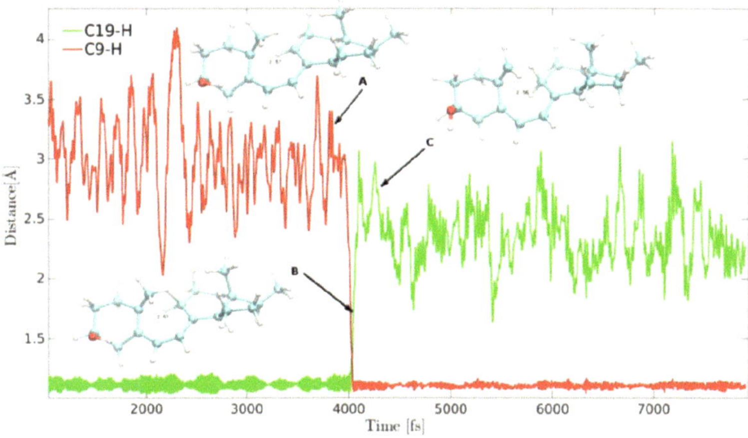

Figure 5.17 Results of a ground state AIMD simulation of the hydrogen transfer dynamics of the reaction previtamin D → vitamin D. Red: distance between C19 and the critical hydrogen atom, green: distance between C9 and the critical hydrogen. Atom numbers as defined in Fig. 5.15. Molecule A represents Pre before hydrogen transfer, Structure B is the transition state, and Structure C is vitamin D after hydrogen transfer. The simulation has been carried out and a temperature of 800 K. The C9-C19 distance has been constrained to (2.56 ± 0.2) Å to keep the molecule in the gZg conformation, mimicking the effect of the biological membrane.

Returning to the purely photochemical reactions in the vitamin D system, the secosteroids Pre and Tachy exhibit an interesting wavelength dependent photochemistry due to their flexible open-ring structure. REMD simulations of Pre exhibit a particular dihedral angle conformation with six high density spots [133] corresponding to the six mentioned minimum energy structures [22, 52]. Similar to the example of tachysterol (Fig. 5.2), also Pre exhibits a conformational dependency of the absorption, which is obtained by computing excitation energies for snapshot structures from REMD simulations (Fig. 5.18). Analogously to the Z-hexatriene derivatives, the wavelength-dependent photochemistry can be probed by correlating the S_1 excitation energy of a snapshot structure from the REMD ensemble with its photoproduct formed in TDDFT-SH simulations. As we can see from Fig. 5.18, the ring-closing reaction forming Lumi occurs mainly for g(+)Zg(+) conformers,

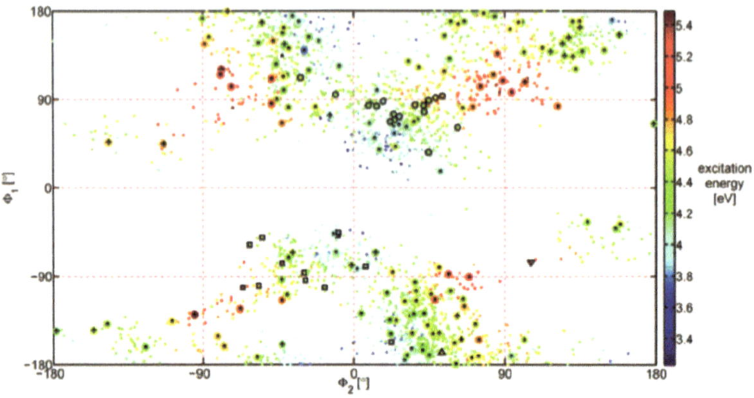

Figure 5.18 Excitation energy and photoproduct as a function of torsional angles Φ_1 and Φ_2 of Pre. The color code indicates the S_1 excitation energy in eV. Unreactive trajectories are indicated by a black dot. Trajectories that led to ring-closure are indicated by squares (provitamin D formation) or by circles (lumisterol formation). Crosses denote tachysterol formation via Z/E isomerization. Δ indicates hydrogen shift, forming vitamin D and ∇ indicates toxisterol formation. Reproduced from Tapavicza, E., Meyer, A. M., and Furche, F. (2011), *Phys. Chem. Chem. Phys.* 13, 20986–20998 with permission from PCCP Owner Societies.

whereas the ring-closing with opposite helicality (forming Pro), occurs mainly for g(-)Zg(-) conformers. The fact that gZg conformers on average exhibit lower S_1 excitation energies explains why ring-closing reaction occur mainly upon excitation at the red side of Pre's spectrum. In contrast, Tachy formation via Z/E-isomerization is prone to occur in t(+)Zt(+) conformers, which exhibit on average higher excitation energies. With this behavior, Pre follows the same trend as the Z-hexatriene derivatives.

While the photochemistry of Pro and Pre have been investigated extensively, the photochemistry and biological role of the E-hexatriene Tachy has attracted less attention. However, biological studies have shown that a variety of Tachy derivatives exhibit binding to the human vitamin D receptor [93, 127]. Furthermore, recently it has been hypothesized that Tachy potentially constitutes a vitamin D reservoir that can be tapped at long wavelength UV radiation and could constitute an important route to produce vitamin D in winter months, where short wavelength UV light is

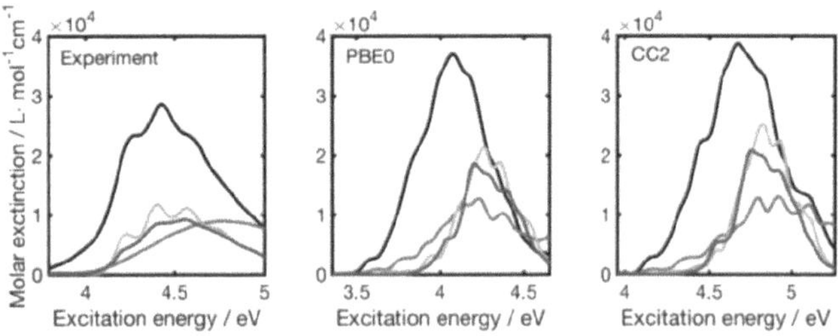

Figure 5.19 Comparison between the experimental absorption spectra measured in ether [91] (left) and the spectra calculated by TDPBE0 (middle) and CC2 (right). Tachysterol: black, provitamin D: green, previtamin D: red, lumisterol: blue. Reproduced from Cisneros, C., Thompson, T., Baluyot, N., Smith, A. C., and Tapavicza, E. (2017), *Phys. Chem. Chem. Phys.* 19, pp. 5763–5777 with permission from the PCCP Owner Societies.

limited [10]. The possibility of inducing photochemical reactions at longer wavelengths in Tachy is already suggested by its absorption spectrum, which extends far more to the red than the spectra of the vitamin D derivatives Pro, Pre, and Lumi (Fig. 5.19). As we can see from Fig. 5.3, the overlap with the solar radiation (brown) is largest with the spectrum of the tEt rotamers; this has a specific impact on photoreactivity of tachysterol. Compared to the other vitamin D derivatives, Tachy's absorption spectrum exhibits also larger extinction coefficients, covering the entire absorption spectra of the other compounds. This behavior is properly described by both theoretical methods, TDDFT and CC2. However, the absorption spectrum alone does not allow to assess its photochemical reactivity.

To investigate the photochemical behavior of Tachy, we carried out TDDFT-SH simulations using initial structures from an ensemble of structures generated by REMD (Fig. 5.20) [45]. Simulations show that to a large part (95.7%), trajectories decay to the ground state without chemical transformation (Fig. 5.21). Interestingly, all trajectories starting from tEt structures decay unreactively and only tEc, cEc, tEc undergo photochemical transformations to a small degree (Fig. 5.21). Pre formation, which is crucial for vitamin D formation, mainly originates from cEc conformers upon excitation at the red side of the spectrum. This finding confirms the possibility that vitamin D could be formed in winter where short-wavelength

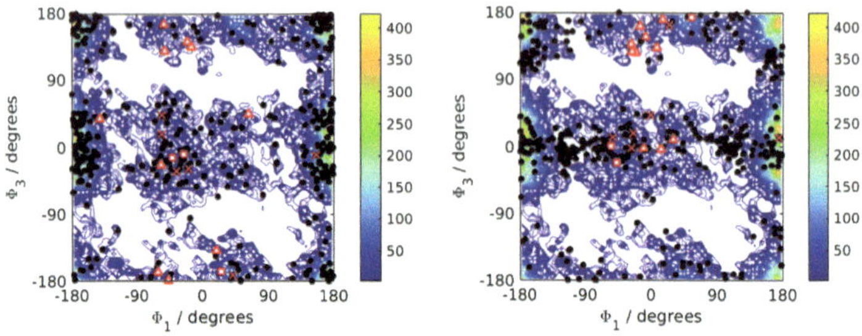

Figure 5.20 Distribution of rotamers as a function of the dihedral angles Φ_1 and Φ_3 in the ground state equilibrium of tachysterol obtained from REMD at 300 K is indicated by the contours left and right. Number of structures per $10^\circ \times 10^\circ$ are indicated by the color code. Left: Position of TDDFT-SH starting structures are indicated by the symbols. Right: Position of TDDFT-SH structures at the time of the surface hop are indicated by the symbols. Black point: unreactive (product Tachy); cross: trans–cis isomerization (product Pre), triangle: hydrogen transfer (toxisterol D1); square: thermal cyclobutene (CB) formation (product: CB-toxisterol). Reproduced from Cisneros, C., Thompson, T., Baluyot, N., Smith, A. C., and Tapavicza, E. (2017), *Phys. Chem. Chem. Phys.* 19, pp. 5763–5777 with permission from the PCCP Owner Societies.

UV light is limited. Furthermore, we find a cyclobutene (CB) derivative formed from cEt and cEc initial structures at the red side of the spectrum. CB toxisterols have been found to thermalize to Pre [33] and therefore they also possibly constitute a source of vitamin D in winter. Furthermore, we also find [1,5]-sigmatropic hydrogen transfer in a few trajectories, which for steric reasons only occurs in the cEt or cEc conformation.

Comparing the dihedral angle conformation of the initial structures of the trajectories with their conformation at the time of the surface hop, we see that the distribution of conformers has changed substantially during the excited state dynamics. Interestingly, we observe one trajectory undergoing [1,5]-sigmatropic hydrogen transfer, in which the initial tEc conformation first transformed to the cEc conformation. For steric reasons, this hydrogen transfer would not have been possible in the tEc conformation. This behavior violates the NEER principle. The anti-NEER behavior can be explained with Tachy's relatively long excited state lifetime (882 fs) compared to the Z-hexatriene derivatives (115–150 fs).

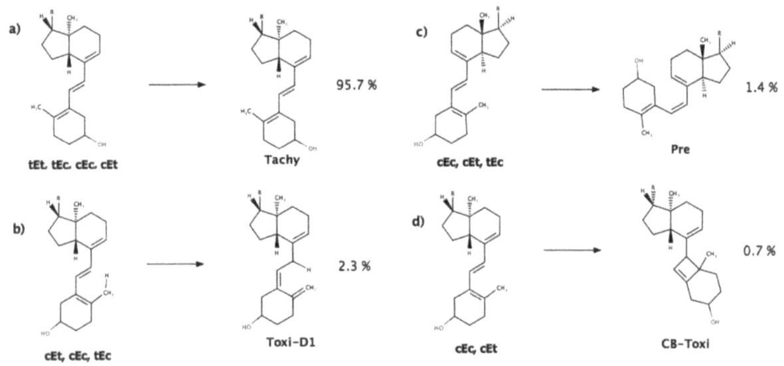

Figure 5.21 Overview of the observed photoreactions of tachysterol: (a) unreactive, (b) [1,5]-sigmatropic hydrogen transfer (toxisterol D1 formation), (c) hula-twist double bond isomerization (previtamin D formation), (d) thermal 2 + 2 electrocyclization (toxisterol-CB formation). Only the main rotamer in which the reaction has been observed is sketched, other rotamers in which the reaction was observed are listed below the reactant. The branching ratios for each reaction channel are given. Reproduced from Cisneros, C., Thompson, T., Baluyot, N., Smith, A. C., and Tapavicza, E. (2017), *Phys. Chem. Chem. Phys.* 19, pp. 5763–5777 with permission from the PCCP Owner Societies.

The long lifetime allows the molecule to adopt an equilibrium of conformers corresponding to the excited state potential energy surface.

5.3.3 Wavelength-Dependent Product Quantum Yields in Z-Hexatriene Derivatives

In the previous sections we have seen the characteristic wavelength dependent photochemistry of 3Z-hexatriene and vitamin D derivatives. Now we will turn to the quantitative prediction of the wavelength dependent PQYs based on TDDFT-SH simulations and electronic spectra predictions. We will focus on the calculations of the ring-closing and Z/E-isomerization PQYs of DMHT and Pre, because for these molecules experimentally determined PQYs are available in the literature for comparison [36, 49]. For the quantitative prediction of the PQY, using Eq. 5.22, we need to group the TDDFT-SH trajectories according to their product formed and compute the electronic absorption spectra averaged over the initial structures of each group. For DMHT, the major reaction channels are ring-closing, Z/E-isomerization, formation of a cyclobutene derivative **11**, and formation of cyclopropan derivatives **8** and **9**, with branching ratios given in Fig. 5.9. We computed the excitation energies and oscillator strengths using TDDFT and the algebraic diagrammatic construction to second order (ADC(2)). Since experimental PQY measurements were carried out in n-pentane, the conductor like screening method (COSMO) [117] was used to mimic the solvent via a dielectric continuum model. Inspecting the distribution of ADC(2) oscillator strengths as a function of the S_1 and S_2 excitation energies (Fig. 5.22), we note that in the region between

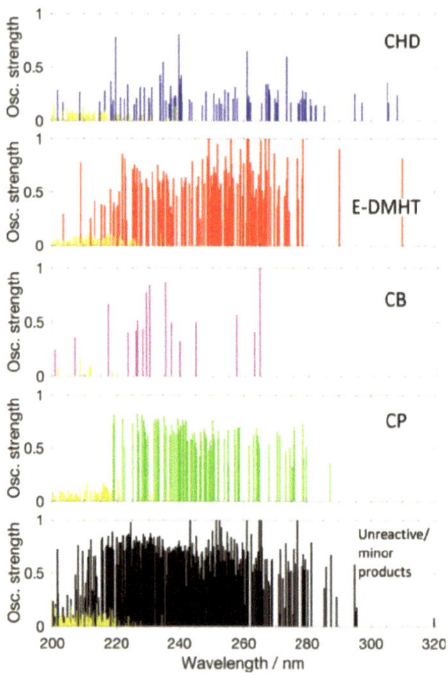

Figure 5.22 Distribution of the ADC(2) oscillator strengths of the first excited singlet state of the initial structures (blue, red, magenta, green, and black) of the Z-DMHT trajectories that form the indicated product. Yellow: oscillator strengths of the second excited singlet state. Reprinted with permission from Thompson and Tapavicza, J. Phys. Chem. Lett. 9, 4758–4764 (2018). Copyright 2019 American Chemical Society.

240 and 320 nm, only oscillator strengths of S_1 are located, whereas S_2 oscillator strengths are located at wavelengths below 240 nm. Thus, the long wavelength photochemistry is mainly governed by S_1 photochemistry. For the ring-closing channel (CHD formation), several structures exhibit excitations in the long-wavelength region between 280 and 320 nm. This is consistent with our previous finding that only gZg conformers undergo ring-closing and that these conformers absorb at the red side of the absorption spectrum. For the Z/E-isomerization (E-DMHT formation) only two structures with excitation energies located in this region are present. To obtain the wavelength dependent PQY, we first converted oscillator strengths of the entire ensemble and for the different reaction channels into absorption spectra (Fig. 5.23, A for ADC(2), and Fig. 5.24, A for TDDFT). Dividing the spectrum of CHD formation by the total spectrum yields the wavelength dependent PQY for CHD formation. Analogously, we obtain the PQYs of the other products. As expected from earlier considerations, we see an increase of CHD formation towards the red side of the spectrum, which is confirmed by the experimental measurements. For Z/E-isomerization, the experimental measurements show a decrease of the PQY with increasing wavelength. This is partially captured by the calculated PQY, which also shows a drop until 300 nm. However, above 300 nm we see an increase, but this increase is possibly an artifact due to the poor statistics in this wavelength region. It is probably caused by only one single trajectory, which shows the isolated excitation energy on the red side in the spectrum (Fig. 5.22). Minor products, **11** and **9** show lower PQY than CHD and E-DMHT, in both, experiment and calculation. While the computed PQY for the cyclobutene product is in the same order of magnitude as in the experimental one, the cyclopropan derivative formation is overestimated by more than 10%. PQYs determined on the basis of TDDFT spectra agree similarly well with experimental results as the PQYs determined based on ADC(2) data. However, we see that the λ_{max} value of the total TDDFT spectrum (Fig. 5.24, A) does not agree as well with experimental value as the λ_{max} value of ADC(2) total spectrum (Fig. 5.23, A). Due to the better agreement of the ADC(2) spectrum with experimental spectrum, one can also expect a more accurate prediction of the PQYs based on ADC(2) computed spectra.

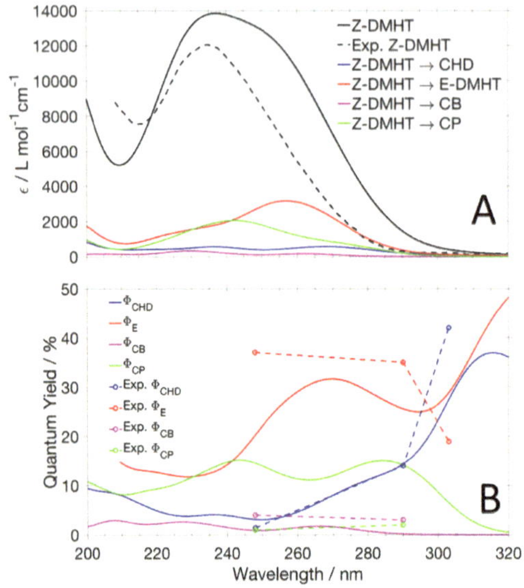

Figure 5.23 (A) ADC(2) Absorption spectra of DMHT. The calculated absorption spectrum of DMHT averaged over all TDDFT-SH initial structures (solid) and experimentally measured spectrum [36] (dashed) are shown in black. The spectra averaged over initial structures of the trajectories leading to the indicated products are given in color. (B) Calculated and experimentally measured [36] (Exp.) wavelength-dependent product quantum yields of E-DMHT (Φ_E), CHD (Φ_{CHD}), CB (Φ_{CB}), and CP (Φ_{CP}). Reprinted with permission from Thompson and Tapavicza, J. Phys. Chem. Lett. 9, 4758–4764 (2018). Copyright 2019 American Chemical Society.

Turning to the calculation of the wavelength-dependent PQY for Pre, we first present the branching rations of the TDDFT-SH simulations (Fig. 5.25). In case of Pre, two distinct ring-closing products, Lumi and Pro are possible, originating from g(+)Zg(+) and g(-)Zg(-) conformers, respectively. Consistent with the dihedral plot (Fig. 5.18), there are only initial structures of trajectories forming Lumi and Pro above 300 nm on the red side of the spectrum (Fig. 5.26). In contrast, Tachy forming trajectories all originate from initial structures with S_1 excitation energies below 300 nm. We also note the larger oscillator strengths of Tachy forming trajectories, compared to Lumi and Pro forming initial structures. The large oscillator strengths are characteristic for the

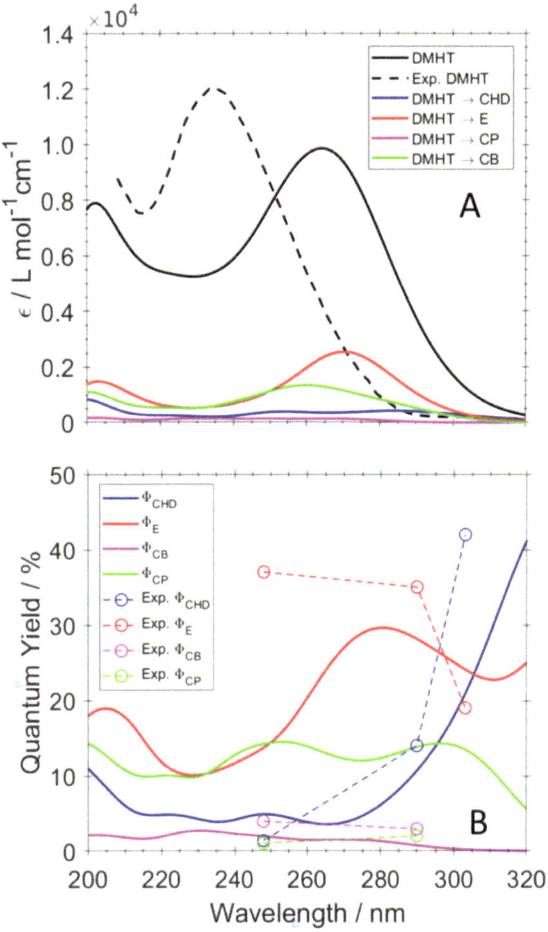

Figure 5.24 (A) TDPBE0 Absorption spectra of DMHT. The calculated absorption spectrum of DMHT averaged over all TDDFT-SH initial structures (solid) and experimentally measured spectrum [36] (dashed) are shown in black. The spectra averaged over initial structures of the trajectories leading to the indicated products are given in color. (B) Calculated and experimentally measured [36] (Exp.) wavelength-dependent product quantum yields of E-DMHT (Φ_E), CHD (Φ_{CHD}), CB (Φ_{CB}), and CP (Φ_{CP}). Reprinted with permission from Thompson and Tapavicza, J. Phys. Chem. Lett. 9, 4758–4764 (2018). Copyright 2019 American Chemical Society.

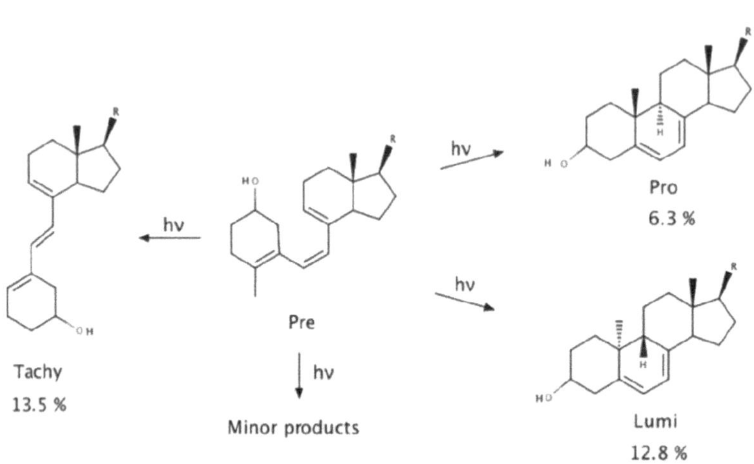

Figure 5.25 Major Reactions and Their Products Observed in TDDFT-SH Simulations of truncated Pre Z/E-isomerization, forming tachysterol, is shown on the left, and ring-closing reactions, forming lumisterol and provitamin D, is shown on the right. Raw branching ratios from simulations are indicated for the products. Reprinted with permission from Thompson and Tapavicza, J. Phys. Chem. Lett. 9, 4758–4764 (2018). Copyright 2019 American Chemical Society.

tZt structures, which are prone to undergo Z/E isomerization. Converting oscillator strengths to absorption spectra and computing the wavelength dependent PQYs, we note a good agreement between calculated and experimentally measured values for Lumi and Pro formation. Not only are calculations able to capture the increased ring-closing towards the red side of the spectrum, they also agree with experiments in the relative magnitudes between Lumi and Pro formation. This is a direct consequence of the higher amount of g(+)Zg(+) structures compared to g(-)Zg(-) structures, suggesting an accurate description of the conformational equilibrium by the REMD simulations.

5.4 Conclusion and Outlook

As we have seen from the comparison between non-adiabatic TSH simulations and experimental results, the simulations are able to

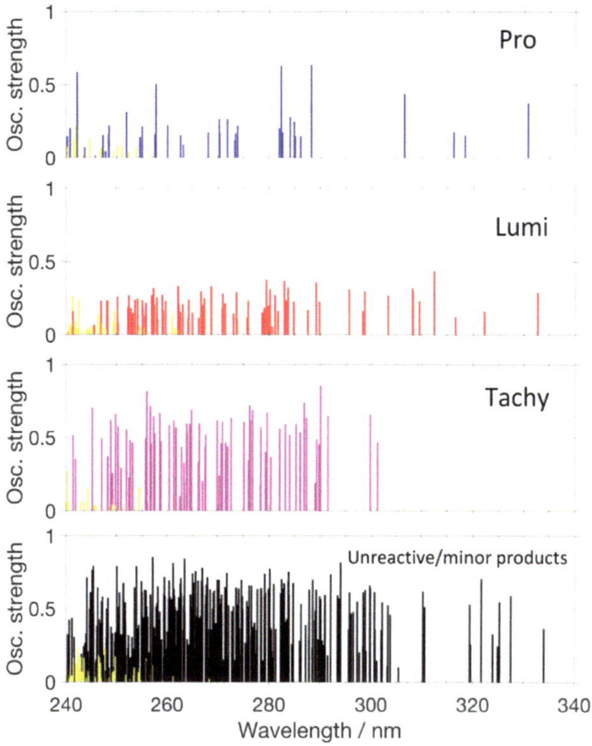

Figure 5.26 Distribution of the ADC(2) oscillator strengths of the first excited singlet state of the initial structures of the Pre trajectories that form the indicated product (blue, red, magenta, and black). Yellow: oscillator strengths of the second excited singlet state. Reprinted with permission from Thompson and Tapavicza, J. Phys. Chem. Lett. 9, 4758–4764 (2018). Copyright 2019 American Chemical Society.

explain conformationally controlled photochemistry in a variety of Z-hexatriene and vitamin D derivatives. An important ingredient to achieve this goal is to draw initial structures for the TSH trajectories from a Boltzmann ensemble of structures that contain the different conformational isomers with the correct statistical distribution. Provided an accurate potential energy function, either from an empirical force field or from ab initio methods is available, enhanced sampling methods such as the presented REMD can be used to obtain an accurate ensemble of structures. Here we have used DFT

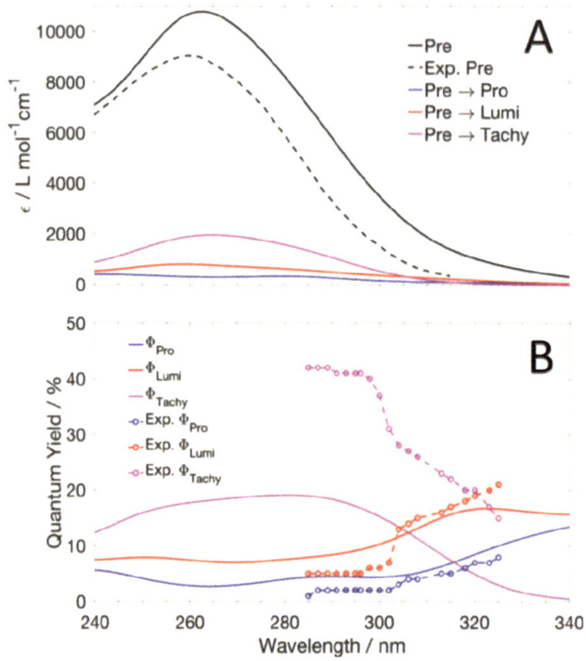

Figure 5.27 (A) Absorption spectra of Pre. Calculated spectrum of all TDDFT-SH initial structures (solid) and experimentally measured spectrum[49] (dashed) are shown in black. The spectra of the initial structures of the trajectories that lead to the indicated reactions are given in color. (B) Calculated and experimentally measured[49] (Exp.) wavelength-dependent product quantum yields of Pre and wavelength-dependent product quantum yields of Lumi (Φ_{Lumi}), Pro (Φ_{Pro}), and Tachy (Φ_{Tachy}). Reprinted with permission from Thompson and Tapavicza, J. Phys. Chem. Lett. 9, 4758–4764 (2018). Copyright 2019 American Chemical Society.

as underlying potential energy function for the initial sampling. However, common approximations to the exchange-correlation functional are known to be problematic in the description of dispersion forces. This could affect the accuracy of the obtained ensemble of structures. Future benchmark studies with ab initio methods that are able to treat dispersion more accurately would be desirable and could possibly lead to a more accurate description of Boltzmann ensembles and in turn positively affect the accuracy of product distributions in TSH simulations.

In the calculation of wavelength dependent PQYs the description of the ground state ensemble of structures affects the accuracy of the PQYs in two ways: firstly, an accurate ensemble is necessary to obtain accurate absorption spectra and secondly, the correct statistical distribution is necessary to obtain correct branching rations of the different photochemical reaction channels. Regarding spectra calculations, we have also seen that highly accurate correlated excited methods such as ADC(2) are often superior to TDDFT. Equally, to obtain good comparability between calculated wavelength dependent PQYs with experimentally measured ones, correlated methods should be chosen for the spectra calculations. It remains the question regarding the accuracy of TDDFT for excited state dynamics. Some issues, such as erroneous description of conical intersections, charge transfer excitations, and double excitations have been mentioned here. These issues are an active field of research [81]. While there have been some studies that employ correlated wavefunction based methods, such as CC2 [112, 135] or CASPT2 [19, 107] in TSH, often the computational resources available only allow TSH simulations based on TDDFT, therefore improvements of TDDFT regarding these issues will increase the accuracy of TSH simulations in the future.

One point that was only briefly mentioned here, is the influence of the solvent the photochemical reactivity. While we used an electrostatic continuum model to account for electronic effects in the calculation of spectra, gauging the influence of solvent viscosity or viscosity of a membrane requires models that explicitly take into account the solvent molecules. This requires a more sophisticated simulation setup. While efficient methods to include the solvent in these calculations have been developed (for instance hybrid quantum mechanical/molecular mechanics, QM/MM methods [89, 101]), they have not yet been applied to compute wavelength dependent PQYs.

Another aspect that possibly affects the outcome of TSH simulations is related to the initial conditions. The cases presented here are all based on initial structures and velocities obtained from MD simulations. Also here, a systematic comparison between TSH using MD-generated initial conditions and initial conditions from a Wigner distribution would be useful to direct future studies.

Besides, it would be interesting to investigate how the inclusion of nuclear quantum effects would affect the TSH simulation of the various photoinduced hydrogen transfer reactions. Methods such as the ring-polymer surface hopping dynamics [121] could be applied to investigate this question.

The large variety of different systems besides HT derivatives that exhibit conformationally controlled photochemistry are still up to study with non-adiabatic simulation techniques. New developments in more efficient excited state methods (e.g. tight binding TDDFT) will allow to also study larger systems and assist the development of photochemical switches and molecular motors in condensed systems.

References

1. Solar position calculator—U.S. Department of Commerce, National Oceanic & Atmospheric Administration, NOAA Research.
2. Ramachandran, G. N., Ramakrishnan, C., and Sasisekharan, V. (1963). Stereochemistry of polypeptide chain configurations, *J. Mol. Biol.* **7**, 1, pp. 95–99.
3. Vroegop, P.J., Lugtenburg, J., and Havinga, E. (1973). Conformational equilibrium and photochemistry of hexa-1,3,5-trienes, *Tetrahedron* **29**, 10, pp. 1393–1398.
4. Dmitrenko, O., Vivian, J .T., Reischl, W., and Frederick, J. H. (1999). Theoretical studies of the first strongly allowed singlet states of 3-desoxy analogs of previtamin D, vitamin D, and their e-isomers, *J. Mol. Struct. THEOCHEM* **467**, 3, pp. 195–210.
5. Levine, B. G., Coe, J. D., Virshup, A. M., and Martínez, T. J.(2008). Implementation of ab initio multiple spawning in the molpro quantum chemistry package, *Chem. Phys.* **347**, 1-3, pp. 3 – 16.
6. Anderson, N. A., Shiang, J. J., and Sension, R. J. (1999). Subpicosecond ring opening of 7-dehydrocholesterol studied by ultrafast spectroscopy, *J. Phys. Chem. A* **103**, 50, p. 10730.
7. Andersson, K., Malmqvist, P.-Å., and Roos, B. O. (1992). Second-order perturbation theory with a complete active space self-consistent field reference function, *J. Chem. Phys.* **96**, 2, pp. 1218–1226.

8. Andersson, K., Malmqvist, P. A., Roos, B. O., Sadlej, A. J., and Wolinski, K. (1990). Second-order perturbation theory with a casscf reference function, *J. Phys. Chem.* **94**, 14, pp. 5483–5488.

9. Andrade, X., Castro, A., Zueco, D., Alonso, J., Echenique, P., Falceto, F., and Rubio, A. (2009). Modified Ehrenfest formalism for efficient large-scale ab initio molecular dynamics, *J. Chem. Theor. Comput.* **5**, 4, pp. 728–742.

10. Andreo, K. (2015). *Generation of previtamin D3 from tachysterol3: A novel approach for producing vitamin D3 in the winter*, Ph.D. thesis, Boston University.

11. Arai, T., Karatsu, T., Sakuragi, H., Tokumaru, K., Tamai, N., and Yamazaki, I. (1989). Highly selective rotational isomerization of 2-vinylanthracene in the excited singlet state. picosecond time-resolved fluorescence study, *Chem. Phys. Lett.* **158**, 5, pp. 429–434.

12. Arai, T., Karatsu, T., Sakuragi, H., Tokumaru, K., Tamai, N., and Yamazaki, I. (1992). Effects of β-substituents on internal rotation of the anthryl group of 2-anthrylethylenes in the excited singlet state, *J. Photochem. Photobiol. A Chem.* **65**, 1-2, pp. 41–51.

13. Baiardi, A., Bloino, J., and Barone, V. (2013). General time dependent approach to vibronic spectroscopy including Franck–Condon, herzberg–teller, and Duschinsky effects, *J. Chem. Theor. Comput.* **9**, pp. 4097–4115.

14. Balasubramani, S. G., Chen, G. P., Coriani, S., Diedenhofen, M., Frank, M. S., Franzke, Y. J., Furche, F., Grotjahn, R., Harding, M. E., Hättig, C., Hellweg, A., Helmich-Paris, B., Holzer, C., Huniar, U., Kaupp, M., Marefat Khah, A., Karbalaei Khani, S., Müller, T., Mack, F., Nguyen, B. D., Parker, S. M., Perlt, E., Rappoport, D., Reiter, K., Roy, S., Rückert, M., Schmitz, G., Sierka, M., Tapavicza, E., Tew, D. P., van Wüllen, C., Voora, V. K., Weigend, F., Wodynski, A., and Yu, J. M. (2020). Turbomole: Modular program suite for ab initio quantum-chemical and condensed-matter simulations, *J. Chem. Phys.* **152**, 18, p. 184107.

15. Baldwin, J. E., and Krueger, S. M. (1969). Stereoselective photochemical electrocyclic valence isomerizations of alpha-phellandrene conformational isomers, *J. Am. Chem. Soc.* **91**, 23, pp. 6444–6447.

16. Barbatti, M. (2011). Nonadiabatic dynamics with trajectory surface hopping method, *Wiley Interdiscip. Rev. Comput. Mol. Sci.* **1**, 4, pp. 620–633.

17. Barbatti, M. (2021). Velocity adjustment in surface hopping: Ethylene as a case study of the maximum error caused by direction choice, *J. Chem. Theor. Comput.* **17**, 5, pp. 3010–3018.

18. Barbatti, M., Aquino, A. J., and Lischka, H. (2010). The UV absorption of nucleobases: Semi-classical ab initio spectra simulations, *Phys. Chem. Chem. Phys.* **12**, 19, pp. 4959–4967.
19. Barbatti, M., Aquino, A. J., Szymczak, J. J., Nachtigallová, D., Hobza, P. and Lischka, H. (2010). Relaxation mechanisms of UV-photoexcited dna and rna nucleobases, *Proc. Natl. Acad. Sci.* **107**, 50, pp. 21453–21458.
20. Barone, V. (2005). Anharmonic vibrational properties by a fully automated second-order perturbative approach, *J. Chem. Phys.* **122**, 1, p. 014108.
21. Bartocci, G., Ciorba, S., Mazzucato, U., and Spalletti, A. (2009). Adiabatic pathways in the conformational and geometrical photoisomerizations of the 1, 2-distyrylbenzene isomers, *J. Phys. Chem. A* **113**, 30, pp. 8557–8568.
22. Bayda, M., Redwood, C. E., Gupta, S., Dmitrenko, O., and Saltiel, J. (2017). Lumisterol to tachysterol photoisomerization in EPA glass at 77 k. A comparative study, *J. Phys. Chem. A* **121**, 12, pp. 2331–2342.
23. Beck, M. H., Jäckle, A., Worth, G. A., and Meyer, H.-D. (2000). The multiconfiguration time-dependent Hartree (MCTDH) method: A highly efficient algorithm for propagating wavepackets, *Phys. Rep.* **324**, 1, pp. 1–105.
24. Begušić, T., Roulet, J., and Vaníček, J. (2018). On-the-fly ab initio semiclassical evaluation of time-resolved electronic spectra, *J. Chem. Phys.* **149**, 24, p. 244115.
25. Begušic, T., and Vanícek, J. (2021). Finite-temperature, anharmonicity, and Duschinsky effects on the two-dimensional electronic spectra from ab initio thermo-field Gaussian wavepacket dynamics, *J. Phys. Chem. Lett.* **12**, 11, pp. 2997–3005.
26. Ben-Nun, M., Quenneville, J., and Martínez, T. J. (2000). Ab initio multiple spawning: Photochemistry from first principles quantum molecular dynamics, *J. Phys. Chem. A* **104**, 22, pp. 5161–5175.
27. Beyer, T., and Swinehart, D. (1973). Algorithm 448: Number of multiply-restricted partitions, *Commun. ACM* **16**, 6, p. 379.
28. Bloino, J., Baiardi, A., and Biczysko, M. (2016). Aiming at an accurate prediction of vibrational and electronic spectra for medium-to-large molecules: An overview, *Int. J. Quantum Chem.* **116**, pp. 1543–1574.
29. Bloino, J., Biczysko, M., Santoro, F., and Barone, V. (2010). General approach to compute vibrationally resolved one-photon electronic spectra, *J. Chem. Theor. Comput.* **6**, 4, pp. 1256–1274.

30. Böckmann, M., Doltsinis, N. L., and Marx, D. (2012). Enhanced photoswitching of bridged azobenzene studied by nonadiabatic ab initio simulation, *J. Chem. Phys.* **137**, 22, p. 22A505.

31. Bonacić-Koutecký, V., and Mitrić, R. (2005). Theoretical exploration of ultrafast dynamics in atomic clusters: Analysis and control, *Chem. Rev.* **105**, 1, pp. 11–66.

32. Boomsma, F., Jacobs, H., Havinga, E., and van der Gen, A. (1975). Studies on vitamin D and related compounds XXIV new irradiation products of previtamin D3. toxisterols. *Tetrahedron Lett.* **16**, 7, p. 427.

33. Boomsma, F., Jacobs, H., Havinga, E., and Van der Gen, A. (1977). The "overirradiation products" of previtamin D and tachysterol: Toxisterols, *Recl. Trav. Chim. Pays-Bas* **96**, 4, pp. 104–112.

34. Braslavsky, S. E. (2007). Glossary of terms used in photochemistry, (IUPAC recommendations 2006), *Pure Appl. Chem.* **79**, 3, pp. 293–465.

35. Brearley, A. M., Flom, S. R., Nagarajan, V., and Barbara, P. F. (1986). Dynamic solvent effects on large-amplitude isomerization rates. 2. 2-(2'-propenyl) anthracene and (e)-2-(but-2'-en-2'-yl) anthracene, *J. Phys. Chem.* **90**, 10, pp. 2092–2099.

36. Brouwer, A., Cornelisse, J., and Jacobs, H. (1987). Wavelength effect on the photochemical reactions of (z)-2, 5-dimethyl-1, 3, 5-hexatrene: selective excitation of rotamers, *Tetrahedron* **43**, 2, pp. 435–438.

37. Browne, W. R., and Feringa, B. L. (2009). Light switching of molecules on surfaces, *Ann. Rev. Phys. Chem.* **60**, pp. 407–428.

38. Browne, W. R., and Feringa, B. L. (2010). Making molecular machines work, in *Nanoscience and Technology: A Collection of Reviews from Nature Journals* (World Scientific), pp. 79–89.

39. Celani, P., Bernardi, F., Robb, M. A., and Olivucci, M. (1996). Do photochemical ring-openings occur in the spectroscopic state? 1b2 pathways for the cyclohexadiene/hexatriene photochemical interconversion, *J. Phys. Chem.* **100**, 50, p. 19364.

40. Ceriotti, M., Bussi, G., and Parrinello, M. (2009). Langevin equation with colored noise for constant-temperature molecular dynamics simulations, *Phys. Rev. Lett.* **102**, 2, p. 020601.

41. Cesaretti, A., Carlotti, B., Elisei, F., Fortuna, C., and Spalletti, A. (2018). Photoinduced ICT vs. excited rotamer intercoversion in two quadrupolar polyaromatic n-methylpyridinium cations, *Phys. Chem. Chem. Phys.* **20**, 4, pp. 2851–2864.

42. Cesaretti, A., Carlotti, B., Elisei, F., Fortuna, C. G., Consiglio, G., and Spalletti, A. (2017). A cationic naphthyl derivative defies the non-

equilibrated excited rotamers principle, *Phys. Chem. Chem. Phys.* **19**, pp. 5262–5272.

43. Chernyak, V., and Mukamel, S. (2000). Density-matrix representation of nonadiabatic couplings in time-dependent density functional (TDDFT) theories, *J. Chem. Phys.* **112**, 8, p. 3572.
44. Christiansen, O., Koch, H., and Jørgensen, P. (1995). The 2nd-order approximate coupled-cluster singles and doubles model cc2, *Chem. Phys. Lett.* **243**, p. 409.
45. Cisneros, C., Thompson, T., Baluyot, N., Smith, A. C., and Tapavicza, E. (2017). The role of tachysterol in vitamin D photosynthesis: A non-adiabatic molecular dynamics study, *Phys. Chem. Chem. Phys.* **19**, pp. 5763–5777.
46. Craig, C. F., Duncan, W. R., and Prezhdo, O. V. (2005). Trajectory surface hopping in the time-dependent Kohn–Sham approach for electron-nuclear dynamics, *Phys. Rev. Lett.* **95**, p. 163001.
47. Craig, I. R., and Manolopoulos, D. E. (2004). Quantum statistics and classical mechanics: Real time correlation functions from ring polymer molecular dynamics, *J. Chem. Phys.* **121**, 8, pp. 3368–3373.
48. Curchod, B. F., and Martínez, T. J. (2018). Ab initio nonadiabatic quantum molecular dynamics, *Chem. Rev.* **118**, 7, pp. 3305–3336.
49. Dauben, W. G., Disanayaka, B., Funhoff, D. J. H., Zhou, B., Kohler, B. E., and Schilke, D. E. (1991). Polyene 21Ag and 11Bu states and the photochemistry of previtamin D3, *J. Am. Chem. Soc.* **113**, 22, p. 8367.
50. De Haan, D. O., Tapavicza, E., Riva, M., Cui, T., Surratt, J. D., Smith, A. C., Jordan, M.-C., Nilakantan, S., Almodovar, M., Stewart, T. N., et al. (2018). Nitrogen-containing, light-absorbing oligomers produced in aerosol particles exposed to methylglyoxal, photolysis, and cloud cycling, *Environ. Sci. Technol.* **52**, 7, pp. 4061–4071.
51. Dierksen, M., and Grimme, S. (2004). The vibronic structure of electronic absorption spectra of large molecules: A time-dependent density functional study on the influence of "exact" Hartree–Fock exchange, *J. Phys. Chem. A* **108**, 46, pp. 10225–10237.
52. Dmitrenko, O., Frederick, J. H., and Reischl, W. (2001). Previtamin D conformations and the wavelength-dependent photoconversions of previtamin D, *J. Photochem. Photobiol. A: Chem.* **139**, 2–3, p. 125.
53. Doltsinis, N. L., and Marx, D. (2002). Nonadiabatic car–parrinello molecular dynamics, *Phys. Rev. Lett.* **88**, 16, p. 166402.

54. Domcke, W., Yarkony, D., and Köppel, H. (2004). *Conical Intersections: Electronic Structure, Dynamics & Spectroscopy*, Advanced Series in Physical Chemistry (World Scientific), ISBN 9789812386724.
55. Dreuw, A., Weisman, J., and Head-Gordon, M. (2003). Long-range charge-transfer excited states in time-dependent density functional theory require non-local exchange, *J. Chem. Phys.* **119**, p. 2943.
56. Duschinsky, F. (1937). The importance of the electron spectrum in multi atomic molecules. concerning the Franck-Condon principle, *Acta Physicochim. URSS* **7**, p. 551.
57. Engel, P. S., and Schexnayder, M. A. (1972). Effect of ring size on the photochemical behavior of beta, gamma-unsaturated ketones, *J. Am. Chem. Soc.* **94**, 26, pp. 9252–9254.
58. Epstein, S. A., Tapavicza, E., Furche, F., and Nizkorodov, S. A. (2013). Direct photolysis of carbonyl compounds dissolved in cloud and fog droplets, *Atmos. Chem. Phys.* **13**, 18, pp. 9461–9477.
59. Filatov, M., Martínez, T. J., and Kim, K. S. (2017). Description of ground and excited electronic states by ensemble density functional method with extended active space, *J. Chem. Phys.* **147**, 6, p. 064104.
60. Flom, S. R., Nagarajan, V., and Barbara, P. F. (1986). Dynamic solvent effects on large-amplitude isomerization rates. 1. 2-vinylanthracene, *J. Phys. Chem.* **90**, 10, pp. 2085–2092.
61. Frenkel, D., and Smit, B. (2001). *Understanding Molecular Simulation: From Algorithms to Applications*, Vol. 1 (Elsevier).
62. Furche, F., and Ahlrichs, R. (2002). Adiabatic time-dependent density functional methods for excited state properties, *J. Chem. Phys.* **117**, p. 7433.
63. Fuss, W., Hofer, T., Hering, P., Kompa, K. L., Lochbrunner, S., Schikarski, T. and Schmid, W. E. (1996). Ring opening in the dehydrocholesterol-previtamin D system studied by ultrafast spectroscopy, *J. Chem. Phys.* **100**, 2, p. 921.
64. Garavelli, M., Page, C. S., Celani, P., Olivucci, M., Schmid, W. E., Trushin, S. A., and Fuss, W. (2001). Reaction path of a sub-200 fs photochemical electrocyclic reaction, *J. Phys. Chem. A* **105**, 18, p. 4458.
65. Garland, C. F., Gorham, E. D., Mohr, S. B., and Garland, F. C. (2009). Vitamin d for cancer prevention: Global perspective, *Ann. Epidemiol.* **19**, 7, p. 468.
66. Ghosh, P., Torner, J., Arora, P. S., and Maayan, G. (2021). Dual control of peptide conformation with light and metal coordination, *Chem. Eur. J.* **27**, 35, pp. 8956–8959.

67. Grathwol, C. W., Wössner, N., Swyter, S., Smith, A. C., Tapavicza, E., Hofstetter, R. K., Bodtke, A., Jung, M., and Link, A. (2019). Azologization and repurposing of a hetero-stilbene-based kinase inhibitor: Towards the design of photoswitchable sirtuin inhibitors, *Beilstein J. Organic Chem.* **15**, 1, pp. 2170–2183.

68. Green, A., Sawada, T., and Shettle, E. (1974). The middle ultraviolet reaching the ground, *Photochem. Photobiol.* **19**, 4, pp. 251–259.

69. Havinga, E. (1973). Vitamin d, example and challenge, *Experientia* **29**, 10, pp. 1181–1193.

70. Havinga, E. (1973). Vitamin d, example and challenge, *Cell. Mol. Life Sci.* **29**, p. 1181.

71. Havinga, E., and Schlatmann, J. L. M. A. (1961). Remarks on the specificities of the photochemical and thermal transformations in the vitamin d field, *Tetrahedron Lett.* **16**, 1-4, p. 146.

72. Heller, D. F., Freed, K. F., and Gelbart, W. M. (1973). The role of accepting modes in the theory of nonradiative transitions, *Chem. Phys. Lett.* **23**, 1, pp. 56–62.

73. Heller, E. J. (1981). Frozen Gaussians: A very simple semiclassical approximation, *J. Chem. Phys.* **75**, 6, pp. 2923–2931.

74. Hilborn, R. C. (1982). Einstein coefficients, cross sections, f values, dipole moments, and all that, *Am. J. Phys.* **50**, 11, pp. 982–986.

75. Holick, M. F. (2003). Vitamin d: A millenium perspective, *J. Cell. Biochem.* **88**, 2, p. 296.

76. Holick, M. F., Tian, X. Q., and Allen, M. (1995). Evolutionary importance for the membrane enhancement of the production of vitamin D3 in the skin of poikilothermic animals. *Proc. Natl. Acad. Sci.* **92**, 8, pp. 3124–3126.

77. Jacobs, H. J. C., and Havinga, E. (1979). *Photochemistry of Vitamin D and Its Isomers and of Simple Trienes* (John Wiley & Sons, Inc.), ISBN 9780470133415.

78. Jankowiak, H.-C., Stuber, J. L., and Berger, R. (2007). Vibronic transitions in large molecular systems: Rigorous prescreening conditions for Franck–Condon factors, *J. Chem. Phys.* **127**, 23, pp. 234101–23.

79. Janoš, J., Madea, D., Mahvidi, S., Mujawar, T., Švenda, J., Suchan, J., Slavícek, P., and Klán, P. (2020). Conformational control of the photodynamics of a bilirubin dipyrrinone subunit: Femtosecond spectroscopy combined with nonadiabatic simulations, *J. Phys. Chem. A,* **124**, 50, 10457–10471.

80. Kubo, R. (1952). Thermal ionization of trapped electrons, *Phys. Rev.* **86**, 6, p. 929.

81. Lee, S., Shostak, S., Filatov, M., and Choi, C. H. (2019). Conical intersections in organic molecules: Benchmarking mixed-reference spin–flip time-dependent dft (MRSF-TD-DFT) vs spin–flip TD-DFT, *J. Phys. Chem. A* **123**, 30, pp. 6455–6462.

82. Levine, B. G., Ko, C., Quenneville, J., and Martinez, T. J. (2006). Conical intersections and double excitations in density functional theory, *Mol. Phys.* **104**, p. 1039.

83. Lewis, F. D., Bassani, D. M., and Reddy, G. D. (1993). Intramolecular photoaddition of secondary alpha-(aminoalkyl) styrenes, *J. Org. Chem.* **58**, 23, pp. 6390–6393.

84. Lewis, F. D., Johnson, R. W., and Johnson, D. E. (1974). Conformational control of photochemical behavior. competitive alpha cleavage and gamma-hydrogen abstraction of alkyl phenyl ketones, *J. Am. Chem. Soc.* **96**, 19, pp. 6090–6099.

85. Li, X., Tully, J. C., Schlegel, H. B., and Frisch, M. J. (2005). Ab initio Ehrenfest dynamics, *J. Chem. Phys.* **123**, 8, p. 084106.

86. Lipson, M., Noh, T., Doubleday, C. E., Zaleski, J. M., and Turro, N. J. (1994). Conformational control of the photochemistry and photophysics of diphenylacetone, *J. Phys. Chem.* **98**, 36, pp. 8844–8850.

87. Lischka, H., Nachtigallova, D., Aquino, A. J., Szalay, P. G., Plasser, F., Machado, F. B., and Barbatti, M. (2018). Multireference approaches for excited states of molecules, *Chem. Rev.* **118**, 15, pp. 7293–7361.

88. Liu, R. S., Turro Jr, N. J., and Hammond, G. S. (1965). Mechanisms of photochemical reactions in solution. xxxi. activation and deactivation of conjugated dienes by energy transfer, *J. Am. Chem. Soc.* **87**, 15, pp. 3406–3412.

89. Liu, X.-Y., Fang, Y.-G., Xie, B.-B., Fang, W.-H., and Cui, G. (2017). Qm/mm nonadiabatic dynamics simulations on photoinduced Wolff rearrangements of 1, 2, 3-thiadiazole, *J. Chem. Phys.* **146**, 22, p. 224302.

90. Liu, Y., Chakraborty, P., Matsika, S., and Weinacht, T. (2020). Excited state dynamics of cis, cis-1, 3-cyclooctadiene: UV pump VUV probe time-resolved photoelectron spectroscopy, *J. Chem. Phys.* **153**, 7, p. 074301.

91. MacLaughlin, J., Anderson, R., and Holick, M. (1982). Spectral character of sunlight modulates photosynthesis of previtamin D3 and its photoisomers in human skin, **216**, 4549, pp. 1001–1003.

92. Madea, D., Mahvidi, S., Chalupa, D., Mujawar, T., Dvorák, A., Muchová, L., Janoš, J., Slavicek, P., Švenda, J., Vítek, L., et al. (2020). Wavelength-dependent photochemistry and biological relevance of a bilirubin dipyrrinone subunit, *J. Org. Chem.* **85**, 20, pp. 13015–13028.

93. Maestro, M. A., Molnár, F., and Carlberg, C. (2019). Vitamin d and its synthetic analogs, *J. Med. Chem.* **62**, 15, pp. 6854–6875.

94. Maitra, N. T., Zhang, F., Cave, R. J., and Burke, K. (2004). Double excitations within time-dependent density functional theory linear response, *J. Chem. Phys.* **120**, 13, pp. 5932–5937.

95. Martineau, A. R., and Forouhi, N. G. (2020). Vitamin d for COVID-19: A case to answer? *Lancet Diab. Endocrinol.* **8**, 9, pp. 735–736.

96. Matsuda, K., Shinkai, Y., Yamaguchi, T., Nomiyama, K., Isayama, M., and Irie, M. (2003). Very high cyclization quantum yields of diarylethene having two n-methylpyridinium ions, *Chem. Lett.* **32**, 12, pp. 1178–1179.

97. Meana-Pañeda, R., and Fernández-Ramos, A. (2012). Tunneling and conformational flexibility play critical roles in the isomerization mechanism of vitamin D, *J. Am. Chem. Soc.* **134**, 1, pp. 346–354.

98. Mebel, A. M., Hayashi, M., Liang, K. K., and Lin, S. H. (1999). Ab initio calculations of vibronic spectra and dynamics for small polyatomic molecules: Role of Duschinsky effect, *J. Phys. Chem. A* **103**, pp. 10674–10690.

99. Meyer-Ilse, J., Akimov, D., and Dietzek, B. (2012). Ultrafast circular dichroism study of the ring opening of 7-dehydrocholesterol, *J. Phys. Chem. Lett.* **3**, 2, pp. 182–185.

100. Micheletti, C., Laio, A., and Parrinello, M. (2004). Reconstructing the density of states by history-dependent metadynamics, *Phys. Rev. Lett.* **92**, 17, p. 170601.

101. Moret, M., Tapavicza, E., Guidoni, L., Rohrig, U., Sulpizi, M., Tavernelli, I. and Rothlisberger, U. (2005). Quantum mechanical/molecular mechanical (QM/MM) Car–Parrinello simulations in excited states, *Chimia* **59**, pp. 493–498.

102. Moriarty, R., and Reardon, R. (1970). The direct and photosensitized decomposition of alkyl azides, *Tetrahedron* **26**, 5, pp. 1379–1392.

103. Mukamel, S., Abe, S., and Islampour, R. (1985). Generating function for electronic spectra of polyatomic molecules, *J. Phys. Chem.* **89**, 2, pp. 201–204.

104. Norval, M., Bjorn, L. O., and de Gruijl, F. R. (2010). Is the action spectrum for the UV-induced production of previtamin D3 in human skin correct? *Photochem. Photobiol. Sci.* **9**, 1, p. 11.

105. Nosé, S. (1984). *Rev. Mod. Phys.* **52**, p. 255.

106. Oppolzer, W. (1982). The intramolecular [2 + 2] photoaddition/cyclobutane-fragmentation sequence in organic synthesis, *Acc. Chem. Res.* **15**, 5, pp. 135–141.

107. Park, J. W., and Shiozaki, T. (2017). On-the-fly caspt2 surface-hopping dynamics, *J. Chem. Theor. Comput.* **13**, 8, pp. 3676–3683.

108. Patoz, A., Begušic, T., and Vanícek, J. (2018). On-the-fly ab initio semiclassical evaluation of absorption spectra of polyatomic molecules beyond the condon approximation, *J. Phys. Chem. Lett.* **9**, 9, pp. 2367–2372.

109. Peng, Q., Niu, Y., Deng, C., and Shuai, Z. (2010). Vibration correlation function formalism of radiative and non-radiative rates for complex molecules, *Chem. Phys.* **370**, 1-3, p. 215.

110. Perdew, J., Burke, K., and Ernzerhof, M. (1996). Generalized gradient approximation made simple, *Phys. Rev. Lett.* **77**, p. 3865.

111. Perdew, J. P., Ernzerhof, M., and Burke, K. (1996). Rationale for mixing exact exchange with density functional approximations, *J. Chem. Phys.* **105**, 22, p. 9982.

112. Plasser, F., Crespo-Otero, R., Pederzoli, M., Pittner, J., Lischka, H., and Barbatti, M. (2014). Surface hopping dynamics with correlated single-reference methods: 9h-adenine as a case study, *J. Chem. Theor. Comput.* **10**, 4, pp. 1395–1405.

113. Ramakrishnan, R., Hartmann, M., Tapavicza, E., and von Lilienfeld, O. A. (2015). Electronic spectra from TDDFT and machine learning in chemical space, *J. Chem. Phys.* **143**, 8, p. 084111.

114. Renth, F., Siewertsen, R., and Temps, F. (2013). Enhanced photoswitching and ultrafast dynamics in structurally modified photochromic fulgides, *Int. Rev. Phys. Chem.* **32**, 1, pp. 1–38.

115. Runge, E., and Gross, E. K. U. (1984). Density-functional theory for time-dependent systems, *Phys. Rev. Lett.* **52**, pp. 997–1000.

116. Schalk, O., Geng, T., Thompson, T., Baluyot, N., Thomas, R. D., Tapavicza, E. and Hansson, T. (2016). Cyclohexadiene revisited: A time-resolved photoelectron spectroscopy and ab initio study, *J. Phys. Chem. A* **120**, 15, p. 2320.

117. Schürmann, A. K. G. (1993). Cosmo: A new approach to dielectric screening in solvents with explicit expressions for the screening energy and its gradient, *J. Chem. Soc. Perkin Trans. 2* **5**, p. 799.

118. Send, R., and Furche, F. (2010). First-order nonadiabatic couplings from time-dependent hybrid density functional response theory:

Consistent formalism, implementation, and performance, *J. Phys. Chem.* **132**, p. 044107.

119. Sharp, T., and Rosenstock, H. (1964). Franck–Condon factors for polyatomic molecules, *J. Chem. Phys.* **41**, pp. 3453–3463.
120. Shemesh, D., and Gerber, R. B. (2018). Molecular dynamics of photoinduced reactions of acrylic acid: Products, mechanisms, and comparison with experiment, *J. Phys. Chem. Lett.* **9**, 3, pp. 527–533.
121. Shushkov, P., Li, R., and Tully, J. C. (2012). Ring polymer molecular dynamics with surface hopping, *J. Chem. Phys.* **137**, 22, p. 22A549.
122. Sneskov, K., and Christiansen, O. (2012). Excited state coupled cluster methods, *Wiley Interdisciplinary Rev. Comput. Mol. Sci.* **2**, 4, pp. 566–584.
123. Sofferman, D. L., Konar, A., Mastron, J. N., Spears, K. G., Cisneros, C., Smith, A. C., Tapavicza, E., and Sension, R. J. (2021). Probing the formation and conformational relaxation of previtamin D3 and analogs in solution and in lipid bilayers, *J. Phys. Chem. B* **125**, 36, pp. 10085–10096.
124. Spangler, C. W., and Hennis, R. P. (1972). The influence of ground state conformation on the photochemical ring-opening of 5-alkylcyclohexa-1, 3-dienes, *J. Chem. Soc. Chem. Commun.* , 1, pp. 24–25.
125. Stein, S. E., and Rabinovitch, B. S. (1973). Accurate evaluation of internal energy level sums and densities including anharmonic oscillators and hindered rotors, *J. Chem. Phys.* **58**, 6, p. 2438.
126. Stendardo, E., Avila Ferrer, F., Santoro, F., and Improta, R. (2012). Vibrationally resolved absorption and emission spectra of dithiophene in the gas phase and in solution by first-principle quantum mechanical calculations, *J. Chem. Theor. Comput.* **8**, pp. 4483–4493.
127. Suda, T., Hallick, R., DeLuca, H. F., and Schnoes, H. K. (1970). 25-hydroxydihydrotachysterol3. Synthesis and biological activity, *Bio-chemistry* **9**, 8, pp. 1651–1657.
128. Sugita, Y., and Okamoto, Y. (1999). Replica-exchange molecular dynamics method for protein folding, *Chem. Phys. Lett.* **314**, 1-2, p. 141.
129. Tang, K.-C., Rury, A., Orozco, M. B., Egendorf, J., Spears, K. G., and Sension, R. J. (2011). Ultrafast electrocyclic ring opening of 7-dehydrocholesterol in solution: The influence of solvent on excited state dynamics, *J. Chem. Phys.* **134**, 10, p. 104503.
130. Tapavicza, E. (2019). Generating function approach to single vibronic level fluorescence spectra, *J. Phys. Chem. Lett.* **10**, pp. 6003–6009.

131. Tapavicza, E., Bellchambers, G. D., Vincent, J. C., and Furche, F. (2013). Ab initio non-adiabatic molecular dynamics, *Phys. Chem. Chem. Phys.* **15**, pp. 18336–18348.

132. Tapavicza, E., Furche, F., and Sundholm, D. (2016). Importance of vibronic effects in the UV-vis spectrum of the 7,7,8,8-tetracyano-quinodimethane anion, *J. Chem. Theor. Comput.* **12**, 10, pp. 5058–5066.

133. Tapavicza, E., Meyer, A. M., and Furche, F. (2011). Unravelling the details of vitamin D photosynthesis by non-adiabatic molecular dynamics simulations, *Phys. Chem. Chem. Phys.* **13**, p. 20986.

134. Tapavicza, E., Tavernelli, I., and Rothlisberger, U. (2007). Trajectory surface hopping within linear response time-dependent density-functional theory, *Phys. Rev. Lett.* **98**, p. 023001.

135. Tapavicza, E., Tavernelli, I., and Rothlisberger, U. (2009). Ab initio excited state properties and dynamics of a prototype σ-bridged-donor-acceptor molecule, *J. Phys. Chem. A* **113**, 35, p. 9595.

136. Tapavicza, E., Tavernelli, I., Rothlisberger, U., Filippi, C., and Casida, M. E. (2008). Mixed time-dependent density-functional theory/classical trajectory surface hopping study of oxirane photochemistry, *J. Chem. Phys.* **129**, 12, p. 124108.

137. Tapavicza, E., Thompson, T., Redd, K., and Kim, D. (2018). Tuning the photoreactivity of z-hexatriene photoswitches by substituents—a non-adiabatic molecular dynamics study, *Phys. Chem. Chem. Phys.* **20**, pp. 24807–24820.

138. Tapavicza, E., von Rudorff, G. F., De Haan, D. O., Contin, M., George, C., Riva, M., and von Lilienfeld, O. A. (2021). Elucidating an atmospheric brown carbon species–toward supplanting chemical intuition with exhaustive enumeration and machine learning, *Environ. Sci. Technol.* **55**, 12, pp. 8447–8457.

139. Tavernelli, I., Röhrig, U. F., and Rothlisberger, U. (2005). Molecular dynamics in electronically excited states using time-dependent density functional theory, *Mol. Phys.* **103**, p. 963.

140. Tavernelli, I., Tapavicza, E., and Rothlisberger, U. (2009). Non-adiabatic dynamics using time-dependent density functional theory: Assessing the coupling strengths, *J. Mol. Struct. THEOCHEM* **914**, 1-3, p. 22.

141. Tavernelli, I., Tapavicza, E., and Rothlisberger, U. (2009). Nonadiabatic coupling vectors within linear response time-dependent density functional theory, *J. Chem. Phys.* **130**, 12, p. 124107.

142. Thompson, T., and Tapavicza, E. (2018). First-principles prediction of wavelength-dependent product quantum yields, *J. Phys. Chem. Lett.* **9**, 16, pp. 4758–4764.

143. Tian, X. Q., and Holick, M. F. (1999). A liposomal model that mimics the cutaneous production of vitamin D3, *J. Biol. Chem.* **274**, 7, p. 4174.
144. Torner, J. M., and Arora, P. S. (2021). Conformational control in a photoswitchable coiled coil, *Chem. Commun.* **57**, 12, pp. 1442–1445.
145. Tully, J. C. (1990). Molecular dynamics with electronic transitions, *J. Chem. Phys.* **93**, p. 1061.
146. van Dijk, A., den Outer, P., van Kranen, H., and Slaper, H. (2016). The action spectrum for vitamin D3: Initial skin reaction and prolonged exposure, *Photochem. Photobiol. Sci.* **15**, pp. 896–909.
147. Vincent, J. C., Muuronen, M., Pearce, K. C., Mohanam, L. N., Tapavicza, E., and Furche, F. (2016). That little extra kick: Nonadiabatic effects in acetaldehyde photodissociation, *J. Phys. Chem. Lett.* **7**, 20, pp. 4185–4190.
148. Wagner, M. (1959). *Z. Naturforschg.* **14a**, p. 81.
149. Wagner, P. J. (1983). Conformational flexibility and photochemistry, *Acc. Chem. Res.* **16**, 12, pp. 461–467.
150. Wagner, P. J., and Stratton, T. J. (1981). Kinetically distinct triplets in the photorearrangement of 2-phenylcyclohexanone to cis-and trans-6-phenyl-5-hexenals, *Tetrahedron* **37**, 19, pp. 3317–3322.
151. Wang, S. (2009). Epidemiology of vitamin D in health and disease, *Nutr. Res. Rev.* **22**, 02, p. 188.
152. Westermayr, J., and Marquetand, P. (2020). Machine learning for electronically excited states of molecules, *Chem. Rev.*.
153. Whitesell, J. K., Minton, M. A., and Tran, V. D. (1989). The non-equilibration of excited rotamers (neer) principle. ground-state conformational bias in triene photocyclizations, *J. Am. Chem. Soc.* **111**, 4, pp. 1473–1476.
154. Wiebeler, C., Bader, C. A., Meier, C., and Schumacher, S. (2014). Optical spectrum, perceived color, refractive index, and non-adiabatic dynamics of the photochromic diarylethene cmte, *Phys. Chem. Chem. Phys.*.
155. Wiebeler, C., and Schumacher, S. (2014). Quantum yields and reaction times of photochromic diarylethenes: Nonadiabatic ab initio molecular dynamics for normal- and inverse-type, *J. Phys. Chem. A* **118**, 36, pp. 7816–7823.
156. Wolf, T. J. A., Kuhlman, T. S., Schalk, O., Martinez, T. J., Moller, K. B., Stolow, A., and Unterreiner, A.-N. (2014). Hexamethylcyclopentadiene: time-resolved photoelectron spectroscopy and ab initio multiple spawning simulations, *Phys. Chem. Chem. Phys.* **16**, pp. 11770–11779.

157. Woodward, R. B., and Hoffmann, R. (1965). Selection rules for sigmatropic reactions, *J. Am. Chem. Soc.* **87**, 11, pp. 2511–2513.

158. Xue, B.-X., Barbatti, M., and Dral, P. O. (2020). Machine learning for absorption cross sections, *J. Phys. Chem. A* **124**, 35, pp. 7199–7210.

159. Yan, Y. J., and Mukamel, S. (1986). Eigenstate-free, green function, calculation of molecular absorption and fluorescence line shapes, *J. Chem. Phys.* **85**, 10, pp. 5908–5923.

160. Yue, L., Yu, L., Xu, C., Zhu, C., and Liu, Y. (2020). Quantum yields of singlet and triplet chemiexcitation of dimethyl 1, 2-dioxetane: ab initio nonadiabatic molecular dynamic simulations, *Phys. Chem. Chem. Phys.* **22**, 20, pp. 11440–11451.

161. Zhu, C., Nobusada, K., and Nakamura, H. (2001). New implementation of the trajectory surface hopping method with use of the Zhu–Nakamura theory, *J. Chem. Phys.* **115**, 7, pp. 3031–3044.

162. Zuehlsdorff, T., and Isborn, C. (2018). Combining the ensemble and Franck–Condon approaches for calculating spectral shapes of molecules in solution, *J. Chem. Phys.* **148**, 2, p. 024110.

163. Zuev, P. S., Sheridan, R. S., Sauers, R. R., Moss, R. A., and Chu, G. (2006). Conformational product control in the low-temperature photochemistry of cyclopropylcarbenes, *Org. Lett.* **8**, 21, pp. 4963–4966.

Chapter 6

Generalized Trajectory-Based Surface-Hopping (GTSH) Nonadiabatic Dynamics with Time-Dependent Density Functional Theory: Methodologies and Applications

Wen-Kai Chen, Xiang-Yang Liu, and Ganglong Cui
Key Laboratory of Theoretical and Computational Photochemistry, Ministry of Education, College of Chemistry, Beijing Normal University, Beijing, China
ganglong.cui@bnu.edu.cn

This chapter is a brief review of the generalized trajectory-based surface hopping nonadiabatic dynamics simulation methods and applications based on time-dependent density functional theory (time and frequency domains) developed by our group and coworkers.

Time-Dependent Density Functional Theory: Nonadiabatic Molecular Dynamics
Edited by Chaoyuan Zhu

ISBN 978-981-4968-42-3 (Hardcover), 978-1-003-31921-4 (eBook)
www.jennystanford.com

6.1 Theoretical Foundation of Nonadiabatic Effects

6.1.1 Breaking Down of Born–Oppenheimer Approximation

The Born–Oppenheimer (BO) approximation is one of the most important approximations in theoretical and computational chemistry. It supposes that nuclei are fixed in a system while electrons are moving around the nuclei. According to this approximation, the wavefunction of a molecular $\Psi(\mathbf{R}, \mathbf{r}, t)$ can be separated into nuclei $\psi(\mathbf{R}, t)$ and electron $\Phi(\mathbf{R}, \mathbf{r})$ parts. Thanks to the BO approximation, it becomes much easier to find an approximate solution of the Schrödinger equation. The BO approximation is reliable when relevant potential energy surfaces (PESs) are far away from each other. However, it breaks down when two or more PESs come close or even crossing, because the nuclear momentum coupling is no longer negligible near the degenerate region of PESs. Unfortunately, photochemical and photophysical problems usually need to deal with close or crossing PESs, thus the BO approximation always breaks down and nonadiabatic effects have to be seriously considered.

In order to illustrate this issue, we begin with the time-dependent Schrödinger equation (TDSE) without the BO approximation:

$$i\hbar \frac{\partial \Psi(\mathbf{R}, \mathbf{r}, t)}{\partial t} = \hat{H}(\mathbf{R}, \mathbf{r})\Psi(\mathbf{R}, \mathbf{r}, t) \tag{6.1}$$

$$\hat{H}(\mathbf{R}, \mathbf{r}) = \hat{T}_n(\mathbf{R}) + \hat{H}_{el}(\mathbf{R}, \mathbf{r}) \tag{6.2}$$

where

$$\hat{T}_n(\mathbf{R}) = -\sum_{\alpha} \frac{\hbar^2}{2m_\alpha} \nabla_\alpha^2 \tag{6.3}$$

is the kinetic energy of nuclei, and

$$\hat{H}_{el}(\mathbf{R}, \mathbf{r}) = -\sum_{i} \frac{\hbar^2}{2m_e} \nabla_i^2 + \hat{V}(\mathbf{R}, \mathbf{r}) \tag{6.4}$$

consists of the kinetic energy of electrons, nuclei-nuclei repulsive interaction, electron-electron repulsive interaction and electron–nuclei attractive interaction. Though the BO expansion [1, 2], the linear combination of product of electronic and nuclear wavefunctions forms a complete set which could represent the total wavefunction exactly:

$$\Psi(\mathbf{R}, \mathbf{r}, t) = \sum_n \psi_n(\mathbf{R}, t)\Phi_n(\mathbf{R}, \mathbf{r}) \tag{6.5}$$

here, Φ_n $(\mathbf{R}, \mathbf{r})$ represents the eigenvectors of time-independent Schrödinger equation (TISE)

$$\hat{H}_{el}(\mathbf{R}, \mathbf{r})|\Phi_n(\mathbf{R}, \mathbf{r}) >= E_n|\Phi_n(\mathbf{R}, \mathbf{r}) > . \tag{6.6}$$

Taking Eq. (6.5) and Eq. (6.6) into Eq. (6.1), after some mathematical derivation, one can get

$$i\hbar\frac{\partial \psi_n(\mathbf{R}, t)}{\partial t} = \left[-\sum_\alpha \frac{\hbar^2}{2m_\alpha}\nabla_\alpha^2 + E_n(\mathbf{R})\right] \psi_n(\mathbf{R}, t) + \sum_m \mathbf{C}_{nm}\psi_m(\mathbf{R}, t) \tag{6.7}$$

in which α represents the index of nuclei. This equation is also known as the coupled-channels equation, which describes the exact expression for vibronic problems, in which

$$\mathbf{C}_{nm} = -\sum_\alpha \frac{\hbar^2}{2m_\alpha} D_{nm}^\alpha(\mathbf{R}, t) + \sum_\alpha \frac{\hbar^2}{m_\alpha}\mathbf{d}_{nm}^\alpha(\mathbf{R}, t)\nabla_\alpha, \tag{6.8}$$

$$\mathbf{d}_{nm}^\alpha(\mathbf{R}, t) = \langle \Phi_n(\mathbf{R}, t) \,|\nabla_\alpha|\, \Phi_m(\mathbf{R}, t)\rangle, \tag{6.9}$$

$$D_{nm}^\alpha(\mathbf{R}, t) = \left\langle \Phi_n(\mathbf{R}, t) \left|\nabla_\alpha^2\right| \Phi_m(\mathbf{R}, t)\right\rangle. \tag{6.10}$$

Here, Eq. (6.9) is called the derivative coupling or first-order nonadiabatic coupling (NAC) while Eq. (6.10) is called kinetic coupling or second-order nonadiabatic coupling. Usually, the first term in Eq. (6.8) is much smaller than the second term, so we can just only consider the second term in Eq. (6.8). We have to mention that $\mathbf{d}_{nm}^\alpha$ can also be expressed as

$$\mathbf{d}_{nm}^\alpha(\mathbf{R}, t) = \frac{\langle \Phi_n(\mathbf{R}, t) \,|\nabla_\alpha H_{el}|\, \Phi_m(\mathbf{R}, t)\rangle}{E_n - E_m}. \tag{6.11}$$

From Eq. (6.11), one can see that when E_n–E_m gets closer to zero (such as conical intersections), $\mathbf{d}_{nm}^\alpha$ goes to infinity. Thus, the nonadiabatic coupling terms become more and more important when the energy of two electronic states becomes close to each other. In the situation such as conical intersections, the

BO approximation breaks down. Due to this issue, we must go beyond the BO approximation and explicitly consider the effects of nonadiabatic couplings in related photophysical and photochemical problems.

6.1.2 Nonadiabatic Molecular Dynamics

As we all know, most of photophysical processes and photochemical reactions not only are ultrafast but also involve multiple, complex electronic states in particular quasi-degenerate regions of PESs. Therefore, nonadiabatic effects are indispensable and in this regard, nonadiabatic molecular dynamics (NAMD) simulation is a powerful tool for shedding important mechanistic light on these ultrafast processes.

NAMD can provide atomic-level insights into excited-state relaxation processes, relationship between structures and properties, timescales of every detailed process, etc. These pieces of information can not only help us understand excited states properties of specific systems but also be used to design compounds and materials better.

Nowadays, a variety of NAMD approaches have been developed to treat various kinds of systems at different simulation levels [3–17]. In the most accurate NAMD method, i.e., quantum wavepacket, etc., both nuclei and electrons are described quantum mechanically; however, this kind of approach is extremely expensive and mainly limited to study small polyatomic systems based on pre-fitted PESs. In other words, full quantum treatment to the nuclear motion is limited to only few nuclear degrees. Instead, mixed quantum-classical (MQC) NAMD methods are more suitable for large systems in various complicated environments, for example, surface hopping, Ehrenfest, and quantum trajectory mean-field methods, etc., which are becoming popular NAMD methods [10, 18, 19]. Obviously, giving a comprehensive introduction of all the NAMD methods is beyond the scope of this chapter and readers who are interested in this field are referred to read related reviews [20–22]. In the following, we will primarily present our developed generalized trajectory surface hopping (GTSH) method and introduce its fundamental equations and representative applications [23].

6.2 Generalized Trajectory Surface Hopping Method

6.2.1 Tully's Fewest Switches Surface Hopping

GTSH method is developed within the framework of the trajectory surface hopping (TSH) proposed by Tully et al. [24, 25], which is one of the most successful and popular mixed quantum-classical NAMD methods. Therefore, it is helpful to briefly discuss this TSH method first.

The basic idea of this approach is that one assumes that the system is propagated on a single potential energy surface (PES) all the time, and has probabilities hopping to other PESs. The motion of nuclei is propagated adiabatically and nonadiabatic effects influence the probabilities of hopping among different PESs.

Here we give a brief introduction and formulation of TSH method. First of all, similar with all other mixed quantum-classical NAMD methods, the motion of nuclei in TSH method is treated classically by Newton's equations

$$F_{\mathbf{R}} - m_{\mathbf{R}} \frac{d^2\mathbf{R}}{dt^2} = 0. \tag{6.12}$$

Here, $F_{\mathbf{R}}$ is the atomic force, which is the negative number of the gradient of the molecule along the PES. In the meantime, electrons of the system are treated by quantum mechanics, i.e., solving the TDSE (see Eq. (6.1)). The time-dependent wavefunction $\Psi(\mathbf{R}, \mathbf{r}, t)$ could be expressed by the linear combination of adiabatic electronic wavefunctions

$$\Psi(\mathbf{R}, \mathbf{r}, t) = \sum_{i=1}^{N} c_i(t)\Phi_i(\mathbf{R}, \mathbf{r}) \tag{6.13}$$

with relationship

$$\left\langle \Phi_j(\mathbf{R}, \mathbf{r}) | \Phi_i(\mathbf{R}, \mathbf{r}) \right\rangle = \delta_{ji}. \tag{6.14}$$

We can take Eq. (6.13) into Eq. (6.1), then multiply $\left\langle \Phi_j(\mathbf{R}, \mathbf{r}) \right|$ from the left side, the TDSE could be rewritten as

$$i\hbar \frac{dc_j(t)}{dt} - c_j(t)E_j(R) + i\hbar \sum_{i}^{N} c_i(t)\mathbf{d}_{ij}^{\alpha} \cdot \mathbf{v}(t) = 0. \tag{6.15}$$

Here, E_j means the jth eigenvalue of the TISE (see Eq. (6.6)) or the energy of the jth adiabatic electronic state; $\mathbf{d}_{ij}^{\alpha}$ is the adiabatic derivative coupling between the ith and jth electronic states. Equation (6.15) is the central equation of propagation of electrons of the system; in some literatures, Eq. (6.15) is also called as the semi-classical time-dependent Schrödinger equation (SC-TDSE) [12].

Except for the propagation of both nuclei and electrons during NAMD, another important issue is that how to distinguish whether hopping occurs or not between different PESs. Nowadays, the algorithm used in the fewest-switches surface hopping (FSSH) method proposed by Tully and colleagues is the most popular algorithm [24, 25]. In this scheme, the number of hoppings between different electronic states, which is supposed to reproduce the statistical distribution of state occupations, is minimized. Here we give the derivation of the transition probability of the FSSH scheme. At time t, the number of trajectories in state i is

$$N_i(t) = \rho_{ii}(t)N. \tag{6.16}$$

Here, N is the total number of trajectories in the simulation, which means N is a constant. ρ_{ii} is the i-th time-dependent diagonal element of density matrix, where density matrix is expressed as

$$\hat{\rho}(t) = |\Psi(\mathbf{R}, \mathbf{r}, t)\rangle\langle\Psi(\mathbf{R}, \mathbf{r}, t)|, \tag{6.17}$$

$$\rho_{ij}(t) = \langle\Phi_i(\mathbf{R}, \mathbf{r})|\hat{\rho}(t)|\Phi_j(\mathbf{R}, \mathbf{r})\rangle = c_i(t)c_j^*(t). \tag{6.18}$$

The occupation of state i at time $t+dt$, similar with Eq. (6.16), is expressed as

$$N_i(t+dt) = \rho_{ii}(t+dt)N. \tag{6.19}$$

Then, $N_i(t) - N_i(t+dt)$ will be the minimal number of transition from state i to all other electronic states in dt when $N_i(t) > N_i(t+dt)$ is assumed. Thus, the transition probability $p_i(t)dt$ from electronic state i to other states in time interval dt could be written as

$$p_i(t)dt = \frac{\rho_{ii}(t) - \rho_{ii}(t+dt)}{\rho_{ii}(t)} \approx -\frac{\dot{\rho}_{ii}dt}{\rho_{ii}}. \tag{6.20}$$

Also, one can get the time derivative of density matrix

$$\dot{\rho}_{ii} = 2Re(c_i^*\dot{c}_i) \tag{6.21}$$

and c_i^*, c_j, E_i are always real numbers. We can take Eq. (6.15) and Eq. (6.21) into Eq. (6.20), then we can get the final expression of the transition probability from electronic state i to electronic state j:

$$p_{ij}(t)dt = 2\frac{Re(c_j^* c_j \mathbf{v} \cdot \mathbf{d}_{ij})}{c_i^* c_i}dt. \tag{6.22}$$

6.2.2 Generalized Trajectory Surface Hopping (GTSH) Method

Although the FSSH approach is becoming popular these years and has a large number of successful applications [16, 26–39], this method still has some limitations and disadvantages. One of the most important limitations of the FSSH approach proposed by Tully et al. is that only internal conversion (IC) processes involving the states with the same multiplicities are considered and formulated in the propagation equations. However, in many photochemistry reactions and photophysical processes, intersystem crossing, which involves states with different multiplicities, also plays a very important role and cannot be neglected. Thus, it is very demanding for developing a generalized NAMD method for considering both internal conversion and intersystem crossing simultaneously.

In order to include the ISC processes into the TSH simulations, spin–orbit interactions must be added into the original FSSH approach. Motivated by this, we developed the generalized trajectory-based surface hopping (GTSH) method several years ago [23].

Figure 6.1 shows the scope of the Tully's original FSSH method and our developed GTSH method. In the GTSH method, intersystem crossing between electronic states with different multiplicities is considered while internal conversion between electronic states with same multiplicities is also included. We have to mention that in the GTSH method, only nonrelativistic adiabatic electronic Hamiltonian is used; while spin–orbit interaction is used to drive intersystem crossing processes. This choice has several advantages: (i) most of popular electronic structure program packages solve the TISE equation with nonrelativistic electronic Hamiltonian; thus, this choice makes maximum use of these available popular packages; (ii) comparing with the original FSSH code, this choice makes minimum

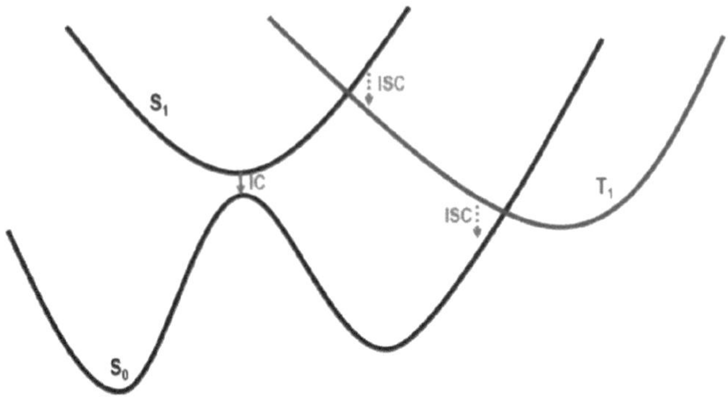

Figure 6.1 The scheme of the original FSSH and our developed GTSH methods. The former only considers the IC processes while the latter one can describe both IC and ISC processes simultaneously within a same framework.

modification of the implementation of GTSH package; (iii) similar with energies, gradients, and nonadiabatic couplings, the spin–orbit couplings (SOCs) also can be calculated "on-the-fly". Next, we give the formulation of the GTSH method.

It is clear that the SOC interaction drives the intersystem crossing process occur. Therefore, we must add the spin–orbit operator as a perturbation to the original zero-order FSSH electronic Hamiltonian

$$\hat{H}(\mathbf{R}, \mathbf{r}, \mathbf{s}) = \hat{H}_0(\mathbf{R}, \mathbf{r}) + \hat{H}_{SO}(\mathbf{R}, \mathbf{r}, \mathbf{s}) \tag{6.23}$$

where $\mathbf{r}$ and $\mathbf{s}$ represent the spatial and spin coordinates of electrons. Insert Eq. (6.23) into the TDSE, then one can get

$$i\hbar\frac{\partial\Psi(\mathbf{R}, \mathbf{r}, \mathbf{s}, t)}{\partial t} = (\hat{H}_0(\mathbf{R}, \mathbf{r}) + \hat{H}_{SO}(\mathbf{R}, \mathbf{r}, \mathbf{s})) \times \Psi(\mathbf{R}, \mathbf{r}, \mathbf{s}, t). \tag{6.24}$$

Here, the time-dependent electronic wavefunction is expanded as linear combination of interested adiabatic zero-order electronic wavefunctions $\Phi_i^0(\mathbf{R}, \mathbf{r}, \mathbf{s})$ with same and difference spins

$$\Psi(\mathbf{R}, \mathbf{r}, \mathbf{s}, t) = \sum_{i=1}^{N} c_i(t)\Phi_i^0(\mathbf{R}, \mathbf{r}, \mathbf{s}). \tag{6.25}$$

It should be noted that only same spin states are considered in the Tully's FSSH method.

These zero-order electronic wavefunctions are eigenfunctions of the zero-order Hamiltonian $\hat{H}_0$ and spin operators S^2 and S_z. They can be separated into spin-free (space) and -relevant (spin) parts. In order to write the whole electronic wavefunction into the sum of products of spin and space parts, a spin factor Θ_{ji} is multiplied to the spin-free function. It is zero for different spins while one for same spins. Based on these considerations, and take Eq. (6.25) into the TDSE and multiply $\langle \Phi_j^0(\mathbf{R}, \mathbf{r}, \mathbf{s})|$ from the left side, one can get

$$i\hbar\frac{dc_j(t)}{dt} - [c_j(t)E_j^0(\mathbf{R}) + \sum_i^N c_i(t)H_{ji}^{SO}(\mathbf{R})] + i\hbar\sum_i^N c_i(t)\mathbf{d}_{ij}^{\alpha}\cdot\mathbf{v}(t)\Theta_{ji} = 0 \qquad (6.26)$$

where H_{ji}^{SO} is the spin–orbit coupling matrix element between state j and i

$$H_{ji}^{SO}(\mathbf{R}) = \langle \Phi_j^0(\mathbf{R}, \mathbf{r}, \mathbf{s}) |\hat{H}_{SO}(\mathbf{R}, \mathbf{r}, \mathbf{s})| \Phi_i^0(\mathbf{R}, \mathbf{r}, \mathbf{s})\rangle . \qquad (6.27)$$

Equation (6.26) is the final electronic propagation equation for the generalized trajectory surface hopping method. If one only wants to consider internal conversion processes between same spin electronic states, the spin–orbit matrix elements are zero and Eq. (6.26) becomes Eq. (6.15). On the other hand, if someone wants to study intersystem crossing processes between different spin electronic states, all of the elements of spin factor Θ_{ji} are zero. Thus, Eq. (6.26) becomes

$$i\hbar\frac{dc_j(t)}{dt} - [c_j(t)E_j^0(\mathbf{R}) + \sum_i^N c_i(t)H_{ji}^{SO}(\mathbf{R})] = 0. \qquad (6.28)$$

Except for the central equation of GTSH method, there is one more important quantity in the GTSH method, which is the transition probability between states i and j. Similar with Eq. (6.22) the fewest-switches of the original FSSH method, the GTSH method introduces the spin–orbit coupling term. Therefore, the transition probability from electronic state i to electronic state j is:

$$p_{ij}(t)dt = 2\frac{Re(c_j^* c_j \mathbf{v}\cdot\mathbf{d}_{ij}) - Im(\hbar^{-1}c_j^* c_j H_{ji}^{SO})}{c_j^* c_j}dt. \qquad (6.29)$$

6.2.3 Generalized Trajectory Surface Hopping Method at QM/MM Level

In order to study nonradiative processes of large systems in complicated environments such as DNA or protein, we have also implemented our GTSH method at the QM/MM level [23, 40, 41]. In the QM/MM implementation, the working equation is similar with Eq. (6.26). However, we must pay attention to three critical quantities: energies, nonadiabatic couplings, and spin–orbit couplings.

First, we introduce the total energy expression of a system at a specific electronic state i, which is written as:

$$\begin{aligned} E_{i,\text{total}}(\mathbf{R}_{\text{QM}}, \mathbf{R}_{\text{MM}}) = E_{i,\text{QM}}(\mathbf{R}_{\text{QM}}) + E_{i,\text{MM}}(\mathbf{R}_{\text{MM}}) \\ + E_{i,\text{QM-MM}}(\mathbf{R}_{\text{QM}}, \text{R}_{\text{MM}}) \end{aligned} \tag{6.30}$$

where $\mathbf{R}_{\text{QM}}$ and $\mathbf{R}_{\text{MM}}$ represent the Cartesian coordinates of atoms in QM and MM regions. $E_{i,\text{total}}$, $E_{i,\text{QM}}$, $E_{i,\text{MM}}$, $E_{i,\text{QM-MM}}$ represent the total energy of the total system, energy of QM region, energy of MM region and energy of QM-MM interaction, respectively.

In our implementation, $E_{i,\text{MM}}$ is always calculated by a standard force field (e.g., AMBER force field or CHARMM force field) no matter which electronic state is. In other word, we use the approximation $E_{i,\text{MM}} \approx E_{\text{MM}}$. This approximation is adoptable because all photo-related regions should be selected into the QM region in a reliable QM/MM simulation. Thus, all atoms in the MM region are supposed to be photo-inert and will not change too much in different electronic states. For similar reasons, van der Waals and bonded QM-MM interactions for the MM region are also supposed to be reliable. On the other hand, the electrostatic QM-MM interaction for the QM regions is used in our implementation [42], which means the MM charges of atoms in MM region are added into the electronic Hamiltonian of QM atoms. The QM-MM boundary is treated by the link atom method combined with a charge-shift scheme, which is implemented in the Chemshell 3.5 program package [41].

For nonadiabatic couplings and spin–orbit couplings, based on the consideration that all photo-related regions are included in the QM region, the influence of surroundings on both SOC and NAC is supposed to be small. Thus, we only take the nonadiabatic couplings and spin–orbit couplings of the QM components and neglected the MM components contributions. This approximation has been used in other FSSH dynamics simulations and conical intersection searching algorithms at the QM/MM level [43–46], which has been demonstrated to work very well.

6.2.4 Algorithm and Implementation of the Generalized Trajectory Surface Hopping Method

Figure 6.2 shows a brief workflow for one NAMD simulation trajectory carried out by our GTSH program package. Here we introduce the protocol of the GTSH method step by step:

(1) Get all initial coordinates and velocities for all trajectories via ground state molecular dynamics simulations or Wigner Sampling [47];
(2) Initialization all coefficients of electronic states that should be considered in the NAMD simulations;
(3) Do electronic structure calculations for the structure at time t of the NAMD simulations. Energies, gradients, nonadiabatic couplings as well as spin–orbit couplings are supposed to be obtained via external quantum chemistry codes;
(4) Get coordinates of the system at time $t + dt$ via the position step of velocity Verlet method;
(5) Do electronic structure calculations for the structure at time $t + dt$ of the NAMD simulations. Energies, gradients, nonadiabatic couplings as well as spin–orbit couplings are supposed to be obtained via external quantum chemistry codes;
(6) Propagate the coefficients of all electronic states from time t to time $t + dt$. This step is implemented by solving the differential equation, i.e., Eq. (6.26) using the fourth-order Runge–Kutta method;
(7) Judge whether surface hopping processes occurred or not from time t to $t + dt$ according to Eq. (6.29);
(8) Write out necessary information to the log files and exchange data between time t and $t + dt$;
(9) Return to step 4 unless time t reach to the total simulation time.

The above simulation protocol is also used for the QM/MM GTSH simulations.

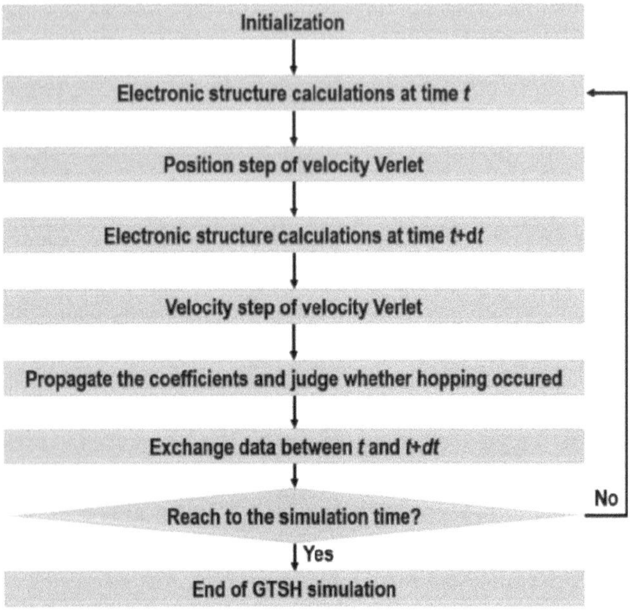

Figure 6.2 Workflow of GTSH method. See Section 6.2.4 for details.

At the beginning of the development of this approach, the GTSH method is used for studying internal conversion and intersystem crossing processes of small organic systems in vacuo (at QM level) or biological systems (at QM/MM level) using multi-configurational ab initio methods (e.g., CASSCF method or OM2/MRCI method) [48–52]. However, these methods are not suitable for large system, such as metal-organic complexes, periodic materials, etc. Three reasons cause this limitation: (i) the computational cost of multi-configurational electronic structure methods increases sharply when the system becomes large; (ii) the number of NACs that are required increases exponentially with the increase of the number of interested excited states (Fig. 6.3); thus, for these kinds of systems, large numbers of electronic states need to be considered, which will cause the calculation of all NACs not affordable; (iii) these multi-configurational electronic structure methods cannot be applied in periodic systems. Thus, developing new methods for solving these three issues is necessary. Motivated by these questions, we have

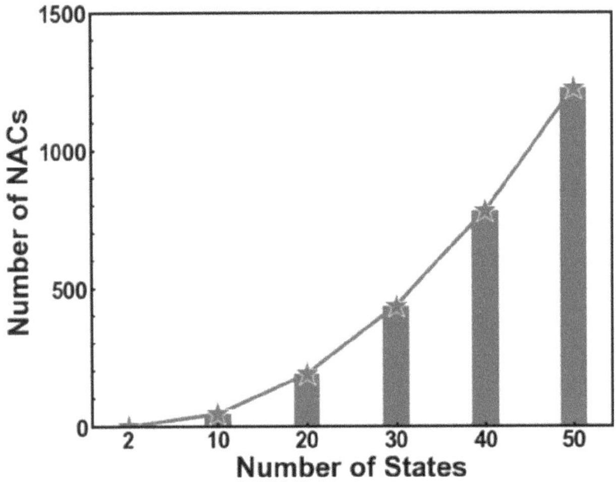

Figure 6.3 Relationship between numbers of NACs needed to be calculated and numbers of interested states.

extended our GTSH method to linear-response (frequency-domain) time-dependent density functional theory (LR-TDDFT) and time-domain density functional theory level with highly efficient NAC and SOC algorithms. In the next section, we will introduce our GTSH method at the linear response time-dependent density functional theory level and their applications for metal-complex systems and zinc phthalocyanine (ZnPc)-fullerene (C_{60}) complexes.

6.3 Generalized Trajectory Surface Hopping Method with Frequency-Domain Time-Dependent Density Functional Theory Method

6.3.1 Linear Response Time-Dependent Density Functional Theory

Nowadays TDDFT is one of the most popular methods for calculating excited state properties of large molecular and biological systems for which high-level multi-configurational or -reference methods cannot afford. A systematic introduction of TDDFT is obviously beyond the scope of this chapter, interested readers are recommended to read relevant books and literatures [53–61]. Here we merely give a brief introduction of the used TDDFT method.

The main idea of TDDFT is similar to the DFT method. In both cases, the electronic density is the key physical quantity, unlike the electronic wavefunction in HF and post-HF electronic structure methods. However, unlike the DFT method focusing on ground-state issues, the TDDFT method can calculate excited-state properties by different formulations in time and frequency domains. In comparison with time-domain ones, frequency-domain LR-TDDFT can calculate state-specific excited-state properties. In addition, LR-TDDFT has a good balance of computational cost and accuracy. Within this LR-TDDFT framework, Casida et al. proposed the widely used Casida equation in 1995 [62]:

$$\begin{pmatrix} \mathbf{A} & \mathbf{K} \\ \mathbf{K} & \mathbf{A} \end{pmatrix} \begin{pmatrix} \mathbf{X} \\ \mathbf{Y} \end{pmatrix} = \Omega \begin{pmatrix} -\mathbf{I} & \mathbf{0} \\ \mathbf{0} & \mathbf{I} \end{pmatrix} \begin{pmatrix} \mathbf{X} \\ \mathbf{Y} \end{pmatrix}. \tag{6.31}$$

Here, the element of matrix **A** and **K** are

$$A_{ia\sigma,i'a'\sigma'}(\Omega) = \delta_{ii'}\delta_{aa'}\delta_{\sigma\sigma'}\omega_{ai\sigma} + K_{ia\sigma,i'a'\sigma'}(\Omega) \tag{6.32}$$

$$K_{ia\sigma,i'a'\sigma'}(\Omega) = \int d^3\mathbf{r} \int d^3\mathbf{r}'\phi^*_{i\sigma}(\mathbf{r})\phi_{a\sigma}(\mathbf{r}) \times f_{Hxc,\sigma\sigma'}(\mathbf{r}, \mathbf{r}', \Omega)\phi^*_{i'\sigma'}(\mathbf{r}')\phi_{a'\sigma'}(\mathbf{r}') \tag{6.33}$$

where i, a are representing the indexes of occupied and unoccupied Kohn–Sham orbitals, and σ represents spin index. Finally, one can get excitation energies of a system through the above Casida equation under approximate exchange-correlation functionals and kernels, adiabatic approximation, etc. [63, 64]. More details about LR-TDDFT and Casida formulations can be found in recent literature [60, 62].

6.3.2 Generalized Trajectory Surface Hopping Method at Linear Response Time-Dependent Density Functional Theory Level

With the LR-TDDFT method, static excited-state calculations of large systems (e.g., metal-organic complexes, metal clusters, etc.) with tens of excited states become possible, which makes ultrafast excited states dynamics simulations using the LR-TDDFT method also feasible. However, a bottleneck of this approach is that the number of nonadiabatic coupling pairs one need to calculate at each dynamic step is large when tens of excited states are considered in the NAMD simulations, as we have mentioned in Section 6.3.1 and Fig. 6.3. Thus, efficient numerical algorithms for nonadiabatic couplings are demanding if one want to do NAMD simulations with tens of excited states. Recently, our group implemented the GTSH method at the LR-TDDFT level with the recently developed algorithm for nonadiabatic couplings [65]. With this method, ultrafast photophysical processes of silicon clusters, metal-organic complex systems are studied via LR-TDDFT-based NAMD simulations. Here we give the algorithm and the final work equation for this efficient nonadiabatic coupling algorithm at the LR-TDDFT level. After that, we present our recently applications with this method.

In the following part of this subsection (Section 6.3.2), we will use *i, j, k*... represent the occupied orbitals; *a, b, c*... represent the unoccupied orbitals (or virtual orbitals); *p, q, r*... represent for any type of orbitals; *K, J* ... represent the indexes of electronic states.

The total time-independent wavefunction of an electronically excited states of a system at LR-TDDFT level can be approximately expressed as the linear combination of the singly excited Slater determinants:

$$\Phi_K = \sum_{ia} \omega_{ia}^K \psi_i^a \tag{6.34}$$

where ω_{ia}^K is the linear combination coefficient of the specific electronic state K of Slater determinant ψ_i^a. The Slater determinants ψ_i^a can be obtained from the ground-state determinant ψ_0 by taking the electron creation and electron annihilation operations on it.

In this framework, the nonadiabatic coupling between electronic states K and J, τ_{KJ}, can be written as

$$\tau_{KJ} = \sum_{ijab} (\omega_{ia}^K \partial_t \omega_{jb}^J \langle \psi_i^a | \psi_j^b \rangle + \omega_{ia}^K \omega_{jb}^J \langle \psi_i^a | \partial_t \psi_j^b \rangle). \tag{6.35}$$

According to the Slater–Condon rules, the first term in Eq. (6.35) can be simplified to $\sum_{ia} \omega_{ia}^K \partial_t \omega_{ia}^J$. And the time differentiation of ψ_j^b is

$$\partial_t \psi_j^b = \sum_{k \neq j} \psi_{jk}^{bk'} + \psi_j^{b'} \tag{6.36}$$

where the meaning of $\psi_j^{b'}$ is that the molecular orbital (MO) ϕ_j is replaced with the time derivative $\partial_t \phi_b$. Therefore, the $\langle \psi_i^a \mid \partial_t \psi_j^b \rangle$ in the second term of Eq. (6.35) becomes

$$\langle \psi_i^a | \partial_t \psi_j^b \rangle = \sum_{k \neq j} \langle \psi_i^a | \psi_{jk}^{bk'} \rangle + \langle \psi_i^a | \psi_j^{b'} \rangle \tag{6.37}$$

and the second term on the right side of Eq. (6.37) can be expressed as

$$\langle \psi_i^a | \psi_j^{b'} \rangle = \delta_{ij} \langle \phi_a | \partial_t \phi_b \rangle. \tag{6.38}$$

Because only one term (i.e., $k = i$ and $a = b$) is not zero due to the orthogonality conditions ($\langle \phi_p \mid \partial_t \phi_p \rangle = 0$ and $\langle \phi_p \mid \phi_q \rangle = \delta_{pq}$). With Eq. (6.37) and Eq. (6.38), we can finally get

$$\langle \psi_i^a | \partial_t \psi_j^b \rangle = \delta_{ij} \langle \phi_a | \partial_t \phi_b \rangle - P_{ij} \delta_{ab} \langle \phi_j | \partial_t \phi_i \rangle \tag{6.39}$$

where P_{ij} is an additional phase factor which is depending on the ordering convention of the molecular orbitals in Slater determinants. Moreover, one can use the finite-difference algorithm to calculate the time differentiation on the molecular orbitals:

$$\langle \phi_p | \partial_t \phi_q \rangle = \frac{\langle \phi_p(t) | \phi_q(t + \Delta t) \rangle}{\Delta t} \tag{6.40}$$

where $\phi_p(t)$ and $\phi_q(t + \Delta t)$ are representing the molecular orbitals at time t and time $t + \Delta t$, respectively. Substituting Eq. (6.36) to Eq. (6.40) into Eq. (6.35), one can get the final work equation for the calculation of nonadiabatic coupling:

$$\begin{aligned} \tau_{KJ} &= \sum_{ia} \omega_{ia}^K \partial_t \omega_{ia}^J + \sum_{iab} \omega_{ia}^K \omega_{ib}^J \langle \phi_a | \partial_t \phi_b \rangle \\ &\quad - \sum_{ija} P_{ij} \omega_{ia}^K \omega_{ja}^J \langle \phi_j | \partial_t \phi_i \rangle \end{aligned} \tag{6.41}$$

More details of this numerical algorithm for nonadiabatic couplings can be found in the original and our recent works [65, 66].

Except for nonadiabatic couplings, spin–orbit couplings are also very important. They have to be added into the total electronic Hamiltonian in NAMD simulations when intersystem crossing processes between singlets and triplets play important roles. In order to calculate SOCs efficiently at the LR-TDDFT level, we have implemented an efficient algorithm for calculating spin–orbit couplings using the single-electron effective operator with an effective charge approximation [67–70]. In this method, the approximate Hamiltonian is written as [69–72]

$$\hat{H}_{SO}^{eff} = \frac{e^2}{2m_e^2c^2}\sum_{i}^{N_{el}}\sum_{\alpha}^{N_A} Z_{\alpha}^{eff}\left(\frac{\mathbf{r}_{i\alpha}}{r_{i\alpha}^3}\times\mathbf{p}_i\right)\cdot\widehat{\mathbf{s}}_i \equiv \sum_{i}^{N_{el}}\xi(r_i)\widehat{\mathbf{l}}_i\cdot\widehat{\mathbf{s}}_i \tag{6.42}$$

where $\xi(r_i) = \frac{e^2}{2m_e^2c^2}\sum_{\alpha}^{N_A}\frac{Z_{\alpha}^{eff}}{r_{i\alpha}^3}$ is a radial function while Z_{α}^{eff} is effective charge of nucleus α. $\hat{\mathbf{s}}_i$ is the spin angular momentum operator of electron i and $\hat{\mathbf{l}}_i = \mathbf{r}_{i\alpha}\times\mathbf{p}_i$ is the orbital angular momentum operator. The effective single-particle Hamiltonian in Eq. (6.42) can be written as following with the terminologies of the second quantization theory:

$$\hat{H}_{SO}^{eff} = \sum_{pq}\sum_{\sigma\sigma'}\langle\phi_{pq}|\xi(r)(\hat{l}_x\hat{s}_x+\hat{l}_y\hat{s}_y+\hat{l}_z\hat{s}_z)|\phi_{q\sigma'}\rangle a_{pq}^{\dagger}a_{q\sigma}. \tag{6.43}$$

Here, $|\phi_{q\sigma}\rangle = |\phi_q\rangle\otimes\sigma$ where $|\phi_q\rangle$ is the Kohn–Sham orbital and σ is the spin orbital (i.e., α spin and β spin). We use the Pauli matrices for the spin angular operator $\hat{\mathbf{s}}$:

$$s_x = \frac{1}{2}\begin{pmatrix}0 & 1\\ 1 & 0\end{pmatrix},\ s_y = \frac{1}{2}\begin{pmatrix}0 & -i\\ i & 0\end{pmatrix},\ s_z = \frac{1}{2}\begin{pmatrix}1 & 0\\ 0 & -1\end{pmatrix} \tag{6.44}$$

After mathematical derivations, the x, y, z components of approximate Hamiltonian $\hat{H}_{SO}^{eff}$ in Eq. (6.43) are expressed as

$$\hat{H}_{SO}^{x} = \frac{1}{2}\sum_{pq}h_{pq}^{x}(a_{p\alpha}^{\dagger}a_{q\beta}+a_{p\beta}^{\dagger}a_{q\alpha}) \tag{6.45}$$

$$\hat{H}_{SO}^{y} = \frac{1}{2i}\sum_{pq}h_{pq}^{y}(a_{p\alpha}^{\dagger}a_{q\beta}-a_{p\beta}^{\dagger}a_{q\alpha}) \tag{6.46}$$

$$\hat{H}_{SO}^{z} = \frac{1}{2}\sum_{pq}h_{pq}^{z}(a_{p\alpha}^{\dagger}a_{q\alpha}-a_{p\beta}^{\dagger}a_{q\beta}) \tag{6.47}$$

where the creation operators $\{a_{p\alpha}^{\dagger}\ldots\}$ and the annihilation operators $\{a_{q\beta}\ldots\}$ has the commutation relation $a_{p\sigma}^{\dagger}a_{q\sigma'} = \delta_{pq}\delta_{\sigma\sigma'}$, and the spin–orbit coupling matrix elements $h_{pq}^{x/y/z} \equiv \langle\phi_p|\xi(r)\hat{l}_{x/y/z}|\phi_q\rangle$. Now we briefly introduce how to build the electronically configuration of excited singlet and triplet states using Kohn–Sham orbitals. The singlet configuration $|S_i^a\rangle$ and triplet configurations

$|T_j^b\rangle_{1,-1/0/1}$ are produced by taking excitation operator on the closed-shell ground state Slater determinant $|GKS\rangle$:

$$|S_i^a\rangle = \frac{1}{\sqrt{2}}(a_{a\alpha}^\dagger a_{i\alpha} + a_{a\beta}^\dagger a_{i\beta})|GKS\rangle \tag{6.48}$$

$$|T_j^b\rangle_{1,1} = -a_{b\alpha}^\dagger a_{j\beta}|GKS\rangle \tag{6.49}$$

$$|T_j^b\rangle_{1,-1} = a_{b\beta}^\dagger a_{j\alpha}|GKS\rangle \tag{6.50}$$

$$|T_j^b\rangle_{1,0} = \frac{1}{\sqrt{2}}(a_{b\alpha}^\dagger a_{j\alpha} - a_{b\beta}^\dagger a_{j\beta}|GKS\rangle \tag{6.51}$$

where $\{i,\ j,\ k\ \ldots\}$ are occupied Kohn–Sham orbitals and $\{a,\ b,\ c\ldots\}$ are unoccupied Kohn–Sham orbitals. In LR-TDDFT framework, the wavefunction of an electronically excited states of a system is approximately written as the linear combination of the singly excited Slater determinants (Eq. (6.34)). With Casida wavefunction ansatz (Eq. (6.31)) [62], $\omega_{ia}^K \equiv N_K^{-1/2}(X+Y)_{ia}^K$. $(X+Y)_{ia}^K$ is the coefficient vector for the K-th excitation in Eq. (6.31) and $N_K = \langle (X+Y)^K \mid (X+Y)^K \rangle$, which is the normalization factor. Here, we have to stress that in the framework of collinear LR-TDDFT, the triplet substates with magnetic quantum numbers $m = 0$ can be obtained while substates with $m = \pm 1$ are approximated [73]. Finally, by using the Wick's theorem [74], the spin–orbit coupling matrix elements between excited singlet and triplet states can be written as

$$\langle S_I|\hat{H}_{SO}|T_J\rangle_{1,1} = \frac{1}{2\sqrt{2}}\sum_{ija}(\omega_{ia}^I)^\dagger \omega_{ja}^J(h_{ji}^x + ih_{ji}^y) - \frac{1}{2\sqrt{2}}\sum_{abi}(\omega_{ia}^I)^\dagger \omega_{ib}^J(h_{ab}^x + ih_{ab}^y) \tag{6.52}$$

$$\langle S_I|\hat{H}_{SO}|T_J\rangle_{1,-1} = -\langle S_I|\hat{H}_{SO}|T_J\rangle_{1,1} \tag{6.53}$$

$$\langle S_I|\hat{H}_{SO}|T_J\rangle_{1,0} = -\frac{1}{2}\sum_{ija}(\omega_{ia}^I)^\dagger \omega_{ja}^J h_{ji}^z + \frac{1}{2}\sum_{abi}(\omega_{ia}^I)^\dagger \omega_{ib}^J h_{ab}^z \tag{6.54}$$

In our implementation, the matrix element of $h_{pq}^d = \sum_{uv}^N c_{pu}^\dagger h_{uv}^d c_{vq}$ between orbitals $|\phi_p\rangle$ and $|\phi_q\rangle$ can be calculated by expanding atomic basis set $\{|\chi_u\rangle, |\chi_v\rangle \ldots\}$ with relationship $|\phi_p\rangle = \sum_u c_{up}|\chi_u\rangle$. h_{uv}^d are calculated by the open-source package MolSOC [75]. More details can be found in our recently published paper [76].

6.3.3 Applications

With the implementation of the above new algorithms and integrating them into our GTSH package, we have studied several complex and large systems which have interesting ultrafast photophysical processes [66, 76–80].

Silicon quantum dots have wide applications in optoelectronics, solar energy, biomedicine, etc. [81, 82]. However, the surface functionalization of this kind of silicon quantum dots may change the nonradiative processes. For an example, saturated dangling bonds with hydrogens or active chromophores always leads to ultrafast decay to the lowest excited state and high photoluminescent quantum yields [83]. Our developed GTSH method is very suitable for studying this interesting ultrafast photophysical processes. Thus, we have selected one chlorine-capped and two nitrogen-capped silicon nanoparticles (i.e., Cl@SiQD, Car@SiQD, Azo@SiQD) reported by Li et al. as our research system (Fig. 6.4) [84]. 500 fs NAMD simulations are performed at the TD-CAM-B3LYP level for these silicon quantum dots [85]. The simulation results show that Cl@SiQD has ultrafast internal conversion processes in <190 fs and decays to the lowest excited singlet state directly. For Car@SiQD system, the S_2 state is trapped >300 fs while for Azo@SiDQ, the S_3 state is trapped >615 fs (see Figs. 6.5 and 6.6). These excited state relaxation behaviors are in excellent agreement with the

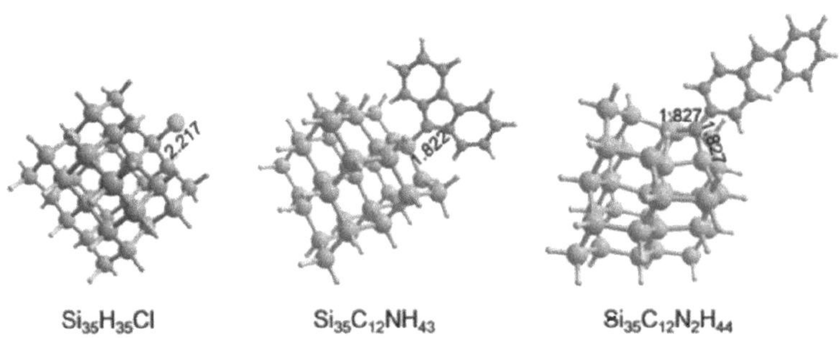

Figure 6.4 Three silicon nanoclusters: (I) $Si_{35}H_{35}Cl$, i.e., Cl@SiQD; (II) $Si_{35}C_{12}NH_{43}$, i.e., Car@SiQD; (III) $Si_{35}C_{12}N_2H_{44}$, i.e., Azo@SiQD. Copyright 2018 Taylor & Francis Group.

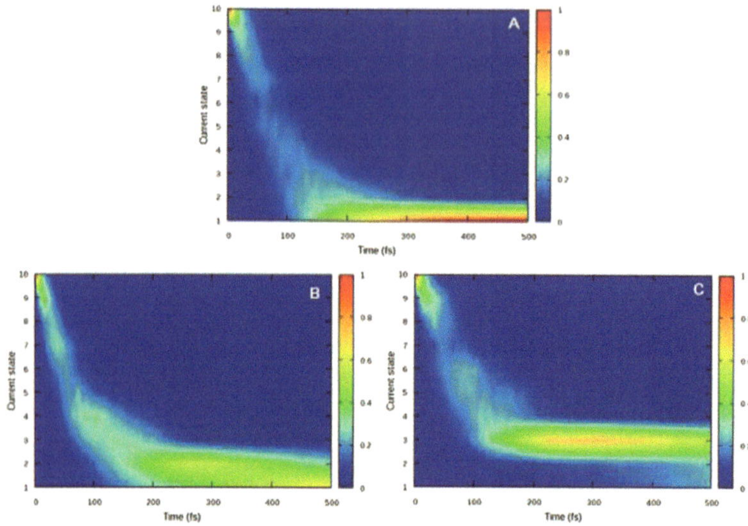

Figure 6.5 Contour plot of time dependent probability density of the current electronic state of (A) Cl@SiQD, (II) Car@SiQD and (III) Azo@SiQD in 500 fs NAMD simulations. Copyright 2018 Taylor & Francis Group.

experimental data and provide new insights for understanding the luminescence properties of silicon quantum dots [66].

Due to the wide range of applications in OLEDs, the iridium (III) complexes have drawn a lot of theoretical and experimental attention [86–91]. Recently, Pomario and coworkers systematically studied the Ir(III) complexes at the femtosecond timescale [92]. However, nonadiabatic dynamics simulations for these complexes have not been reported. We have used the GTSH method to study these Ir complexes (Fig. 6.7). In the NAMD simulations, 5 lowest singlet excited states and 10 lowest triplet excited states are involved. From the time-dependent population of these three Ir complexes (Fig. 6.8), we can see that all three complexes have ultrafast internal conversion and intersystem crossing processes between singlets and triplets, which are caused by large nonadiabatic couplings, spin–orbit couplings and small energy gaps among excited states. Although all these complexes have ultrafast ISC processes, the relaxation processes are different. For **Ir1**, the system decays to T_1 state very fast after the ISC process. For **Ir2**, system decays to

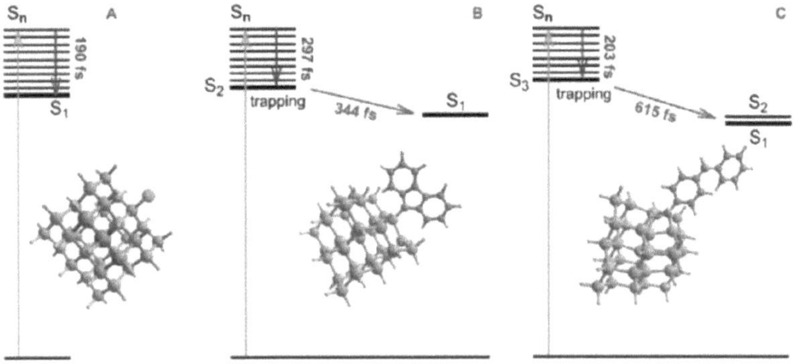

Figure 6.6 Schematic relaxation mechanisms of (A) Cl@SiQD, (B) Car@SiQD and (C) Azo@SiQD based on NAMD simulations. Copyright 2018 Taylor & Francis Group.

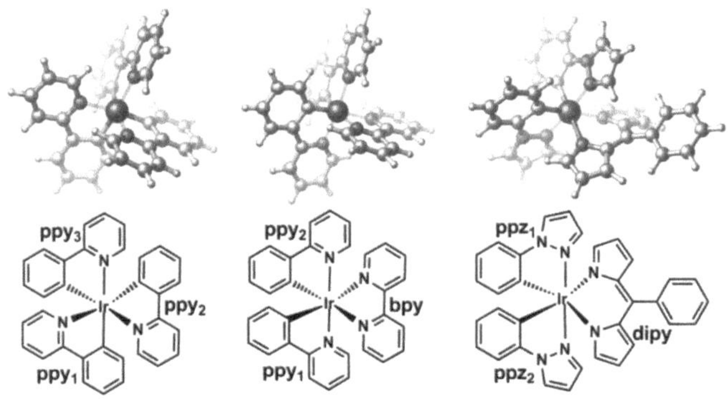

Figure 6.7 Three Ir complexes studied by nonadiabatic molecular dynamics simulations, referred as **Ir1**, **Ir2** and **Ir3**. Copyright 2018 American Chemical Society.

T_2 state after the ISC process, then the T_2-S_1 intersystem crossing process occurs, followed by the S_1-T_1 intersystem crossing process happened after the ISC process between T_2-S_1. For **Ir3**, the system traps at T_2 state for 500 fs timescale (Fig. 6.8). With our NAMD simulations, we gave the useful insights of the relaxation processes of Ir(III) complexes with different ligands, which could be helpful for the design of new Ir(III) complexes based materials [78].

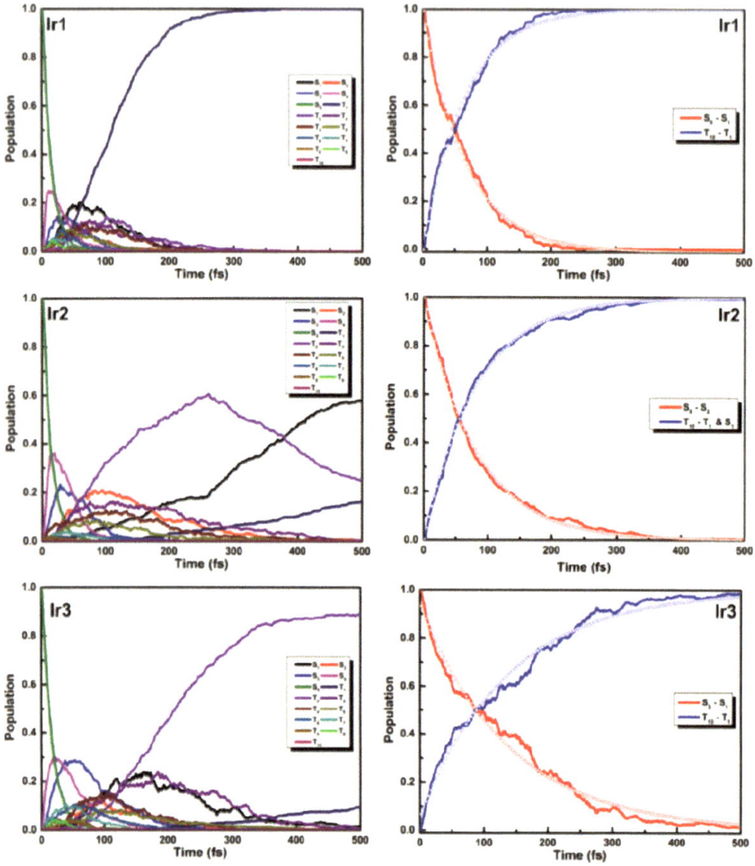

Figure 6.8 Time dependent populations **Ir1**, **Ir2** and **Ir3**. Copyright 2018 American Chemical Society.

Au(I)-C σ bond could significantly change the excited states electronic structures and excited states relaxation dynamics [93–97]. Recently, Crespo-Hernández et al. have studied relaxation dynamics of three similar Au(I) complexes by the femtosecond transient absorption spectroscopic techniques (Fig. 6.10) [98]. Their experiments show that these Au(I) complexes have ultrafast ISC processes around 350 fs. However, the details of the relaxation mechanism remain unclear. Thus, we performed the NAMD simulations at the TDDFT level for these three complexes. The results show

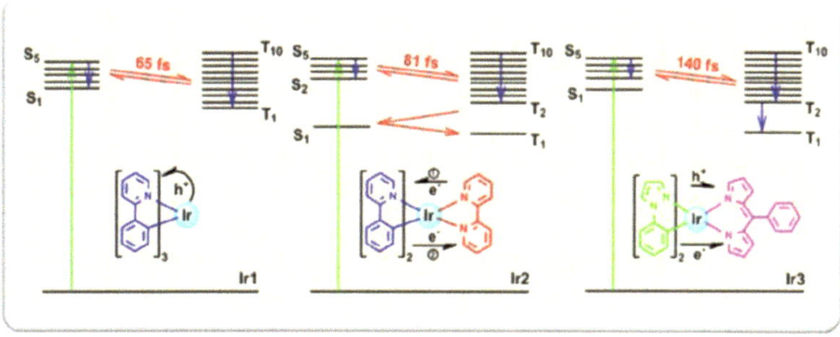

Figure 6.9 Excited-state relaxation mechanisms of **Ir1**, **Ir2**, and **Ir3** based on NAMD simulations. Copyright 2018 American Chemical Society.

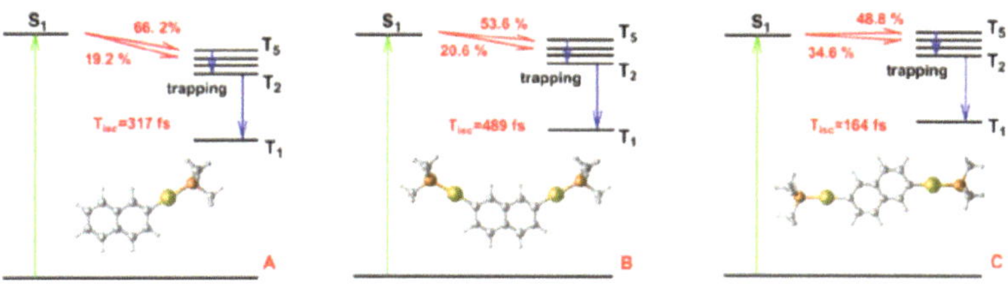

Figure 6.10 Three gold(I) complexes **Au1**, **Au2** and **Au3** studied by nonadiabatic molecular dynamics simulations and proposed relaxation mechanism. Copyright 2018 American Institute of Physics.

that S_1 state decays with time constants of 317, 489 and 164 fs for **Au1**, **Au2** and **Au3**, respectively (Fig. 6.11). Further analysis shows that **Au1** and **Au2** mainly hop from S_1 state to T_5 state (66.2% and 53.6%) while the S_1-T_4 ISC processes are much less (19.2% and 20.6%). In contrast, for **Au3**, the contributions of S_1-T_5 and S_1-T_4 transitions are similar (34.6% and 48.8%). We further analyze the energy gaps and spin–orbit couplings between S_1 and T_n states ($n = 3, 4, 5$) (Figs. 6.12 and 6.13). These data show that the smaller energy gaps and larger SOC cause the faster S_1-T_n ISC processes for **Au3** (164 fs). The smaller SOC values make the ISC processes of **Au2** slower (Fig. 6.10). These results and conclusions may provide new insights for understanding the photophysical properties and designing new Au(I) compounds with better properties [77].

The photoinduced charge separation dynamics of zinc phthalocyanine (ZnPc)-fullerene (C_{60}) complexes can be influenced

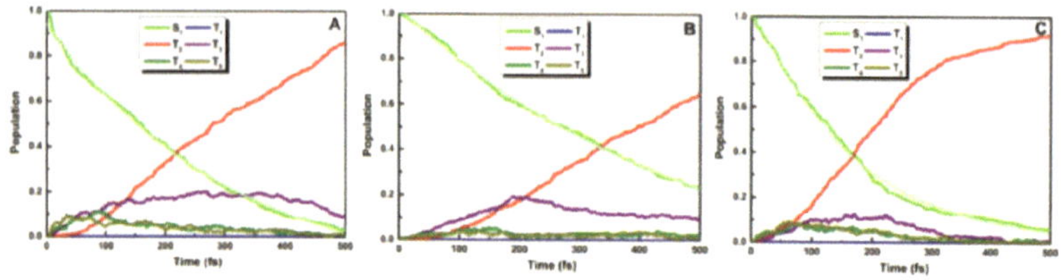

Figure 6.11 Time dependent populations of (A) **Au1**, (B) **Au2** and (C) **Au3** based on NAMD simulations. Copyright 2018 American Institute of Physics.

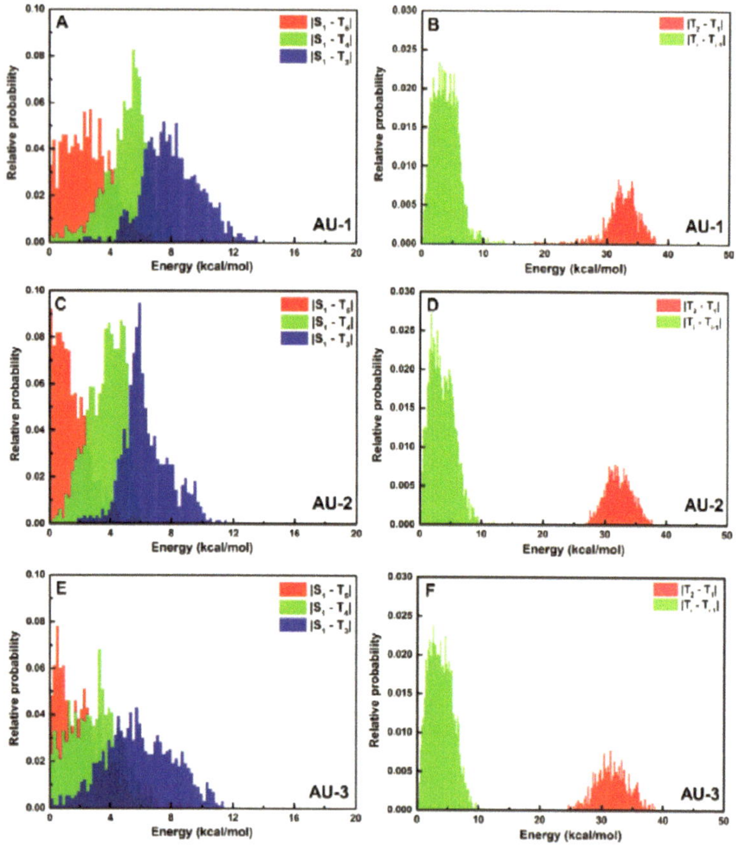

Figure 6.12 Distributions of energy gaps (left panels) between S_1 and T_n (n = 3–5) and (right panels) between different triplet states of (A and B) **Au1**, (C and D) **Au2** and (E and F) **Au3** in NAMD simulations. Copyright 2018 American Institute of Physics.

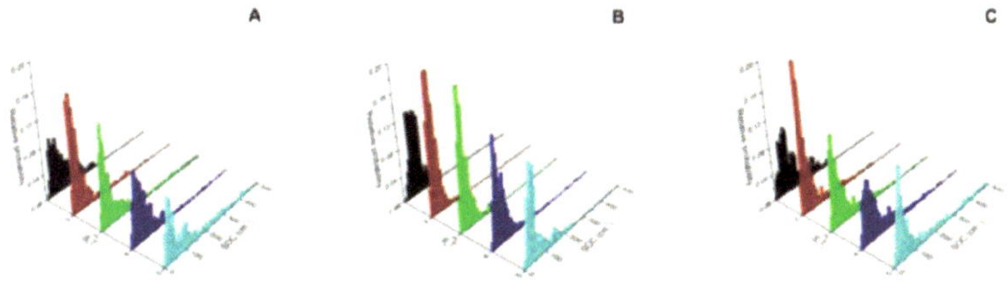

Figure 6.13 Distributions of SOC (in cm^{-1}) between S_1 and T_n ($n = 1$–5) states of (A) **Au1**, (B) **Au2** and (C) **Au3** in NAMD simulations. Copyright 2018 American Institute of Physics.

remarkably by the spatial orientations and the bonding configurations between ZnPc and C_{60} [99–103]. However, the underlying physical origin is unclear. Recently, we have employed the TDDFT-based NAMD simulations with two kinds of ZnPc-C_{60} complexes (i.e., noncovalently bonded ZnPc-C_{60} complexes with different space orientations and covalently bonded ZnPc-C_{60} dyads with different bonding configurations).

For the noncovalently ZnPc-C_{60} complexes, we have selected two different space orientations, which are called face-on and edge-on configurations, respectively (Fig. 6.14). With electronic structure calculations at the TDDFT level, we find that both face-on and edge-on ZnPc-C_{60} complexes are excited to local excitation (LE) states within ZnPc. However, the LE states of the face-on configuration are the S_3 and S_4 states while the edge-on configuration are the S_1 and S_2 states. For the face-on configuration, the two lowest singlet

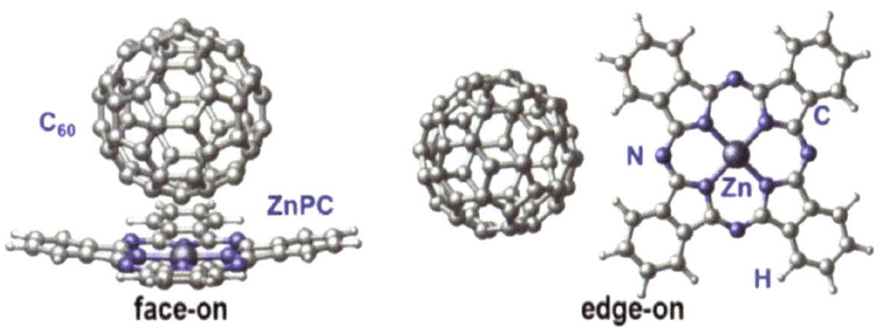

Figure 6.14 Structures of ZnPc and C_{60} with (left) face-on and (right) edge-on configurations. Copyright 2020 American Chemical Society.

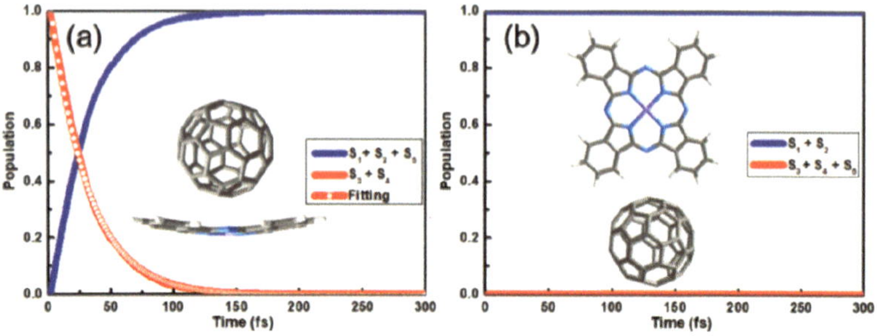

Figure 6.15 Time dependent state populations of (a) face-on and (b) edge-on configurations ZnPc-C_{60}. Copyright 2020 American Chemical Society.

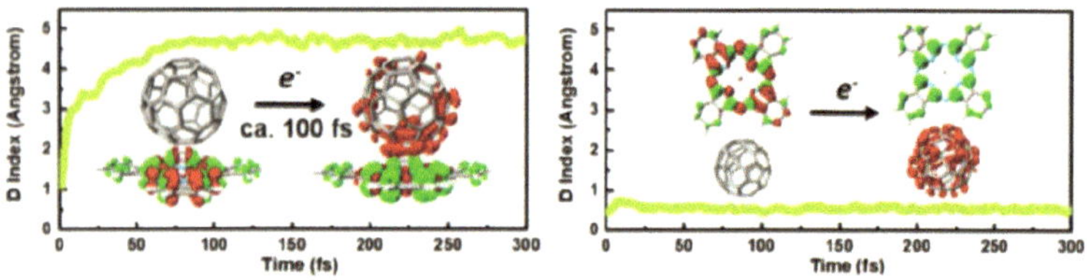

Figure 6.16 Time dependent distance of centroids of electron and hole of (a) face-on and (b) edge-on configurations ZnPc-C_{60}. Copyright 2020 American Chemical Society.

excited states S_1 and S_2 are charge transfer (CT) states. These CT states are supposed to be populated during the relaxation processes of the face-on ZnPc-C_{60} complex. For the edge-on configuration, the energies of the CT states are much higher than the LE states, so the charge separation dynamics is not favorable. Moreover, 300 fs NAMD simulations at the TD-CAM-B3LYP level are performed for both face-on and edge-on ZnPc-C_{60} (Fig. 6.15). The NAMD results are consistent with the electronic structure calculations. In the face-on configuration, exciton transfer from LE states to CT states occurs within ca. 100 fs. On the other hand, for the edge-on configuration, no exciton transfer and charge separation processes happen in 300 fs NAMD simulations (Fig. 6.16) [80].

Except for different space orientations of ZnPc and C_{60}, we have also studied the influence of different bonding configurations of ZnPc-C_{60} dyads. Herein, we have selected the covalently bonded ZnPc-C_{60} dyads with 5-6 and 6-6 bonding configurations to illustrate

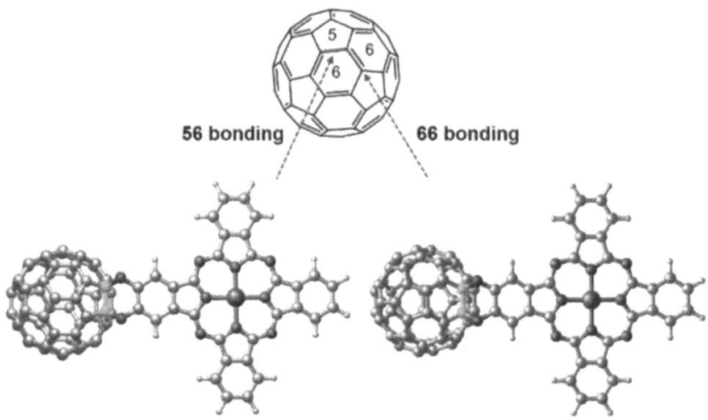

Figure 6.17 Structures of ZnPc and C_{60} with (left) 5-6 bonding and (right) 6-6 bonding configurations. 5-6 bonding means two oxygen atoms of ZnPc are bonded to carbons by both hexagonal and pentagonal rings while 6-6 bonding means two oxygen atoms of ZnPc are bonded to carbons by two hexagonal rings. Copyright 2020 Wiley-VCH GmbH.

the effects of bonding configurations (Fig. 6.17). Similar with the noncovalently bonded ZnPc-C_{60} complexes, the LE states within ZnPc of both 5-6 bonding and 6-6 bonding ZnPc-C_{60} dyads are spectroscopically bright states. For 5-6 bonding ZnPc-C_{60}, the energies of the CT states are lower than the LE states. On the contrary, the energies of the CT states are much higher than the LE states within ZnPc in 6-6 bonding configuration (Fig. 6.18). This makes that no exciton transfer processes occur for the 6-6 bonding ZnPc-C_{60} and the exciton will be trapped at the LE states within ZnPc, which is similar with the edge-on orientation ZnPc-C_{60} complex. However, in 500 fs NAMD simulations, the charge separation process and the following energy transfer to fullerene occurs in an ultrafast way for the 5-6 bonding configuration (Fig. 6.19). In one word, these theoretical findings of different behaviors of ZnPc-C_{60} dyads could provide valuable knowledge to help design new heterojunctions with different target properties in the future [79].

From these applications, one can see that the GTSH method at the LR-TDDFT level is reliable and very useful to simulate many ultrafast photophysical processes of large interesting molecules and clusters, etc.

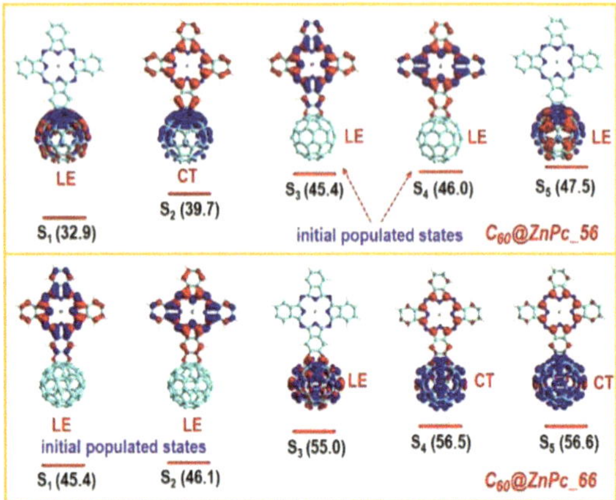

Figure 6.18 Electronic transition densities and excitation energies (in kcal/mol) of S_1-S_5 states of the (up) 5-6 bonding and (bottom) 6-6 bonding configurations. Copyright 2020 Wiley-VCH GmbH.

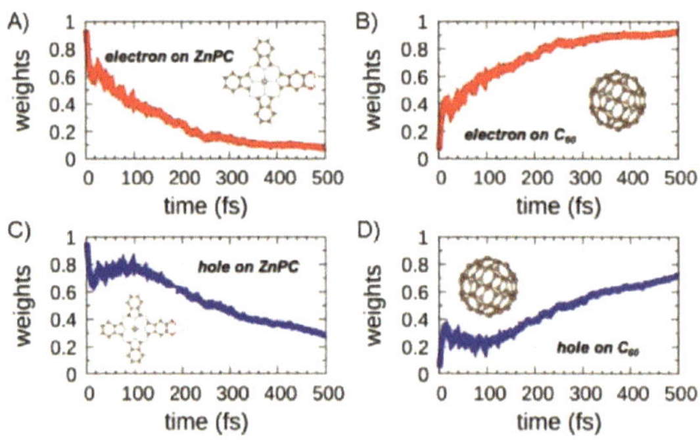

Figure 6.19 Time dependent electron and hole weights based on NAMD simulations of ZnPc and C_{60} fragments of 5-6 configuration. Copyright 2020 Wiley-VCH GmbH.

6.4 Generalized Trajectory-Based Surface-Hopping Method with Time-Domain Time-Dependent Density Functional Theory Method

Although we have developed the GTSH method at the LR-TDDFT level, which has the ability to conduct nonadiabatic dynamics simulations for the systems within 100-200 atoms, it still has limitations. For example, the behavior of photoinduced charge-transfer processes at the interface of heterojunctions is very important in materials, but the size of this kind of materials is huge (up to several μm or more). In addition, theoretical studies of these materials must be considered under the periodic boundary condition (PBC). Obviously, nowadays, the periodic LR-TDDFT method is not a very suitable electronic structure method for large materials simulations. Instead, an efficient NAMD method with plane-wave basis sets at the time-domain TDDFT level has been proposed by Prezhdo et al., which is beneficial for simulating photoinduced carrier dynamics (electron and hole) of periodic complex systems [104, 105]. Similarly, in recent years, we have also developed and implemented this kind of NAMD method but with localized Gaussian-like basis sets and have extended it to include the SOC effects to treat metal-containing periodic systems with remarkable SOC effects using collinear and noncollinear DFT methods. Moreover, we have also made several successful applications to manifest the methodological advantages. Next, we will introduce briefly the algorithms and applications of the developed method.

6.4.1 Time-Domain Time-Dependent Density Functional Theory

In the theoretical framework of TDDFT, the most important quantity is the electronic density. The time-dependent electron density $\rho(\mathbf{r}, t)$ can be expressed as the summation of all the occupied time-dependent single-electron Kohn–Sham orbitals $\phi(\mathbf{r}, t)$ [61]:

$$\rho(\mathbf{r}, t) = \sum_{n=1}^{N_e} |\phi_n(\mathbf{r}, t)|^2 \tag{6.55}$$

where N_e represents the number of electrons and $\mathbf{r}$ represents the coordinates of electrons. Meanwhile, for $n = 1, 2, \ldots, N_e$, the single-electron Kohn–Sham orbitals $\phi_n(\mathbf{r}, t)$ is satisfied with the time-domain single particle Kohn–Sham equation:

$$i\hbar\frac{\partial\phi_n(\mathbf{r}, t)}{\partial t} = \hat{H}(\mathbf{r}, \mathbf{R}, \phi)\phi_n(\mathbf{r}, t). \tag{6.56}$$

The time-dependent Kohn–Sham orbitals $\phi_n(\mathbf{r}, t)$ can be expanded by a set of adiabatic Kohn–Sham orbitals

$$\phi_n(\mathbf{r}, t) = \sum_k c_k(t)\chi_k(\mathbf{r}, \mathbf{R}) \tag{6.57}$$

Take Eq. (6.57) into Eq. (6.56) and one can get the equations of motion (EOM) of the time-dependent expansion coefficients $c_j(t)$:

$$i\hbar\frac{\partial c_j(t)}{\partial t} = \sum_k c_k(t)(\epsilon_k\delta_{jk} - i\hbar d_{jk}) \tag{6.58}$$

in which the ϵ_k is the energy of the k-th adiabatic state of the system, d_{jk} is the nonadiabatic coupling (NAC) between adiabatic states j and k. The NAC term arises because the adiabatic Kohn–Sham orbitals depend on the nuclear coordinates. One can find that the above EOM is very similar to Eq. (6.15) for electronically excited and ground states except that the Kohn–Sham orbital energies and NAC terms between Kohn–Sham orbitals are used in Eq. (6.58).

6.4.2 Generalized Trajectory-Based Surface-Hopping at Time-Domain Time-Dependent Density Functional Theory Level

In previous works, Prezhdo et al. have derived the above EOM and implemented the algorithm with plane-wave basis sets [6, 104, 106–108]. Nonadiabatic couplings are calculated by the real-space grid integration. Here, we have developed and implemented this kind of method with localized Gaussian-like basis sets. In such case, nonadiabatic coupling d_{jk} in Eq. (6.58) is calculated numerically as overlaps of wavefunctions of adiabatic states at time t and $t + dt$ by a finite difference method:

$$\begin{aligned} d_{jk} &= \langle\phi_j(\mathbf{r}, \mathbf{R})|\frac{\partial\phi_k(\mathbf{r}, \mathbf{R})}{\partial t}\rangle \\ &\approx \frac{\langle\phi_j(t)|\phi_k(t+\Delta t)\rangle - \langle\phi_j(t+\Delta t)|\phi_k(t)\rangle}{2\Delta t} \end{aligned} \tag{6.59}$$

Importantly, in some materials systems, spin–orbit coupling (SOC) effects play important roles in optoelectronic properties, e.g., Pb-containing perovskite materials [109–112]. In such situation, the SOC effects must be included in the DFT calculations. In order to consider the SOC effects, we have used the two-component noncollinear DFT method to calculate related physical quantities for propagating the EOM, Eq. (6.58) in NAMD simulations [113–116]. In this method, an orbital ϕ_k is represented using two spinors, which naturally mix $\alpha(\phi_k^\alpha)$ and $\beta(\phi_k^\beta)$ components:

$$\phi_k = \begin{pmatrix} \phi_k^\alpha \\ \phi_k^\beta \end{pmatrix}. \tag{6.60}$$

The expression of overlaps of wavefunctions of adiabatic states within the framework of two-component noncollinear DFT method is now written as:

$$\begin{aligned}\langle \phi_j(t)|\phi_k(t+\Delta t)\rangle &= \begin{pmatrix} \phi_j^\alpha(t) & \phi_j^\beta(t) \end{pmatrix} \begin{pmatrix} \phi_k^\alpha(t+\Delta t) \\ \phi_k^\beta(t+\Delta t) \end{pmatrix} \\ &= \langle \phi_j^\alpha(t)|\phi_k^\alpha(t+\Delta t)\rangle + \langle \phi_j^\beta(t)|\phi_k^\beta(t+\Delta t)\rangle.\end{aligned} \tag{6.61}$$

On the other hand, in order to measure charge-transfer processes quantitatively between different components or fragments during NAMD simulations, a density-matrix-based method has been proposed and developed by us [117]. First of all, a density matrix D is defined as:

$$D_{\mu v i}(t) = p_i(t)\chi_{\mu i}\chi_{v i}^* \tag{6.62}$$

where $\chi_{\mu i}$ is the atomic orbital coefficient of μ-th atom of the i-th adiabatic state and the $p_i(t)$ represents the occupation number of the i-th adiabatic state based on the time-dependent coefficients $c_i(t)$. Then, we redefine a population matrix P, which is similar to the Mulliken charge analysis, using the density-matrix D and the atomic overlap matrix S:

$$P_{\mu v i} = D_{\mu v i}S_{\mu v} \tag{6.63}$$

The a-th atomic charge P_a is the summation of basis functions of the a-th atom and all involved adiabatic electronic states i

$$P_a = \sum_i \left(\sum_{\mu\in a, v\in a} P_{\mu v i} + \frac{1}{2}\left(\sum_{\mu\in a, v\notin a} P_{\mu v i} + \sum_{\mu\notin a, v\in a} P_{\mu v i} \right)\right) \tag{6.64}$$

Once one has the atomic charges, the charges of the specific fragment A can be obtained by summing all atomic charges of the atoms in fragment A:

$$P_A = \sum_{a\in A} P_a. \tag{6.65}$$

If we expand Eq. (6.65) by taking Eq. (6.63)–(6.64) into it, we can rewrite P_A as

$$P_A = \sum_i p_i(t) \left(\sum_{a \in A} \left(\sum_{\mu \in a, \nu \in a} \chi_{\mu i} \chi_{\nu i}^* S_{\mu\nu} \right) \right.$$
$$\left. + \frac{1}{2} \left(\sum_{\mu \in a, \nu \notin a} \chi_{\mu i} \chi_{\nu i}^* S_{\mu\nu} + \sum_{\mu \notin a, \nu \in a} \chi_{\mu i} \chi_{\nu i}^* S_{\mu\nu} \right) \right) \quad (6.66)$$

The differential of P_A is written as

$$dP_A = d(\sum_i c_i^* c_i P_A) = \sum_i (d(c_i^* c_i) P_{Ai} + c_i^* c_i dP_{Ai}). \quad (6.67)$$

Here, the first term represents the hoppings between different electronic states, which corresponds to the nonadiabatic electronic transfer, and the second term represents the electron population changes in one adiabatic state induced by motions of atoms, which corresponds to the adiabatic electronic transfer.

6.4.3 Applications with Collinear and Noncollinear DFT Methods

Using the time-domain TDDFT-based GTSH method with collinear and noncollinear DFT approaches, we have studied several kinds of ultrafast photoinduced charge transfer processes of interesting heterojunctions, including two-dimensional transition metal dichalcogenides (TMDs), fullerenes, carbon nanotubes, perovskites, etc. [117–125]. Next we will briefly discuss the important findings gained by the NAMD simulations.

TMDs can form mix-dimensional heterostructures with zero- or one-dimensional organic or inorganic molecules. This kind of mixed dimensional organic-inorganic heterostructures possess novel and excellent optoelectronic properties and have many potential applications in light-emitting diodes, photodetectors, and photovoltaic devices [126–129]. Due to its tunable optoelectronic properties, molybdenum disulfide (MoS_2) is one of the most studied TMDs experimentally and theoretically in the past several years [130–133].

Recently, ZnPc-MoS_2 has been studied experimentally by Chan et al. via ultrafast photoluminescence spectroscopic techniques

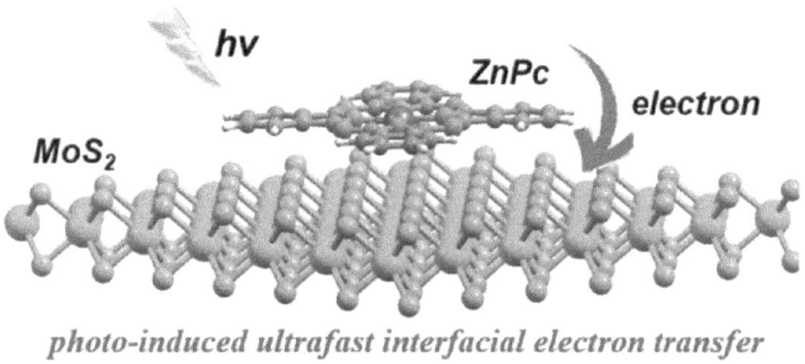

Figure 6.20 Structure of ZnPc-MoS_2 complex. Copyright 2018 American Chemical Society.

(Fig. 6.20) [134]. Their work shows that ultrafast electron transfer process from ZnPc to MoS_2 occurs within 100 fs. Obviously, nonadiabatic dynamics simulations could give more details of the mechanism and dynamical processes. Thus, we performed the NAMD simulations at the PBE+D3 level. The ZnPc is adsorbed to the monolayer MoS_2 in a parallel orientation, which gives ca. 3 eV adsorption energy. Our calculations show that ZnPc is excited by photons, which agrees with the experimental data. Then, an ultrafast electron transfer from the LUMO of ZnPc to conduction band of MoS_2 occurs in ca. 10 fs (Fig. 6.21). In the electron transfer process, we have found that the low-frequency out-of-plane vibrational motion of MoS_2, the in-plane and out-of-plane vibrational motion at low- and high-frequency of ZnPc are playing important roles. Moreover, the change of adiabatic states, which is induced by the atomic motions during the NAMD simulations, contributes most in the electron transfer process [117].

As we have mentioned, the photoelectronic properties of MoS_2 based TMDs are easy to tune. Thus, we have also studied MoS_2 with another organic molecule, i.e., pentacene (Fig. 6.22). Opposite to the MoS_2-ZnPc complex, the MoS_2 part is excited in the MoS_2-Pentacene complex, then an ultrafast hole transfer from MoS_2 to pentacene happens with time constant 611 fs. A hole cooling process happens in ca. 15 ps. After that, we have also studied the electron-hole recombination process by NAMD simulations, which is about 10.2 ns (Fig. 6.23) [120].

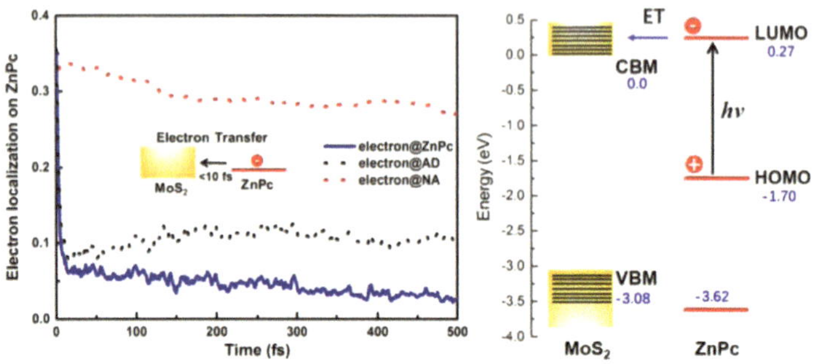

Figure 6.21 (left) Time-dependent electron weights on ZnPc based on NAMD simulations and (right) energy profile of VBM/CBM of MoS_2 and HOMO/LUMO of ZnPc. Copyright 2018 American Chemical Society.

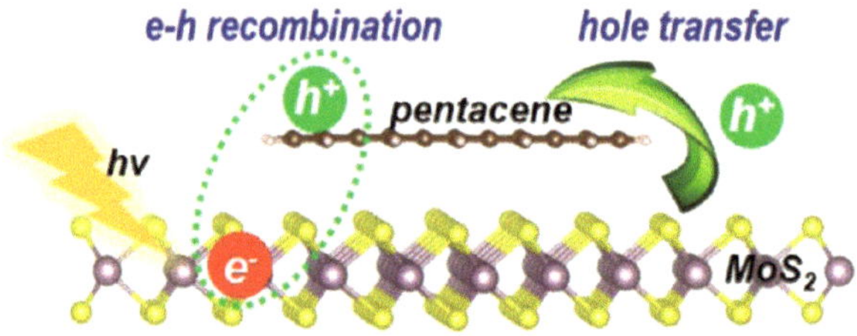

Figure 6.22 Structure of MoS_2-pentacene complex. Copyright 2019 American Chemical Society.

Moreover, we have also found that the thickness of the organic layer plays an important role in regulating interfacial photoinduced charge transfer dynamics. As one example, we have studied the different carrier dynamics of PTB7-1L@MoS_2 and PTB7-5L@MoS_2 (Fig. 6.24). The calculated DOS shows that the occupied states of PTB7-5L@MoS_2 is denser between CBM and VBM of the MoS_2 than PTB7-1L@MoS_2. Moreover, the energy gap of PTB7-5L@MoS_2 between the lowest occupied orbitals and VBM is smaller than PTB7-1L@MoS_2, which indicates that the hole transfer process of PTB7-5L@MoS_2 is much easier than PTB7-1L@MoS_2. Our NAMD simulations have as well demonstrated this viewpoint. The nonadiabatic dynamics simulations show that the adiabatic electron transfer happens in <10 fs for both PTB7-1L@MoS_2 and PTB7-5L@MoS_2. However, the time constant of hole transfer of PTB7-

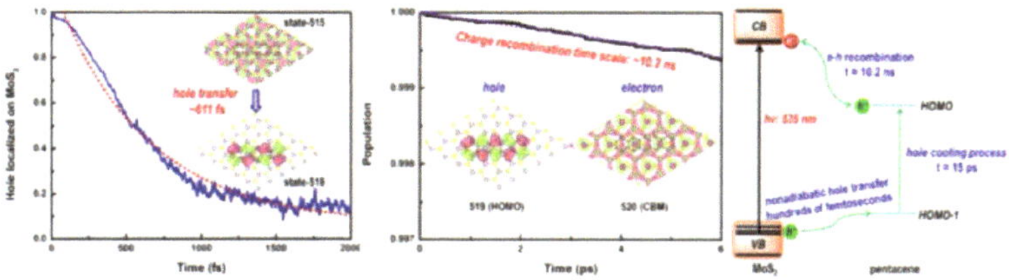

Figure 6.23 (left) Time-dependent hole weights on MoS_2 based on NAMD simulations, (middle) time-dependent electron-hole recombination amount based on NAMD simulations and (right) energy profile of VBM/CBM of MoS_2 and HOMO/HOMO-1 of pentacene and schematic diagram of photoinduced carrier dynamics of MoS_2-pentacene complex. Copyright 2019 American Chemical Society.

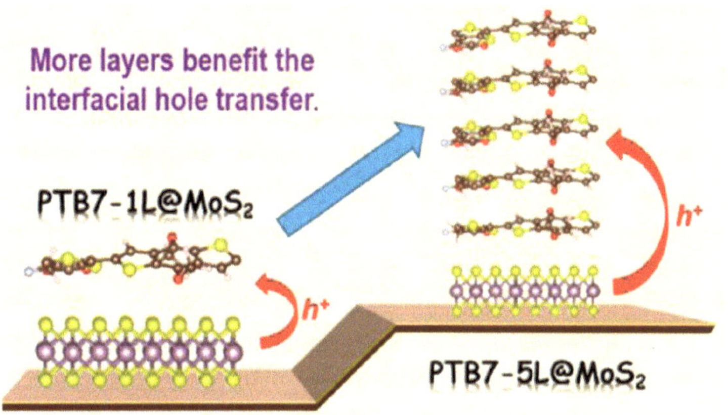

Figure 6.24 Structures of (left) PTB7-1L@MoS_2 and (right) PTB7-5L@MoS_2. Copyright 2019 American Chemical Society.

1L@MoS_2 is ca. 70 ps while the time constant of hole transfer of PTB7-5L@MoS_2 is within 1 ps (Fig. 6.25). From our calculations, we have shown that the hole transfer process of PTB7-MoS_2 complex is heavily dependent on the number of PTB7 layers. These new findings are useful for designing new TMD materials [121].

The spin–orbit coupling (SOC) effects are important for electronic structures and thereto optoelectronic properties of perovskites [109–112]. With our developed GTSH method with the two-component noncollinear DFT method, we have studied the

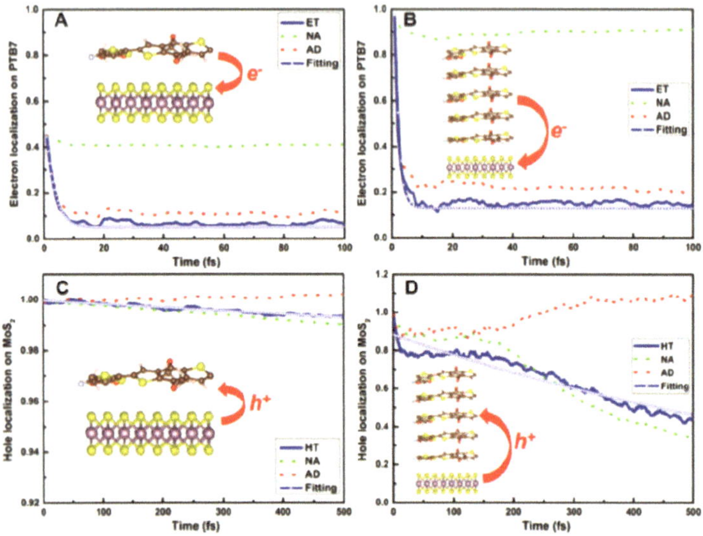

Figure 6.25 Time-dependent electron weights on PTB7 of (A) PTB7-1L@MoS_2 and (B) PTB7-5L@MoS_2; time-dependent hole weights on MoS_2 of (C) PTB7-1L@MoS_2 and (D) PTB7-5L@MoS_2. Copyright 2019 American Chemical Society.

importance of the SOC effects of perovskite materials on their optoelectronic properties.

A typical example is C_{60}-Py@$MAPbI_3$ heterostructure (Fig. 6.26), Liu et al. have found that this heterostructure has achieved the power conversion efficiencies of ca. 18.27% [135]. However, the interfacial charge transfer properties of C_{60}-Py@$MAPbI_3$ are still unclear. Motivated by this issue, we have studied the interfacial electron dynamics with and without SOC effects. The dynamics simulation shows totally different results with and without SOC effects. When considering SOC effects, the electron transfer is ultrafast within 1 ps from $MAPbI_3$ to C_{60}-Py (Fig. 6.27). However, when SOC effects are absent, no electron transfer process is observed. With these results, we have demonstrated that the SOC effects are non-negligible in interfacial charge carrier dynamics of perovskites [118].

We have also studied the photoinduced charge transfer dynamics of recently popular two-dimensional Ruddlesden–Popper (RP)

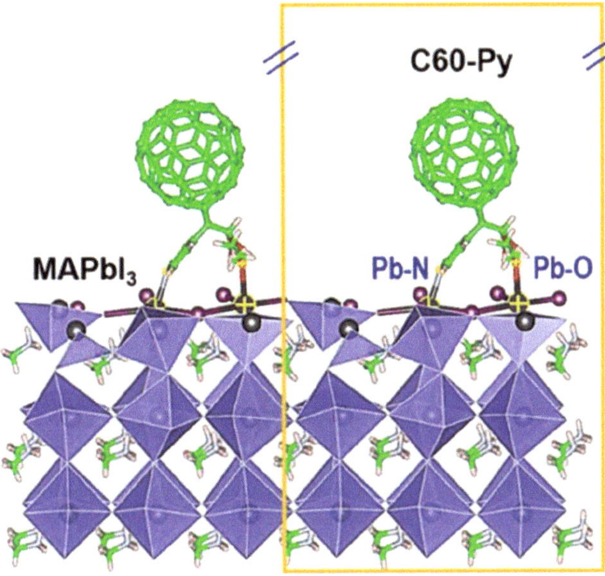

Figure 6.26 Structure of C_{60}-Py@$MAPbI_3$ heterostructure. Copyright 2021 American Chemical Society.

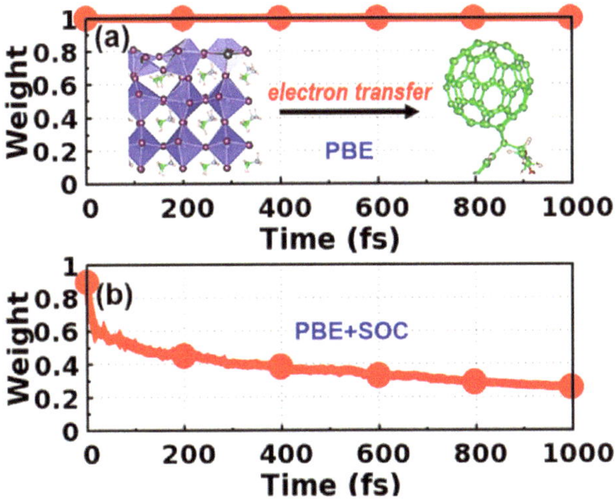

Figure 6.27 Time-dependent electron weights on $MAPbI_3$ based on NAMD simulations at (a) PBE+D3 and (b) PBE+D3+SOC level. Copyright 2021 American Chemical Society.

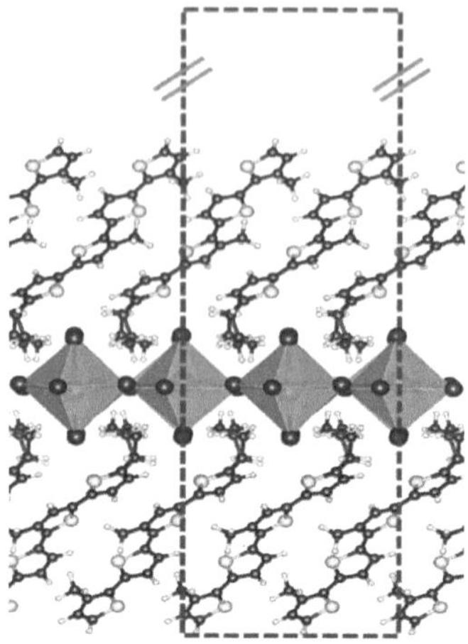

Figure 6.28 Structure of $(4Tm)_2PbI_4$. Copyright 2021 American Chemical Society.

phase perovskites, $(4Tm)_2PbI_4$, in order to help the community to design and synthesize better 2D perovskites materials. $(4Tm)_2PbI_4$ has recently synthesized and the interfacial carrier dynamics have been studied by Dou et al. [136, 137]. However, the detailed mechanism of interfacial charge transfer processes remains to be clarified. We have calculated the PDOS of $(4Tm)_2PbI_4$ with and without considering SOC effect (Fig. 6.29). Only with taking the SOC effects into accounts, the $(4Tm)_2PbI_4$ performs as the Type-II heterojunction, which agrees with the experimental data. When SOC effect is neglected, the $(4Tm)_2PbI_4$ shows properties as Type-I heterojunctions, which is contradictory to the reported experiments. Thus, NAMD simulations are performed at PBE+D3+SOC level. Electron transfer from 4Tm to PbI_4 has two phases, with time constants of 53 fs and 64.5 ps. We can find that the slower one dominates the electron transfer process. Hole transfer from PbI_4 to 4Tm is much faster than electron transfer process, which has the time constant of 1.4 ps (Fig. 6.30). Through this work, we have provided the new insights of the interfacial carrier dynamics of 2D perovskite materials. Meanwhile, we have shown that explicitly including the SOC effects for the theoretical studies for perovskite materials is quite necessary [125].

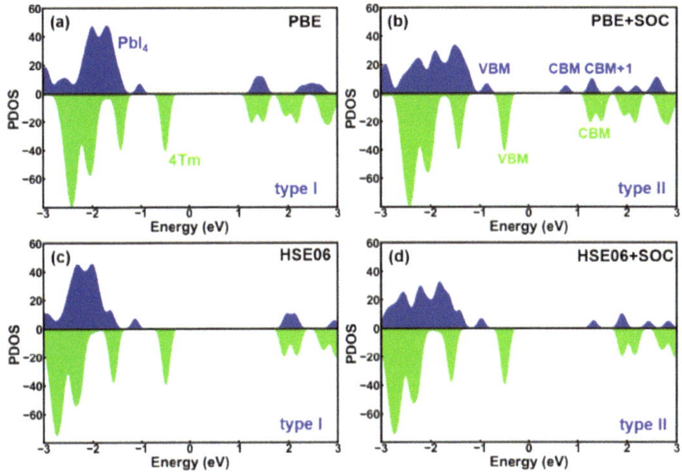

Figure 6.29 PDOS of $(4Tm)_2PbI_4$ at (a) PBE+D3, (b) PBE+D3+SOC, (c) HSE06 and (d) HSE06+D3+SOC level. Copyright 2021 American Chemical Society.

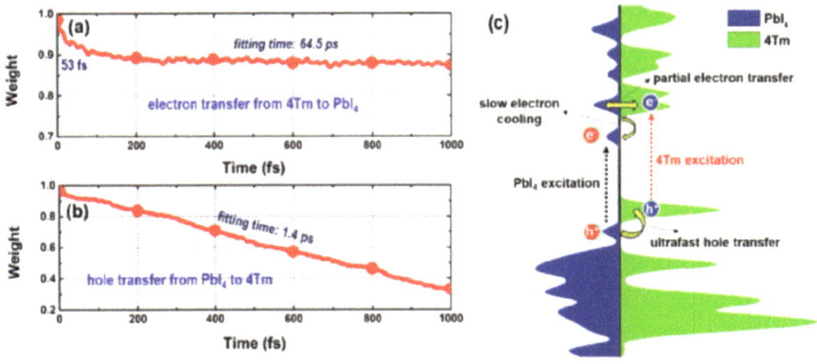

Figure 6.30 Time-dependent (a) electron weight on 4Tm and (b) hole weight on PbI_4 based on NAMD simulations. (c) Proposed mechanism of charge carrier transfer processes of $(4Tm)_2PbI_4$. Copyright 2021 American Chemical Society.

6.5 Conclusion and Perspective

In this chapter we have briefly introduced our recently developed GTSH nonadiabatic molecular dynamics methods with TDDFT methods (time- and frequency-domain) for simulating both ultrafast excited-state relaxation dynamics of large organometallic molecules and nanoclusters (including both IC and ISC processes), and ultrafast photoinduced interfacial carrier dynamics of periodic materials systems (with and without the SOC effects). In addition, with our developed simulation and analysis methods, we have successfully simulated a lot of ultrafast internal conversion and intersystem crossing processes, exciton transfer and dissociation, excitation energy transfer, and photoinduced interfacial carrier dynamics of several metal-containing systems, nanoclusters, donor-acceptor dyads, TMDs, and perovskite materials, etc. These pioneering simulations provide new insights for understanding ultrafast photoinduced physical and chemical processes of these isolated and extended systems. In addition, methodologically, we have established practical nonadiabatic simulation tools for simulating various ultrafast processes of molecules, nanoclusters, biological systems, and materials simulations. Their extended applications in related research fields will be explored in near future.

However, one must see there are enough room to improve the accuracy and efficiency of the GTSH method for large systems in several directions. Firstly, different exchange-correlation-functionals in TDDFT calculations always give more or less dissimilar excited-state properties. To avoid this issue, more robust electronic structure methods, e.g., GW-BSE, etc., should be combined with the GTSH method, which is already under the progress in our group. Secondly, how to go beyond the single-particle approximation in simulating photoinduced interfacial carrier dynamics? To do that, one must consider electron-hole interaction to simulate exciton transfer and dissociation, etc. In this regard, periodic time-dependent density functional theory method could be a potential method, however, which is very expensive for large periodic systems. Finally, how to avoid the classical path approximation in materials simulations? So that one can simulate photoinduced chemical and physical processes with large conformational changes and chemical bond-breaking and -making. All these questions motivate us to move one step further in developing novel TDDFT based nonadiabatic dynamics simulation methods, which is our primary focus in the next decade.

References

1. Born, M., Huang, K., and Lax, M. (1955). Dynamical theory of crystal lattices. *Am. J. Phys.*, **23**, 474–474.
2. Born, M., and Oppenheimer, R. (1927). Zur quantentheorie der molekeln. *Ann. Phys.*, **389**, 457–484.
3. Mai, S., Marquetand, P., and Gonzalez, L. (2015). A general method to describe intersystem crossing dynamics in trajectory surface hopping. *Int. J. Quantum Chem.*, **115**, 1215–1231.
4. Du, L., and Lan, Z. (2015). An on-the-fly surface-hopping program JADE for nonadiabatic molecular dynamics of polyatomic systems: implementation and applications. *J. Chem. Theory Comput.*, **11**, 1360–1374.
5. Barbatti, M., Granucci, G., Persico, M., Ruckenbauer, M., Vazdar, M., Eckert-Maksic, M., and Lischka, H. (2007). The on-the-fly surface-hopping program system Newton-X: application to ab initio simulation of the nonadiabatic photodynamics of benchmark systems. *J. Photochem. Photobiol. A*, **190**, 228–240.
6. Akimov, A. V., and Prezhdo, O. V. (2013). The PYXAID program for non-adiabatic molecular dynamics in condensed matter systems. *J. Chem. Theory Comput.*, **9**, 4959–4972.
7. Ben-Nun, M., Quenneville, J., and Martinez, T. J. (2000). Ab initio multiple spawning: photochemistry from first principles quantum molecular dynamics. *J. Phys. Chem. A*, **104**, 5161–5175.
8. Doltsinis, N. L., and Marx, D. (2002). Nonadiabatic Car–Parrinello molecular dynamics. *Phys. Rev. Lett.*, **88**, 166402.
9. Worth, G. A., Hunt, P., and Robb, M. A. (2003). Nonadiabatic dynamics: a comparison of surface hopping direct dynamics with quantum wavepacket calculations. *J. Phys. Chem. A*, **107**, 621–631.
10. Li, X., Tully, J. C., Schlegel, H. B., and Frisch, M. J. (2005). Ab initio Ehrenfest dynamics. *J. Chem. Phys.*, **123**, 084106.
11. Mitric, R., Werner, U., Wohlgemuth, M., Seifert, G., and Bonacic-Koutecky, V. (2009). Nonadiabatic dynamics within time-dependent density functional tight binding method. *J. Phys. Chem. A*, **113**, 12700–12705.
12. Barbatti, M. (2011). Nonadiabatic dynamics with trajectory surface hopping method. *WIREs Comput. Mol. Sci.*, **1**, 620–633.
13. Worth, G. A., Meyer, H. D., Koeppel, H., Cederbaum, L. S., and Burghardt, I. (2008). Using the MCTDH wavepacket propagation method to

describe multimode non-adiabatic dynamics. *Int. Rev. Phys. Chem.*, **27**, 569–606.

14. Wang, H. B., and Thoss, M. (2003). Multilayer formulation of the multiconfiguration time-dependent Hartree theory. *J. Chem. Phys.*, **119**, 1289–1299.
15. Beck, M. H., Jäckle, A., Worth, G. A., and Meyer, H.-D. (2000). The multiconfiguration time-dependent Hartree (MCTDH) method: a highly efficient algorithm for propagating wavepackets. *Phys. Rep.*, **324**, 1–105.
16. Yu, L., Xu, C., Lei, Y., Zhu, C., and Wen, Z. (2014). Trajectory-based nonadiabatic molecular dynamics without calculating nonadiabatic coupling in the avoided crossing case: trans <-> cis photoisomerization in azobenzene. *Phys. Chem. Chem. Phys.*, **16**, 25883–25895.
17. Yu, L., Xu, C., and Zhu, C. (2015). Probing the $\pi \rightarrow \pi^*$ photoisomerization mechanism of cis-azobenzene by multi-state ab initio on-the-fly trajectory dynamics simulation. *Phys. Chem. Chem. Phys.*, **17**, 17646–17660.
18. Tully, J. C. (2012). Perspective: nonadiabatic dynamics theory. *J. Chem. Phys.*, **137**, 22A301.
19. Feng, W., Xu, L., Li, X.-Q., Fang, W., and Yan, Y. (2012). Nonadiabatic molecular dynamics simulation: an approach based on quantum measurement picture. *AIP Adv.*, **4**, 077131.
20. Crespo-Otero, R., and Barbatti, M. (2018). Recent advances and perspectives on nonadiabatic mixed quantum-classical dynamics. *Chem. Rev.*, **118**, 7026–7068.
21. Curchod, B. F. E., and Martinez, T. J. (2018). Ab initio nonadiabatic quantum molecular dynamics. *Chem. Rev.*, **118**, 3305–3336.
22. Nelson, T. R., White, A. J., Bjorgaard, J. A., Sifain, A. E., Zhang, Y., Nebgen, B., Fernandez-Alberti, S., Mozyrsky, D., Roitberg, A. E., and Tretiak, S. (2020). Non-adiabatic excited-state molecular dynamics: theory and applications for modeling photophysics in extended molecular materials. *Chem. Rev.*, **120**, 2215–2287.
23. Cui, G. L., and Thiel, W. (2014). Generalized trajectory surface-hopping method for internal conversion and intersystem crossing. *J. Chem. Phys.*, **141**, 124101.
24. Tully, J. C., and Preston, R. K. (1971). Trajectory surface hopping approach to nonadiabatic molecular collisions: the reaction of H^+ with D_2. *J. Chem. Phys.*, **55**, 562–572.

25. Hammes-Schiffer, S., and Tully, J. C. (1994). Proton-transfer in solution—molecular-dynamics with quantum transitions. *J. Chem. Phys.,* **101**, 4657–4667.
26. Richter, M., Marquetand, P., Gonzalez-Vazquez, J., Sola, I., and Gonzalez, L. (2011). SHARC: ab initio molecular dynamics with surface hopping in the adiabatic representation including arbitrary couplings. *J. Chem. Theory Comput.,* **7**, 1253–1258.
27. Granucci, G., Persico, M., and Spighi, G. (2012). Surface hopping trajectory simulations with spin-orbit and dynamical couplings. *J. Chem. Phys.,* **137**, 22A501.
28. Jaeger, H. M., Fischer, S., and Prezhdo, O. V. (2012). Decoherence-induced surface hopping. *J. Chem. Phys.,* **137**, 22A545.
29. Plasser, F., Granucci, G., Pittner, J., Barbatti, M., Persico, M., and Lischka, H. (2012). Surface hopping dynamics using a locally diabatic formalism: charge transfer in the ethylene dimer cation and excited state dynamics in the 2-pyridone dimer. *J. Chem. Phys.,* **137**, 22A514.
30. Kelly, A., and Markland, T. E. (2013). Efficient and accurate surface hopping for long time nonadiabatic quantum dynamics. *J. Chem. Phys.,* **139**, 014104.
31. Subotnik, J. E., Ouyang, W., and Landry, B. R. (2013). Can we derive Tully's surface-hopping algorithm from the semiclassical quantum Liouville equation? Almost, but only with decoherence. *J. Chem. Phys.,* **139**, 211101.
32. Tempelaar, R., van der Vegte, C. P., Knoester, J., and Jansen, T. L. C. (2013). Surface hopping modeling of two-dimensional spectra. *J. Chem. Phys.,* **138**, 164106.
33. Tapavicza, E., Tavernelli, I., and Rothlisberger, U. (2007). Trajectory surface hopping within linear response time-dependent density-functional theory. *Phys. Rev. Lett.,* **98**, 023001.
34. Gao, X., and Thiel, W. (2017). Non-Hermitian surface hopping. *Phys. Rev. E,* **95**, 013308.
35. Fazzi, D., Barbatti, M., and Thiel, W. (2016). Unveiling the role of hot charge-transfer states in molecular aggregates via nonadiabatic dynamics. *J. Am. Chem. Soc.,* **138**, 4502–4511.
36. Barbatti, M., and Lischka, H. (2008). Nonadiabatic deactivation of 9H-adenine: a comprehensive picture based on mixed quantum-classical dynamics. *J. Am. Chem. Soc.,* **130**, 6831–6839.
37. Barbatti, M. (2014). Photorelaxation induced by water-chromophore electron transfer. *J. Am. Chem. Soc.,* **136**, 10246–10249.

38. Ha, J.-K., Lee, I. S., and Min, S. K. (2018). Surface hopping dynamics beyond nonadiabatic couplings for quantum coherence. *J. Phys. Chem. Lett.,* **9**, 1097–1104.
39. Mai, S., Marquetand, P., and Gonzalez, L. (2016). Intersystem crossing pathways in the noncanonical nucleobase 2-thiouracil: a time-dependent picture. *J. Phys. Chem. Lett.,* **7**, 1978–1983.
40. Warshel, A., and Levitt, M. (1976). Theoretical studies of enzymic reactions—dielectric, electrostatic and steric stabilization of carbonium-ion in reaction of lysozyme. *J. Mol. Bio.,* **103**, 227–249.
41. Senn, H. M., and Thiel, W. (2009). QM/MM methods for biomolecular systems. *Angew. Chem., Int. Ed.,* **48**, 1198–1229.
42. Bakowies, D., and Thiel, W. (1996). Hybrid models for combined quantum mechanical and molecular mechanical approaches. *J. Phys. Chem.,* **100**, 10580–10594.
43. Toniolo, A., Granucci, G., and Martinez, T. J. (2003). Conical intersections in solution: a QM/MM study using floating occupation semiempirical configuration interaction wave functions. *J. Phys. Chem. A,* **107**, 3822–3830.
44. Virshup, A. M., Punwong, C., Pogorelov, T. V., Lindquist, B. A., Ko, C., and Martinez, T. J. (2009). Photodynamics in complex environments: ab initio multiple spawning quantum mechanical/molecular mechanical dynamics. *J. Phys. Chem. B,* **113**, 3280–3291.
45. Lan, Z., Lu, Y., Fabiano, E., and Thiel, W. (2011). QM/MM nonadiabatic decay dynamics of 9H-adenine in aqueous solution. *ChemPhysChem,* **12**, 1989–1998.
46. Eckert-Maksic, M., Vazdar, M., Ruckenbauer, M., Barbatti, M., Mueller, T., and Lischka, H. (2010). Matrix-controlled photofragmentation of formamide: dynamics simulation in argon by nonadiabatic QM/MM method. *Phys. Chem. Chem. Phys.,* **12**, 12719–12726.
47. Wigner, E. (1932). On the quantum correction for thermodynamic equilibrium. *Phys. Rev.,* **40**, 0749–0759.
48. Wang, Y. T., Liu, X. Y., and Fang, W. H. (2019). Mechanism of the O_2 ($^1\Delta_g$) generation from the Cl_2/H_2O_2 basic aqueous solution explored by the combined ab initio calculation and nonadiabatic dynamics simulation. *J. Comput. Chem.,* **40**, 447–455.
49. Chen, W. K., Fang, W. H., and Cui, G. L. (2019). Integrating machine learning with the multilayer energy-based fragment method for excited states of large systems. *J. Phys. Chem. Lett.,* **10**, 7836–7841.

50. Liu, X. Y., Fang, Y. G., Xie, B. B., Fang, W. H., and Cui, G. L. (2017). QM/MM nonadiabatic dynamics simulations on photoinduced Wolff rearrangements of 1,2,3-thiadiazole. *J. Chem. Phys.,* **146**, 224302.
51. Li, C. X., Guo, W. W., Xie, B. B., and Cui, G. L. (2016). Photodynamics of oxybenzone sunscreen: nonadiabatic dynamics simulations. *J. Chem. Phys.,* **145**, 074308.
52. Xia, S. H., Liu, X. Y., Fang, Q., and Cui, G. L. (2015). Photodissociation dynamics of $CH_3C(O)SH$ in argon matrix: a QM/MM nonadiabatic dynamics simulation. *J. Chem. Phys.,* **143**, 194303.
53. Adamo, C., and Jacquemin, D. (2013). The calculations of excited-state properties with time-dependent density functional theory. *Chem. Soc. Rev.,* **42**, 845–856.
54. Laurent, A. D., Adamo, C., and Jacquemin, D. (2014). Dye chemistry with time-dependent density functional theory. *Phys. Chem. Chem. Phys.,* **16**, 14334–14356.
55. Runge, E., and Gross, E. K. U. (1984). Density-functional theory for time-dependent systems. *Phys. Rev. Lett.,* **52**, 997–1000.
56. Caillie, C. V., and Amos, R. D. (1999). Geometric derivatives of excitation energies using SCF and DFT. *Chem. Phys. Lett.,* **308**, 249–255.
57. Caillie, C. V., and Amos, R. D. (2000). Geometric derivatives of density functional theory excitation energies using gradient-corrected functionals. *Chem. Phys. Lett.,* **317**, 159–164.
58. Scalmani, G., Frisch, M. J., Mennucci, B., Tomasi, J., Cammi, R., and Barone, V. (2006). Geometries and properties of excited states in the gas phase and in solution: theory and application of a time-dependent density functional theory polarizable continuum model. *J. Chem. Phys.,* **124**, 94107.
59. Grimme, S., Antony, J., Ehrlich, S., and Krieg, H. (2010). A consistent and accurate ab initio parametrization of density functional dispersion correction (DFT-D) for the 94 elements H-Pu. *J. Chem. Phys.,* **132**, 154104.
60. Casida, M. E., and Huix-Rotllant, M. (2012). Progress in time-dependent density-functional theory. *Annu. Rev. Phys. Chem.,* **63**, 287–323.
61. Marques, M. A. L., and Gross, E. K. U. (2004). Time-dependent density functional theory. *Annu. Rev. Phys. Chem.,* **55**, 427–455.
62. Casida, M. E. (1995). Time-dependent density functional response theory for molecules, in *Recent Advances in Density Functional Methods: Part I,* World Scientific, pp. 155–192.

63. Hirata, S., and Head-Gordon, M. (1999). Time-dependent density functional theory within the Tamm-Dancoff approximation. *Chem. Phys. Lett.*, **314**, 291–299.

64. Cui, G. L., and Yang, W. (2010). Challenges with range-separated exchange-correlation functionals in time-dependent density functional theory calculations. *Mol. Phys.*, **108**, 2745–2750.

65. Ryabinkin, I. G., Nagesh, J., and Izmaylov, A. F. (2015). Fast numerical evaluation of time-derivative nonadiabatic couplings for mixed quantum-classical methods. *J. Phys. Chem. Lett.*, **6**, 4200–4203.

66. Liu, X.-Y., Xie, X.-Y., Fang, W.-H., and Cui, G. L. (2018). Photoinduced relaxation dynamics of nitrogen-capped silicon nanoclusters: a TD-DFT study. *Mol. Phys.*, **116**, 869–884.

67. Hess, B. A., Marian, C. M., Wahlgren, U., and Gropen, O. (1996). A mean-field spin-orbit method applicable to correlated wavefunctions. *Chem. Phys. Lett.*, **251**, 365–371.

68. Fedorov, D. G., Koseki, S., Schmidt, M. W., and Gordon, M. S. (2003). Spin-orbit coupling in molecules: chemistry beyond the adiabatic approximation. *Int. Rev. Phys. Chem.*, **22**, 551–592.

69. Koseki, S., Gordon, M. S., Schmidt, M. W., and Matsunaga, N. (1995). Main group effective nuclear charges for spin-orbit calculations. *J. Phys. Chem.*, **99**, 12764–12772.

70. Koseki, S., Schmidt, M. W., and Gordon, M. S. (1998). Effective nuclear charges for the first- through third-row transition metal elements in spin-orbit calculations. *J. Phys. Chem. A*, **102**, 10430–10435.

71. Chiodo, S. G., and Russo, N. (2009). One-electron spin–orbit contribution by effective nuclear charges. *J. Comput. Chem.*, **30**, 832–839.

72. Gao, X., Bai, S., Fazzi, D., Niehaus, T., Barbatti, M., and Thiel, W. (2017). Evaluation of spin-orbit couplings with linear-response time-dependent density functional methods. *J. Chem. Theory. Comput.*, **13**, 515–524.

73. de Carvalho, F. F., Curchod, B. F. E., Penfold, T. J., and Tavernelli, I. (2014). Derivation of spin-orbit couplings in collinear linear-response TDDFT: a rigorous formulation. *J. Chem. Phys.*, **140**, 144103.

74. Szabo, A., and Ostlund, N. S. (2012). *Modern Quantum Chemistry: Introduction to Advanced Electronic Structure Theory*. Courier Corporation.

75. Chiodo, S. G., and Leopoldini, M. (2014). MolSOC: a spin–orbit coupling code. *Comput. Phys. Commun.*, **185**, 676–683.

76. Fang, Y.-G., Peng, L.-Y., Liu, X.-Y., Fang, W.-H., and Cui, G. L. (2019). QM/MM nonadiabatic dynamics simulation on ultrafast excited-state

relaxation in osmium(II) compounds in solution. *Comput. Theor. Chem.*, **1155**, 90–100.

77. Liu, X. Y., Li, Z. W., Fang, W. H., and Cui, G. L. (2018). Nonadiabatic dynamics simulations on internal conversion and intersystem crossing processes in gold(I) compounds. *J. Chem. Phys.*, **149**, 044301.
78. Liu, X. Y., Zhang, Y. H., Fang, W. H., and Cui, G. L. (2018). Early-time excited-state relaxation dynamics of iridium compounds: distinct roles of electron and hole transfer. *J. Phys. Chem. A*, **122**, 5518–5532.
79. Li, Z. W., Yang, J. J., Liu, X. Y., Fang, W. H., Wang, H., and Cui, G. L. (2021). Chemical bonding as a new avenue for controlling excited-state properties and excitation energy-transfer processes in zinc phthalocyanine-fullerene dyads. *Chem. Eur. J.*, **27**, 4159–4167.
80. Liu, X. Y., Li, Z. W., Fang, W. H., and Cui, G. L. (2020). Nonadiabatic exciton and charge separation dynamics at interfaces of zinc phthalocyanine and fullerene: orientation does matter. *J. Phys. Chem. A*, **124**, 7388–7398.
81. Montalti, M., Cantelli, A., and Battistelli, G. (2015). Nanodiamonds and silicon quantum dots: ultrastable and biocompatible luminescent nanoprobes for long-term bioimaging. *Chem. Soc. Rev.*, **44**, 4853–4921.
82. Cheng, X., Lowe, S. B., Reece, P. J., and Gooding, J. J. (2014). Colloidal silicon quantum dots: from preparation to the modification of self-assembled monolayers (SAMS) for bio-applications. *Chem. Soc. Rev.*, **43**, 2680–2700.
83. Li, Q., He, Y., Chang, J., Wang, L., Chen, H., Tan, Y.-W., Wang, H., and Shao, Z. (2013). Surface-modified silicon nanoparticles with ultra-bright photoluminescence and single-exponential decay for nanoscale fluorescence lifetime imaging of temperature. *J. Am. Chem. Soc.*, **135**, 14924–14927.
84. Li, Q., Luo, T.-Y., Zhou, M., Abroshan, H., Huang, J., Kim, H. J., Rosi, N. L., Shao, Z., and Jin, R. (2016). Silicon nanoparticles with surface nitrogen: 90% quantum yield with narrow luminescence bandwidth and the ligand structure based energy law. *ACS Nano*, **10**, 8385–8393.
85. Yanai, T., Tew, D. P., and Handy, N. C. (2004). A new hybrid exchange-correlation functional using the coulomb-attenuating method (CAM-B3LYP). *Chem. Phys. Lett.*, **393**, 51–57.
86. King, K., Spellane, P., and Watts, R. J. (1985). Excited-state properties of a triply ortho-metalated iridium (III) complex. *J. Am. Chem. Soc.*, **107**, 1431–1432.

87. King, K., and Watts, R. (1987). Dual emission from an ortho-metalated iridium (III) complex. *J. Am. Chem. Soc.*, **109**, 1589–1590.

88. Tamayo, A. B., Alleyne, B. D., Djurovich, P. I., Lamansky, S., Tsyba, I., Ho, N. N., Bau, R., and Thompson, M. E. (2003). Synthesis and characterization of facial and meridional tris-cyclometalated iridium (III) complexes. *J. Am. Chem. Soc.*, **125**, 7377–7387.

89. Schulze, M., Steffen, A., and Würthner, F. (2015). Near-IR phosphorescent ruthenium (II) and iridium (III) perylene bisimide metal complexes. *Angew. Chem., Int. Ed.*, **54**, 1570–1573.

90. Brahim, H., and Daniel, C. (2014). Structural and spectroscopic properties of Ir (III) complexes with phenylpyridine ligands: absorption spectra without and with spin–orbit-coupling. *Comput. Theor. Chem.*, **1040**, 219–229.

91. Penfold, T. J., Gindensperger, E., Daniel, C., and Marian, C. M. (2018). Spin-vibronic mechanism for intersystem crossing. *Chem. Rev.*, **118**, 6975–7025.

92. Pomarico, E., Silatani, M., Messina, F., Braem, O., Cannizzo, A., Barranoff, E., Klein, J., Lambert, C., and Chergui, M. (2016). Dual luminescence, interligand decay, and nonradiative electronic relaxation of cyclometalated iridium complexes in solution. *J. Phys. Chem. C*, **120**, 16459–16469.

93. Lima, J. C., and Rodriguez, L. (2011). Applications of gold (I) alkynyl systems: a growing field to explore. *Chem. Soc. Rev.*, **40**, 5442–5456.

94. Chao, H.-Y., Lu, W., Li, Y., Chan, M. C., Che, C.-M., Cheung, K.-K., and Zhu, N. (2002). Organic triplet emissions of arylacetylide moieties harnessed through coordination to $[Au(PCy_3)]^+$. Effect of molecular structure upon photoluminescent properties. *J. Am. Chem. Soc.*, **124**, 14696–14706.

95. Partyka, D. V., Zeller, M., Hunter, A. D., and Gray, T. G. (2006). Relativistic functional groups: aryl carbon–gold bond formation by selective transmetalation of boronic acids. *Angew. Chem., Int. Ed.*, **45**, 8188–8191.

96. Lu, W., Kwok, W.-M., Ma, C., Chan, C. T.-L., Zhu, M.-X., and Che, C.-M. (2011). Organic triplet excited states of gold (I) complexes with oligo (o- or m-phenyleneethynylene) ligands: conjunction of steady-state and time-resolved spectroscopic studies on exciton delocalization and emission pathways. *J. Am. Chem. Soc.*, **133**, 14120–14135.

97. Vogt, R. A., Peay, M. A., Gray, T. G., and Crespo-Hernandez, C. E. (2010). Excited-state dynamics of (organophosphine) gold (I) pyrenyl isomers. *J. Phys. Chem. Lett.*, **1**, 1205–1211.

98. Vogt, R. A., Gray, T. G., and Crespo-Hernandez, C. E. (2012). Subpicosecond intersystem crossing in mono-and di (organophosphine) gold (I) naphthalene derivatives in solution. *J. Am. Chem. Soc.,* **134**, 14808–14817.

99. Guldi, D. M., Gouloumis, A., Vázquez, P., and Torres, T. (2002). Charge-transfer states in strongly coupled phthalocyanine fullerene ensembles. *Chem. Commun.*, 2056–2057.

100. Javaid, S., and Javed Akhtar, M. (2015). An ab-initio density functional theory investigation of fullerene/Zn-phthalocyanine (C_{60}/ZnPc) interface with face-on orientation. *J. Appl. Phys.,* **118**, 045305.

101. Javaid, S., and Javed Akhtar, M. (2016). Orientation dependant charge transfer at fullerene/Zn-phthalocyanine (C_{60}/ZnPc) interface: implications for energy level alignment and photovoltaic properties. *Appl. Phys. Lett.,* **109**, 051605.

102. Javaid, S., and Akhtar, M. J. (2016). Adsorption and electronic properties of fullerene/Zn-phthalocyanine (C_{60}/ZnPc) interface with face-on orientation: a van der Waals corrected density functional theory investigation. *Chem. Phys. Lett.,* **649**, 73–77.

103. Santos, E. J., and Wang, W. (2016). Ultrafast charge-transfer in organic photovoltaic interfaces: geometrical and functionalization effects. *Nanoscale,* **8**, 15902–15910.

104. Wang, L., Long, R., and Prezhdo, O. V. (2015). Time-domain ab initio modeling of photoinduced dynamics at nanoscale interfaces. *Annu. Rev. Phys. Chem.,* **66**, 549–579.

105. Habenicht, B. F., Craig, C. F., and Prezhdo, O. V. (2006). Time-domain ab initio simulation of electron and hole relaxation dynamics in a single-wall semiconducting carbon nanotube. *Phys. Rev. Lett.,* **96**, 187401.

106. Akimov, A. V., and Prezhdo, O. V. (2014). Advanced capabilities of the PYXAID program: Integration schemes, decoherence: effects, multiexcitonic states, and field-matter interaction. *J. Chem. Theory Comput.,* **10**, 789–804.

107. Akimov, A. V., Neukirch, A. J., and Prezhdo, O. V. (2013). Theoretical insights into photoinduced charge transfer and catalysis at oxide interfaces. *Chem. Rev.,* **113**, 4496–4565.

108. Zheng, Q., Chu, W., Zhao, C., Zhang, L., Guo, H., Wang, Y., Jiang, X., and Zhao, J. (2019). Ab initio nonadiabatic molecular dynamics investigations on the excited carriers in condensed matter systems. *WIREs Comput. Mol. Sci.,* **9**, e1411.

109. Zhang, L., Zhang, X., and Lu, G. (2020). Intramolecular band alignment and spin-orbit coupling in two-dimensional halide perovskites. *J. Phys. Chem. Lett.,* **11**, 6982–6989.

110. Even, J., Pedesseau, L., Jancu, J.-M., and Katan, C. (2013). Importance of spin–orbit coupling in hybrid organic/inorganic perovskites for photovoltaic applications. *J. Phys. Chem. Lett.,* **4**, 2999–3005.

111. Du, M. H. (2015). Density functional calculations of native defects in $CH_3NH_3PbI_3$: effects of spin-orbit coupling and self-interaction error. *J. Phys. Chem. Lett.,* **6**, 1461–1466.

112. Li, W., Zhou, L., Prezhdo, O. V., and Akimov, A. V. (2018). Spin–orbit interactions greatly accelerate nonradiative dynamics in lead halide perovskites. *ACS Energy Lett.,* **3**, 2159–2166.

113. von Barth, U., and Hedin, L. (1972). A local exchange-correlation potential for the spin polarized case. I. *J. Phys. C: Solid State Phys.,* **5**, 1629.

114. Kubler, J., Hock, K.-H., Sticht, J., and Williams, A. (1988). Density functional theory of non-collinear magnetism. *J. Phys. F: Met. Phys.,* **18**, 469.

115. Sticht, J., Höck, K., and Kübler, J. (1989). Non-collinear itinerant magnetism: the case of Mn_3Sn. *J. Phys.: Condens. Matter,* **1**, 8155.

116. Oda, T., Pasquarello, A., and Car, R. (1998). Fully unconstrained approach to noncollinear magnetism: application to small Fe clusters. *Phys. Rev. Lett.,* **80**, 3622.

117. Liu, X. Y., Xie, X. Y., Fang, W. H., and Cui, G. L. (2018). Theoretical insights into interfacial electron transfer between zinc phthalocyanine and molybdenum disulfide. *J. Phys. Chem. A,* **122**, 9587–9596.

118. Liu, X. Y., Yang, J. J., Chen, W. K., Akimov, A. V., Fang, W. H., and Cui, G. L. (2021). Spin-orbit coupling accelerates the photoinduced interfacial electron transfer in a fullerene-based perovskite heterojunction. *J. Phys. Chem. Lett.,* **12**, 1131–1137.

119. Yang, J. J., Liu, X. Y., Fang, W. H., Xiao, D., and Cui, G. L. (2019). Photoinduced carrier dynamics at the interface of black phosphorus and bismuth vanadate. *J. Phys. Chem. A,* **123**, 10019–10029.

120. Xie, X. Y., Liu, X. Y., Fang, Q., Fang, W. H., and Cui, G. L. (2019). Photoinduced carrier dynamics at the interface of pentacene and molybdenum disulfide. *J. Phys. Chem. A,* **123**, 7693–7703.

121. Liu, X. Y., Chen, W. K., Fang, W. H., and Cui, G. L. (2019). Nonadiabatic dynamics simulations reveal distinct effects of the thickness of PTB7

on interfacial electron and hole transfer dynamics in PTB7@MoS_2 heterostructures. *J. Phys. Chem. Lett.,* **10**, 2949–2956.

122. Yang, J. J., Li, Z. W., Liu, X. Y., Fang, W. H., and Cui, G. L. (2020). Photoinduced electron transfer from carbon nanotubes to fullerenes: C_{60} versus C_{70}. *Phys. Chem. Chem. Phys.,* **22**, 19542–19548.

123. Yang, J. J., Liu, X. Y., Li, Z. W., Frauenheim, T., Yam, C., Fang, W. H., and Cui, G. L. (2021). The spin-orbit interaction controls photoinduced interfacial electron transfer in fullerene-perovskite heterojunctions: C_{60} versus C_{70}. *Phys. Chem. Chem. Phys.,* **23**, 6536–6543.

124. Xie, X. Y., Yang, J. J., Liu, X. Y., Fang, Q., Fang, W. H., and Cui, G. L. (2021). Interfacial photoinduced carrier dynamics tuned by polymerization of coronene molecules encapsulated in carbon nanotubes: bridging type-I and type-II heterojunctions. *Phys. Chem. Chem. Phys.,* **23**, 13503–13511.

125. Yang, J.-J., Chen, W.-K., Liu, X.-Y., Fang, W.-H., and Cui, G. L. (2021). Spin-orbit coupling is the key to promote asynchronous photoinduced charge transfer of two-dimensional perovskites. *JACS Au,* **1**, 1178–1186.

126. Britnell, L., Ribeiro, R. M., Eckmann, A., Jalil, R., Belle, B. D., Mishchenko, A., Kim, Y.-J., Gorbachev, R. V., Georgiou, T., and Morozov, S. (2013). Strong light-matter interactions in heterostructures of atomically thin films. *Science,* **340**, 1311–1314.

127. Fiori, G., Bonaccorso, F., Iannaccone, G., Palacios, T., Neumaier, D., Seabaugh, A., Banerjee, S. K., and Colombo, L. (2014). Electronics based on two-dimensional materials. *Nat. Nanotechnol.,* **9**, 768–779.

128. Chhowalla, M., Shin, H. S., Eda, G., Li, L.-J., Loh, K. P., and Zhang, H. (2013). The chemistry of two-dimensional layered transition metal dichalcogenide nanosheets. *Nat. Chem.,* **5**, 263–275.

129. Withers, F., Del Pozo-Zamudio, O., Mishchenko, A., Rooney, A., Gholinia, A., Watanabe, K., Taniguchi, T., Haigh, S. J., Geim, A., and Tartakovskii, A. (2015). Light-emitting diodes by band-structure engineering in van der Waals heterostructures. *Nat. Mater.,* **14**, 301–306.

130. Mak, K. F., Lee, C., Hone, J., Shan, J., and Heinz, T. F. (2010). Atomically thin MoS_2: a new direct-gap semiconductor. *Phys. Rev. Lett.,* **105**, 136805.

131. Nie, Z., Long, R., Teguh, J. S., Huang, C.-C., Hewak, D. W., Yeow, E. K., Shen, Z., Prezhdo, O. V., and Loh, Z.-H. (2015). Ultrafast electron and hole relaxation pathways in few-layer MoS_2. *J. Phys. Chem. C,* **119**, 20698–20708.

132. Mannebach, E. M., Li, R., Duerloo, K.-A., Nyby, C., Zalden, P., Vecchione, T., Ernst, F., Reid, A. H., Chase, T., and Shen, X. (2015). Dynamic structural response and deformations of monolayer MoS_2 visualized by femtosecond electron diffraction. *Nano Lett.,* **15**, 6889–6895.

133. Mai, C., Barrette, A., Yu, Y., Semenov, Y. G., Kim, K. W., Cao, L., and Gundogdu, K. (2014). Many-body effects in valleytronics: direct measurement of valley lifetimes in single-layer MoS_2. *Nano Lett.,* **14**, 202–206.

134. Kafle, T. R., Kattel, B., Lane, S. D., Wang, T., Zhao, H., and Chan, W.-L. (2017). Charge transfer exciton and spin flipping at organic–transition-metal dichalcogenide interfaces. *ACS Nano,* **11**, 10184–10192.

135. Liu, H.-R., Li, S.-H., Deng, L.-L., Wang, Z.-Y., Xing, Z., Rong, X., Tian, H.-R., Li, X., Xie, S.-Y., and Huang, R.-B. (2019). Pyridine-functionalized fullerene electron transport layer for efficient planar perovskite solar cells. *ACS Appl. Mater. Interfaces,* **11**, 23982–23989.

136. Shi, E., Yuan, B., Shiring, S. B., Gao, Y., Guo, Y., Su, C., Lai, M., Yang, P., Kong, J., and Savoie, B. M. (2020). Two-dimensional halide perovskite lateral epitaxial heterostructures. *Nature,* **580**, 614–620.

137. Gao, Y., Shi, E., Deng, S., Shiring, S. B., Snaider, J. M., Liang, C., Yuan, B., Song, R., Janke, S. M., and Liebman-Peláez, A. (2019). Molecular engineering of organic–inorganic hybrid perovskites quantum wells. *Nat. Chem.,* **11**, 1151–1157.

Chapter 7

Multistate Nonadiabatic Molecular Dynamics: The Role of Conical Intersection between the Excited States

Panwang Zhou[a] and Keli Han[a,b]

[a] *Institute of Molecular Sciences and Engineering, Institute of Frontier and Interdisciplinary Science, Shandong University, Qingdao 266237, P. R. China*

[b] *State Key Laboratory of Molecular Reaction Dynamics, Dalian Institute of Chemical Physics, Chinese Academy of Science, Dalian 116023, Liaoning, China*

klhan@dicp.ac.cn

7.1 Introduction

In the condensed phase, depending on the wavelength of excitation light, the molecules can be excited to the lowest or upper excited states. However, the fluorescence and phosphorescence usually can be emitted only from the lowest singlet and triplet excited state, respectively. That is, the emission wavelength is independent of the excitation wavelength. This rule is known as Kasha's rule, and its strict statement is "*polyatomic molecules generally luminesce with appreciable yield only from the lowest excited state of a given*

Time-Dependent Density Functional Theory: Nonadiabatic Molecular Dynamics
Edited by Chaoyuan Zhu

ISBN 978-981-4968-42-3 (Hardcover), 978-1-003-31921-4 (eBook)
www.jennystanford.com

multiplicity" [1]. The validity of Kasha's rule is based on the fact that for most emissive molecules of a given multiplicity, the energy gaps between upper excited states are usually far smaller than those between the ground state and the lowest excited state. Therefore, the internal conversion (IC) from the upper excited state to the lowest excited state is much fast than the rate of fluorescence or phosphorescence. The fluorescence and phosphorescence from upper excited states are referred to as anti-Kasha's rule and have been observed for a few organic molecules [2]. Extending Kasha's rule to common excited-state events indicated that the photochemical reaction should also occur from the lowest excited state and the exceptions of this phenomenon are known as "anti-Kasha photochemistry" [3].

As a corollary, Vavilov's rule states that *the quantum yield of luminescence is generally independent of the wavelength of excitation* [1]. Although anti-Kasha's rule and anti-Kasha photochemistry have drawn much attention and some excellent reviews have been published [2, 3], the studies on anti-Vavilov's rule have been reported rarely. Experimentally, the wavelength-dependent photoisomerization quantum yield has been observed for some molecules, i.e., photoactive yellow protein (PYP) chromophore [4], trans-urocanic acid [5, 6], arylazoimidazoles [7, 8], and 8-oxo-deoxyguanosine [9, 10]. Our recent theoretical studies [11–15] have demonstrated that photoisomerization of these molecules still occur on S_1 state and thereby obey Kasha's rule. However, these results violate Vavilov's rule and can be regarded as anti-Vavilov photochemistry. The experimentally observed wavelength-dependent photoisomerization quantum yield may result from minimal energy conical intersection (MECI) between S_1 and S_2 states, which opens a new nonradiative decay path that is lacked when excited directly to the S_1 state [11–15].

MECI between S_1 and S_2 states may also play an important role in excited-state decays of molecules in which the energies of S_1 and S_2 states are close to each other [16–18]. Under the circumstances, S_1 state is usually a bright and locally excited (LE) $\pi\pi^*$ state, and S_2 is a dark state, i.e., nπ^* state, or charge transfer (CT) state. The molecule will be excited firstly to S_1 state, and then relax to the dark state S_2 through MECI between S_1 and S_2 states. After that, the order of S_1 and S_2 states will be changed, and the S_1 state becomes the dark state. Lastly, the molecule returns to the ground state by nonradiative decay and thereby quenching the fluorescence.

In this chapter, we will focus on how MECI between S_1 and S_2 states affects the excited-state decays and quantum yield of fluorescence and photoisomerization of organic molecules. We will discuss this topic from the point of view of theoretical studies. That is, how to describe accurately the excited-state dynamics of organic molecules with state-of-art calculations when interactions between multi excited states are involved.

7.2 Theory and Methods

Owning to the balance between computational efficiency and accuracy of results, time-dependent density functional theory (TDDFT) has been used widely in investigating the excited-state properties of medium and large molecules [19–25]. Nevertheless, it is a great challenge for TDDFT to accurately describe CT state [26, 27] and MECI [28–30] for a long time. With the development of range-separated functional [26, 27] and spin-flip DFT [31–33], TDDFT can now deal with CT state and MECI to a large extent accurately, making the TDDFT becomes the first choice to study excited-state [19–33].

Alternatively, the CASPT2//CASSCF protocol is a typical theoretical method in studying the photochemistry involving the MECI [11–15, 17, 18, 21, 23, 30, 34]. This protocol employs complete-active-space self-consistent-field (CASSCF) to optimize geometries and corrects the energy with second-order perturbation theory (CASPT2) to account for dynamic electron correlation effects external to the active space.

Both TDDFT and CASPT2//CASSCF were employed in our static calculations. Moreover, we have also performed nonadiabatic ab initio dynamics based on Zhu–Nakamura theory [35, 36] to capture crucial dynamical information that may be overlooked by static calculations. A summary of the surface hopping dynamics method based on Zhu–Nakamura's theory [35, 36] is provided as follows. This method is applied in Landau–Zener (LZ) type of nonadiabatic transition [36]. By monitoring the adiabatic energy gap between two involved states, a hopping will be initiated when the trajectory propagates to a local minimum, in which the potential energy surface (PES) can be reduced to a one-dimensional problem. The surface hopping probability is computed firstly by employing the appropriate component of the Zhu–Nakamura formulas [35, 36], and then compared to a generated random number (0–1) to determine whether the hopping occurs. If the hopping is allowed, the velocities need to be adjusted to ensure energy conservation.

The minimum separation r_0 between the two adiabatic PES $V_1(r)$ and $V_2(r)$ is firstly determined, and then the crossing energy is given by $E_0 = (V_1(r_0) + V_2(r_0))/2$. With E_0 we can determine $r_0{}^+$ and $r_0{}^-$ by $V_1(r_0{}^+) = V_2(r_0{}^-) = E_0$. If $E > V_1(r_0)$, the nonadiabatic transition probability p_{ZN} is computed by using effective coupling constant a^2 and effective collision energy b^2 as

$$p_{ZN} = \exp\left[-\frac{\pi}{4a}\left(\frac{2}{b^2 + \sqrt{b^4 + 0.4a^2 + 0.7}}\right)^{\frac{1}{2}}\right] \tag{7.1}$$

where

$$a^2 = \frac{\hbar^2\sqrt{d^2 - 1}}{\mu\left(r_0^+ + r_0^-\right)^2\left[V_2\left(r_0\right) - V_1\left(r_0\right)\right]} \tag{7.2}$$

$$b^2 = \frac{\left(E - E_0\right)\sqrt{d^2 - 1}}{\left[V_2\left(r_0\right) - V_1\left(r_0\right)\right]/2} \tag{7.3}$$

with

$$d^2 = \frac{\left[V_2\left(r_0^-\right) - V_1(r_0^-)\right]\left[V_2\left(r_0^+\right) - V_1(r_0^+)\right]}{\left[V_2\left(r_0\right) - V_1\left(r_0\right)\right]^2} \tag{7.4}$$

If $E < V_1(r_0)$, the transition probability in the deep tunneling region is

$$P_{12} = 4p\left(1 - p\right)\sin^2\left(\psi\right) \tag{7.5}$$

where

$$p = \left[1 + B\left(\frac{\sigma}{\pi}\right)e^{2\delta} - g_2\sin^2\left(\sigma\right)\right]^{-1} \tag{7.6}$$

and

$$\psi = \arg(U_1) \tag{7.7}$$

The parameters in Eqs. (7.6) and (7.7) can be approximately computed by using a^2 and b^2.

We have implemented the above method into a homemade program NAIMD-DICP [11–15, 37–39] and applied it to investigate the nonadiabatic dynamics of several organic molecules.

7.3 Results and Discussion

7.3.1 Wavelength-Dependent Photoisomerization Quantum Yield

Owning to ultrafast and reversible *trans-cis* isomerization process upon photoexcitation, azobenzene (AB) and its analogs have been applied widely in molecular switches, storage devices, molecular motors, etc. [40–43]. Phenylazoimidazole is one type of AB-based compound and has a heterocyclic ring in place of the phenyl group. Experimental studies reveal that the isomerization quantum yield of 1-N-methyl-2-(tolyazo)-imidazole (Tai-Me, Fig. 7.1) is wavelength-dependent, which is 0.35 when excited to S_1 state and will decrease to 0.21 when excited to the S_2 state [44, 45]. When N-methyl or the methyl on six-membered is moved, wavelength-dependent photoisomerization quantum yield is not observed for 2-(tolyazo)imidazole (Tai-H, Fig. 7.1) and 2-(phenylazo)imidazole (Pai-H, Fig. 7.1) [44, 45]. The static calculations and on-the-fly surface hopping excited-state dynamical simulations have been performed to uncover the underlying mechanism for the wavelength-dependent photoisomerization quantum yield of Tai-Me and the substituent effect [11, 12].

At the CASSCF(12,10)/6-31G(d) level, one MECI between S_1 and S_2 states and five MECIs between S_0 and S_1 states were located for Pai-H, Tai-H, and Tai-Me [11, 12]. The MECI between S_1 and S_2 states possesses a planar structure and is denoted as $\text{Planar}_{S2/S1}$. The MECIs between S_1 and S_0 can be divided into two groups. The first group involves one MECI and is related to a broadening of the C2N3N4 and N3N4C5 (Fig. 7.1) bond angles. The second group involves four MECIs and is related to a twisting motion around the NH bond. The first and second groups of MECI are denoted as $\text{NNC}_{S1/S0}$ and $\text{TWIST}i_{S1/S0}$ ($i = 1, 2, 3, 4$), respectively.

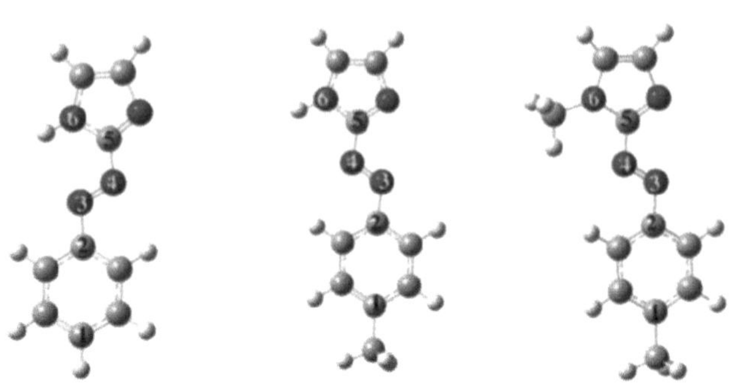

Figure 7.1 The numbering scheme for the three stable arylazoimidazoles structures on the S_0 state. left: Pai-H; middle: Tai-H, and right: Tai-Me. Here and in other figures: white: H; grey: C; blue: N. Reprinted with permission from Ref. [12]. Copyright 2017 American Chemical Society.

TWIST1(3)$_{S1/S0}$ is a mirror structure of TWIST2(4)$_{S1/S0}$ with an opposite rotation direction.

For Pai-H, on-the-fly nonadiabatic molecular dynamic (MD) simulation demonstrated that, upon photoexcitation to S_1 state, only 30% of molecules return to S_0 through the NNC$_{S1/S0}$ MECI, and 70% of molecules still remain on S_1 [12]. The results of nonadiabatic MD simulation of Pai-H from S_2 state are similar to those of from S_1 state, except that the system will first decay to S_1 state through Planar$_{S2/S1}$ within approximately 100 fs. The twisting MECIs are not involved in the excited-state decays of Pai-H from either S_2 or S_1 state, and no *trans-cis* photoisomerization process is observed within our simulation time (1600 fs). These results agree well with the experimental observations that Pai-H has a much lower photoisomerization quantum yield than azobenzene. The excited-state decays of Pai-H from S_2 and S_1 states are summarized in Scheme 7.1.

The non-radiative decays of Tai-H from S_1 and S_2 states are obviously faster than those of Pai-H. Within 1600 fs, all the included trajectories in our nonadiabatic MD simulations from S_1 and S_2 states have returned to S_0 state [12]. The estimated lifetimes of the S_2 and S_1 states were approximately 50 and 798 fs, respectively. Upon photoexcitation to S_2 state, Tai-H will also first return to S_1

Pai-H-Decay Scheme

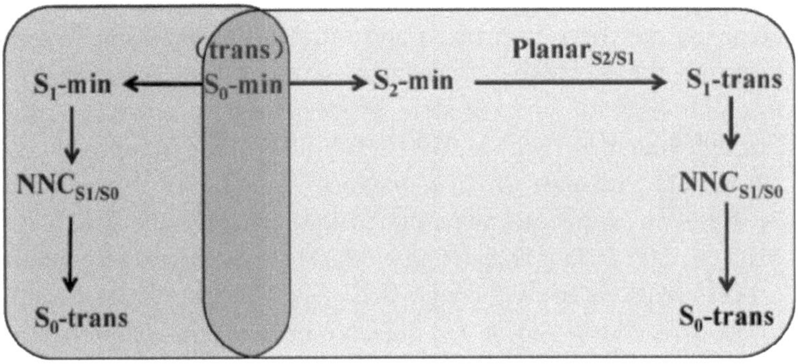

Scheme 7.1 The decay scheme of Pai-H following excitation to the S_1 and S_2 states, respectively. Reprinted with permission from Ref. [12]. Copyright 2017 American Chemical Society.

state through Planar$_{S2/S1}$. After that, the excited-state decay from S_1 to S_0 is similar to that of directly excited to the S_1 state. Two excited-state decay branches are involved, one around the NNC$_{S1/S0}$ MECI, and the other one around the twisting-motion-characterized MECIs. The fractions of the latter are larger than the former. The excited-state decays of Tai-H from S_2 and S_1 states are summarized in Scheme 7.2.

Tai-H-Decay Scheme

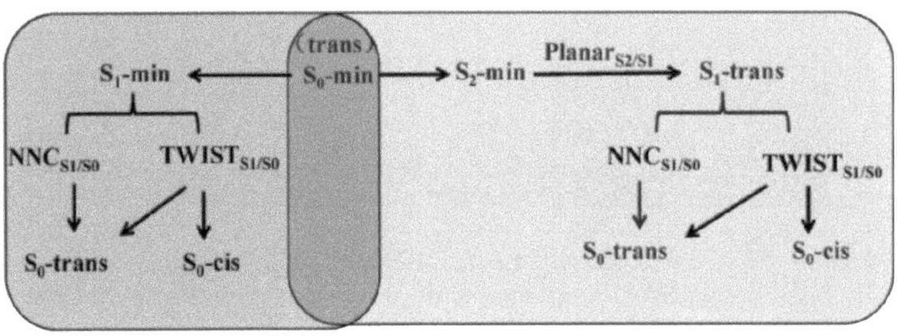

Scheme 7.2 The decay scheme of Tai-H following excitation to the S_1 and S_2 states, respectively. Reprinted with permission from Ref. [12]. Copyright 2017 American Chemical Society.

Upon photoexcitation to S_1 state, Tai-Me can only decay to the ground state through the twisting-motion-characterized MECIs, involving the *trans-cis* process and returning to the original *trans*-form. All the included 50 trajectories return to S_0 before 975 fs. Upon photoexcitation to S_2 state, Tai-Me still first return to S_1 state through $Planar_{S2/S1}$ within 150 fs, and the average is approximately 54 fs [11]. However, in this instance, Tai-Me can return to the S_0 state through both twisting-motion-characterized MECIs and $NNC_{S1/S0}$ MECI. That is, exciting of Tai-Me to S_2 opens a new decay process from S_1 to S_0 through $NNC_{S1/S0}$. This process will return to the original *trans*-form and does not involve photoisomerization, thereby decreasing the quantum yield of photoisomerization. The excited-state decays of Tai-Me from S_2 and S_1 states are summarized in Scheme 7.3, which can explain reasonably the experimentally observed wavelength-dependent photoisomerization quantum yield of Tai-Me [44, 45]. Comparing the excited-state decays of Pai-H, Tai-H, and Tai-Me, we can conclude that the methyl group on the imidazole ring would make the NNC bending become an unfavorable decay channel, whereas the methyl group on the phenyl ring could make the N=N twisting rotation become a favorable decay channel [11, 12].

Tai-Me-Decay Scheme

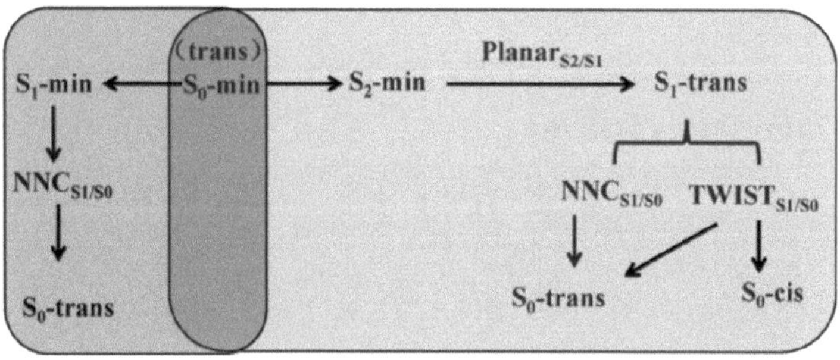

Scheme 7.3 The decay scheme of Tai-Me following excitation to the S_1 and S_2 states, respectively. Reprinted with permission from Ref. [12]. Copyright 2017 American Chemical Society.

Combined static electronic structure calculations and on-the-fly dynamics simulations based on the CASPT2//CASSCF method have also been performed to investigate the wavelength-dependent photoisomerization process of the neutral form of the photoactive yellow protein (PYP) chromophore [14] and trans-urocanic acid [13]. The obtained conclusions are similar to those of arylazoimidazoles [11, 12]. Photoexcitation of neutral form of the PYP chromophore to S_2 state activates a new decay channel from S_1 to S_0 through a MECI characterized by a puckering distortion of the ring. This decay pathway does not involve the *trans-cis* process and decreases the photoisomerization quantum yield [14]. The excited state decays of trans-urocanic acid are more sophisticated than arylazoimidazoles and neutral form of the PYP chromophore, which involves two excited-state decays from S_2 to S_1 state [13]. An overwhelming majority of molecules decay to S_1 by a planar or pucker characterized MECI, and a very small fraction of molecules decay to S_1 through a twisting-motion-characterized MECI around the CC double bond. Only the latter channel can drive the *trans-cis* photoisomerization process, whereas the former channel will drive the molecule to return to the original trans-form [13]. These results can explain the reasons why almost no photoisomerization is observed when trans-urocanic acid is excited to the S_2 state.

In summary, MECI between S_1 and S_2 states plays a significant role in excited-state decays of molecules and may open a new decay pathway from S_1 to S_0 that is lacked when the molecule is excited directly to the S_1 state.

7.3.2 Vibronic Interaction between the Close-Lying $\pi\pi^*$ and nπ^* States

Existence of energetically close S_1 ($\pi\pi^*$) and S_2 (nπ^*) states in nitrogen-heterocyclic and aromatic carbonyl compounds can lead to a significant fluorescence quenching. The "proximity effect" has been proposed to explain this phenomenon [47–49]. That is, the pseudo-Jahn–Teller effect between $S_1(\pi\pi^*)$ and $S_2(n\pi^*)$ states will make the out-of-plane bending mode act as an efficient accepting mode for the nonradiative transition from $S_1(\pi\pi^*)$ to the ground state, leading to an increasing of nonradiative transition rate. An alternative explanation indicated that, although the molecule was firstly excited to the $\pi\pi^*$ state, it will relax to the nπ^* state through a

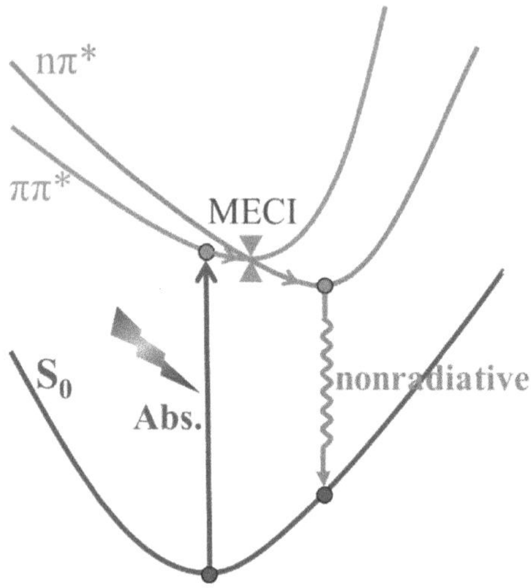

Figure 7.2 Schematic diagram of fluorescence quenching mechanism by nπ* state through MECI between $\pi\pi$* and nπ* states.

MECI between $\pi\pi$* and nπ* states (see Fig. 7.2). After that, the order of $\pi\pi$* and nπ* states will be changed and the nπ* state become the S_1 state. Then the molecule returns to the ground state from nπ* state through nonradiative decay. Recent theoretical studies on the latter mechanism demonstrated that accurately describing the vibronic interaction between $\pi\pi$* and nπ* states represent a major challenge for the widely used TDDFT method [16–18].

By employing the TD-B3LYP/def2-TZVP method, Su and coworkers [16] investigated the fluorescence quenching of 2,5-dimethoxybenzaldehyde (DMBA) in chloroform. At the optimized S_0 geometry, the S_1 and S_2 states were $\pi\pi$* and nπ* states, respectively (see Fig. 7.3). At the optimized nπ* geometry, the order of $\pi\pi$* and nπ* states was inverted, indicating that a MECI between $\pi\pi$* and nπ* states was involved (see Fig. 7.3). Although the energy of $\text{MECI}_{\pi\pi^*/n\pi^*}$ of DMBA was only larger than the Franck–Condon (FC) point approximately 3 kcal/mol, the excited-state decay through this MECI was still energetically unfavorable, because the energy decreased gradually from FC point to the optimized geometry of $\pi\pi$*. Further extensive theoretical studies on 8-methoxy-4-methyl-2H-benzo[g]chromen-2-one (MMBC) by Tang et al. [17]

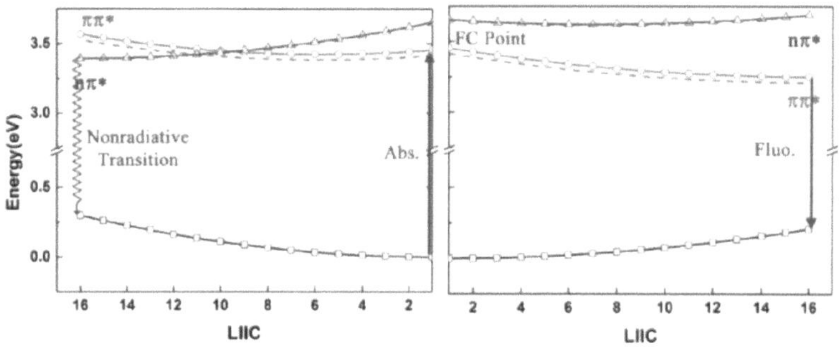

Figure 7.3 TD-B3LYP/def2-TZVP Computed energy profiles along the constructed LIIC pathways between the optimized ground and $\pi\pi^*$ (right) and $n\pi^*$ (left) states. Both the nonequilibrium (solid line) and equilibrium (dashed line) solvation models are used. Reprinted with permission from Ref. [16]. Copyright 2019 American Chemical Society.

demonstrated that the carbonyl stretch can act as an FC active mode and provide a driving force for excited-state decay from $n\pi^*$ states.

In non-hydrogen-bonding solvents, MMBC is almost nonfluorescent. TDDFT calculations with different functionals (PBE0, B3LYP, CAM-B3LYP, M062X, and ωB97XD) reveal that the energy of $n\pi^*$ state is always larger than that of $\pi\pi^*$ state. The small energy gap between these two states is observed at the optimized $n\pi^*$ geometry, which is approximately 0.11 eV at the TD-PBE0/TZVP level with the corrected linear response (cLR) scheme (top of Fig. 7.4). When the CASSCF is applied, the energy gap between $\pi\pi^*$ and $n\pi^*$ states significantly decreases and inversion of these two states occurs at the optimized $n\pi^*$ geometry, indicating that $\text{MECI}_{\pi\pi^*/n\pi^*}$ is involved. However, extended multiconfiguration quasidegenerate perturbation theory at the second-order of perturbation theory expansion (XMCQDPT2) computed energy profiles along the constructed linearly interpolated internal coordinate (LIIC) pathways between CASSCF optimized geometries are significantly different from those of CASSCF results (middle and bottom of Fig. 7.4). These results demonstrated that neither TDDFT nor CASSCF can provide an accurate description of energy profiles between energetically close $\pi\pi^*$ and $n\pi^*$ states.

When the extended multistate CASPT2 (XMS-CASPT2) method was used to optimize the geometries of MMBC, some unexpected

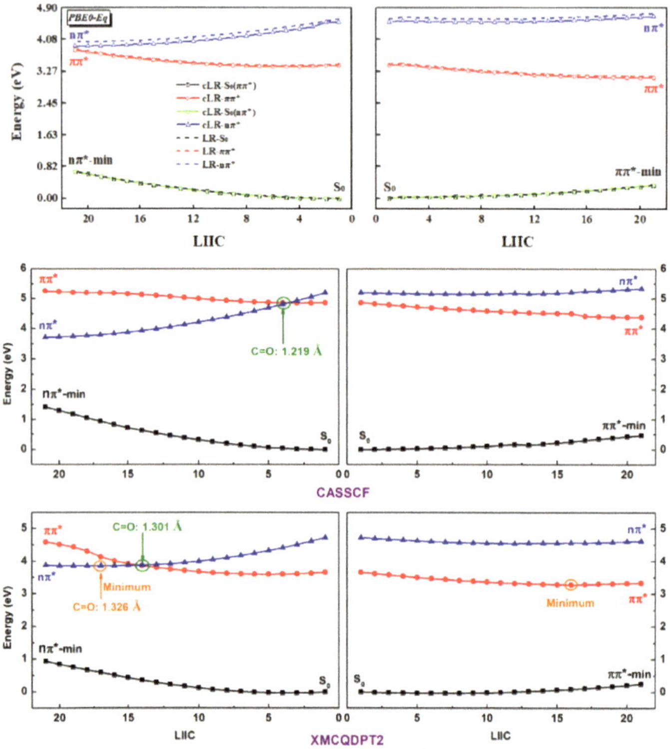

Figure 7.4 The PBE0/ZTVP (top) computed energy profiles along the constructed LIIC pathways between the optimized S_0 and first $\pi\pi^*$ (right) and $n\pi^*$ (left) states with the LR and cLR equilibrium solvation model. CASSCF (middle) and XMCQDPT2 (bottom) computed energy profiles along the constructed LIIC pathways between the SA4-CASSCF/6-31g(d,p) optimized S_0 and the first excited $\pi\pi^*$ (right) and $n\pi^*$ (left) states. Reprinted with permission from Ref. [17]. Copyright 2020 American Chemical Society.

results that were not obtained at the TDDFT and CASSCF levels were revealed. First, and most importantly, the optimized geometry of $\pi\pi^*$ state was no longer planar with some dihedral angles increasing to approximately 8°. This allowed the carbonyl stretch vibration mode to be coupled with the stretch mode of the neighboring C–C bonds of the pyran ring, and the two coupled normal modes were 70 and 71 (inset of Fig. 7.5b). The determined energy barrier from FC point to $\mathrm{MECI}_{\pi\pi^*/n\pi^*}$ at the XMS-CASPT2 level was approximately 6 kcal/mol (Fig. 7.5a). Therefore, the excited decay from the nπ^* state was still energetically unfavorable. The vibrationally resolved absorption spectrum of MMBC was then computed to determine which modes were excited during the excitation process. The stick bands were denoted as n^x, where n was the index of the excited normal mode and x was the quantum number. The results reveal that the second strongest stick band was 71^1 and most of the stick bands in the range of 3.5 to 4.2 eV involve the contribution from 71^x (Fig. 7.5b). The normal mode 71 involved the carbonyl stretch mode and the stretch mode of the neighboring C–C bonds of the pyran ring and correspond to the reaction coordinate from FC to the equilibrium geometry of the nπ^* state. Therefore, based on the XMS-CASPT2 results, the fluorescence quenching of MMBC in non-hydrogen-bonding solvents can be explained as follows. The vibronic interaction between the $\pi\pi^*$ and nπ^* states distort the geometry of $\pi\pi^*$ state along the out-of-plane mode, thereby leading to a coupling between the stretch mode of the carbonyl and its neighboring C–C bond in the pyran ring. Upon photoexcitation, the coupled stretch mode is vibronically active and provides the driving force for excited-state from the nπ^* state through $\mathrm{MECI}_{\pi\pi^*/n\pi^*}$. Moreover, the nature of the vibronic interaction between the $\pi\pi^*$ and nπ^* states can only be captured by the XMS-CASPT2 method, but not the popular TDDFT and CASSCF methods.

7.3.3 Minimal Energy Conical Intersection between Locally Excited and Charge Transfer States

Intramolecular photoinduced electron transfer (PET) can lead to the formation of a CT state with radical-ion-pair nature. Once the energy of the CT state was lower than that of the first LE state, a significant fluorescence quenching can be observed because the formed CT state was a dark state. Akin to the energetically close $\pi\pi^*$ and nπ^* states, the order of the LE and CT states may be changed from the FC point to the equilibrium geometry of the CT state. Under such circumstances, the MECI between LE and CT states ($\mathrm{MECI}_{\mathrm{LE/CT}}$) was involved and both the geometries of LE and CT states needed

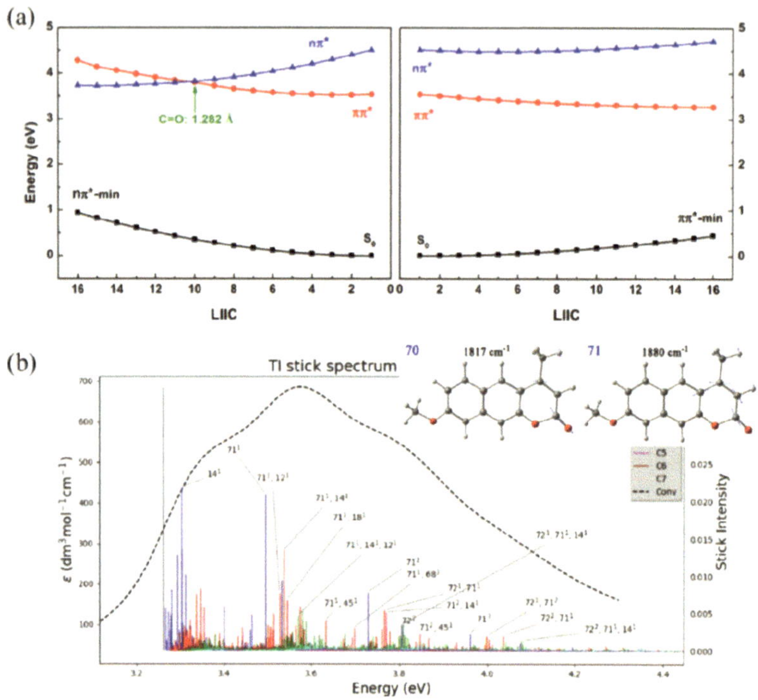

Figure 7.5 (a) The XMS-CASPT2//SA4-CASSCF(14,12)/SVP computed energy profiles along the constructed LIIC pathways between the XMS-CASPT2 optimized S0 and the first excited $\pi\pi^*$ (right) and $n\pi^*$ (left) states. (b) The XMS-CASPT2 computed absorption spectrum of MMBC in gas phase including only the FC contribution, as well as the stick spectra and the assignment for main bands. The XMS-CASPT2//SA3-CASSCF(10,9)/SVP computed normal vibration mode of MMBC in the first excited $\pi\pi^*$ state are shown as insets. Reprinted with permission from Ref. [17]. Copyright 2020 American Chemical Society.

to be optimized. Recently, the fluorescence quenching by CT state involving $MECI_{LE/CT}$ has been reported for BODIPYs [50] and a new imidazo[1,5-α] pyridine-based fluorescent probe [20]. Because the fluorescence quenching mechanisms by PET and $n\pi^*$ state were similar to each other, further discussion on mechanisms will not be presented.

It is well known that the vertical excitation energies (VEEs) of CT state were usually underestimated by TDDFT with certain conventional approximate functional [51, 52]. Although TDDFT with range-separated functional can provide an accurate description of CT state, which may overestimate the VEEs of LE state [53–57]. Finally, the optimally tuned range-separated TDDFT can accurately compute the VEEs of LE and CT states simultaneously [58, 59]. Therefore, one can deal with the PET mechanism well within the TDDFT framework, and the multi-reference methods had been used rarely.

Owning to high fluorescence quantum yield (Φ_{fl}) in aqueous solution and tolerance to photobleaching, fluorescein and its derivatives have been applied to detect several molecules that are of considerable chemical and biological importance [60–63]. Recent extensive theoretical studies based on the XMCQDPT2//CASSCF method have proposed a new set of mechanisms that can explain well the fluorescence behavior of fluorescein derivatives in gas phase, acidic and alkaline solutions, as well as the effects of substitutes on the benzene ring, the intermolecular hydrogen bonding, and temperature [18]. Here we only focus on the fluorescence quenching mechanism of fluorescein monoanion in the gas phase, in which the orientation-controlled PET was involved [18].

The dominant form of fluorescein in the gas phase is fluorescein monoanion (see Scheme 7.4), which was nearly non-fluorescent and showed a very weak and broadband with maximum emission between 550 nm (2.25 eV) and 650 nm (1.91 eV) [64, 65].

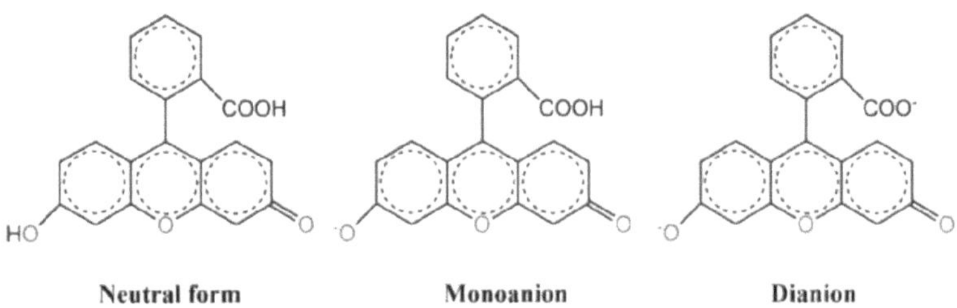

Scheme 7.4 Chemical structures of neutral, monoanionic, and dianionic form of fluorescein. Reprinted with permission from Ref. [18]. Copyright 2021 American Chemical Society.

Depending on the orientation of OH of carboxyl, there exist three different conformers (see Fig. 7.6). The previous study has

Figure 7.6 The different conformers of fluorescein monoanion. Reprinted with permission from Ref. [18]. Copyright 2021 American Chemical Society.

demonstrated that the relative populations of the FL_MonoX1, FL_MononX2, and FL_MonoX3 should be 85:14:1.35 [66].

At both the TD-CAMB3LYP/TZVP and SA3-CASSCF(12,12)/6-31g(d) levels, the S_1 and S_2 states of FL_MonoX1 were LE and CT states, respectively, with large energy gaps of approximately 0.66 eV, indicating that the CT state could not be involved in the excited-state decays (see Fig. 7.7). The CT state energies of FL_MononX2 and FL_MonoX3 significantly decreased and were lower than those of LE states at the TD-CAMB3LYP/TZVP level, indicating that the fluorescence quenching of FL_MonoX2 and FL_MonoX3 resulted from the dark CT state. The CT state involved PET from xanthene moiety to benzene ring, whereas the involved transition orbitals of LE state were all localized on xanthene moiety (see Fig. 7.8c). At the CASSCF level, S_1 and S_2 states were superposition states involving the contribution from LE and CT states, making the geometry optimization of pure LE and CT states difficult. Figure 7.8b shows the XMS-CASPT2 computed LIIC pathways of FL_MonoX2 between the CAM-B3LYP/TZVP optimized S_0 and CT (left) (LE (right)) geometries. The results revealed that, at the FC point, the order of states computed by XMS-CASPT2 was opposite to that of TD-CAM-B3LYP. XMS-CASPT2 calculated S_1 state should be regarded as LE state with the weights of LE and CT configurations approximately 0.68 and 0.08, respectively, whereas the TD-CAM-B3LYP calculated S_1 state was a pure CT state. A similar conclusion has also been drawn for the S_2 state, which was CT state (LE (0.09)/CT (0.70)) at XMS-CASPT2 level and pure LE state at TD-CAM-B3LYP level, respectively. These results indicated that the TD-CAM-B3LYP might not describe accurately the coupling between LE and CT states of FL_MonoX2 (and FL_MonoX3). From the FC point to the TD-CAM-

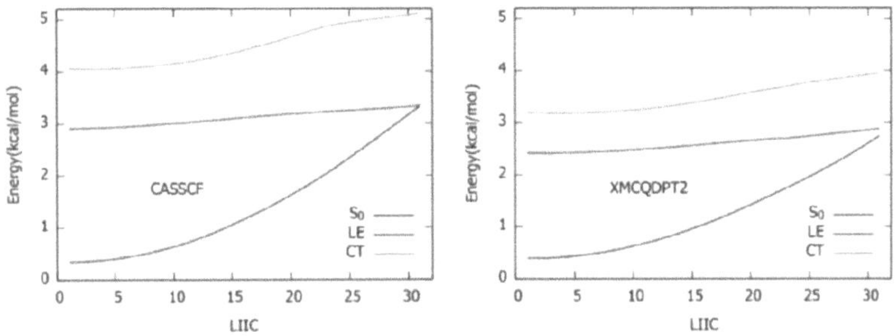

Figure 7.7 The CASSCF (left) and XMCQDPT2 (right) computed energy profiles along the constructed LIIC pathways between the SA3-CASSCF(12,12)/6-31G(d) optimized S_1 and MECI geometries of fluorescein monoanion (FL_MonoX1). Reprinted with permission from Ref. [18]. Copyright 2021 American Chemical Society.

B3LYP optimized geometry of the CT state, the weights of CT and LE configurations of XMS-CASPT2 computed S_1 (S_2) state increase (decrease) and decrease (increase), respectively, making the XMS-CASPT2 computed order of LE and CT states has been changed (see left of Fig. 7.8b). From the FC point to the TD-CAM-B3LYP optimized geometry of LE state, the XMS-CASPT2 computed order of LE and CT states was always opposite to that of TD-CAM-B3LYP (see right of Fig. 7.8b). Finally, the XMS-CASPT2 computed LIIC pathways in Fig. 7.8b revealed that the nonradiative transition from the CT state to the ground state was energetically favorable for FL_MonoX2. These results were in accordance with the weak emission of fluorescein monoanion. Similar results have also been obtained for FL_MonoX3 expect that the energy of CT state of FL_MonoX3 was slightly lower than that of FL_MonoX2.

7.4 Summary and Outlook

In summary, in this chapter, we discuss the role of MECI between S_1 and S_2 states in tuning the photophysical and photochemical properties of organic molecules. The combined static calculations based on CASPT2//CASSCF and on-the-fly surface hopping excited-state dynamical simulations based on CASSCF reveal that the MECI between S_1 and S_2 states may open a new IC pathway from S_1 state

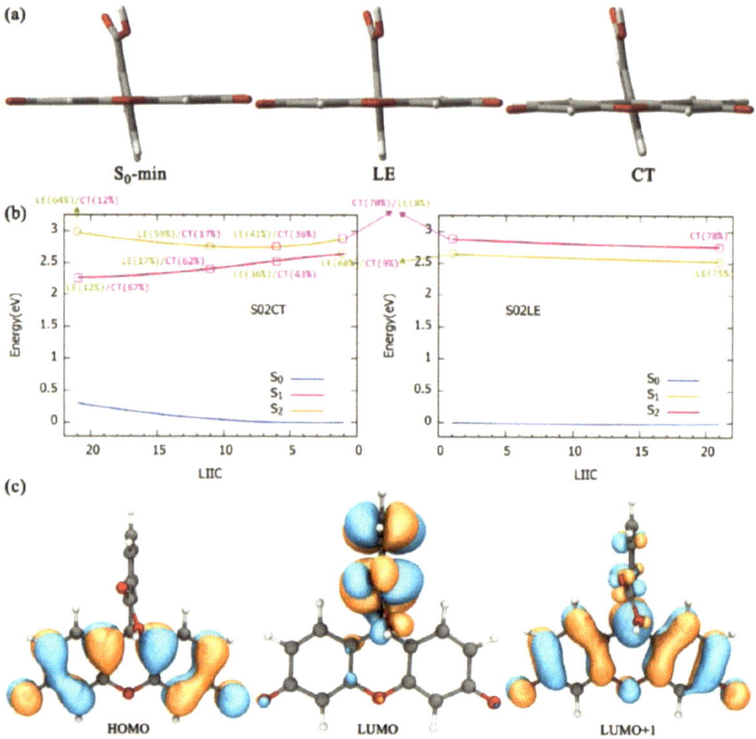

Figure 7.8 (a) The CAM-B3LYP/TZVP optimized geometries of S_0, LE, and CT states of fluorescein monoanion (FL_MonoX2) in gas phase. (b) The XMS-CASPT2//CASSCF(12,12)/6-31g(d) computed energy profiles along the constructed LIIC pathways based on the CAM-B3LYP/TZVP optimized geometries. (c) The CAM-B3LYP computed HOMO, LUMO, and LUMO+1 of fluorescein monoanion (FL_MonoX2) at the optimized S_0 geometry. Reprinted with permission from Ref. [18]. Copyright 2021 American Chemical Society.

to S_0 state, which will be active when the molecule is excited to S_2 state and is inactive when the molecule is excited to the S_1 state. The new IC pathway may not involve the photoisomerization process and thereby lead to wavelength-dependent photoisomerization quantum yield [11–15].

MECI between energetically close $\pi\pi^*$ and $n\pi^*$ states (or LE and CT states) play a vital role in fluorescence quenching by dark $n\pi^*$ state (or CT state) [16–18]. The popular TDDFT method cannot capture the nature of the strong coupling between $\pi\pi^*$ and nπ^* states (or LE and CT states) accurately, which may overestimate the energy gap between $\pi\pi^*$ and nπ^* states [17] and underestimate the energy of CT state even with the range-separated functional CAM-B3LYP [18]. The vibronic interaction between the $\pi\pi^*$ and nπ^* states can only be described accurately when the XMS-CASPT2 method was used to optimize geometries [17]. The XMS-CASPT2 computed energy profiles of FL_MonoX2 and FL_MonoX3 were significantly different from those of TD-CAM-B3LYP [18]. All these results demonstrate the failure of TDDFT in dealing with the excited-state decays involving energetically close $\pi\pi^*$ and nπ^* states (or LE and CT states). Further theoretical methods within the TDDFT framework are needed to develop to tackle these problems.

References

1. Klán, P., and Wirz, J. (2009). *Photochemistry of Organic Compounds: From Concepts to Practice*, A John Wiley and Sons, Ltd, Publication.
2. Itoh, T. (2012). Fluorescence and phosphorescence from higher excited states of organic molecules. *Chem. Rev.* **112**, 4541–4568.
3. Demchenko, A. P., Tomin, V. I., and Chou, P.-T. (2017). Breaking the Kasha rule for more efficient photochemistry. *Chem. Rev.* **117,** 13353–13381.
4. Ryan, W. L., Gordon, D. J., and Levy, D. H. (2002). Gas-Phase photochemistry of the photoactive yellow protein chromophore trans-p-coumaric acid. *J. Am. Chem. Soc.* **124**, 6194–6201.
5. Ryan, W. L., and Levy, D. H. (2001). Electronic spectroscopy and photoisomerization of trans-urocanic acid in a supersonic jet. *J. Am. Chem. Soc.* **123**, 961–966.
6. Morrison, H., Bernasconi, C., and Pandey, G. (1984). A wavelength effect on urocanic acid E/Z photoisomerization. *Photochem. Photobiol.* **40**, 549–550.
7. Otsuki, J., Suwa, K., Narutaki, K., Sinha, C., Yoshikawa, I., and Araki, K. (2005). Photochromism of 2-(phenylazo)imidazoles. *J. Phys. Chem. A* **109**, 8064–8069.
8. Otsuki, J., Suwa, K., Sarker, K. K., and Sinha, C. (2007). Photoisomerization and thermal isomerization of arylazoimidazoles. *J. Phys. Chem. A* **111**, 1403–1409.

9. Changenet-Barret, P., Gustavsson, T., Improta, R., and Markovitsi, D. (2015). Ultrafast excited-state deactivation of 8-hydroxy-2′-deoxyguanosine studied by femtosecond fluorescence spectroscopy and quantum-chemical calculations. *J. Phys. Chem. A* **119**, 6131–6139.
10. Lu, Z., Beckstead, A. A., Kohler, B., and Matsika, S. (2015). Excited state relaxation of neutral and basic 8-oxoguanine. *J. Phys. Chem. B* **119**, 8293–8301.
11. Zhao, L., Zhou, P.-W., and Zhao, G.-J. (2016). Non-adiabatic dynamics simulation exploration of the wavelength-dependent photoinduced relaxation mechanism of trans-N-1-methyl-2-(tolylazo) imidazole in the gas phase. *RSC Adv.* **6**, 64323–64331.
12. Zhao, L., Liu, J.-Y., and Zhou, P.-W. (2017). Effect of Methylation on the photodynamical behavior of arylazoimidazoles: new insight from theoretical ab initio potential energy calculations and molecular dynamics simulations. *J. Phys. Chem. A* **121**, 141–150.
13. Zhao, L., Zhou, P.-W., and Zhao, G.-J. (2016). Non-adiabatic dynamics investigation of the radiationless decay mechanism of trans-urocanic acid in the S_2 state. *J. Chem. Phys.* **145**, 044316.
14. Zhao, L., Liu, J.-Y., and Zhou, P.-W. (2019). Does the wavelength dependent photoisomerization process of the *p*-coumaric acid come out from the electronic state dependent pathways? *Spectrochim. Acta A* **211**, 203–211.
15. Zhao, L., Zhou, P.-W., Liu, X.-X., Zheng, H.-X., Zhan, K.-Y., Luo, J.-H., and Liu, B. (2021). Theoretical studies of the ultrafast deactivation mechanism of 8-oxo-guanine on the S_1 and S_2 electronic states in gas phase. *Spectrochim. Acta A* **244**, 118884.
16. Su, Y., Li, K., and Yu, X. (2019). Theoretical studies on the fluorescence enhancement of benzaldehydes by intermolecular hydrogen bonding. *J. Phys. Chem. B* **123**, 884–890.
17. Tang, Z., Li, P., Liu, J.-Y., and Zhou, P.-W. (2020). Carbonyl stretch as a Franck–Condon active mode and driving force for excited-state decay of 8-methoxy-4-methyl-2H-benzo[g]chromen-2-one from $n\pi^*$ State. *J. Phys. Chem. B* **124**, 11472–11480.
18. Zhou, P.-W., Tang, Z., Li, P., and Liu, J.-Y. (2021). Unraveling the mechanism for tuning the fluorescence of fluorescein derivatives: the role of conical intersection and $n\pi^*$ state. *J. Phys. Chem. Lett.* **12**, 6478–6485.
19. Zhou, P.-W., and Han, K.-L. (2018). Unraveling the detailed mechanism of excited-state proton transfer. *Acc. Chem. Res.* **51**, 1681–1690.

20. Tang, Z., Bai, T.-X., and Zhou, P.-W. (2020). Sensing mechanism of a fluorescent probe for cysteine: photoinduced electron transfer and invalidity of excited-state intramolecular proton transfer. *J. Phys. Chem. A* **124**, 6920–6927.

21. Lou, Z.-R., Zhou, X.-Y., Tang, Z., and Zhou, P.-W. (2020). Theoretical insights into the excited state decays of a donor-acceptor dyad: is the twisted and rehybridized intramolecular charge-transfer state involved? *J. Phys. Chem. B* **124**, 4564–4572.

22. Tang, Z., and Zhou, P.-W. (2020). New insights into the excited state dynamics of quinoline-pyrazole isomerism. *J. Phys. Chem. B* **124**, 3400–3407.

23. Zhou, P.-W., Li, P., Zhao, Y.-L., and Han K.-L. (2019). Restriction of flip-flop motion as a mechanism for aggregation-induced emission. *J. Phys. Chem. Lett.* **10**, 6929–6935.

24. Hubin, P. O., Laurent, A. D., Vercauteren, D. P., and Jacquemin, D. (2014). Investigation of ESIPT in a panel of chromophores presenting N–H··· N intramolecular hydrogen bonds. *Phys. Chem. Chem. Phys.* **46**, 25288–25295.

25. Azarias, C., Budzák, Š., Laurent, A. D., Ulrich, G., and Jacquemin, D. (2016). Tuning ESIPT fluorophores into dual emitters. *Chem. Sci.* **7**, 3763–3774.

26. Dreuw, A., and Head-Gordon, M. (2004). Failure of time-dependent density functional theory for long-range charge-transfer excited states: the zincbacteriochlorin–bacteriochlorin and bacteriochlorophyll—spheroidene complexes. *J. Am. Chem. Soc.* **126**, 4007–4016.

27. Autschbach, J. (2009). Charge-transfer excitations and time-dependent density functional theory: problems and some proposed solutions. *ChemPhysChem* **10**, 1757–1760.

28. Casida, M. E., and Huix-Rotllant, M. (2012). Progress in time-dependent density functional theory. *Annu. Rev. Phys. Chem.* **63**, 287–323.

29. González, L., Escudero, D., and Serrano-Andrés, L. (2012). Progress and challenges in the calculation of electronic excited states. *ChemPhysChem.* **13**, 28–51.

30. Zhou, P.-W., Liu, J.-Y., Han, K.-L., and He, G.-Z. (2014). The photoisomerization of 11-*cis*-retinal protonated Schiff base in gas phase: insight from spin–flip density functional theory. *J. Comput. Chem.* **35**, 109–120.

31. Minezawa, N., and Gordon, M. S. (2009). Optimizing conical intersections by spin–flip density functional theory: application to ethylene. *J. Phys. Chem. A* **113**, 12749–12753.

32. Minezawa, N., and Gordon, M. S. (2011). Photoisomerization of stilbene: a spin–flip density functional theory approach. *J. Phys. Chem. A* **115**, 7901–7911.

33. Minezawa, N., and Gordon, M. S. (2012). Optimizing conical intersections of solvated molecules: the combined spin–flip density functional theory/effective fragment potential method. *J. Chem. Phys.* **137**, 034116.

34. Zhou, P.-W., and Zhao, L. (2019). How does the O^6-methylation regulate the excited-state decay of guanine monomers. *J. Phys. Chem. B* **123**, 201–206.

35. Zhu, C.-Y., and Nakamura, H. (1994). Theory of nonadiabatic transition for general two-state curve crossing problems. I. Nonadiabatic tunneling case. *J. Chem. Phys.* **101**, 10630–10647.

36. Zhu, C.-Y., Nobusada, K., and Nakamura, H. (2001). New implementation of the trajectory surface hopping method with use of the Zhu–Nakamura theory. *J. Chem. Phys.* **115**, 3031–3044.

37. Gao, A.-H., Li, B., Zhang, P.-Y., and Han, K. L. (2012). Nonadiabatic ab initio molecular dynamics of photoisomerization in bridged azobenzene. *J. Chem. Phys.* **137**, 204305:1–7.

38. Zhao, L., Zhou, P.-W., Li, B., Gao, A.-H., and Han, K.-L. (2014). Non-adiabatic dynamics of isolated green fluorescent protein chromophore anion. *J. Chem. Phys.* **141**, 235101:1–9.

39. Zhao, L., Liu, J.-Y., and Zhou, P.-W. (2016). New insight into the photoisomerization process of the salicylidene methylamine under vacuum. *J. Phys. Chem. A* **120**, 7419–7426.

40. Puntoriero, F., Ceroni, P., Balzani, V., Bergamini, G., and Vögtle, F. (2007). Photoswitchable dendritic hosts: a dendrimer with peripheral azobenzene groups. *J. Am. Chem. Soc.* **129**, 10714–10719.

41. Wang, L., Xu, J., Zhou, H., Yi, C., and Xu, W. (2009). *cis-trans* isomerization mechanism of 4-aminoazobenzene in the S_0 and S_1 states: a CASSCF and DFT study. *J. Photochem. Photobiol., A* **205**, 104–108.

42. Ciminelli, C., Granucci, G., and Persico, M. (2004). The photoisomerization mechanism of azobenzene: a semiclassical simulation of non-adiabatic dynamics. *Chem. Eur. J.* **10**, 2327–2341.

43. Samanta, S., Beharry, A. A., Sadovski, O., McCormick, T. M., Babalhavaeji, A., Tropepe, V., and Woolley, G. A. (2013). Photoswitching azo compounds in vivo with red light. *J. Am. Chem. Soc.* **135**, 9777–9784.

44. Otsuki, J., Suwa, K., Narutaki, K., Sinha, C., Yoshikawa, I., and Araki, K. (2005). Photochromism of 2-(Phenylazo)imidazoles. *J. Phys. Chem. A* **109**, 8064–8069.

45. Otsuki, J., Suwa, K., Sarke, K. K., and Sinha, C. (2007). Photoisomerization and thermal isomerization of arylazoimidazoles. *J. Phys. Chem. A* **111**, 1403–1409.

46. Ryan, W. L., and Levy D. H. (2001). Electronic spectroscopy and photoisomerization of trans-urocanic acid in a supersonic jet. *J. Am. Chem. Soc.* **123**, 961–966.

47. Wassam, W. A., and Lim, E. C. (1978). "Proximity effect" in radiationless transitions. *J. Chem. Phys.* **68**, 433–454.

48. Wassam, W. A., and Lim, E. C. (1978). "Proximity effect" and radiationless transitions in aromatic molecules with non-bonding electrons. *J. Mol. Struct.* **47**, 129–198.

49. Lim, E. C. (1986). Proximity effect in molecular photophysics: dynamical consequences of pseudo-Jahn–Teller interaction. *J. Phys. Chem. A* **90**, 6770–6777.

50. Chi, W., Chen, J., Liu, W., Wang, C., Qi, Q., Qiao, Q., Tan, T. M., Xiong, K., Liu, X., Kang, K., Chang, Y.-T., Xu, Z., and Liu, X. (2020). A general descriptor ΔE enables the quantitative development of luminescent materials based on photoinduced electron transfer. *J. Am. Chem. Soc.* **142**, 6777–6785.

51. Dreuw, A., and Head-Gordon, M. (2004). Failure of time-dependent density functional theory for long-range charge-transfer excited states: the Zincbacteriochlorin–Bacteriochlorin and Bacteriochlorophyll–Spheroidene complexes. *J. Am. Chem. Soc.* **126**, 4007–4016.

52. Autschbach, J. (2009). Charge-transfer excitations and time-dependent density functional theory: problems and some proposed solutions. *ChemPhysChem* **10**, 1757–1760.

53. Leang, S. S., Zahariev, F., and Gordon, M. S. (2012). Benchmarking the performance of time-dependent density functional methods. *J. Chem. Phys.* **136**, 104101.

54. Laurent, A. D., and Jacquemin, D. (2013). TD-DFT benchmarks: a review. *Int. J. Quantum Chem.* **113**, 2019–2039.

55. Zhou, P.-W. (2018). Why the lowest electronic excitations of rhodamines are overestimated by time-dependent density functional theory? *Int. J. Quantum Chem.* **118**, e25780.

56. Zhou, P.-W., Liu, J.-Y., Yang, S.-Q., Chen, J.-S., Han, K.-L., and He, G. Z. (2012). The invalidity of the photo-induced electron transfer mechanism for fluorescein derivatives. *Phys. Chem. Chem. Phys.* **14**, 15191–15198.

57. Zhou, P.-W., Ning, C., Alsaedi, A., and Han, K.-L. (2016). The effects of heteroatoms Si and S on tuning the optical properties of rhodamine-

and fluorescein-based fluorescence probes: a theoretical analysis. *ChemPhysChem* **17**, 3139–3145.

58. Zhou, P.-W., and Zhao, L. (2018). Accurate description of excited state intramolecular proton transfer that involves zwitterionic state using optimally tuned range-separated time-dependent density functional theory. *Int. J. Quantum Chem.* **118**, e25618.
59. Kuritz, N., Stein, T., Baer, R., and Kronik, L. (2011). Charge-transfer-like $\pi \rightarrow \pi^*$ excitations in time-dependent density functional theory: a conundrum and its solution. *J. Chem. Theory Comput.* **7**, 2408–2415.
60. Kobayashi, H., Ogawa, M., Alford, R., Choyke, P. L., and Urano, Y. (2010). New strategies for fluorescent probe design in medical diagnostic imaging. *Chem. Rev.* **110**, 2620–2640.
61. Zheng, H., Zhan, X.-Q., Bian, Q.-N., and Zhang, X.-J. (2013). Advances in modifying fluorescein and rhodamine fluorophores as fluorescent chemosensors. *Chem. Commun.* **49**, 429–447.
62. Urano, Y., Kamiya, M., Kanda, K., Ueno, T., Hirose, K., and Nagano, T. (2005). Evolution of fluorescein as a platform for finely tunable fluorescence probes. *J. Am. Chem. Soc.* **127**, 4888–4894.
63. Kamiya, M., Urano, Y., Ebata, N., Yamamoto, M., Kosuge, J., and Nagano, T. (2005). Extension of the applicable range of fluorescein: a fluorescein-based probe for western blot analysis. *Angew. Chem., Int. Ed.* **44**, 5439–5441.
64. Yao, H., and Jockush, R. (2013). Fluorescence and electronic action spectroscopy of mass-selected gas-phase fluorescein, 2′,7′-dichloro-fluorescein, and 2′,7′-difluorofluorescein ions. *J. Phys. Chem. A* **117**, 1351–1359.
65. Kjær, C., Nielsen, S. B., and Stockett, M. H. (2017). Sibling rivalry: intrinsic luminescence from two xanthene dye monoanions, resorufin and fluorescein, provides evidence for excited-state proton transfer in the latter. *Phys. Chem. Chem. Phys.* **19**, 24440–24444.
66. Yao, H., Steill, J. D., Oomens, J., and Jockusch, R. A. (2011). Infrared multiple photon dissociation action spectroscopy and computational studies of mass-selected gas-phase fluorescein and 2′,7′-dichlorofluorescein ions. *J. Phys. Chem. A* **115**, 9739–9747.

Chapter 8

Excited Carrier Dynamics in Condensed Matter Systems Investigated by ab initio Nonadiabatic Molecular Dynamics

Qijing Zheng,[a] Weibin Chu,[b] Xiang Jiang,[a] Lili Zhang,[a] Yunzhe Tian,[a] Hongli Guo,[a] and Jin Zhao[a,b,c]

[a] *ICQD/Hefei National Laboratory for Physical Sciences at Microscale, CAS Key Laboratory of Strongly-Coupled Quantum Matter Physics, and Department of Physics, University of Science and Technology of China, Hefei, Anhui 230026, China*
[b] *Department of Physics and Astronomy, University of Pittsburgh, Pittsburgh, Pennsylvania 15260, USA*
[c] *Synergetic Innovation Center of Quantum Information and Quantum Physics, University of Science and Technology of China, Hefei, Anhui 230026, China*
zhaojin@ustc.edu.cn

8.1 Introduction

The excited carrier dynamics generally exist in different physical and chemical processes in condensed matter systems, and it plays crucial role in different research areas such as solar energy conversion and optoelectronics (Hong et al., 2014; Withers et al., 2015; Wang and Xia, 2015; Britnell et al., 2013; Yu et al., 2013; Wang et al.,

Time-Dependent Density Functional Theory: Nonadiabatic Molecular Dynamics
Edited by Chaoyuan Zhu

ISBN 978-981-4968-42-3 (Hardcover), 978-1-003-31921-4 (eBook)
www.jennystanford.com

2012; Feng et al., 2012; Mak et al., 2012; Eda et al., 2010). For example, in photovoltaic systems, the lifetime of the excited carriers, the relaxation of the hot carriers and the charge transfer efficiency at the interface affect the efficiency of the device collectively. In the photocatalytic processes at the surfaces, the photoexcited carriers need to have long-enough lifetime during which they need to be trapped by the absorbed molecules efficiently to stimulate the following chemical reactions. In optoelectronics, photoexcited carrier dynamics is linked to several different figures of merit (Kumar et al., 2014; Korn et al., 2011; Docherty et al., 2014; Shi et al., 2013; Nie et al., 2014; Wang et al., 2015). For instance, it is important to suppress the nonradiative electron–hole (*e-h*) recombination for light emitting devices. By contrast, for optical switches, the excited carrier lifetime needs to be short in order to have a fast response. Thus, for the design of new solar energy and optoelectronic devices, it is important to understand the excited carrier dynamics in condensed matter systems at the atomic level.

The challenges for the excited carrier dynamics in condensed matter systems exist in several different aspects. First, different with the small molecular systems, the condensed matter systems usually contain large amount of electrons and atoms. Second, there are different quasi-particles (QP) as well as defects, impurities, crystal boundaries existing in the condensed matter systems, which will interact with the excited carriers during their relaxation process. Finally, different interactions, such as electron–phonon (*e-ph*) interaction, spin-orbit coupling (SOC) and the many-body effects of *e-h* pair come into play collectively. To understand such complex excited carrier dynamics in condensed matter systems, the ab initio investigations are essential.

In the recent 20 years, ab initio nonadiabatic molecular dynamics (NAMD) methods for the condensed matter systems have been developed. In particular, if the nuclear quantum effects are known to be unimportant, then one can reduce the computational cost significantly via the mixed quantum-classical (MQC) method, where only the electrons (quantum subsystem) are treated quantum mechanically while the nuclei (classical subsystem) are considered as classical particles . The MQC method includes quantum–classical Liouville dynamics (Mac Kernan et al., 2008; Kelly and Markland,

2013), nonadiabatic Bohmian dynamics (Curchod et al., 2011), coupled-trajectories mixed quantum–classical method (Agostini et al., 2016), Ehrenfest dynamics (also known as mean-field method) (Tully, 1998; Ma et al., 2016; Bircher et al., 2017) and trajectory surface hopping (TSH) method (Tully, 1990; Craig et al., 2005). Among them, the last two are perhaps the most widely used ones. In the mean-field approach, the classical subsystem evolves on a single potential energy surface (PES), which is weight-averaged over the quantum states. Trajectory surface hopping method was proposed to deal with the quantum-classical correlation by introducing stochastic hops between different PESs (Barbatti, 2011; Tully, 1990). The hopping algorithm is not unique and the most popular one is Tully's fewest-switches algorithm (Tully, 1990). For the evolution of the quantum subsystem, a general way is to expand the electronic wavefunction in a basis set (representation), the Schrödinger equation then becomes a set of equations for the expanding coefficients. Under this scheme, the knowledge of the ground state, excited states and nonadiabatic couplings (NACs) between them are required. General quantum chemistry method can be used to generate all the information needed and it has been implemented in a few codes (Barbatti, 2011; Akimov and Prezhdo, 2013, 2014; Barbatti et al., 2007; Du and Lan, 2015) Tavernelli and coworkers(Tapavicza et al., 2007). showed that the necessary quantities can also be rigorously derived from LR-TDDFT. Prezhdo and coworkers (Craig et al., 2005) develop a scheme where adiabatic Kohn–Sham (KS) orbitals are used to expand the electronic wavefunction. It has been shown that the adiabatic KS orbitals can be seen as zeroth-order approximation to LR-TDDFT adiabatic states (Fischer et al., 2011). This expansion along with the so-called classical-path approximation (CPA) (Wang et al., 2016b; Akimov and Prezhdo, 2013; Long et al., 2017), in which the nuclear dynamics of the system remains unaffected by the dynamics of the electronic degrees of freedom, significantly reduces the computational cost and can be applied to condensed matter systems with large number of atoms.

In Section 8.2 we will introduce the method combining time-dependent Kohn–Sham equation with the surface hopping based on classical-path approximation (CPA). (TDKS–NAMD). In Section 8.3

and 8.4, the applications based on TDKS–NAMD method on interfacial charge transfer and *e*-*h* recombination are discussed. In Section 8.5 the new development of ab initio NAMD based on $GW +$ real-time BSE ($GW +$ rtBSE–NAMD) and its application on the spin-valley exciton dynamics in transition metal dichalcogenide (TMD) is introduced.

8.2 Time-Dependent Kohn–Sham Equation Combined with Surface Hopping (TDKS–NAMD)

In nonadiabatic molecular dynamics, the nuclear motions drive the transitions between the electronic states, which in turn influence the evolution of the nulcei. The classical-path approximation (CPA) (Wang et al., 2016b; Akimov and Prezhdo, 2013; Long et al., 2017) reduces the computational cost significantly by assuming that the nuclear trajectory remains unaffected by the electronic dynamics, while the electronic Hamiltonian still depends parametrically on the nuclear coordinates. Under this approximation, the nulcear motions can be obtained by ground state molecular dynamics at some finite temperature. CPA is valid when electron–electron interactions are much stronger than electron–phonon interactions, or if the kinetic energy of nuclei is significantly greater than the amount of energy exchanged during transitions (Akimov and Prezhdo, 2013). This condition is satisfied in most condensed matter systems, where there is a large amount of electrons in the system and moderate photoexcitation barely affects the system.

In TDKS–NAMD as implemented in Hefei–NAMD (Zheng et al., 2019a), the excited state wavefunction is $\psi_e(\mathbf{r}, t)$ is expanded in the basis of instantaneous adiabatic KS orbitasl $\phi_j(\mathbf{r}, \mathbf{R}(t))$, which are obtained by solving the time-independent KS equation at configuration $\mathbf{R}(t)$, i.e.,

$$\psi_e(\mathbf{r}, t) = \sum_j c_j(t)\phi_j(\mathbf{r}, \mathbf{R}(t)) \tag{8.1}$$

by inserting Eq. (8.1) into the time-dependent Kohn–Sham equation, a set of differential equations for the coefficients $c_j(t)$ is produced

$$i\hbar\frac{\partial}{\partial t}c_j(t) = \sum_k c_k(t)\left[\varepsilon_k\delta_{jk} - i\hbar\mathbf{d}_{jk}(t)\right] \tag{8.2}$$

The square modulus of the coefficients, $|c_j(t)|^2$, can be explained as the population of the corresponding KS orbitals. Here, ε_k is the KS energy of the k-th adiabatic KS states and $\mathbf{d}_{jk}$ is the NACs between adiabatic KS states j and k. The NAC can be written as:

$$\mathbf{d}_{jk} = \langle\phi_j|\frac{\partial}{\partial t}|\phi_k\rangle = \sum_I \frac{\langle\phi_j|\nabla_{\mathbf{R}_I}\mathcal{H}|\phi_k\rangle}{\varepsilon_k - \varepsilon_j}\cdot\dot{\mathbf{R}}_I \tag{8.3}$$

where $\mathcal{H}$ is the KS Hamiltonian, ϕ_j, ϕ_k, ε_j and ε_k are the wavefunctions and eigenvalues for electronic states j and k. $\dot{\mathbf{R}}_I$ is the velocity of the nuclei. Thus, the NAC mainly depend on the energy difference term $\varepsilon_k - \varepsilon_j$, the electron–phonon (*e-ph*) coupling term $\langle\phi_j|\nabla_{\mathbf{R}_I}\mathcal{H}|\phi_k\rangle$ and the nuclear velocity term $\dot{\mathbf{R}}_I$.

The NACs can be calculated numerically using finite-difference method (Billeter and Curioni, 2005; Hammes Schiffer and Tully, 1994).

$$\begin{aligned}\mathbf{d}_{jk} &= \langle\phi_j|\frac{\partial}{\partial t}|\phi_k\rangle \\ &\approx \frac{\langle\phi_j(t)|\phi_k(t+\Delta t)\rangle - \langle\phi_j(t+\Delta t)|\phi_k(t)\rangle}{2\Delta t}\end{aligned} \tag{8.4}$$

In plane-wave basis DFT code (e.g., VASP, Quantum Espresso), the KS wavefunctions are expanded as superpositions of plane-waves, which are themselves delocalized non-centered orthonormalized functions. As a result, the overlap in Eq. (8.4) can be simply written as the inner product of the corresponding plane-wave coefficients. It has to stressed that in most plane-wave DFT codes pseudopotential is used, which means that the KS wavefunctions are pseudo-wavefunctions. Chu et al. proposed a method to compute the all-electron NACs based on projected augmented wave (PAW) method (Chu et al., 2020b). Moreover, since arbitrary phase can be assigned to the KS wavefunctions, phase correction must be applied in the numerical evaluation of NACs (Billeter and Curioni, 2005; Akimov, 2018; Westermayr et al., 2019).

With the coefficients $c_j(t)$ and the NACs, hopping probabilities between the adiabatic KS states according the Tully's fewest-switches algorithm can be obtained as: (Tully, 1990)

$$P_{j\to k}(t, \Delta t) = \frac{2\Re[c_j^* c_k \mathbf{d}_{jk}]\Delta t}{c_j^* c_j} \tag{8.5}$$

Further, the probabilities are multiplied by a Boltzmann factor within the CPA. Besides FSSH, the decoherence-induced surface hopping (DISH)(Jaeger et al., 2012) was also implemented in Hefei-NAMD.

The time-dependent real-space information of the photocarrier can be obtained as the norm-square of the time-dependent KS orbitals $|\psi_e(\mathbf{r}, t)|^2$. By expanding using Eq. (8.1), we have

$$\begin{aligned}\rho(\mathbf{r}, t) &= |\psi_e(\mathbf{r}, t)|^2 \\ &= \sum_j c_j^*(t) c_j(t)\, \phi_j^*(\mathbf{r}, \mathbf{R}(t))\, \phi_j(\mathbf{r}, \mathbf{R}(t))\end{aligned} \tag{8.6}$$

A further separation of the above quantity can be obtained by taking the time derivative of Eq. (8.6)

$$\begin{aligned}\frac{\mathrm{d}\rho(\mathbf{r}, t)}{\mathrm{d}t} &= \sum_j \left[\frac{\mathrm{d}}{\mathrm{d}t} c_j^*(t) c_j(t)\right] \phi_j^*(\mathbf{r}, \mathbf{R}(t))\phi_j(\mathbf{r}, \mathbf{R}(t)) \\ &\quad + c_j^*(t) c_j(t) \left[\frac{\mathrm{d}}{\mathrm{d}t}\phi_j^*(\mathbf{r}, \mathbf{R}(t))\phi_j(\mathbf{r}, \mathbf{R}(t))\right]\end{aligned} \tag{8.7}$$

As can be seen from the Eq. (8.7), the time variation can be clearly divided into two parts: a change primarily due to the variation of the KS states populations $c_j^*(t)c_j(t)$, which comes from the first part of the equation and we denotes it as the nonadiabatic contribution (NA). For the second part of the equation, the time variation derives from the change of the adiabatic KS orbitals and as a result, we refer to this part as the adiabatic contribution (AD).

8.3 Interfacial Charge Transfer Dynamics

8.3.1 Charge Transfer at Molecule/Semiconductor

Since the seminal discovery of water splitting on TiO_2 surface by Fujishima and Honda in 1972, to understand the mechanism and improve the efficiency of photocatalytic reactions on TiO_2 surface has long been great interest for many researchers. In the photocatalytic reactions, primary reagents at high potentials are created by photoexcited electrons and holes at a photocatalytic interface, and the subsequent redox chemistry can be driven. Photoexcited carriers are generated in the bulk of TiO_2 and they will migrate to the surface, where they will be trapped by the adsorbed molecules and stimulate the following chemical reactions. Such charge transfer dynamics at the molecule/TiO_2 interface plays a crucial role in photocatalytic reactions. Chu et al. have chosen CH_3OH/TiO_2 and CO_2/TiO_2 as two prototypical systems to investigate the photoexcited hole and electron trap processes by the molecules at TiO_2 surface (Chu et al., 2016, 2020c).

8.3.1.1 Ultrafast photoexcited hole transfer at CH_3OH/TiO_2 interface

Chu et al. chose the most intensively studied rutile $TiO_2(110)$ surface to perform our investigation. In photocatalytic reactions, hole scavenger has been used to trap photoexcited holes and prevent the *e-h* recombination (Sakai et al., 2003; Nakamura and Nakato, 2004; Shkrob and Sauer, 2004; Thompson and Yates, 2005; Kawai and Sakata, 1980). In the photocatalytic investigations on TiO_2 surface, CH_3OH has been known as a hole scavenger for long time (Thompson and Yates, 2005; Tamaki et al., 2006; Henderson, 2011). After trapping a photoexcited hole, further photoreaction can be driven to form CH_2O (Henderson, 2011; Shen and Henderson, 2011). Yet, there have been debates on the hole trapping ability of CH_3OH. Some experimental investigations suggest that instead of CH_3OH, CH_3O formed by thermal deprotonation is the active species for the hole trapping. Chu et al. have used ab initio TDKS–NAMD simulation to clarify this problem (Chu et al., 2016).

As indicated in Fig. 8.1, Chu et al. have investigated three different dynamical processes at CH_3OH/TiO_2 interface, which includes (i) the forward hole transfer from TiO_2 to CH_3OH, e.g., the hole trapping process; (ii) the reverse hole transfer from CH_3OH to TiO_2; (iii) the hot hole relaxation process. As shown in Fig. 8.2a, using (2×1) unit cell, Chu et al. have investigated three different adsorption

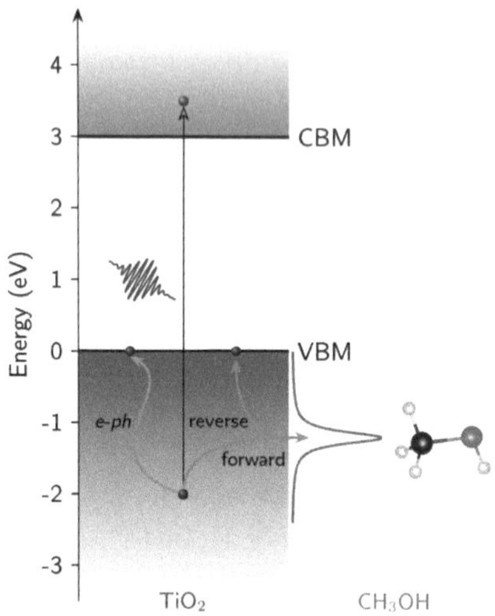

Figure 8.1 Schematic diagram of photoexcited hole dynamics at the CH_3OH/TiO_2 interface.

structures with 1 monolayer coverage including the molecular adsorption (labeled as M), half-dissociated adsorption (labeled as HD) and dissociated adsorption (labeled as D). Figure 8.2b shows the forward hole transfer process. It can be seen that for M structure there is no forward hole transfer. Only when CH_3OH dissociates to form CH_3O, the hole transfer starts to occur. For D structure, where both of the CH_3OH molecules dissociate, around 20% of the photoexcited hole can transfer from TiO_2 to CH_3O. Such results suggest that CH_3O is the hole scavenger instead of CH_3OH. The major reason is the energy level alignment, namely, the highest occupied molecular orbital (HOMO) of CH_3O is closer to the valence band maximum (VBM). In addition, the forward hole transfer process is temperature dependent, suggesting a significant role of phonon excitation and *e-ph* coupling. The backward hole transfer from molecule to TiO_2 behaves similarly for the three structures (Fig. 8.2c), which all happen within 10 fs. However, for D structure, there is around 20% hole remains on CH_3O, which can promote the following photoreaction. Finally, it is found that the hot hole relaxation process is significantly accelerated by molecular

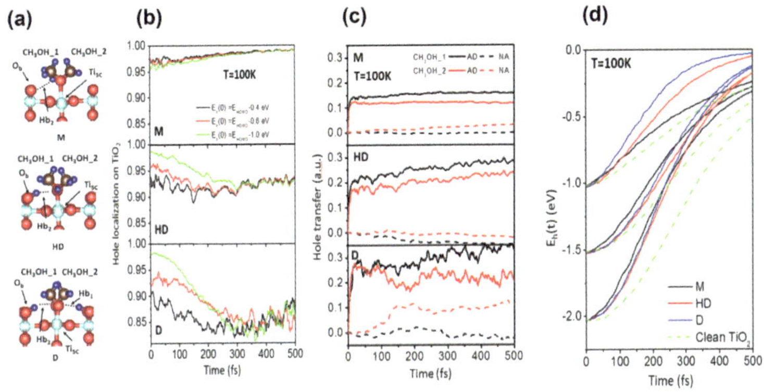

Figure 8.2 (a) The top and side views of the most stable molecular (M), half-dissociated (HD), and fully dissociated (D) adsorption structures. M1 and M2 label different methanol molecules in the unit cell and Hb_1 and Hb_2 their hydrogen bonds. The cyan, red, brown, and blue balls represent Ti, O, C, and H atoms, respectively. (b) Averaged forward hole transfer from TiO_2 bulk states to adsorbed molecules at 100 K. (c) Averaged hole reverse transfer from TiO_2 bulk states to adsorbed molecules at 100 K d)The time dependence of energy relaxation of the photogenerated holes with different initial energies at 100 K.

adsorption. It was attributed to the strong *e-ph* coupling of the HOMOs of adsorbed molecules (Fig. 8.2d).

Our results suggest that at CH_3OH/TiO_2 interface, there are two crucial factors affect the photoexcited hole dynamics. The first one is the energy level alignment. If the HOMO of the orbital is close or even higher than VBM of TiO_2, it will be able to trap the photoexcited holes. The second crucial factor is the phonon excitation and *e-ph* interaction. At higher temperature, more phonons are excited and the hole trapping process is accelerated. The *e-ph* interaction also accelerates the hot hole relaxation process.

8.3.1.2 CO_2 photoreduction on TiO_2 driven by transient capture of photoexcited electron

The photoreduction of CO_2 is one of the most challenging photocatalytic reactions on TiO_2 surface. The most critical bottleneck lies in the activation of CO_2. Different with CH_3OH/TiO_2, the energy level alignment at CO_2/TiO_2 interface is unfavorable for hot electron trapping, since the lowest unoccupied molecular orbital (LUMO) of CO_2 is 2.2 to 4.0 eV above the conduction band minimum (CBM) (Tan et al., 2011; Lee et al., 2011; Sorescu et al., 2011). Thus, to stabilize the LUMO of CO_2 close to CBM is the critical step to break through the bottleneck of CO_2 photoreduction.

Chu et al. also chose rutile $TiO_2(110)$ surface to study the photoreduction of CO_2(Chu et al., 2020c). They found that if a CO_2 adsorbs in an oxygen vacancy (O_v) traps a photoexcited electron, a $CO_2^{\cdot-}$ can be formed. Since the LUMO of CO_2 is much higher than CBM of TiO_2, such $CO_2^{\cdot-}$ has a certain lifetime, within which the electron will relax back to the conduction band of TiO_2. Therefore, such $CO_2^{\cdot-}$ can be understood as a transient $CO_2^{\cdot-}$. Although the transient $CO_2^{\cdot-}$ has a limited lifetime, it is found to affect the CO_2 photoreduction notably. The ab initio TDKS–NAMD simulation proves that if the lifetime of the transient $CO_2^{\cdot-}$ is longer than 13 fs, the bending and antisymmetric stretching modes of CO_2 will be successfully excited and its LUMO will decrease below the CBM within 20 fs. After the energy stabilization, CO_2 is able to trap the photoexcited hot electron within 100 fs, forming a stable $CO_2^{\cdot-}$, and will be reduced to CO within 30 to 40 fs. Such a photoreduction of CO_2 induced by the formation of transient $CO_2^{\cdot-}$ is summarized in Fig. 8.3.

In the whole photoreduction process, the excitation of bending and antisymmetric stretching vibrational modes are essential for the stabilization of CO_2 LUMO. In Chu et al.'s work, they propose that the formation of a transient $CO_2^{\cdot-}$ with lifetime longer than 13 fs can stimulate the vibration modes excitation and the following photoreduction. However, their ab initio TDKS–NAMD simulations also show that the lifetime of falls in to 10 to 15 fs, which is just around the threshold of efficient vibration mode excitation. They propose that there are other methods to excite the bending and anti-symmetric stretching modes, such as using different semiconductor surfaces, deposit metal atoms and clusters, as well as using the newly developed techniques such as THz photoexcitation, which can excite vibrational modes selectively (Hu et al., 2014; Gierz et al., 2015; Pomarico et al., 2017; Mitrano et al., 2016; Wang et al., 2013). In this work, Chu et al. shows that for photocatalytic reactions, if the band alignment is not favorable for charge carrier trapping, the phonon/vibrational modes excitation may be able to change

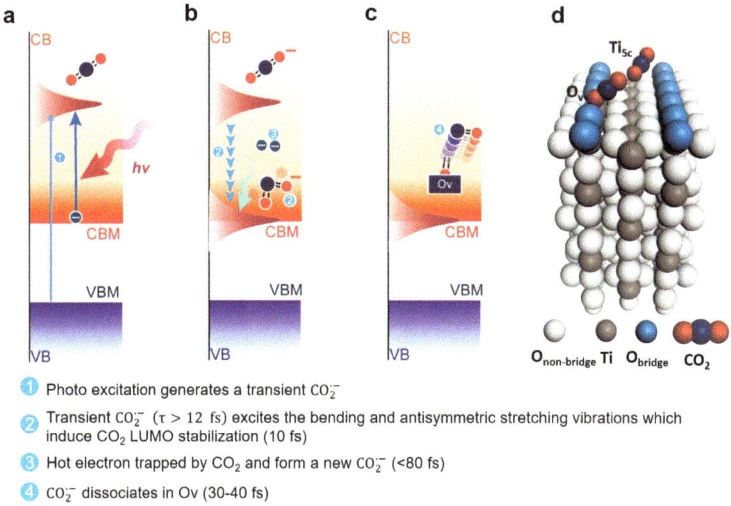

Figure 8.3 Diagram of CO_2 photoreduction on the TiO_2 surface. The four steps of CO_2 photoreduction and the corresponding time scales are schematically indicated in (a)–(c). (d) Schematic showing the CO_2 molecule adsorbed in the O_V and Ti_{5C} sites on the $TiO_2(110)$ surface.

the energy level alignment and stimulate the photoreaction, which provide valuable insights into the design of efficient photocatalysts.

8.3.2 Charge Transfer at van der Waals Heterostructure

8.3.2.1 Phonon-assisted ultrafast charge transfer at MoS_2/WS_2

The newly developed two-dimensional (2D) materials have brought new concepts for the design of solar energy conversion and optoelectronic devices. Especially, different 2D materials can form van der Waals (vdW) heterostructures with various functionalities. For 2D vdW heterostructure, the interlayer interaction is expected to be weak. However, the ultrafast charge transfer dynamics at the interface has been observed by time-resolved measurements. It was urgent and essential to understand the physical mechanism of the ultrafast charge dynamics at the vdW heterostructure interface.

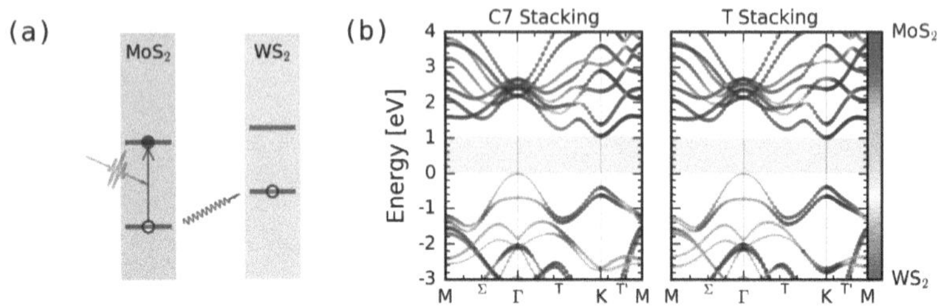

Figure 8.4 (a) Schematic of the photoexcitation and hole transfer in a MoS_2/WS_2 heterostructure. (b) Band structures of a MoS_2/WS_2 heterostructure with C7 and T stackings.

Zheng et al. used ab initio TDKS–NAMD simulation to investigate the photoexcited hole transfer at MoS_2/WS_2 interface, which is a prototypical TMD system (Gong et al., 2013). As shown in Fig. 8.4, they form type-II band alignment. The most stable C7 and T stackings show similar band structures. The photoexcited hole is initially generated at $MoS_2@K$. It is also possible to be scattered to $MoS_2@\Gamma$ through *e-ph* interaction. They have studied the charge transfer dynamics with initially excited hole at $MoS_2@K$ or $MoS_2@\Gamma$ with C7 and T stackings. As shown in Fig. 8.5a–d, the ultrafast interface charge transfer with a timescale of 20 fs have been observed in both C7 and T stackings at 300 K. In momentum space, the charge transfer occurs through $MoS_2@\Gamma$ – $WS_2@\Sigma$ – $WS_2@\Gamma$ or $MoS_2@K$ – $WS_2@\Sigma$ – $WS_2@\Gamma$, where Σ locates between Γ and M points. The $MoS_2@\Gamma/MoS_2@K$ – $WS_2@\Sigma$ process corresponds to the ultraface interface charge transfer and $WS_2@\Sigma$ – $WS_2@\Gamma$ corresponds to a slower charge relaxation. More analysis shows that such an ultrafast charge transfer is assisted by the intralayer optical phonon mode A_1, which has a wave number of 400 cm^{-1} and period of 80 fs. The 20 fs ultrafast charge transfer is a quarter of its period. To certify the phonon-assisted ultrafast charge transfer dynamics mechanism, we have performed the temperature dependent investigation. As shown in Fig. 8.5e–h, when the temperature is decreased to 100 K, such an ultrafast interlayer charge transfer is suppressed. Similar phonon-assisted ultrafast charge transfer was also found at lateral MoS_2/WS_2 heterostructure (Zheng et al., 2019b).

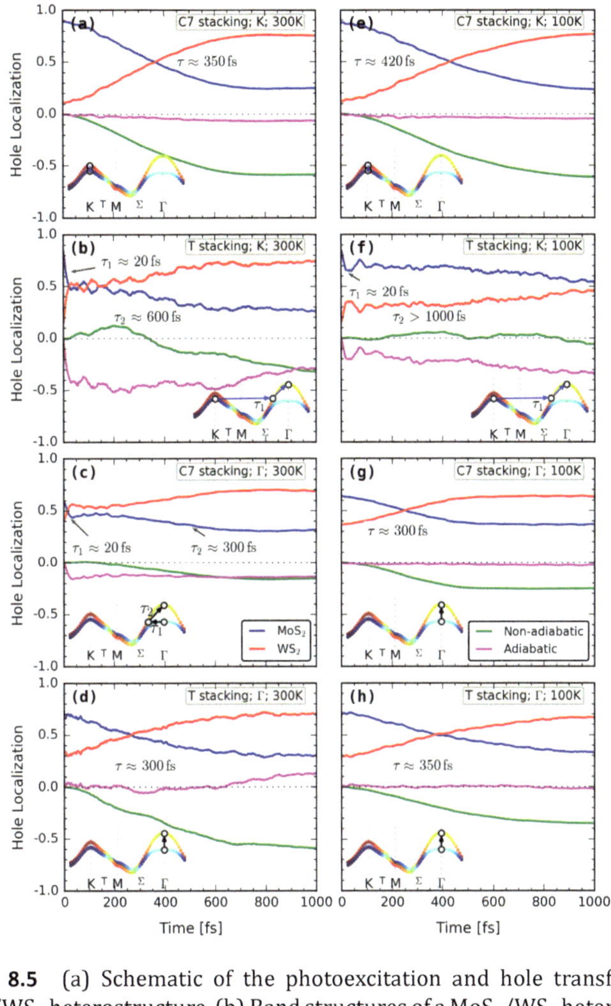

Figure 8.5 (a) Schematic of the photoexcitation and hole transfer in a MoS_2/WS_2 heterostructure. (b) Band structures of a MoS_2/WS_2 heterostructure with C7 and T stackings.

8.3.2.2 Phonon-coupled charge oscillation at $MoSe_2/WSe_2$

Based on the phonon-assisted charge transfer perspective, Zheng et al. further proposed that the phonon excitation and *e-ph* coupling can be modified by changing the elements of the TMD heterostructure. Therefore, they have further investigated the charge transfer dynamics at $MoSe_2/WSe_2$ (Zheng et al., 2018). When the S element

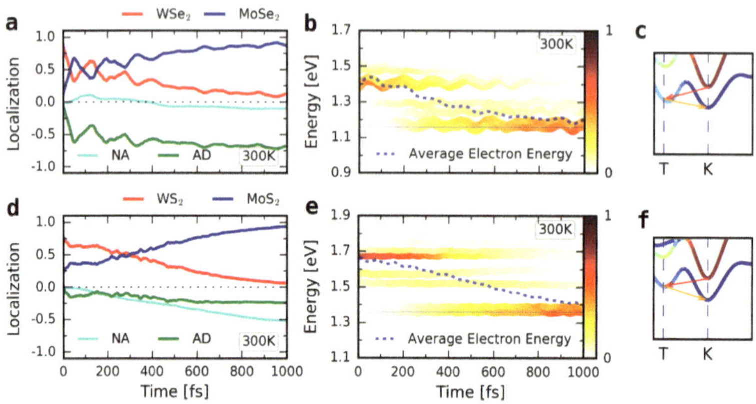

Figure 8.6 Nonadiabatic molecular dynamics results of $MoSe_2/WSe_2$ and MoS_2/WS_2 and 300 K, respectively. (a, d) Time-dependent electron spatial localization. (b, e) Time-dependent electron energy change. (c, f) Schematics of the electron relaxation route in the momentum space.

is replaced by Se with larger mass, the wave number of A_1 mode is decreased from 400 to 230 cm^{-1}. Therefore, comparing with MoS_2/WS_2, there are more phonons excited in $MoSe_2/WSe_2$ at 300 K. Different with the hole transfer, the electron transfer in both MoS_2/WS_2 and $MoSe_2/WSe_2$ occurs through $WX_2@K$ – $MoX_2@T$ – MoX2@K (X=S, Se) route. And the stronger *e-ph* coupling leads to the phonon coupled interlayer electron oscillation in $MoSe_2/WSe_2$ as shown in Fig. 8.6a–b. If the temperature is decreased to 100 K, such phonon-coupled charge oscillation will be suppressed.

8.3.2.3 Control the charge transfer dynamics at MoS_2/WS_2 by external stress

Besides changing the elements, applying external stress may also change the *e-ph* coupling and the band structure in TMD heterostructure and affect the charge transfer dynamics. Tian et al. found that external tensile stress can suppress the electron transfer at MoS_2/WS_2 interface (Tian et al., 2020). As shown in Fig. 8.7, if applying a tensile strain up to 4%, the hole transfer can always happen but the electron transfer is significantly suppressed. As shown in Fig. 8.7d–f, for the electron transfer at MoS_2/WS_2 interface,

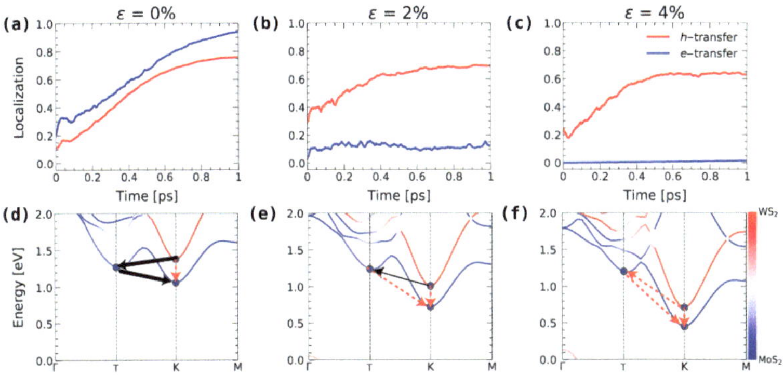

Figure 8.7 Time-dependent spatial charge localization of h-transfer and e-transfer under 0 (a), 2 (b), and 4% (c) biaxial tensile strain. The red line represents the hole localization on WS_2 in the h-transfer process. The blue line represents the electron localization on MoS_2 in the e-transfer process. Major electron relaxation route in momentum space under 0 (d), 2 (e), and 4% (f) biaxial tensile strain.. The red dashed line represents the blocked charge-transfer path. The black thick line represents the sufficient charge transfer path, while the black thin line represents the partially suppressed charge-transfer path.

the donor is the $WS_2@K$ while the acceptor is the $MoS_2@K$ states. Both the donor and acceptor states are localized in the single layer and their NAC is rather weak. Therefore, the direct charge transfer between them is rather difficult. Without external strain, the electron transfer can happen through $WS_2@K$ – $MoS_2@T$ – $MoS_2@K$ route, where $MoS_2@T$ behaves as an intermediate state as indicated in Fig. 8.7d. However, if external tensile strain is applied, the energy of the intermediate state $MoS_2@T$ will increase and the charge transfer from $WS_2@K$ to $MoS_2@T$ is not favorable. In this case, the electron transfer is suppressed.

It is also easy to understand why the hole transfer is not affected by the external tensile strain. For the hole transfer, the charge acceptor is the valance band maximum (VBM) $WS_2@\Gamma$ which has the delocalized character over the two TMD layers. And such character does not change with the tensile strain. Therefore, it is always able to accept the holes.

8.3.2.4 Comparing with other works

The phonon assisted ultrafast interlayer charge transfer mechanism proposed by Zheng et al. has been certified by different experiments (Zhou et al., 2019; Liu et al., 2020a; Wallauer et al., 2020; Aeschlimann et al., 2020). For example, X. Y. Zhu has observed the phonon assisted charge transfer between K and Q valleys at MoS_2/WS_2 interface using time- and angle-resolved photoemission spectroscopy (Liu et al., 2020a). However, the proposed temperature dependence has not been observed by experiments. It can be attributed to the nuclear quantum effects or exciton effects.

It should be noted that the interlayer charge transfer dynamics at TMD interface has been intensively studied and different theoretical perspectives have been proposed (Wang et al., 2016a; Ji et al., 2017). For example, using Erhenfest NAMD simulation, Wang et al. found the hole transfer at MoS_2/WS_2 happens within 100 fs, accompanying charge oscillation between the two layers. They bring forward a model that the charge oscillation is coupled with the dipole moment (Wang et al., 2016a). Ji et al. studied this process using the same method (Ji et al., 2017), different with Zheng et al. they found the hole transfer dynamics depends on the stacking significantly, providing different charge transfer channels. They proposed that in the experiments, the hole transfer happens through the fastest channel. One can see that the Erhenfest dynamics and surface hopping scheme gave different charge transfer timescale and physical pictures. Benchmarks based on advanced theoretical or time-resolved experimental investigations are necessary in the future.

8.4 Electron–Hole Recombination in Semiconductors

In the solar energy conversion process, the nonradiative *e-h* recombination process transfers the photo-energy to heat, which is widely known as an energy waste which will decrease the solar energy conversion efficiency. Such nonradiative *e-h* recombination often occurs through the defect/impurity states, namely, the defects

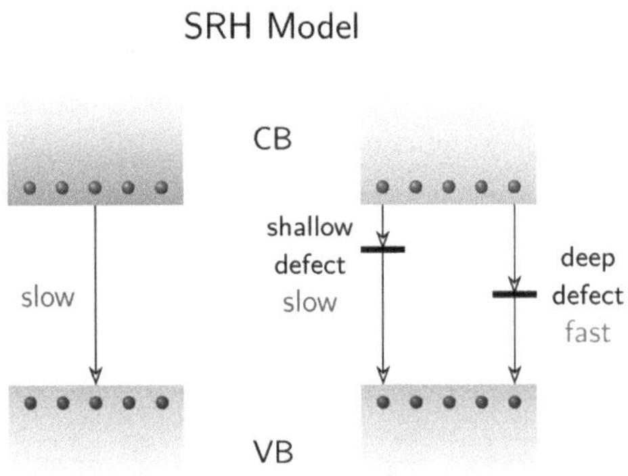

Figure 8.8 SRH mechanism for *e-h* recombination in semiconductors with defects.

and impurities often behave as *e-h* recombination centers. How to reduce the defect/impurity mediated *e-h* recombination is a primary concern.

As early as the 1950s, the Shockley–Read–Hall (SRH) mechanism, which considers the localized defect states in the band gap, was proposed. As schematically shown in Fig. 8.8, the localized defect state with its energy close to the middle of the band gap is defined as a deep defect level, which was proposed to issue in fast *e-h* recombination and behave as recombination center. By contrast, the defect state close to VBM or CBM is known as shallow defect level, which was believed not to accelerate the *e-h* recombination significantly. SRH model has been widely used for decades. However, in this model only the charge-carrier concentration and the energy level positions are considered. The lifetime of photoexcited hole and electron is treated as an empirical parameter. In the nonradiative process, the energy of photoexcited carrier transfers to the phonon system. Thus, the *e-ph* plays the dominating role. In SRH model, neither the phonon excitation nor the *e-ph* coupling are considered. Fortunately, in ab initio TDKS–NAMD, the *e-ph* coupling and phonon excitation is simulated by combining the TDKS equation with the AIMD. It is of great interest to compare the SRH model with TDKS–NAMD simulation.

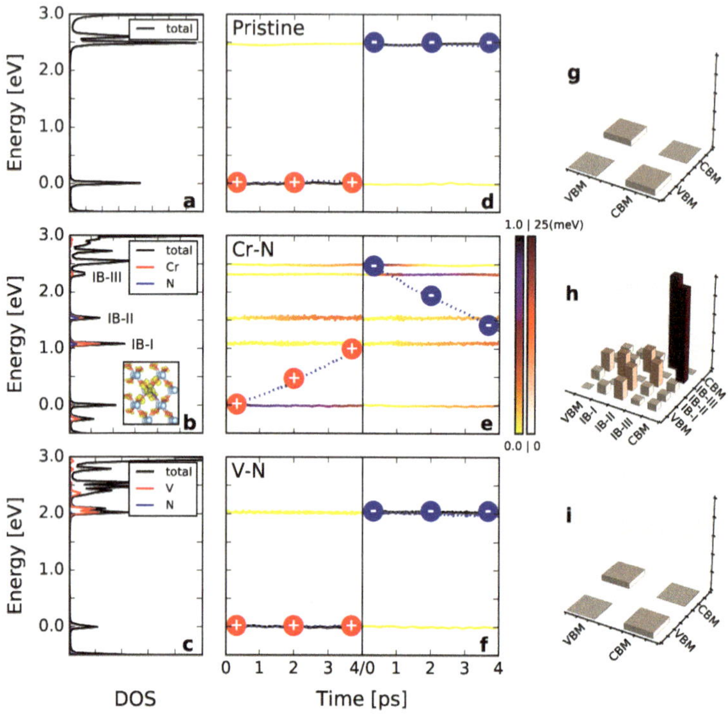

Figure 8.9 Electronic structures and the time-dependent electron/hole (*e/h*) dynamics in undoped, Cr–N- and V–N-doped TiO_2. (a-c) The total and partial DOS. (d-f) The averaged time-dependent e/h energy relaxation at 300 K. The color strip indicates the *e/h* distribution on different energy states, and the dashed line represents the averaged *e/h* energy. The energy reference is the average VBM energy. (g-i) The averaged NAC elements in undoped and Cr–N- and V–N-doped TiO_2 at 300 K. The inset in panel b shows the spatial distribution of the excess charge induced by Cr–N codoping, in which the Ti, O, Cr, and N atoms are marked by large light blue, small red, large deep blue, and small purple balls, respectively.

8.4.1 Electron–Hole Recombination in TiO_2

Zhang et al. choose TiO_2 as a prototypical system to study (Zhang et al., 2018). As a wide band gap semiconductor, TiO_2 can only absorb the ultraviolet light which is only 5% of the solar energy. Element doping is known as an effective method to reduce the band gap of TiO_2 and improve its solar energy conversion efficiency.

Comparing with the single element doping, the *n-p* co-doping was proposed to be more energetically stable. Noncompensated Cr–N and compensated V–N co-doping were chosen for contrast. For noncompensated Cr–N co-doping, one excess electron is introduced to TiO_2 and a deep defect level is formed. For the compensated V–N co-doping, there are no excess charges and only shallow defect level is formed. As shown in Fig. 8.9, the TDKS–NAMD simulation shows that for Cr–N co-doped TiO_2, the *e-h* recombination occurs within several ps, and the deep defect level behaves as an *e-h* recombination center. For V–N co-doped TiO_2, the *e-h* recombination happens within a ns timescale, which is comparable to the undoped TiO_2. The results in TiO_2 are in good agreement with SRH model.

8.4.2 Electron–Hole Recombination in Halide Perovskite

Although SRH model works well in TiO_2, Chu et al. found that nonradiative *e-h* recombination in halide perovskite $MAPbI_3$ behaves distinctly. They have chosen five different defect structures. The structures and projected density of states (PDOS) are shown in Fig. 8.10. It can be seen that two of them have deep defect states while the other three has shallow defect states. As shown in Fig. 8.11, out of expectation, it was found that no matter the defect state is

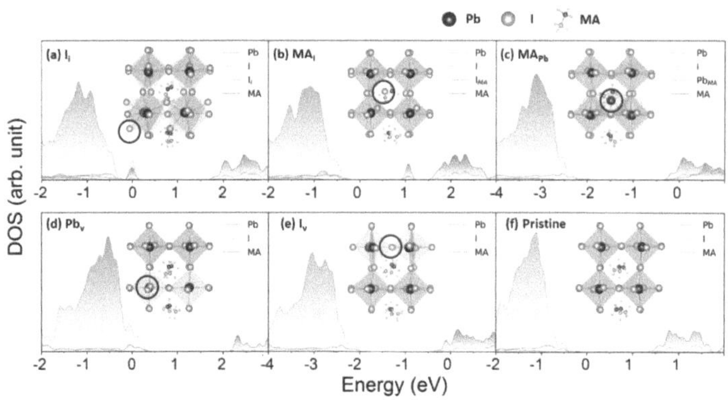

Figure 8.10 Atom-projected DOS for different defective and pristine $MAPbI_3$. (A to F) Defective and Pristine systems of $MAPbI_3$. The energy reference is located at the Fermi level. Inset shows corresponding atomic structure, with blue circle indicating the defect location.

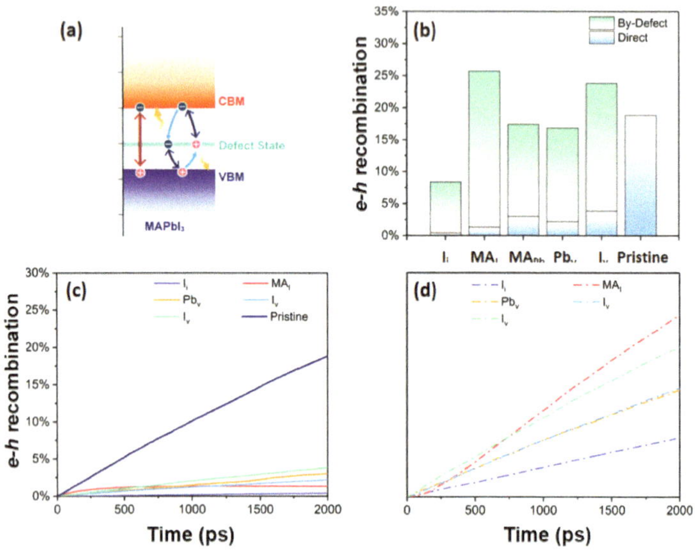

Figure 8.11 The *e-h* recombination process in $MAPbI_3$ systems. (A) Schematic map of the direct and by-defect *e-h* recombination processes. (B) *e-h* recombined percentage for different systems after 2 ns. The direct and by-defect *e-h* recombined percentages are shown by blue and green color bars. (C and D) Time-dependent *e-h* recombined percentage for different systems. The direct and by-defect *e-h* recombined percentage is shown in (C) and (D), respectively.

deep or shallow, the *e-h* recombination is not accelerated notably. For I interstitial (I_i) defect, the defect state even suppress the *e-h* recombination. Such phenomenon was known as defect tolerance, which is in good agreement with the experimental observation (Akkerman et al., 2018; Kovalenko et al., 2017; Wei et al., 2019).

In NAMD simulations, the *e-h* recombination rate is strongly dependent on the NAC between the donor and acceptor states. As expressed in Eq. (8.3), NAC is determined by (i) the overlap between the donor and acceptor orbitals; (ii) energy difference between the donor and acceptor states; and the nuclear velocity term. All of them are related to the *e-ph* coupling. Figure 8.12 presents the Fourier transform (FT) of the autocorrelation function of VBM, CBM and the defect state energies for the 5 different defective structures and the pristine system. One can see the frequencies of the dominate

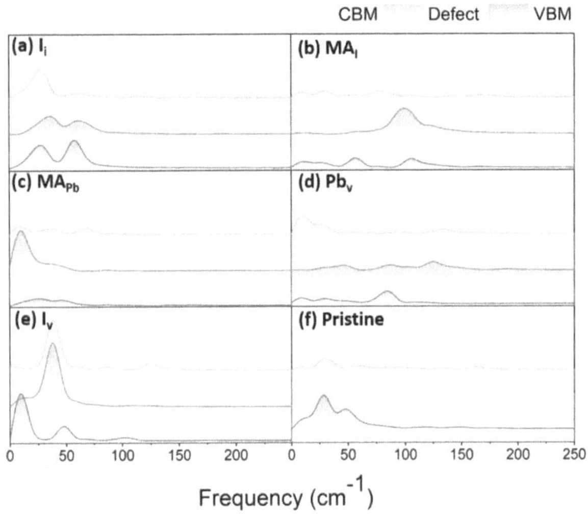

Figure 8.12 The Fourier transform spectra of the autocorrelation function of the VBM, the CBM, and the defect state energies. (a to f) Defective and Pristine systems of $MAPbI_3$.

phonon modes are all below 100 cm^{-1}. Such low phonon frequencies are correlated with atoms with small nuclear velocities, leading to small NAC. These low-frequency phonon modes are contributed by the inorganic Pb-I lattice, which was attributed to the small bulk modulus of $MAPbI_3$. Chu et al. also revealed similar defect tolerance due to low-frequency phonons exist in $CsPbI_3$ (Chu et al., 2020a).

The results by Chu et al. suggest the break-down of SRH model in halide perovskites is attributed to the *e-ph* coupling with the low-frequency lattice phonons, suggesting that soft lattice efficiently suppresses the charge recombination.

8.4.3 Electron–Hole Recombination in 2D Materials

The idea of *e-ph* coupling with the low-frequency lattice phonons leads to slow *e-h* recombination also applies in 2D black phosphorene (BP). Zhang et al. have studied different defects and oxidation in BP (Zhang et al., 2019a,b). They found that the defects and oxidation can suppress the *e-h* recombination. Similar results were reported by Long et al. (Long et al., 2016). Analysis shows that introducing the defects can excite more phonons with low-frequencies, which will reduce the recombination rate of charge carriers. In addition to the role of low-frequency phonons, the phonon induced electronic decoherence also contribute to the suppression of *e-h* recombination. In the defective or oxidized BP, the phonons are scattered by the defects or impurity, generate multiple phonon peaks, which will reduce the decoherence time and further suppress the *e-h* recombination.

From the ab initio NAMD simulation, one can see that the major factor affects the nonradiative *e-h* recombination is the *e-ph* coupling. Therefore, different schemes were proposed to tune the *e-h* recombination rate by changing the *e-ph* coupling. Guo et al. proposed that applying 1% tensile strain will excite the ZA mode in BP, which is strongly coupled with VBM. And the ZA phonon excitation will accelerate the *e-h* recombination. They also proposed that the *e-h* recombination timescale, namely, the lifetime of the photoexcited carriers, can be tuned by family atom doping (Guo et al., 2021). It was found the carrier lifetime has a strong correlation with the mass of the doping atoms. A doping atom with larger mass results in a longer lifetime. Combining with the ultrafast spectroscopy, Nie et al. proved the photoexcited carrier lifetime in $MoSe_2$ can be engineered by changing different substrate (Nie et al., 2019). NAMD simulations show that the interfacial *e-ph* coupling is the leading mechanism for the lifetime tuning.

8.5 Exciton Dynamics

As discussed in Sections 8.3 and 8.4, by combining AIMD with TDKS and surface hopping scheme (TDKS–NAMD), the *e-ph* coupling can be included in the excited carrier dynamics simulation at the ab initio level. However, there are other aspects which will affect the excited carrier dynamics are not included in the TDKS–NAMD method. For example, when the *e-h* pair is generated by photoexcitation, they can form an exciton through the Coulomb interaction. The many-body effects in the exciton formation raised by *e-h* interaction is one of the most important aspects which can affect the excited carrier dynamics. Especially, in 2D material, when the screening is weak, the exciton effects cannot be neglected.

The real-time ab initio methods for exciton dynamics are still in their infancy. The conventional ab initio approach to study the excitons in 2D materials is the GW plus Bethe–Salpeter equation (GW + BSE) method (Hybertsen and Louie, 1986; Rohlfing and Louie, 2000; Onida et al., 2002; Ramasubramaniam, 2012). Although Ismail-Beigi and Louie developed a method to calculate the excited state forces based on GW + BSE, providing a possibility to simulate the excited state dynamics. Still it is mostly used to study the "static" exciton properties such as exciton binding energy (E_b) and wavefunction rather than dynamics. The TDKS–NAMD discussed above is based on single-particle KS basis sets. Recently, the linear response time-dependent functional theory (LR-TDDFT) has been combined with NAMD (Zhang et al., 2020; Liu et al., 2020b). It has been used to investigate the exciton dynamics at TMD heterostructure interface. However, it is known that LR-TDDFT has problems in dealing with excitonic states in condensed matter systems. In addition, SOC is not included in that approach. An ab initio method to describe the spin- and time-resolved exciton dynamics is urgently needed.

8.5.1 *GW*+Real-Time BSE NAMD Method

Jiang et al. have extended one-body TDKS–NAMD to many-body GW plus real-time propagation of BSE (GW + rtBSE–NAMD) for the investigation of exciton dynamics (Jiang et al., 2021). The BSE Hamiltonian used for the real-time propagation is expressed by spinor notation as:

$$\langle \mathbf{k}cv|\mathcal{H}|\mathbf{k}'c'v'\rangle = (E^{\mathrm{QP}}_{\mathbf{k}c} - E^{\mathrm{QP}}_{\mathbf{k}v})\delta_{\mathbf{k}\mathbf{k}'}\delta_{cc'}\delta_{vv'} + W^{\mathbf{k}cv}_{\mathbf{k}'c'v'} + V^{\mathbf{k}cv}_{\mathbf{k}'c'v'} \quad (8.8)$$

Here v/c represents the index of the hole and electron. $E^{\mathrm{QP}}_{\mathbf{k}c}$ and $E^{\mathrm{QP}}_{\mathbf{k}c}$ are the quasi-particle (QP) energies for the electron and hole, respectively. $\langle \mathbf{k}cv\rangle \equiv \psi_{\mathbf{k}c}(\mathbf{r}_e)\psi^*_{\mathbf{k}v}(\mathbf{r}_h)$ are the electron–hole-pair basis sets based on spinor formalism, which is constructed as

$$\psi_{kn}(\mathbf{r}) = \psi_{\uparrow kn}(\mathbf{r})|\uparrow\rangle + \psi_{\downarrow kn}(\mathbf{r})|\downarrow\rangle \equiv \sum_{\sigma=\uparrow,\downarrow} |\sigma\mathbf{k}n\rangle \otimes |\sigma\rangle \quad (8.9)$$

where $\psi_{\sigma kn}(\mathbf{r}) = e^{i\mathbf{k}\cdot\mathbf{r}}u_{\sigma\mathbf{k}n}(\mathbf{r})$ is the spin up/down Bloch-like wavefunctions that can be obtained from the DFT calculations. W and V represent the screened Coulomb and exchange interaction between the electron and hole, which can be written as

$$W^{\mathbf{k}cv}_{\mathbf{k}'c'v'} = \frac{1}{\Omega}\sum_{\mathbf{G}\mathbf{G}'}\frac{4\pi\,\varepsilon^{-1}_{\mathbf{G}\mathbf{G}}(\mathbf{k}-\mathbf{k}')}{|\mathbf{k}-\mathbf{k}'+\mathbf{G}|\,|\mathbf{k}-\mathbf{k}'+\mathbf{G}'|} \times \left[B^{\uparrow\mathbf{k}c}_{\uparrow\mathbf{k}'c'}(\mathbf{G}) + B^{\downarrow\mathbf{k}c}_{\downarrow\mathbf{k}'c'}(\mathbf{G})\right] \times \left[B^{\uparrow\mathbf{k}v}_{\uparrow\mathbf{k}'v'}{}^{*}(\mathbf{G}') + B^{\downarrow\mathbf{k}v}_{\downarrow\mathbf{k}'v'}{}^{*}(\mathbf{G}')\right] \tag{8.10}$$

$$V^{\mathbf{k}cv}_{\mathbf{k}'c'v'} = \frac{1}{\Omega}\sum_{\mathbf{G}\neq 0}\frac{4\pi}{|\mathbf{G}|^2}\left[B^{\uparrow\mathbf{k}c}_{\uparrow\mathbf{k}'c'}(\mathbf{G}) + B^{\downarrow\mathbf{k}c}_{\downarrow\mathbf{k}'c'}(\mathbf{G})\right] \times \left[B^{\uparrow\mathbf{k}v}_{\uparrow\mathbf{k}'v'}{}^{*}(\mathbf{G}) + B^{\downarrow\mathbf{k}v}_{\downarrow\mathbf{k}'v'}{}^{*}(\mathbf{G})\right] \tag{8.11}$$

Here $\varepsilon^{-1}_{\mathbf{G}\mathbf{G}'}(\mathbf{k}-\mathbf{k}')$ is the inverse dielectric function that can be obtained from the random phase approximation in the GW calculation. The Bloch integral B defined as $B^{\sigma\mathbf{k}n}_{\sigma\mathbf{k}'n'}(\mathbf{G}) = \langle u_{\sigma\mathbf{k}n}|e^{i\mathbf{G}\cdot\mathbf{r}}|u_{\sigma\mathbf{k}'n'}\rangle$ can be calculated from the spinor basis sets, where **G** is the reciprocal lattice vector.

Using the BSE Hamiltonian of Eq. (8.8), the time-dependent two-particle Schrödinger equation follows

$$i\hbar\frac{\partial\Psi(\mathbf{r}_e,\mathbf{r}_h,t)}{\partial t} = \mathcal{H}(\mathbf{R};t)\Psi(\mathbf{r}_e,\mathbf{r}_h,t) \tag{8.12}$$

The CPA approximation is used following the TDKS–NAMD method, where the nulcear coordinates $\mathbf{R}(t)$ in the Hamiltonian is described by AIMD simulation. Combining with the expansion of the exciton state

$$\Psi(\mathbf{r}_e,\mathbf{r}_h,t) = \sum_{\mathbf{k}}\sum_{c}^{\text{elec}}\sum_{v}^{\text{hole}} A_{\mathbf{k}cv}(t)\psi_{\mathbf{k}c}(\mathbf{r}_e)\psi_{\mathbf{k}v}(\mathbf{r}_h)^* \tag{8.13}$$

where $\psi_{\mathbf{k}c}(\mathbf{r}_e)$ and $\psi_{\mathbf{k}v}(\mathbf{r}_h)$ are the spinor wavefunctions defined in Eq. (8.9). The time propagation of the expanding coefficients $A_{\mathbf{k}cv}$ can be written as,

$$i\hbar\dot{A}_{\mathbf{k}cv} = \sum_{\mathbf{k}'}\sum_{c'}^{\text{elec}}\sum_{v'}^{\text{hole}}\left[\langle\mathbf{k}cv|\mathcal{H}|\mathbf{k}'c'v'\rangle - i\hbar\langle\mathbf{k}cv|\frac{\partial}{\partial t}|\mathbf{k}'c'v'\rangle\right]A_{\mathbf{k}'c'v'} \tag{8.14}$$

Coherent time evolution of excitonic states coupled to the nuclear subsystem is described in Eq. (8.14). Here it is expressed by a superposition of the adiabatic basis sets, which is deviated from the laws of quantum mechanics since the system can only exist in one pure state if been measured. In this work we have used the fewest switches surface hopping (FSSH) scheme developed by Tully (Tully, 1990) to overcome this limitation.

Different with TDKS–NAMD method, in GW + rtBSE–NAMD method, the NAC includes off the diagonal elements of $\langle \mathbf{k}cv|\mathcal{H}|\mathbf{k}'c'v'\rangle - i\hbar\langle \mathbf{k}cv|\frac{\partial}{\partial t}|\mathbf{k}'c'v'\rangle$ in Eq. (8.14), where

$$\langle \mathbf{k}cv|\frac{\partial}{\partial t}|\mathbf{k}'c'v'\rangle = \left\{ \left[\langle\uparrow \mathbf{k}c|\frac{\partial}{\partial t}|\uparrow \mathbf{k}c'\rangle + \langle\downarrow \mathbf{k}c|\frac{\partial}{\partial t}|\downarrow \mathbf{k}c'\rangle \right] \delta_{vv'} + \left[\langle\uparrow \mathbf{k}v|\frac{\partial}{\partial t}|\uparrow \mathbf{k}v'\rangle + \langle\downarrow \mathbf{k}v|\frac{\partial}{\partial t}|\downarrow \mathbf{k}v'\rangle \right]^* \delta_{cc'} \right\} \delta_{\mathbf{k}\mathbf{k}'} \tag{8.15}$$

This part contains the *e-ph* interaction contributed by both the electron and the hole, which is an extension of NAC in TDKS–NAMD. Besides the *e-ph* part, $\langle \mathbf{k}cv|\mathcal{H}|\mathbf{k}'c'v'\rangle$ contains the off-diagonal contribution by W and V (Eq. (8.10) and (8.11)) that is also included in NAC.

The most challenging part for the application of GW + rtBSE–NAMD method is the GW calculation for each time step. For example, for a simulation of a ps timescale trajectory, 10^3 times of GW calculation is needed, the ultra-expensive computational cost makes the simulation to be difficult. To overcome such difficulty, the rigid-dielectric function approximation is proposed where the dielectric function is assumed not to change during the AIMD simulation. This approximation is reasonable if the temperature is far away from the phase transition point. Based on this approximation, only a single point G_0W_0 is needed to achieve the dielectric function. The QP energies can be calculated using the scissor operator by adding $\Delta E_{\text{GW-PBE}}$ to the PBE KS energies. Then the W and V can be constructed by time-varying spinor basis sets as stated in Eqs. (8.10) and (8.11).

The workflow of GW + rtBSE–NAMD can be summarized as follows: (i) Perform an AIMD simulation at a certain temperature;

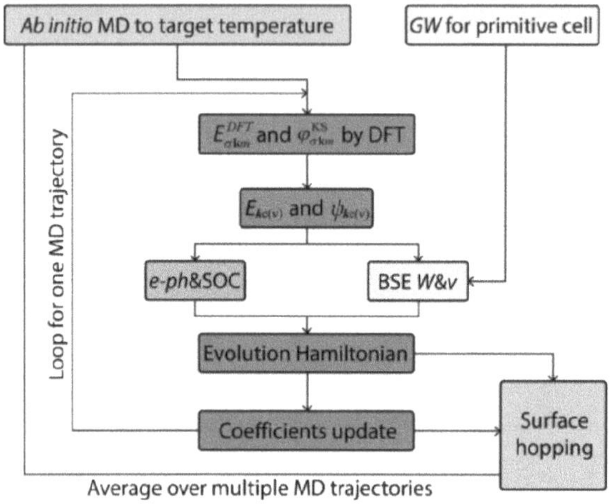

Figure 8.13 Work chart of current scheme for GW + rtBSE–NAMD method. The inner loop works on one AIMD trajectory and the outer loop represents the statistics on samplings of different initial structures.

(ii) get the time-varying KS orbitals based on DFT calculations on each structure along the AIMD trajectory; (iii) get the time-varying spinor-wavefunction basis sets by applying the SOC on the KS orbitals; (iv) perform a single G_0W_0 calculation on the optimized structure to get the QP energies and dielectric function; (v) calculate W and V based on the time-varying spinor basis sets and dielectric function and construct the BSE Hamiltonian together with the QP energies; (vi) solve the time-dependent Schrödinger equation using the BSE Hamiltonian; (vii) perform NAMD simulation by surface hopping scheme based on sampling of different initial structures and exciton trajectories and obtain the dynamics behavior based on ensemble average. A flowchart of the method is shown in Fig. 8.13.

8.5.2 Spin Valley Exciton Dynamics in MoS_2

To test the validity of GW + rtBSE–NAMD, Jiang et al. chose a prototypical TMD material MoS_2 to investigate (Jiang et al., 2021).

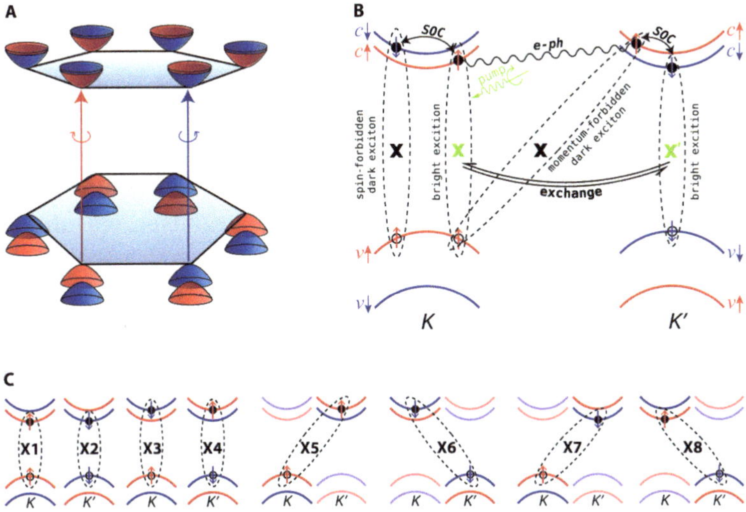

Figure 8.14 Schematic showing spin-valley dynamics in TMD systems. (A) The band structure at the band edges near K and K'. (B) The intervalley bright exciton transition and bright-to-dark exciton transition processes are shown. They are induced by *e-h* exchange, SOC, and *e-ph* coupling, respectively. (C) The *e-h* pairs involved during the exciton dynamics.

Due to its hexagonal structure, there are six valleys formed at K and K' points at the corner of the Brillouin zone as shown in Fig. 8.14a. The valley states are the basic ingredients for valleytronics, which has the opportunity to offer information processing schemes that are superior to charge- and spin-based semiconductor technologies (Cao et al., 2012; Mak et al., 2012; Zeng et al., 2012; Schaibley et al., 2016; Mai et al., 2014). In monolayer MoS_2, the broken inversion symmetry and strong SOC lead to spin-valley locking where the spin splitting has opposite signs at K and K' valleys (Cao et al., 2012; Xiao et al., 2012). The spin-valley excitons formed by parallel can be optically excited by circularly polarized light at K or K' valleys, which play a vital role in the optically driven valleytronic devices based on TMDs.

The timescale to keep the valley polarization is known as the valley lifetime, which is affected by different factors as *e-h* interaction, *e-ph* interaction and SOC et al. After the spin-valley exciton formation, there can be intervalley bright exciton scattering between K and K' valleys, which is known as valley depolarization. This process was expected to be very slow due to the spin-valley locking, because spin flip of both the electron and hole together with a large momentum transfer (from K to K') is needed for a valley pseudospin change. Yet, previous studies reported relatively short valley lifetimes of picosecond (ps) timescales (Cao et al., 2012; Schaibley et al., 2016; Lagarde et al., 2014). Several studies proposed that the intervalley bright exciton scattering is induced by *e-h* exchange interaction (Schaibley et al., 2016; Yu and Wu, 2014), which leads to the short valley lifetime. While the others debate the *e-ph* scattering is the dominate mechanism. Further, a bright valley exciton can transfer into a dark exciton. Through the spin flip of the electron or hole, which can be realized through an external magnetic field or internal SOC, a spin-forbidden dark exciton can be formed. Otherwise, a momentum-forbidden dark exciton can be formed with the electron and hole located in different valleys through the scattering with defects or phonons. The GW + rtBSE–NAMD method can help people to understand such complex spin-valley exciton dynamics at the ab initio level.

Jiang et al. first checked the QP energies and the E_b to verify the validity of the method. A single G_0W_0 calculation based on Heyd-Scuseria-Ernzerhof (HSE) (Heyd et al., 2003; Heyd and Scuseria, 2004; Heyd et al., 2006; Krukau et al., 2006) functional (G_0W_0@HSE) using a primitive cell is performed. And the 2×3 orthogonal unit cell is used to perform the AIMD simulation, in which both the K and K' are folded into the Γ point of the supercell. Figure 8.15 shows the time evolution of QP energies close to the CBM and the VBM of MoS_2. The averaged QP band gaps (E_g) for MoS_2 is 2.66 eV, in good agreement with previous investigations (Qiu et al., 2013; Ramasubramaniam, 2012). The degeneracy of $K_\uparrow$ vs. $K'_\downarrow$ and $K'_\uparrow$ vs. $K_\downarrow$ is kept since the phonon excitation simulated by AIMD does not break the time-reversal symmetry. As shown in Fig. 8.15a, the conduction band (CB) splitting between CB_$K_\uparrow$/CB_$K'_\downarrow$ and CB_$K'_\uparrow$/CB_$K_\downarrow$ is smaller than 20 meV. By contrast, the splitting of valance band (VB), namely VB_$K_\uparrow$/VB_$K'_\downarrow$, is higher than VB_$K'_\uparrow$/VB_$K_\downarrow$ by 160 meV due to the strong SOC of the VB. The large spin splitting in the VB ensures that the spin-valley locking is preserved with phonon excitations.

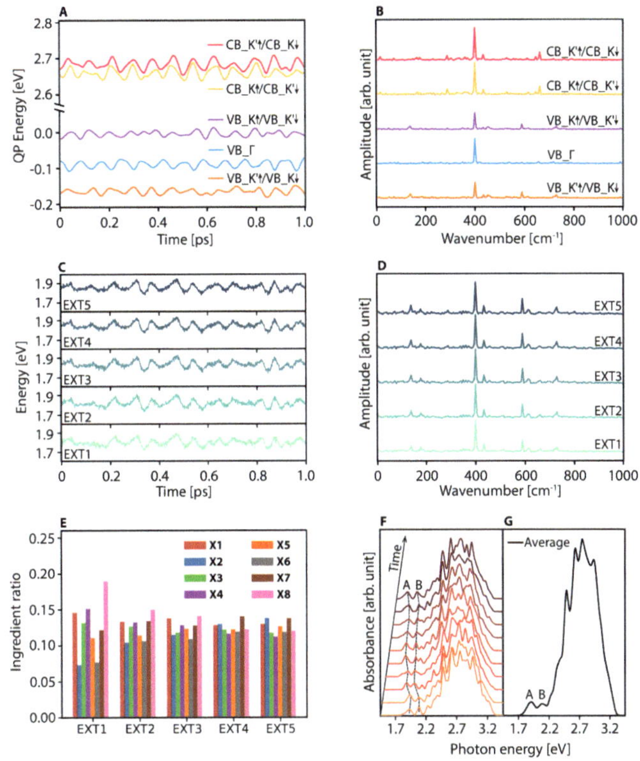

Figure 8.15 Time-dependent QP and exciton energies. (A and B) Time evolutions of the QP energies close to CBM/VBM and their Fourier transforms (FTs). (C and D) Time-dependent energies of the lowest five exciton states (EXT1 to EXT5) and their FTs. (E) The averaged ingredients over the MD trajectory for the lowest five exciton states contributed by X1-X8 configurations shown in Fig. 8.14. F Snapshots of absorption spectrum along the MD trajectory from $t = 0$ to $t = 200$ fs. The spectrum is plotted in every 20 fs. (G) Averaged absorption spectrum along the MD trajectory. The peaks for (A) and (B) excitons are marked.

The eigenstates obtained from the diagonalization of time-dependent BSE Hamiltonian (labeled as EXT) are the superpositions of different *e-h* pairs. In MoS_2, due to the large SOC induced energy splitting in VB states, the VB_$K'_{\uparrow}$/VB_$K_{\downarrow}$ and the VB_Γ states barely contribute to the low energy EXTs. While the two states at the VB (VB_$K_{\uparrow}$ and VB_$K'_{\downarrow}$) and the four states at the CB (CB_$K_{\uparrow}$, CB_$K'_{\downarrow}$

and CB_$K'_\uparrow$, CB_$K_\downarrow$) form eight *e-h* pairs [schematically shown in Fig. 8.14] that contribute to the low energy EXTs. The *e-h* pairs with parallel spin in the same valley are referred as bright excitons ($K_\uparrow K_\uparrow$ labeled as X1 and $K'_\downarrow K'_\downarrow$ labeled as X2). The other six *e-h* pairs are referred as dark excitons, in which $K_\downarrow K_\downarrow$ and $K'_\uparrow K'_\uparrow$ are intravalley spin forbidden excitons (labeled as X3 and X4); $K'_\uparrow K_\uparrow$ and $K_\downarrow K'_\downarrow$ are intervalley momentum forbidden excitons (labeled as X5 and X6); $K'_\downarrow K_\uparrow$ and $K_\uparrow K'_\downarrow$ are the intervalley excitons with both spin and momentum forbidden (labeled as X7 and X8).

The time-dependent energy of the lowest five EXTs (labeled as EXT1-EXT5) is plotted in Fig. 8.15c. Their energy difference is smaller than 100 meV. In Fig. 8.15e the averaged contribution from X1-X8 to the lowest five EXTs over 5 ps is presented. The bright and dark components are mixed in all the EXTs. In Fig. 8.15f 10 absorption spectrum snapshots are plotted from $t = 0$ to $t = 200$ fs, and the averaged absorption spectrum over the AIMD trajectory is shown in Fig. 8.15g. The averaged energy of the lowest A exciton is 1.91 eV, with a binding energy of 0.75 eV. These results are in good agreement with previous theoretical and experimental results (Qiu et al., 2013; Ramasubramaniam, 2012). The *e-ph* coupling contributed by the electron/hole and the exciton can be understood from the fluctuation of the QP and EXT energies as shown in Fig. 8.15b,d, respectively. From the FT spectra, it can be seen the major phonon coupled with the exciton is the optical A_1 mode at 400 cm^{-1} and the acoustic phonon mode at 200 cm^{-1}. In addition, there are additional minor peaks around 600 and 700 cm^{-1}, which is a combination of the optical A_1 mode and the acoustic mode.

In our simulation, the bright exciton X1 which can be excited by the circularly polarized light is set as the initial excitation. The time-dependent population on different exciton configurations are shown in Fig. 8.16a. The fluctuations of these curves implies the exciton dynamics couples with phonon excitation. FT spectra gives a peak at 600 cm^{-1}, corresponding to the combination of the optical A_1 and acoustic modes. After the excitation, the population of X1 decays from 98% to 68% within 30 fs, and the population of dark excition X5 increases from 0% to 30%. Such a timescale corresponds half of the period of phonon peak at 600 cm^{-1}. Such an ultrafast process corresponds to the intervalley *e-ph* scattering of

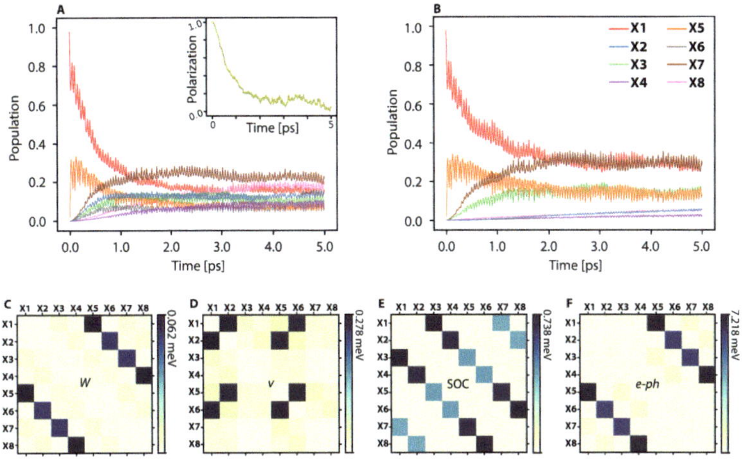

Figure 8.16 Dynamics results and nonadiabatic couplings. (A and B) Time evolution of the population on X1 to X8 (A) with and (B) without the e-h interaction *W* and *V* in the NAMD simulation. The time-dependent valley polarization is inserted in (A). (C to F) Averaged NACs contributed by *W*, *V*, SOC, and *e-ph*, respectively.

the electrons between $CB\_K_{\uparrow}$ and $CB\_K'_{\uparrow}$. After that, the populations of X1 and X5 start to decay in a slower manner and simultaneously, the populations on X7, X8 and X2 start to increase. After around 4 ps, the exciton population reaches the equilibrium and X7 and X8 has the highest population, which are 22% and 18%, respectively. It is because these two exciton states have the lowest energies due to the SOC induced energy splitting at the CB and the lacking of exchange interaction. The two bright excitons X1 and X3 host 18% and 15% population, respectively. The remaining population distributes over the other four dark excitons.

The intervalley bright exciton transition, namely, the X1-X2 process corresponds to the valley depolarization process observed by the experiments. It can be calculated as:

$$\eta(t) = \frac{n_{X1}(t) - n_{X2}(t)}{n_{X1}(t) + n_{X2}(t)} \tag{8.16}$$

where $n_{X1}(t)$ and $n_{X2}(t)$ are the time dependent population of X1 and X2, respectively. The results are shown in the inset of Fig. 8.16a. The depolarization timescale is around 5 ps, in good agreement with

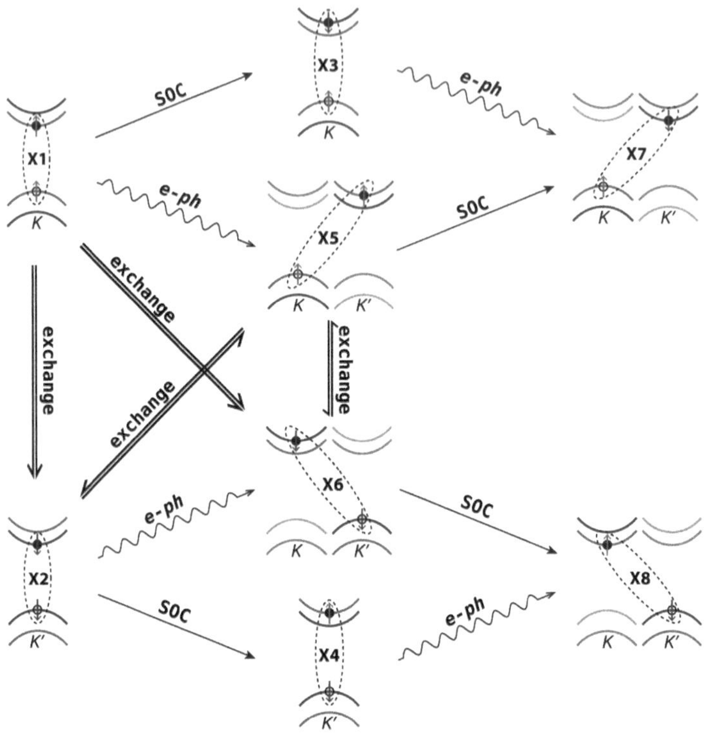

Figure 8.17 Schematic map of the exciton dynamics channels and the correlated mechanisms. The excitons can be divided into two groups with different hole status: X1, X3, X5, and X7 with $K_\uparrow$ hole and X2, X4, X6, and X8 with $K'_\downarrow$ hole. The SOC and *e-ph* contribute to the exciton transition within one group, and the exchange interaction contributes to the exciton transition between different groups.

experimental results (Mai et al., 2014; Lagarde et al., 2014; Yu and Wu, 2014).

The exciton transition mechanism is schematically shown in Fig. 8.17. The eight excitons are divided into two groups. X1, X3, X5 and X7 belong to the first group, where the hole locates in K valley. X2, X4, X6 and X8 belongs to the second group, where the hole locates in K' valley. The ultrafast X1-X5 transition is due to the *e-ph* interaction. The SOC is responsible for the coupling between the bright and spin-forbidden (dark) exciton states, e.g., X1-X3 and X5-X7, which happens with $\sim$ ps timescale. The most interesting, the transition between the two groups is mediated *e-h* exchange interaction. If we use the TDKS–NAMD simulation, the exciton transition can only occur among the first group, where the hole always locates in K valley (as shown in Fig. 8.16b.

In the GW + rtBSE–NAMD simulation, the *e-ph*, SOC and the *e-h* interaction (W and V) all contribute to the NAC elements. In Fig. 8.16c–f, the NAC contributed by *e-ph*, SOC, W and V are plotted. One can see that *e-ph* is one order of magnitude larger than SOC and V, which is in line with the ultrafast *e-ph* scattering within several tens of fs between X1 and X5. The SOC and V has the same magnitude, corresponding to dynamical processes with ps timescale. The contribution of W to the NACs is the smallest, which is one order of magnitude smaller than SOC and V. It suggests that although W modifies the diagonal elements of BSE Hamiltonian significantly, which is the major contribution of the exciton binding energy, it barely influences the spin-valley exciton dynamics in MoS_2.

8.6 Summary and Perspectives

This chapter has reviewed recent theoretical works from J. Zhao's group of excited carrier dynamics in condensed matter systems using ab initio time-dependent NAMD simulation, including (i) charge transfer dynamics at interfaces, (ii) *e-h* recombination in semiconductors, and (iii) spin-valley exciton dynamics in 2D TMD material.

The charge transfer dynamics at different interfaces have been studied using TDKS–NAMD method. It is found that phonon excitation and *e-ph* coupling play crucial role in these dynamical processes. At semiconductor/molecule interfaces, the energy level alignment and *e-ph* coupling determine the photoexcited carrier dynamics. Further, it was found that specific vibrational modes/phonon excitation can change the energy level alignment and enhance the interfacial charge transfer. At vdW interfaces, it was proposed that phonon excitation can assist the ultrafast interfacial charge transfer. Moreover, if the *e-ph* coupling is large enough, there can be phonon-couple charge oscillation.

TDKS–NAMD method was also used to study the *e-h* recombination in semiconductors, and was compared with the widely used SRH model. It was found that in TiO_2 the deep defect state behaves as an *e-h* recombination center, while the shallow defect state does not accelerate the *e-h* recombination. This is in good agreement with the SRH model. However, in halide perovskite and 2D BP, the defect tolerance was reported even for the deep defect level. It was attributed to the excitation and coupling with the low-frequency phonon mode, which is originated from the soft lattice.

Finally, the newly developed GW + rtBSE–NAMD method was introduced, in which the two-body real-time BSE is combined with the NAMD scheme. This method was applied to investigate the spin-valley exciton dynamics in 2D TMD material MoS_2. It is found that the fast bright valley exciton scattering is due to the *e-h* exchange interaction, while the SOC and *e-ph* coupling are responsible to the bright-to-dark exciton transition. This method can be widely used to investigate the exciton dynamics in different solids.

In the future, we propose that ab initio NAMD method can be applied to broader research area based on the further development. We plan to develop several new aspects in Hefei-NAMD including:

(1) *The nuclear quantum effects (NQEs)*. NQEs are important for systems containing light elements like hydrogen. For example, at many molecule/semiconductor interfaces, the hydrogen bond network is naturally formed and the NQEs may play an important role to affect the interfacial charge transfer dynamics. We plan to combine the scheme of Feynman path integrals with the NAMD to investigate how the NQEs affect the charge transfer dynamics.

(2) *The dynamics in momentum space.* In the current approach, the *e-ph* coupling is calculated by NAC, which is simulated using AIMD. To sample different K-points in the Brillouin zone, a supercell is needed. Therefore, the density of K-points is limited by the size of the supercell. In the future, we plan to use the perturbation theory to calculate the e-ph coupling elements (EPCs) and use EPCs to calculate the NAC in NAMD simulation. In this case, the dynamics with denser K-points can be simulated.

(3) *The improvement of CPA approximation*. In the current approach, CPA is applied and the nuclear dynamics is not affected by the excited carrier. In the newly developed $GW+$ rtBSE–NAMD method, for each simulation, several thousands of exciton state energies can be obtained. We propose to construct the potential energy surface of the excited states based on these exciton state energies through the machine learning approach. If the potential energy surface of the excited states can be constructed, we can then calculate the atomic forces in the excited states and improve the CPA approximation.

It should be mentioned that the ab initio simulation of excited carrier dynamics in condensed matter systems is a newly emerging research frontier. There are several other groups have accomplished outstanding research works in this area. For example, Prezhdo and Akimov have developed PYXAID code, which have been widely used (Akimov and Prezhdo, 2013, 2014). Lu, Akmiov, and Cui et al. have combined NAMD with LR-TDDFT to include the exciton effects (Liu et al., 2020b; Li et al., 2018; Cui and Thiel, 2014). Lu et al. have also applied LR-TDDFT to calculate the atomic forces in large systems. Meng and Kaxiras have developed TDAP code based on Erhenfest dynamics (Meng and Kaxiras, 2008; Ma et al., 2016; Lian et al., 2018). Similar method was also applied in PWmat code developed by L. Wang (Jia et al., 2013b,a). We expect this research area will attract more and more interests from scientists in physics and chemistry in the future.

References

Aeschlimann, S., Rossi, A., Chávez-Cervantes, M., Krause, R., Arnoldi, B., Stadtmüller, B., Aeschlimann, M., Forti, S., Fabbri, F., Coletti, C. and Gierz, I. (2020). Direct evidence for efficient ultrafast charge separation in epitaxial ws_2/graphene heterostructures, *Science Advances* **6**, 20, doi: 10.1126/sciadv.aay0761.

Agostini, F., Min, S. K., Abedi, A. and Gross, E. K. (2016). Quantum-classical nonadiabatic dynamics: coupled- vs independent-trajectory methods,

Journal of Chemical Theory and Computation **12**, 5, pp. 2127–2143, doi: 10.1021/acs.jctc.5b01180.

Akimov, A. V. (2018). A simple phase correction makes a big difference in nonadiabatic molecular dynamics, *Journal of Physical Chemistry Letters* **9**, 20, pp. 6096–6102, doi:10.1021/acs.jpclett.8b02826.

Akimov, A. V. and Prezhdo, O. V. (2013). The PYXAID program for non-adiabatic molecular dynamics in condensed matter systems, *Journal of Chemical Theory and Computation* **9**, 11, pp. 4959–4972, doi:10.1021/ct400641n.

Akimov, A. V. and Prezhdo, O. V. (2014). Advanced capabilities of the PYXAID program: Integration schemes, decoherence effects, multiexcitonic states, and field-matter interaction, *Journal of Chemical Theory and Computation* **10**, 2, pp. 789–804, doi:10.1021/ct400934c.

Akkerman, Q. A., Rainò, G., Kovalenko, M. V. and Manna, L. (2018). Genesis, challenges and opportunities for colloidal lead halide perovskite nanocrystals, *Nature Materials* **17**, 5, pp. 394–405, doi: 10.1038/s41563-018-0018-4, URL http://www.nature.com/articles/s41563-018-0018-4.

Barbatti, M. (2011). Nonadiabatic dynamics with trajectory surface hopping method, *Wiley Interdisciplinary Reviews: Computational Molecular Science* **1**, 4, pp. 620–633.

Barbatti, M., Granucci, G., Persico, M., Ruckenbauer, M., Vazdar, M., Eckert-Maksić, M. and Lischka, H. (2007). The on-the-fly surface-hopping program system Newton-X: Application to ab initio simulation of the nonadiabatic photodynamics of benchmark systems, *Journal of Photochemistry and Photobiology A: Chemistry* **190**, 2-3, pp. 228–240, doi:10.1016/j.jphotochem.2006.12.008.

Billeter, S. R. and Curioni, A. (2005). Calculation of nonadiabatic couplings in density-functional theory, *Journal of Chemical Physics* **122**, 3, p. 034105, doi:10.1063/1.1834562.

Bircher, M. P., Liberatore, E., Browning, N. J., Brickel, S., Hofmann, C., Patoz, A., Unke, O. T., Zimmermann, T., Chergui, M., Hamm, P., Keller, U., Meuwly, M., Woerner, H.-J., Vaníček, J. and Rothlisberger, U. (2017). Nonadiabatic effects in electronic and nuclear dynamics, *Structural Dynamics* **4**, 6, p. 061510, doi:10.1063/1.4996816.

Britnell, L., Ribeiro, R. M., Eckmann, A., Jalil, R., Belle, B. D., Mishchenko, A., Kim, Y. J., Gorbachev, R. V., Georgiou, T., Morozov, S. V., Grigorenko, A. N., Geim, A. K., Casiraghi, C., Castro Neto, A. H. and Novoselov, K. S. (2013). Strong light-matter interactions in heterostructures of atomically thin films, *Science* **340**, 6138, pp. 1311–1314, doi:10.1126/science.1235547.

Cao, T., Wang, G., Han, W., Ye, H., Zhu, C., Shi, J., Niu, Q., Tan, P., Wang, E., Liu, B. and Feng, J. (2012). Valley-selective circular dichroism of monolayer molybdenum disulphide, *Nature Communications* **3**, 1, p. 887, doi:10.1038/ncomms1882.

Chu, W., Saidi, W. A., Zhao, J. and Prezhdo, O. V. (2020a). Soft lattice and defect covalency rationalize tolerance of β-$CsPbI_3$ perovskite solar cells to native defects, *Angewandte Chemie - International Edition* **59**, 16, pp. 6435–6441, doi:10.1002/anie.201915702.

Chu, W., Saidi, W. A., Zheng, Q., Xie, Y., Lan, Z., Prezhdo, O. V., Petek, H. and Zhao, J. (2016). Ultrafast dynamics of photongenerated holes at a CH_3OH/TiO_2 Rutile Interface, *Journal of the American Chemical Society* **138**, 41, pp. 13740–13749, doi:10.1021/jacs.6b08725.

Chu, W., Zheng, Q., Akimov, A. V., Zhao, J., Saidi, W. A. and Prezhdo, O. V. (2020b). Accurate computation of nonadiabatic coupling with projector augmented-wave pseudopotentials, *Journal of Physical Chemistry Letters* **11**, 23, pp. 10073–10080, doi:10.1021/acs.jpclett.0c03080.

Chu, W., Zheng, Q., Prezhdo, O. V. and Zhao, J. (2020c). CO_2 photoreduction on metal oxide surface is driven by transient capture of hot electrons: *Ab Initio* quantum dynamics simulation, *Journal of the American Chemical Society* **142**, 6, pp. 3214–3221, doi:10.1021/jacs.9b13280.

Craig, C. F., Duncan, W. R. and Prezhdo, O. V. (2005). Trajectory surface hopping in the time-dependent Kohn-Sham approach for electron-nuclear dynamics, *Physical Review Letter* **95**, 16, p. 163001, doi:10.1103/PhysRevLett.95.163001.

Cui, G. and Thiel, W. (2014). Generalized trajectory surface-hopping method for internal conversion and intersystem crossing, *Journal of Chemical Physics* **141**, 12, doi:10.1063/1.4894849.

Curchod, B. F. E., Tavernelli, I. and Rothlisberger, U. (2011). Trajectory-based solution of the nonadiabatic quantum dynamics equations: an on-the-fly approach for molecular dynamics simulations, *Physical Chemistry Chemical Physics* **13**, 8, pp. 3231–3236, doi:10.1039/C0CP02175J.

Docherty, C. J., Parkinson, P., Joyce, H. J., Chiu, M. H., Chen, C. H., Lee, M. Y., Li, L. J., Herz, L. M. and Johnston, M. B. (2014). Ultrafast transient terahertz conductivity of monolayer MoS_2 and WSe_2 grown by chemical vapor deposition, *ACS Nano* **8**, 11, pp. 11147–11153, doi:10.1021/nn5034746.

Du, L. and Lan, Z. (2015). An on-the-fly surface-hopping program JADE for nonadiabatic molecular dynamics of polyatomic systems: Implementation and applications, *Journal of Chemical Theory and Computation* **11**, 4, pp. 1360–1374, doi:10.1021/ct501106d.

Eda, G., Yamaguchi, H., Voiry, D., Fujita, T., Chen, M. and Chhowalla, M. (2010). Photoluminescence from chemically exfoliated MoS_2, *Nano Letters* **11**, 12, pp. 5111–5116, doi:10.1021/nl201874w.

Feng, J., Qian, X., Huang, C. W. and Li, J. (2012). Strain-engineered artificial atom as a broad-spectrum solar energy funnel, *Nature Photonics* **6**, 12, pp. 866–872, doi:10.1038/nphoton.2012.285.

Fischer, S. A., Habenicht, B. F., Madrid, A. B., Duncan, W. R. and Prezhdo, O. V. (2011). Regarding the validity of the time-dependent kohn-sham approach for electron-nuclear dynamics via trajectory surface hopping, *Journal of Chemical Physics* **134**, 2, p. 024102.

Gierz, I., Calegari, F., Aeschlimann, S., Chávez Cervantes, M., Cacho, C., Chapman, R. T., Springate, E., Link, S., Starke, U., Ast, C. R. and Cavalleri, A. (2015). Tracking primary thermalization events in graphene with photoemission at extreme time scales, *Physical Review Letter* **115**, p. 086803, doi:10.1103/PhysRevLett.115.086803.

Gong, C., Zhang, H., Wang, W., Colombo, L., Wallace, R. M. and Cho, K. (2013). Band alignment of two-dimensional transition metal dichalcogenides: Application in tunnel field effect transistors, *Applied Physics Letters* **103**, 5, p. 053513, doi:10.1063/1.4817409.

Guo, H., Chu, W., Prezhdo, O. V., Zheng, Q. and Zhao, J. (2021). Strong modulation of band gap, carrier mobility and lifetime in two-dimensional black phosphorene through acoustic phonon excitation, *Journal of Physical Chemistry Letters* **12**, pp. 3960–3967, doi:10.1021/acs.jpclett.1c00747.

Hammes Schiffer, S. and Tully, J. C. (1994). Proton transfer in solution: Molecular dynamics with quantum transitions, *Journal of Chemical Physics* **101**, 6, pp. 4657–4667, doi:10.1063/1.467455.

Henderson, M. A. (2011). A surface science perspective on tio2 photocatalysis, *Surface Science Reports* **66**, 6-7, pp. 185–297.

Heyd, J. and Scuseria, G. E. (2004). Efficient hybrid density functional calculations in solids: Assessment of the Heyd-Scuseria-Ernzerhof screened Coulomb hybrid functional, *Journal of Chemical Physics* **121**, 3, pp. 1187–1192, doi:10.1063/1.1760074.

Heyd, J., Scuseria, G. E. and Ernzerhof, M. (2003). Hybrid functionals based on a screened Coulomb potential, *Journal of Chemical Physics* **118**, 18, pp. 8207–8215, doi:10.1063/1.1564060.

Heyd, J., Scuseria, G. E. and Ernzerhof, M. (2006). Erratum: Hybrid functionals based on a screened Coulomb potential (Journal of Chemical Physics (2003) 118 (8207)), *Journal of Chemical Physics* **124**, 21, p. 219906, doi:10.1063/1.2204597.

Hong, X., Kim, J., Shi, S. F., Zhang, Y., Jin, C., Sun, Y., Tongay, S., Wu, J., Zhang, Y. and Wang, F. (2014). Ultrafast charge transfer in atomically thin MoS2/WS2 heterostructures, *Nature Nanotechnology* **9**, 9, pp. 682–686, doi:10.1038/nnano.2014.167.

Hu, W., Kaiser, S., Nicoletti, D., Hunt, C. R., Gierz, I., Hoffmann, M. C., Le Tacon, M., Loew, T., Keimer, B. and Cavalleri, A. (2014). Optically enhanced coherent transport in $YBa_2Cu_3O_{6.5}$ by ultrafast redistribution of interlayer coupling, *Nature Materials* **13**, 7, pp. 705–711, doi:10.1038/nmat3963.

Hybertsen, M. S. and Louie, S. G. (1986). Electron correlation in semiconductors and insulators: Band gaps and quasiparticle energies, *Physical Review B* **34**, pp. 5390–5413, doi:10.1103/PhysRevB.34.5390.

Jaeger, H. M., Fischer, S. and Prezhdo, O. V. (2012). Decoherence-induced surface hopping, *Journal of Chemical Physics* **137**, 22, p. 22A545, doi:10.1063/1.4757100.

Ji, Z., Hong, H., Zhang, J., Zhang, Q., Huang, W., Cao, T., Qiao, R., Liu, C., Liang, J., Jin, C., Jiao, L., Shi, K., Meng, S. and Liu, K. (2017). Robust Stacking-Independent Ultrafast Charge Transfer in MoS_2/WS_2 Bilayers, *ACS Nano* **11**, 12, pp. 12020–12026, doi:10.1021/acsnano.7b04541.

Jia, W., Cao, Z., Wang, L., Fu, J., Chi, X., Gao, W. and Wang, L. W. (2013a). The analysis of a plane wave pseudopotential density functional theory code on a GPU machine, *Computer Physics Communications* **184**, 1, pp. 9–18, doi:10.1016/j.cpc.2012.08.002.

Jia, W., Fu, J., Cao, Z., Wang, L., Chi, X., Gao, W. and Wang, L. W. (2013b). Fast plane wave density functional theory molecular dynamics calculations on multi-GPU machines, *Journal of Computational Physics* **251**, pp. 102–115, doi:10.1016/j.jcp.2013.05.005.

Jiang, X., Zheng, Q., Lan, Z., Saidi, W. A., Ren, X. and Zhao, J. (2021). Real-time GW-BSE investigations on spin-valley exciton dynamics in monolayer transition metal dichalcogenide, *Science Advances* **7**, 10, p. eabf3759, doi:10.1126/sciadv.abf3759.

Kawai, T. and Sakata, T. (1980). Photocatalytic hydrogen production from liquid methanol and water, *Journal of the Chemical Society, Chemical Communications*, 15, pp. 694–695.

Kelly, A. and Markland, T. E. (2013). Efficient and accurate surface hopping for long time nonadiabatic quantum dynamics, *Journal of Chemical Physics* **139**, 1, p. 014104, doi:10.1063/1.4812355.

Korn, T., Heydrich, S., Hirmer, M., Schmutzler, J. and Schller, C. (2011). Low-temperature photocarrier dynamics in monolayer MoS_2, *Applied Physics Letters* **99**, 10, p. 102109, doi:10.1063/1.3636402.

Kovalenko, M. V., Protesescu, L. and Bodnarchuk, M. I. (2017). Properties and potential optoelectronic applications of lead halide perovskite nanocrystals, *Science* **358**, 6364, pp. 745–750, doi:10.1126/science.aam7093, URL https://www.sciencemag.org/lookup/doi/10.1126/science.aam7093.

Krukau, A. V., Vydrov, O. A., Izmaylov, A. F. and Scuseria, G. E. (2006). Influence of the exchange screening parameter on the performance of screened hybrid functionals, *Journal of Chemical Physics* **125**, 22, p. 224106, doi:10.1063/1.2404663.

Kumar, N., Cui, Q., Ceballos, F., He, D., Wang, Y. and Zhao, H. (2014). Exciton-exciton annihilation in $mose_2$ monolayers, *Physical Review B* **89**, p. 125427, doi:10.1103/PhysRevB.89.125427.

Lagarde, D., Bouet, L., Marie, X., Zhu, C. R., Liu, B. L., Amand, T., Tan, P. H. and Urbaszek, B. (2014). Carrier and polarization dynamics in monolayer MoS_2, *Physical Review Letters* **112**, 4, p. 47401, doi:10.1103/PhysRevLett.112.047401.

Lee, J., Sorescu, D. C. and Deng, X. (2011). Electron-induced dissociation of CO_2 on TiO_2(110), *Journal of the American Chemical Society* **133**, 26, pp. 10066–10069, doi:10.1021/ja204077e.

Li, W., Zhou, L., Prezhdo, O. V. and Akimov, A. V. (2018). Spin-orbit interactions greatly accelerate nonradiative dynamics in lead halide perovskites, *ACS Energy Letters* **3**, 9, pp. 2159–2166, doi:10.1021/acsenergylett.8b01226.

Lian, C., Guan, M., Hu, S., Zhang, J. and Meng, S. (2018). Photoexcitation in solids: first-principles quantum simulations by real-time TDDFT, *Advanced Theory and Simulations* **1**, 8, p. 1800055, doi:10.1002/adts.201800055.

Liu, F., Li, Q. and Zhu, X.-Y. (2020a). Direct determination of momentum-resolved electron transfer in the photoexcited van der Waals heterobilayer WS_2/MoS_2, *Physical Review B* **101**, p. 201405, doi:10.1103/PhysRevB.101.201405.

Liu, J., Zhang, X. and Lu, G. (2020b). Excitonic effect drives ultrafast dynamics in van der Waals heterostructures, *Nano Letters* **20**, 6, pp. 4631–4637, doi:10.1021/acs.nanolett.0c01519.

Long, R., Fang, W. and Akimov, A. V. (2016). Nonradiative electron-hole recombination rate is greatly reduced by defects in monolayer black phosphorus: ab initio time domain study, *Journal of Physical Chemistry Letters* **7**, 4, pp. 653–659, doi:10.1021/acs.jpclett.6b00001.

Long, R., Prezhdo, O. V. and Fang, W. (2017). Nonadiabatic charge dynamics in novel solar cell materials, *Wiley Interdisciplinary Reviews: Computational Molecular Science* **7**, 3, p. e1305.

Ma, W., Zhang, J., Yan, L., Jiao, Y., Gao, Y. and Meng, S. (2016). Recent progresses in real-time local-basis implementation of time dependent density functional theory for electron-nucleus dynamics, *Computational Materials Science* **112**, pp. 478–486, doi:10.1016/j.commatsci.2015.08.056.

Mac Kernan, D., Ciccotti, G. and Kapral, R. (2008). Trotter-based simulation of quantum-classical dynamics, *Journal of Physical Chemistry B* **112**, 2, pp. 424–432, doi:10.1021/jp0761416.

Mai, C., Barrette, A., Yu, Y., Semenov, Y. G., Kim, K. W., Cao, L. and Gundogdu, K. (2014). Many-body effects in valleytronics: Direct measurement of valley lifetimes in single-layer MoS_2, *Nano Letters* **14**, 1, pp. 202–206, doi:10.1021/nl403742j.

Mak, K. F., He, K., Shan, J. and Heinz, T. F. (2012). Control of valley polarization in monolayer MoS_2 by optical helicity, *Nature Nanotechnology* **7**, 8, pp. 494–498, doi:10.1038/nnano.2012.96.

Meng, S. and Kaxiras, E. (2008). Real-time, local basis-set implementation of time-dependent density functional theory for excited state dynamics simulations, *Journal of Chemical Physics* **129**, 5, p. 054110, doi:10.1063/1.2960628.

Mitrano, M., Cantaluppi, A., Nicoletti, D., Kaiser, S., Perucchi, A., Lupi, S., Di Pietro, P., Pontiroli, D., Riccò, M., Clark, S. R., Jaksch, D. and Cavalleri, A. (2016). Possible light-induced superconductivity in K3 C60 at high temperature, *Nature* **530**, 7591, pp. 461–464, doi:10.1038/nature16522.

Nakamura, R. and Nakato, Y. (2004). Primary intermediates of oxygen photoevolution reaction on tio2 (rutile) particles, revealed by in situ ftir absorption and photoluminescence measurements, *Journal of the American Chemical Society* **126**, 4, pp. 1290–1298.

Nie, Z., Long, R., Sun, L., Huang, C. C., Zhang, J., Xiong, Q., Hewak, D. W., Shen, Z., Prezhdo, O. V. and Loh, Z. H. (2014). Ultrafast carrier thermalization and cooling dynamics in few-layer MoS2, *ACS Nano* **8**, 10, pp. 10931–10940, doi:10.1021/nn504760x.

Nie, Z., Shi, Y., Qin, S., Wang, Y., Jiang, H., Zheng, Q., Cui, Y., Meng, Y., Song, F., Wang, X., Turcu, I. C., Wang, X., Xu, Y., Shi, Y., Zhao, J., Zhang, R. and Wang, F. (2019). Tailoring exciton dynamics of monolayer transition Metal dichalcogenides by interfacial electron-phonon coupling, *Communications Physics* **2**, 1, p. 103, doi:10.1038/s42005-019-0202-0.

Onida, G., Reining, L. and Rubio, A. (2002). Electronic excitations: density-functional versus many-body green's-function approaches, *Review of Modern Physics* **74**, pp. 601–659, doi:10.1103/RevModPhys.74.601.

Pomarico, E., Mitrano, M., Bromberger, H., Sentef, M. A., Al-Temimy, A., Coletti, C., Stöhr, A., Link, S., Starke, U., Cacho, C., Chapman, R., Springate, E., Cavalleri, A. and Gierz, I. (2017). Enhanced electron-phonon coupling in graphene with periodically distorted lattice, *Physical Review B* **95**, p. 024304, doi:10.1103/PhysRevB.95.024304.

Qiu, D. Y., da Jornada, F. H. and Louie, S. G. (2013). Optical spectrum of mos_2: Many-body effects and diversity of exciton states, *Physical Review Letter* **111**, p. 216805, doi:10.1103/PhysRevLett.111.216805.

Ramasubramaniam, A. (2012). Large excitonic effects in monolayers of molybdenum and tungsten dichalcogenides, *Physical Review B* **86**, p. 115409, doi:10.1103/PhysRevB.86.115409.

Rohlfing, M. and Louie, S. G. (2000). Electron-hole excitations and optical spectra from first principles, *Physical Review B* **62**, pp. 4927–4944, doi:10.1103/PhysRevB.62.4927.

Sakai, N., Fujishima, A., Watanabe, T. and Hashimoto, K. (2003). Quantitative evaluation of the photoinduced hydrophilic conversion properties of tio2 thin film surfaces by the reciprocal of contact angle, *The Journal of Physical Chemistry B* **107**, 4, pp. 1028–1035.

Schaibley, J. R., Yu, H., Clark, G., Rivera, P., Ross, J. S., Seyler, K. L., Yao, W. and Xu, X. (2016). Valleytronics in 2D materials, *Nature Reviews Materials* **1**, 11, p. 16055, doi:10.1038/natrevmats.2016.55.

Shen, M. and Henderson, M. A. (2011). Identification of the active species in photochemical hole scavenging reactions of methanol on tio2, *The Journal of Physical Chemistry Letters* **2**, 21, pp. 2707–2710.

Shi, H., Yan, R., Bertolazzi, S., Brivio, J., Gao, B., Kis, A., Jena, D., Xing, H. G. and Huang, L. (2013). Exciton dynamics in suspended monolayer and few-layer MoS_2 2D crystals, *ACS Nano* **7**, 2, pp. 1072–1080, doi:10.1021/nn303973r.

Shkrob, I. A. and Sauer, M. C. (2004). Hole scavenging and photo-stimulated recombination of electron-hole pairs in aqueous tio2 nanoparticles, *The Journal of Physical Chemistry B* **108**, 33, pp. 12497–12511.

Sorescu, D. C., Lee, J., Al-Saidi, W. A. and Jordan, K. D. (2011). CO_2 adsorption on TiO_2(110) rutile: Insight from dispersion-corrected density functional theory calculations and scanning tunneling microscopy experiments, *The Journal of Chemical Physics* **134**, 10, p. 104707, doi: 10.1063/1.3561300.

Tamaki, Y., Furube, A., Murai, M., Hara, K., Katoh, R. and Tachiya, M. (2006). Direct observation of reactive trapped holes in tio2 undergoing photocatalytic oxidation of adsorbed alcohols: evaluation of the reaction rates and yields, *Journal of the American Chemical Society* **128**, 2, pp. 416–417.

Tan, S., Zhao, Y., Zhao, J., Wang, Z., Ma, C., Zhao, A., Wang, B., Luo, Y., Yang, J. and Hou, J. (2011). CO_2 dissociation activated through electron attachment on the reduced rutile $TiO_2(110)$-1×1 surface, *Physical Review B* **84**, 15, p. 155418, doi:10.1103/PhysRevB.84.155418.

Tapavicza, E., Tavernelli, I. and Rothlisberger, U. (2007). Trajectory surface hopping within linear response time-dependent density-functional theory, *Physical Review Letters* **98**, 2, p. 023001.

Thompson, T. L. and Yates, J. T. (2005). Monitoring hole trapping in photoexcited tio2(110) using a surface photoreaction, *The Journal of Physical Chemistry B* **109**, 39, pp. 18230–18236, doi:10.1021/jp0530451.

Tian, Y., Zheng, Q. and Zhao, J. (2020). Tensile strain-controlled photogenerated carrier dynamics at the van der Waals heterostructure interface, *Journal of Physical Chemistry Letters* **11**, 3, pp. 586–590, doi:10.1021/acs.jpclett.9b03534.

Tully, J. C. (1990). Molecular dynamics with electronic transitions, *The Journal of Chemical Physics* **93**, 2, pp. 1061–1071, doi:10.1063/1.459170.

Tully, J. C. (1998). Mixed quantum-classical dynamics, *Faraday Discussions* **110**, 0, pp. 407–419, doi:10.1039/a801824c.

Wallauer, R., Marauhn, P., Reimann, J., Zoerb, S., Kraus, F., Güdde, J., Rohlfing, M. and Höfer, U. (2020). Momentum-resolved observation of ultrafast interlayer charge transfer between the topmost layers of Mos_2, *Physical Review B* **102**, p. 125417, doi:10.1103/PhysRevB.102.125417.

Wang, H., Bang, J., Sun, Y., Liang, L., West, D., Meunier, V. and Zhang, S. (2016a). The role of collective motion in the ultrafast charge transfer in van der Waals heterostructures, *Nature Communications* **7**, 1, p. 11504, doi:10.1038/ncomms11504.

Wang, H., Zhang, C., Chan, W., Tiwari, S. and Rana, F. (2015). Ultrafast response of monolayer molybdenum disulfide photodetectors, *Nature Communications* **6**, 1, p. 8831, doi:10.1038/ncomms9831.

Wang, L., Akimov, A. and Prezhdo, O. V. (2016b). Recent progress in surface hopping: 2011-2015, *Journal of Physical Chemistry Letters* **7**, 11, pp. 2100–2112.

Wang, Q. H., Kalantar-Zadeh, K., Kis, A., Coleman, J. N. and Strano, M. S. (2012). Electronics and optoelectronics of two-dimensional transition metal dichalcogenides, *Nature Nanotechnology* **7**, 11, pp. 699–712, doi: 10.1038/nnano.2012.193.

Wang, X. and Xia, F. (2015). Van der Waals heterostructures: Stacked 2D materials shed light, *Nature Materials* **14**, 3, pp. 264–265, doi:10.1038/nmat4218.

Wang, Y. H., Steinberg, H., Jarillo-Herrero, P. and Gedik, N. (2013). Observation of floquet-bloch states on the surface of a topological insulator, *Science* **342**, 6157, pp. 453–457, doi:10.1126/science.1239834.

Wei, Y., Cheng, Z. and Lin, J. (2019). An overview on enhancing the stability of lead halide perovskite quantum dots and their applications in phosphor-converted LEDs, *Chemical Society Reviews* **48**, 1, pp. 310–350, doi:10.1039/C8CS00740C.

Westermayr, J., Gastegger, M., Menger, M. F., Mai, S., González, L. and Marquetand, P. (2019). Machine learning enables long time scale molecular photodynamics simulations, *Chemical Science* **10**, 35, pp. 8100–8107, doi:10.1039/c9sc01742a.

Withers, F., Del Pozo-Zamudio, O., Mishchenko, A., Rooney, A. P., Gholinia, A., Watanabe, K., Taniguchi, T., Haigh, S. J., Geim, A. K., Tartakovskii, A. I. and Novoselov, K. S. (2015). Light-emitting diodes by band-structure engineering in van der Waals heterostructures, *Nature Materials* **14**, 3, pp. 301–306, doi:10.1038/nmat4205.

Xiao, D., Liu, G.-B., Feng, W., Xu, X. and Yao, W. (2012). Coupled spin and valley physics in monolayers of mos_2 and other group-vi dichalcogenides, *Physical Review Letter* **108**, p. 196802, doi:10.1103/PhysRevLett.108.196802.

Yu, T. and Wu, M. W. (2014). Valley depolarization due to intervalley and intravalley electron-hole exchange interactions in monolayer mos_2, *Physical Review B* **89**, p. 205303, doi:10.1103/PhysRevB.89.205303.

Yu, W. J., Liu, Y., Zhou, H., Yin, A., Li, Z., Huang, Y. and Duan, X. (2013). Highly efficient gate-tunable photocurrent generation in vertical heterostructures of layered materials, *Nature Nanotechnology* **8**, 12, pp. 952–958, doi:10.1038/nnano.2013.219.

Zeng, H., Dai, J., Yao, W., Xiao, D. and Cui, X. (2012). Valley polarization in MoS_2 monolayers by optical pumping, *Nature Nanotechnology* **7**, 8, pp. 490–493, doi:10.1038/nnano.2012.95.

Zhang, L., Chu, W., Zheng, Q., Benderskii, A. V., Prezhdo, O. V. and Zhao, J. (2019a). Suppression of electron-hole recombination by intrinsic defects in 2D monoelemental material, *Journal of Physical Chemistry Letters* **10**, 20, pp. 6151–6158, doi:10.1021/acs.jpclett.9b02620.

Zhang, L., Vasenko, A. S., Zhao, J. and Prezhdo, O. V. (2019b). Mono-elemental properties of 2D black phosphorus ensure extended charge carrier lifetimes under oxidation: time-domain *Ab Initio* analysis, *Journal of Physical Chemistry Letters* **10**, 5, pp. 1083–1091, doi:10.1021/acs.jpclett.9b00042.

Zhang, L., Zheng, Q., Xie, Y., Lan, Z., Prezhdo, O. V., Saidi, W. A. and Zhao, J. (2018). Delocalized impurity phonon induced electron-hole recombination in doped semiconductors, *Nano Letters* **18**, 3, pp. 1592–1599, doi:10.1021/acs.nanolett.7b03933.

Zhang, X., Lu, G., Baer, R., Rabani, E. and Neuhauser, D. (2020). Linear-response time-dependent density functional theory with stochastic range-separated hybrids, *Journal of Chemical Theory and Computation* **16**, 2, pp. 1064–1072, doi:10.1021/acs.jctc.9b01121.

Zheng, Q., Chu, W., Zhao, C., Zhang, L., Guo, H., Wang, Y., Jiang, X. and Zhao, J. (2019a). Ab initio nonadiabatic molecular dynamics investigations on the excited carriers in condensed matter systems, *Wiley Interdisciplinary Reviews: Computational Molecular Science* **9**, 6, p. e1411, doi:10.1002/wcms.1411.

Zheng, Q., Xie, Y., Lan, Z., Prezhdo, O. V., Saidi, W. A. and Zhao, J. (2018). Phonon-coupled ultrafast interlayer charge oscillation at van der Waals heterostructure interfaces, *Physical Review B* **97**, 20, p. 205417, doi: 10.1103/PhysRevB.97.205417.

Zheng, Z., Zheng, Q. and Zhao, J. (2019b). Ultrafast electron transfer dynamics in lateral transition-metal dichalcogenide heterostructures, *Electronic Structure* **1**, 3, p. 034001, doi:10.1088/2516-1075/ab3b28.

Zhou, H., Zhao, Y. and Zhu, H. (2019). Dielectric environment-robust ultrafast charge transfer between two atomic layers, *The Journal of Physical Chemistry Letters* **10**, 2, pp. 150–155, doi:10.1021/acs.jpclett.8b03596.

Chapter 9

Time-Dependent Density Matrix Renormalization Group for Quantum Chemistry

Xiaoyu Xie,[a] Yao Yao,[b] and Haibo Ma[a]

[a] *School of Chemistry and Chemical Engineering, Nanjing University, No. 163 Xianlin Avenue, Nanjing 210023, Jiangsu Province, China*

[b] *Department of Physics and State Key Laboratory of Luminescent Materials and Devices, South China University of Technology, No. 381 Wushan Road, Guangzhou 510640, Guangdong Province, China*

haibo@nju.edu.cn

Quantum chemistry methods, by approximately solving the Schrödinger equation, have been widely used to investigate various chemical processes and understand corresponding chemical phenomena. In particular, dynamics methods, including full quantum dynamics methods, mixed quantum-classical dynamics and quantum dissipation methods, focus on the time evolution of microscopic processes in these phenomena within different time scales ranging from attosecond to nanosecond. Among these methods, full quantum dynamics approaches try to solve the time-dependent Schrödinger equation (TDSE) of the total chemical systems consisting of both electrons and nucleuses as accurately as possible. Since the degrees of freedom of total chemical systems are immense to be

Time-Dependent Density Functional Theory: Nonadiabatic Molecular Dynamics
Edited by Chaoyuan Zhu

ISBN 978-981-4968-42-3 (Hardcover), 978-1-003-31921-4 (eBook)
www.jennystanford.com

treated directly, approximations are required to solve TDSE. Matrix product state (MPS) is one efficient way to compress quantum states and reduce computational cost. Consequently, nowadays the time-dependent density matrix renormalization group (tDMRG) approaches within the MPS framework have become powerful tools to study the dynamic behaviors of quantum chemistry systems.

In this chapter, an introduction is given in Section 9.1 to present the application of MPS framework as well as tDMRG variants in chemical systems. In Sections 9.2 and 9.3, the elementary principles of MPS approaches are provided, including the basic concept of MPS and matrix product operator (MPO), the density matrix renormalization group (DMRG) method and tDMRG approaches based on the MPS/MPO framework to solve the time-independent and dependent Schrödinger equation, respectively. Then, several examples of tDMRG applications in chemical systems investigated by the authors, including charge carrier dynamics in polymer chain, charge transfer and dissociation in organic systems, nonadiabatic dynamics of excited states S_1/S_2 in pyrazine molecule and singlet fission processes in organic systems, are displayed in Section 9.4. Finally, summary and outlook are given in Section 9.5.

9.1 Introduction

Theoretically, solving the time-dependent or independent Schrödinger equation of a given non-relativistic quantum system is the most straightforward and intuitive idea to investigate its out-of-equilibrium behaviors of dynamical processes or static properties of stationary states, respectively. However, the configuration space of a chemical system expands exponentially with the increasing size of the system. More importantly, there is ubiquitous disorder in spatial and/or temporal domains, which results in the breaking of the Bloch and/or Floquet theorem, thence that the spatial translation symmetry is normally not held in realistic chemical systems leading to great difficulties for obtaining exact solutions of large chemical systems. To deal with this so-called curse of dimensionality, many theoretical solutions have been proposed by applying different approximations. Among these methods, tensor product methods [9, 17] treat quantum states, in terms of the products of local basis, as high-order tensors with an exponentially increasing number of coefficients and compress the information of the high-dimensional coefficient set by decomposing these high-order tensors into products of several localized low-rank and low-order tensors via different tensor decomposition algorithms. For instance, the Tucker decomposition [40] decomposes a high-order and high-rank tensor into the product of a set of matrices and one small Tucker core tensor with the same order but low rank. This well-known decomposition algorithm can be considered as a high-order single value decomposition (HOSVD) [7, 8] and has been applied in the multi-configuration time-dependent Hartree (MCTDH) methods [3, 27]. Obviously, this algorithm suffers from the curse of dimensionality for a system with large number of local sites since the size of Tucker core still grows exponentially with respect to the order of the core. The hierarchical Tucker (HT) decomposition was introduced to overcome this problem and the multi-layer MCTDH (ML-MCTDH) [46] was implemented.

Alternatively, the tensor train decomposition (TT; in the mathematical literature) [30] or matrix product state representation (MPS; in the physical literature) [35, 43] provides another way for the decomposition of quantum states, which decomposes a high-order tensor with high rank into a product of local low-order (order 2 or 3) tensors with a one-dimension (1D) topology. Furthermore, Hamiltonian operators can also be decomposed as products of localized components (i.e., matrix product operator, MPO) and make it possible to solve the Schrödinger equation within the localized representation of both state and Hamiltonian. This decomposition algorithm has been generalized to tensor network states (TNS) such as projected entangled-pair states (PEPS) [41, 43] or tensor tree networks (TTN) [13, 29, 36, 38] for non-1D systems. Among various tensor product methods, DMRG based on the tensor train decomposition has been widely recognized as the most accurate numerical tool for calculating 1D strongly correlated systems with low computational cost, taking advantage of the efficient compression of quantum states and localized representations of states and operators with its underlying MPS/TT formulation.

Following the success of DMRG for describing static quantities of systems, various time-dependent variants (i.e., tDMRG) have been developed and implemented over the last years, extending MPS/TT framework to explore the real-time quantum dynamics of chemical systems by solving the TDSE. In 2004, White and Feiguin, Daley et al. [6] and Verstraete et al. [42] proposed algorithms based on the time-evolving block decimation (TEBD) algorithm of Vidal [44, 45]. They all use the Suzuki–Trotter decomposition [35, 39] of the Hamiltonian into commutative sets of two-body local terms. Therefore, their real-time evolution operator in MPO format can be efficiently constructed and applied to the time-dependent state in MPS format. Unfortunately, this approach is not easy applicable to the systems whose Hamiltonian contains long-range interactions, which is common for quantum chemistry applications as well as system-bath problems, but only after the usage of numerous time-consuming site-swapping operations ("swap gates"). In order to handle such general Hamiltonians, one can apply global time integration solvers (e.g. Runge–Kutta [10, 33, 34] and Krylov [31] approaches) of TDSE to the compressed MPS wavefunction directly without explicitly constructing the time-dependent state. To exploit the second advantage of MPS/MPO framework (i.e., the localized representation and treatment of states/Hamiltonian in MPS/MPO format), several local tDMRG methods, including local Krylov [11, 31] and time-dependent variational principle (TDVP) [14, 15, 21] approaches, were also proposed, which solve a sequence of localized effective TDSEs by introducing appropriate projectors of MPS/TT and inserting the projectors into the original TDSE. Moreover, the MPS/TT structure has also been successfully utilized in describing the mean-field operators in MCTDH [16, 18], the reduced/auxiliary density operators or density vectors in hierarchical equations of motion (HEOM) simulations [4, 37], and the grid-based wavepacket in split-operator Fourier transform (SOFT). [12] In the recent years, tDMRG has been successfully applied for studying various realistic chemical problems, ranging from modeling exciton dynamics in photovoltaic and photosynthetic systems [5, 53] to simulating vibrationally resolved one-dimensional and two-dimensional spectroscopy in molecular aggregates [2, 33] as well as computing the response functions in ab initio systems [1, 34].

9.2 Matrix Product State, Density Matrix Renormalization Group

In this section and the next one, we present a brief introduction to the basic knowledge of DMRG and tDMRG methods, respectively. For more detailed understanding of these methods, we recommend to the readers some review and practical papers [20, 22, 31, 35].

9.2.1 Matrix Product State and Matrix Product Operator

Considering a system with n sites (e.g., electrons as fermions, phonons and photons as bosons, and so on), a general quantum state $|\psi\rangle$ can be written as a linear combination of the orthonormal complete basis set consisting of the product of local configuration $\{\sigma_i\}$ in n-dimension Hilbert space,

$$|\psi\rangle = \sum_{\{\sigma_i\}} c_{\sigma_1\sigma_2\cdots\sigma_n} |\sigma_1\sigma_2\cdots\sigma_n\rangle . \tag{9.1}$$

The combination coefficient $c_{\sigma_1\sigma_2\cdots\sigma_n}$ could be regarded as an element of a rank-n tensor, which can be rewritten as a tensor train (TT) structure by performing sequential singular value decomposition (SVD) processes. Consequently, the quantum state is reformulated as a matrix product state (MPS, see Fig. 9.1),

$$\begin{aligned}|\psi\rangle &= \sum_{\{\sigma_i\}} \sum_{\{\alpha_i=1\}}^{\{m_i\}} A[1]^{\sigma_1}_{1,\alpha_1} A[2]^{\sigma_2}_{\alpha_1,\alpha_2} \cdots A[n]^{\sigma_n}_{\alpha_{n-1},1} |\sigma_1\sigma_2\cdots\sigma_n\rangle \\ &= \sum_{\{\sigma_i\}} \boldsymbol{A}^{\sigma_1}_1 \boldsymbol{A}^{\sigma_2}_2 \cdots \boldsymbol{A}^{\sigma_n}_n |\sigma_1\sigma_2\cdots\sigma_n\rangle. \end{aligned} \tag{9.2}$$

Here, $A[i]^{\sigma_i}_{\alpha_{i-1},\alpha_i}$ (shorten as $A^{\sigma_i}_{\alpha_{i-1},\alpha_i}$ for convenience) is an element of the rank-3 tensor $A[i]$, which has one physical leg (σ_i) and two bond legs (α_{i-1} and α_i) with dimension of m_{i-1} and m_i individually ($m_0 = m_n = 1$). $\boldsymbol{A}^{\sigma_i}_i$ is the matrix slide of the tensor for a given local configuration of site i (σ_i).

Moreover, $m = \max(m_i)$ is defined as the bond dimension of the state $|\psi\rangle$ and related to the amount of entanglement in the states. m can be reduced during the SVD processes (i.e., state $|\psi\rangle$ could be compressed) by setting some thresholds of singular values and discarding corresponding components, which is identical to

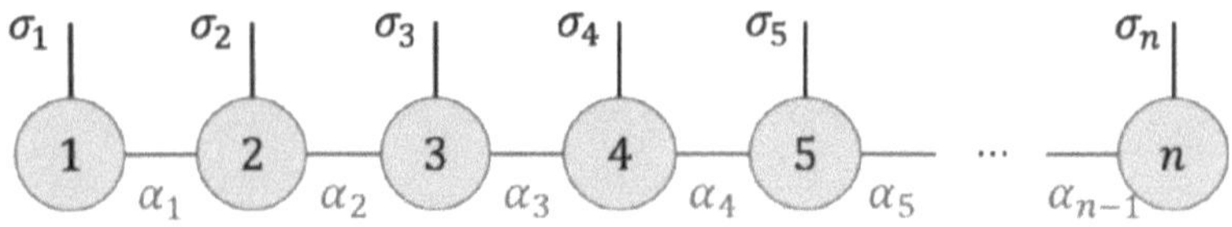

Figure 9.1 Illustration of an MPS. Each circle represents an MPS local tensor component $\boldsymbol{A}\,[\boldsymbol{i}]^{\sigma_i}_{\alpha_{i-1},\alpha_i}$.

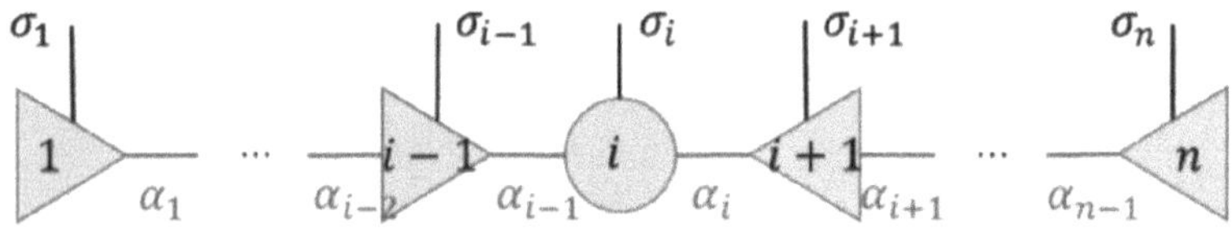

Figure 9.2 Illustration of a mixed-canonical MPS. Each left (right) triangle represents a left (right) orthonormal local tensor component, while the circle represents the local tensor component of the active site.

discarding components with small eigenvalues of the reduce density matrix in traditional DMRG algorithm. In practice, this truncation can be performed via either discarding components whose singular values are below a setting threshold (i.e., dynamical block state selection, DBSS [19]) or fixing the maximal number m of retained components.

Due to the gauge transformation symmetry ($\boldsymbol{A}_1^{\sigma_i}\boldsymbol{A}_{i+1}^{\sigma_{i+1}} = \boldsymbol{A}_1^{\sigma_i}\boldsymbol{X}^{-1}\boldsymbol{X}\boldsymbol{A}_{i+1}^{\sigma_{i+1}}$) of MPS, a state may have an arbitrary MPS structure. For convenience, the left (right) orthonormal rank-3 tensor $L^{\sigma_i}_{\alpha_{i-1},\alpha_i}(R^{\sigma_i}_{\alpha_{i-1},\alpha_i})$ is used to rebuild the structure of MPS as shown below:

$$\sum_{\sigma_i,\alpha_{i-1}} L^{\sigma_i}_{\alpha_{i-1},\alpha_i}\left(L^{\sigma_i}_{\alpha_{i-1},\alpha_i'}\right)^* = \delta_{\alpha_i,\alpha_i'}$$

$$\sum_{\sigma_i,\alpha_i}\left(R^{\sigma_i}_{\alpha_{i-1},\alpha_i}\right)^* R^{\sigma_i}_{\alpha_{i-1}',\alpha_i} = \delta_{\alpha_{i-1},\alpha_{i-1}'}. \tag{9.3}$$

Then, mixed-canonical MPS (see Fig. 9.2) form can be defined as

$$|\psi\rangle = \sum_{\{\sigma_i\}} \boldsymbol{L}_1^{\sigma_1}\boldsymbol{L}_2^{\sigma_2}\cdots\boldsymbol{M}_i^{\sigma_i}\cdots\boldsymbol{R}_{n-1}^{\sigma_{n-1}}\boldsymbol{R}_n^{\sigma_n}\,|\sigma_1\sigma_2\cdots\sigma_i\cdots\sigma_{n-1}\sigma_n\rangle$$

$$= \sum_{\sigma_i,\alpha_{i-1},\alpha_i} M^{\sigma_i}_{\alpha_{i-1},\alpha_i}\left|\mathcal{I}^{[1:i-1]}_{\alpha_{i-1}}\right\rangle|\sigma_i\rangle\left|\mathcal{R}^{[i+1:n]}_{\alpha_i}\right\rangle. \tag{9.4}$$

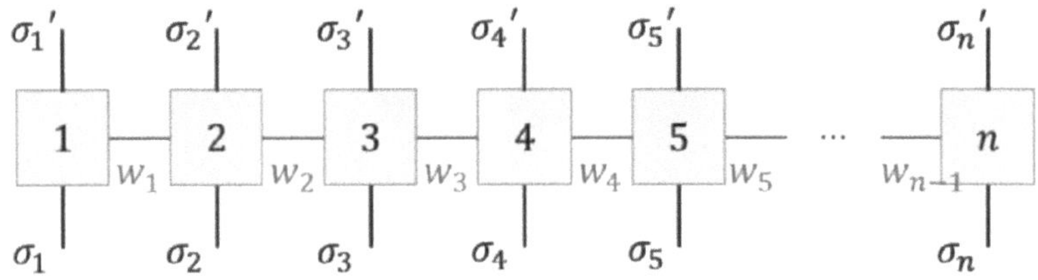

Figure 9.3 Illustration of an MPO. Each square represents an MPO local tensor component $\boldsymbol{W[i]}^{\sigma_i}_{\alpha_{i-1},\alpha_i}$.

where $\left|\mathcal{I}^{[1:i-1]}_{\alpha_{i-1}}\right\rangle$ $(\left|\mathcal{R}^{[i+1:n]}_{\alpha_i}\right\rangle)$ are block configurations consisting of the left (right) orthonormal basis and the site i is called active site or orthogonality center. Particularly, an MPS is called a left (right) canonical MPS when the active site $i = n$ $(i = 1)$.

Analogously, a matrix product operator (MPO, see Fig. 9.3) form is available for an operator which can be expressed as a linear combination in the local configuration space,

$$\hat{O} = \sum_{\{\sigma_i\},\{\sigma_i'\}} \sum_{\{w_i=1\}}^{\{s_i\}} W[1]^{\sigma_1,\sigma_1'}_{1,w_1} \cdots W[n]^{\sigma_n,\sigma_n'}_{w_{n-1},1} \left|\sigma_1 \cdots \sigma_n\right\rangle \left\langle \sigma_1' \cdots \sigma_n'\right|. \tag{9.5}$$

Here, the local component $W[i]$ is a rank-4 tensor with two band legs (w_{i-1} and w_i, $s_0 = s_n = 1$) and two physical legs (σ_n and σ_n').

Within the framework of MPS/MPO, there are two operations (as shown in Fig. 9.4) playing important roles in DMRG/tDMRG algorithms as well as further state analyzation, which are the overlap between two states and the application of an operator to a state.

The calculation of two MPSs is quite straightforward because of the orthonormality of the basis,

$$\begin{aligned}\langle \phi \mid \psi \rangle = &\sum_{\{\sigma_i'\},\{\alpha_i'\}} \left(A'^{\sigma_1'}_{1,\alpha_1'} A'^{\sigma_2'}_{\alpha_1',\alpha_2'} \cdots A'^{\sigma_n'}_{\alpha_{n-1}',1}\right)^* \left\langle \sigma_1'\sigma_2' \cdots \sigma_n'\right| \\ &\times \sum_{\{\sigma_i\},\{\alpha_i\}} A^{\sigma_1}_{1,\alpha_1} A^{\sigma_2}_{\alpha_1,\alpha_2} \cdots A^{\sigma_n}_{\alpha_{n-1},1} \left|\sigma_1\sigma_2 \cdots \sigma_n\right\rangle \\ = &\sum_{\{\sigma_i\}} \left(\boldsymbol{A}'^{\sigma_n}_n\right)^\dagger \cdots \left(\boldsymbol{A}'^{\sigma_1}_1\right)^\dagger \boldsymbol{A}^{\sigma_1}_1 \cdots \boldsymbol{A}^{\sigma_n}_n.\end{aligned} \tag{9.6}$$

Practically, the calculation of the final expression has to be ordered with a suitable sequence considering the computational efficiency.

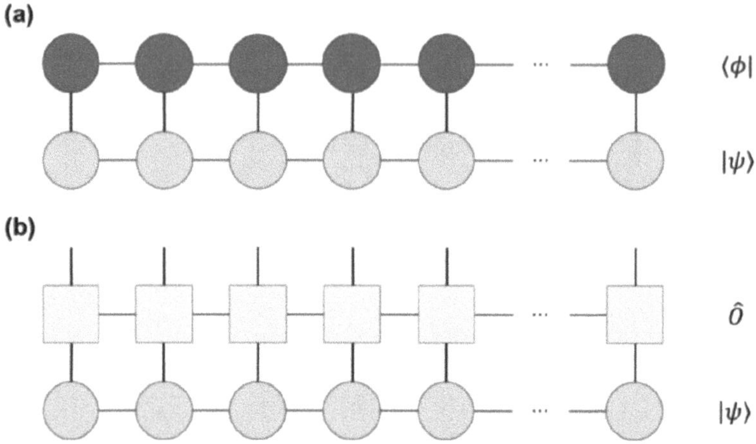

Figure 9.4 Illustration of the two important applications of the MPS/MPO framework. (a) Overlap of two MPSs, resulting to a scale value. (b) Acting of an MPO on an MPS, resulting to another MPS.

For the application operation, there are two ways. One is directly acting the MPO on the MPS. According to the orthonormality of the basis,

$$\begin{aligned}\hat{O}\left|\psi\right\rangle = & \sum_{\{\sigma_i\},\{\sigma_i'\},\{w_i\}} W_{1,w_1}^{\sigma_1,\sigma_1'} \cdots W_{w_{n-1},1}^{\sigma_n,\sigma_n'} \left|\sigma_1\cdots\sigma_n\right\rangle \left\langle\sigma_1'\cdots\sigma_n'\right| \\ & \times \sum_{\{\sigma_i''\},\{\alpha_i\}} A_{1,\alpha_1}^{\sigma_1''} \cdots A_{\alpha_{n-1},1}^{\sigma_n''} \left|\sigma_1''\cdots\sigma_n''\right\rangle \\ = & \sum_{\{\sigma_i\},\{w_i,\alpha_i\}} A_{1,w_1\alpha_1}^{\prime\sigma_1} \cdots A_{w_{n-1}\alpha_{n-1},1}^{\prime\sigma_n} \left|\sigma_1\cdots\sigma_n\right\rangle, \end{aligned} \tag{9.7}$$

where the new MPS local tensor components are given as

$$A_{w_{i-1}\alpha_{i-1},w_i\alpha_i}^{\prime\sigma_i} = \sum_{\sigma_i'} W_{w_{i-1},w_i}^{\sigma_i\sigma_i'} A_{\alpha_{i-1},\alpha_i}^{\sigma_i'}. \tag{9.8}$$

Another way is based on the calculus of variations, i.e., the target state would be obtained by minimizing the distance between $\hat{O}\left|\psi\right\rangle$ and a guess state $\left|\phi\right\rangle$,

$$\begin{aligned} & \min_{\left|\phi\right\rangle} \left\| \left|\phi\right\rangle - \hat{O}\left|\psi\right\rangle \right\|^2 \\ & = \min_{\left|\phi\right\rangle} \left(\left\langle\phi \mid \phi\right\rangle - \left\langle\phi \mid \hat{O}\psi\right\rangle - \left\langle\hat{O}\psi \mid \phi\right\rangle + \left\langle\hat{O}\psi \mid \hat{O}\psi\right\rangle \right). \end{aligned} \tag{9.9}$$

To deal with this minimization, variational method is applied for the local tensor components of $\langle\phi|$ in a mixed canonical form,

$$\frac{\partial}{\partial\left(M^{\sigma_i}_{\alpha_{i-1},\alpha_i}\right)^*}\left(\langle\phi\mid\phi\rangle-\left\langle\phi\mid\hat{O}\psi\right\rangle\right)=0. \tag{9.10}$$

Within the mixed canonical form,

$$\begin{aligned}\langle\phi\mid\phi\rangle &= \sum_{\sigma_i,\alpha_{i-1},\alpha_i}\left(M^{\sigma_i}_{\alpha_{i-1},\alpha_i}\right)^*\left\langle R^{[i+1:n]}_{\alpha_i}\right|\langle\sigma_i|\left\langle I^{[1:i-1]}_{\alpha_{i-1}}\right| \\ &\quad\times\sum_{\sigma_i',\alpha_{i-1}',\alpha_i'}M^{\sigma_i'}_{\alpha_{i-1}',\alpha_i'}\left|I^{[1:i-1]}_{\alpha_{i-1}'}\right\rangle|\sigma_i'\rangle\left|R^{[i+1:n]}_{\alpha_i'}\right\rangle \\ &= \sum_{\substack{\sigma_i,\alpha_{i-1},\alpha_i,\\ \sigma_i',\alpha_{i-1}',\alpha_i'}}\left(M^{\sigma_i}_{\alpha_{i-1},\alpha_i}\right)^* M^{\sigma_i'}_{\alpha_{i-1}',\alpha_i'}\delta_{\sigma_i,\sigma_i'}\delta_{\alpha_{i-1},\alpha_{i-1}'}\delta_{\alpha_i,\alpha_i'} \\ &= \sum_{\sigma_i,\alpha_{i-1},\alpha_i}\left(M^{\sigma_i}_{\alpha_{i-1},\alpha_i}\right)^* M^{\sigma_i}_{\alpha_{i-1},\alpha_i}.\end{aligned} \tag{9.11}$$

Therefore,

$$\frac{\partial\langle\phi\mid\phi\rangle}{\partial\left(M^{\sigma_i}_{\alpha_{i-1},\alpha_i}\right)^*}=M^{\sigma_i}_{\alpha_{i-1},\alpha_i}. \tag{9.12}$$

Thus,

$$M^{\sigma_i}_{\alpha_{i-1},\alpha_i}=\frac{\partial}{\partial\left(M^{\sigma_i}_{\alpha_{i-1},\alpha_i}\right)^*}\left\langle\phi\mid\hat{O}\psi\right\rangle. \tag{9.13}$$

i.e., the local tensor component $M^{\sigma_i}_{\alpha_{i-1},\alpha_i}$ can be computed via Eq. 9.13 (also see Fig. 9.5) by fixing other local components of $|\phi\rangle$ and the target state tend to be converged by sweeping the calculation of $M^{\sigma_i}_{\alpha_{i-1},\alpha_i}$ with i from 1 to n and backward.

9.2.2 Density Matrix Renormalization Group

For the time-independent Schrödinger equation (TISE), calculus of variations could be applied to minimize the expectation energy of system $\left\langle\psi\left|\hat{H}\right|\psi\right\rangle$ and obtain the ground state $|\psi\rangle$.

Given the normalization of target state, we can define the Lagrange multiplier for the TISE as

$$L(\lambda)=\left\langle\psi\left|\hat{H}\right|\psi\right\rangle-\lambda\left(\langle\psi\mid\psi\rangle-1\right). \tag{9.14}$$

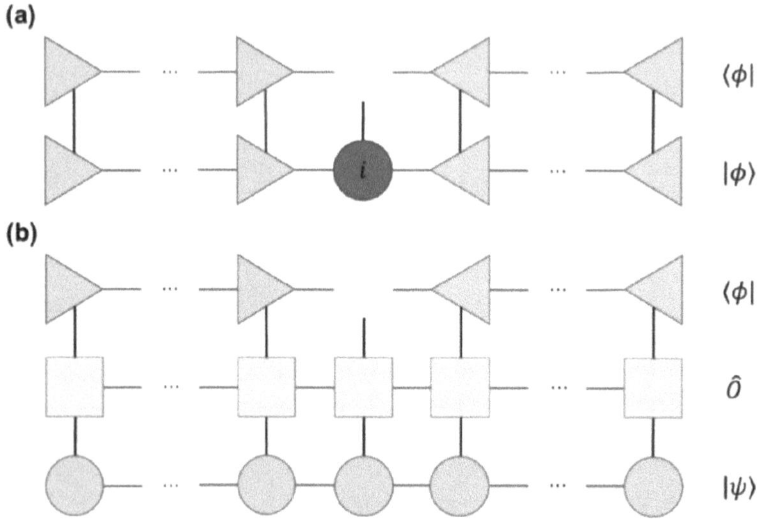

Figure 9.5 Illustration for the application operation $|\boldsymbol{\phi}\rangle = \hat{\boldsymbol{O}}\,|\boldsymbol{\psi}\rangle$ via variational method. (a) Left hand side of Eq. 9.13, i.e., $\boldsymbol{M}^{\sigma_i}_{\alpha_{i-1},\alpha_i}$. (b) Right hand side of Eq. 9.13.

Within the MPS/MPO framework, the energy minimization means that the derivative of the Lagrange multiplier with respect to the conjugation of each tensor component $\left(M^{\sigma_i}_{\alpha_{i-1},\alpha_i}\right)^*$ is zero,

$$0 = \frac{\partial L}{\partial\left(M^{\sigma_i}_{\alpha_{i-1},\alpha_i}\right)^*} = \frac{\partial\left\langle\psi\left|\hat{H}\right|\psi\right\rangle}{\partial\left(M^{\sigma_i}_{\alpha_{i-1},\alpha_i}\right)^*} - \lambda\frac{\partial\left\langle\psi\mid\psi\right\rangle}{\partial\left(M^{\sigma_i}_{\alpha_{i-1},\alpha_i}\right)^*}. \tag{9.15}$$

To calculate the first term of Eq. 9.15, the MPO of Hamiltonian operator is rewritten as a product of three blocks for an active site i,

$$\hat{H} = \sum_{\sigma_i,\sigma_i',w_{i-1},w_i} \hat{H}^{L,[1:i-1]}_{w_{i-1}} W^{\sigma_i,\sigma_i'}_{w_{i-1},w_i} \left|\sigma_i\right\rangle\left\langle\sigma_i'\right| \hat{H}^{R,[i+1:n]}_{w_i}. \tag{9.16}$$

Consequently, the left and right blocks of total $\left\langle\psi\left|\hat{H}\right|\psi\right\rangle$ are defined as

$$\begin{aligned} \hat{H}^{L,[1:i]}_{w_i,\alpha_i,\alpha_i'} &= \left\langle I^{[1:i]}_{\alpha_i}\right| \hat{H}^{L,[1:i]}_{w_i} \left| I^{[1:i]}_{\alpha_i'}\right\rangle \\ \hat{H}^{R,[i:n]}_{w_{i-1},\alpha_{i-1},\alpha_{i-1}'} &= \left\langle R^{[i:n]}_{\alpha_{i-1}}\right| \hat{H}^{R,[i:n]}_{w_{i-1}} \left| R^{[i:n]}_{\alpha_{i-1}'}\right\rangle. \end{aligned} \tag{9.17}$$

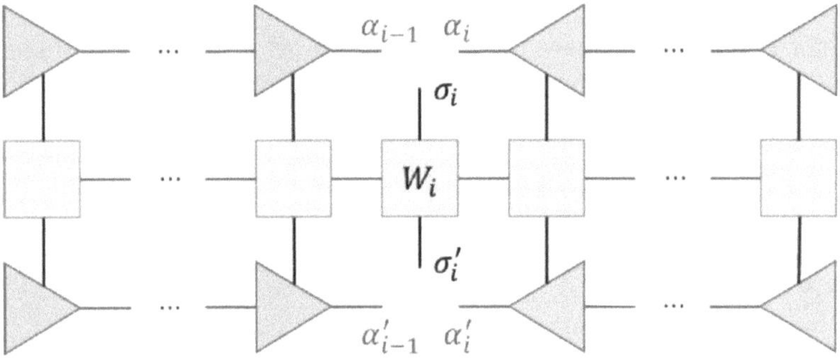

Figure 9.6 Illustration of the one-site effective Hamiltonian $\hat{\boldsymbol{H}}_{\mathrm{i}}^{\mathrm{eff},1}$ at site $\boldsymbol{i}$ with its six legs.

And a rank-6 one-site effective Hamiltonian $\hat{\boldsymbol{H}}_i^{\mathrm{eff},1}$ (Fig. 9.6) for site i is defined as,

$$\left(\hat{\boldsymbol{H}}_i^{\mathrm{eff},1}\right)_{\alpha_{i-1},\alpha_i,\alpha_{i-1}',\alpha_i'}^{\sigma_i,\sigma_i'} = \sum_{w_{i-1},w_i} \hat{H}_{w_{i-1},\alpha_{i-1},\alpha_{i-1}'}^{L,[1:i-1]} W_{w_{i-1},w_i}^{\sigma_i,\sigma_i'} \hat{H}_{w_i,\alpha_i,\alpha_i'}^{R,[i+1:n]}. \quad (9.18)$$

Finally,

$$\langle \psi | \hat{H} | \psi \rangle = \sum_{\substack{\sigma_i,\alpha_{i-1},\alpha_i,\\ \sigma_i',\alpha_{i-1}',\alpha_i'}} \left(M_{\alpha_{i-1},\alpha_i}^{\sigma_i}\right)^* \left(\hat{\boldsymbol{H}}_i^{\mathrm{eff},1}\right)_{\alpha_{i-1},\alpha_i,\alpha_{i-1}',\alpha_i'}^{\sigma_i,\sigma_i'} M_{\alpha_{i-1}',\alpha_i'}^{\sigma_i'}. \quad (9.19)$$

Therefore,

$$\frac{\partial \langle \psi | \hat{H} | \psi \rangle}{\partial \left(M_{\alpha_{i-1},\alpha_i}^{\sigma_i}\right)^*} = \sum_{\sigma_i',\alpha_{i-1}',\alpha_i'} \left(\hat{\boldsymbol{H}}_i^{\mathrm{eff},1}\right)_{\alpha_{i-1},\alpha_i,\alpha_{i-1}',\alpha_i'}^{\sigma_i,\sigma_i'} M_{\alpha_{i-1}',\alpha_i'}^{\sigma_i'}. \quad (9.20)$$

Combining with Eq. 9.12, we have

$$0 = \sum_{\sigma_i',\alpha_{i-1}',\alpha_i'} \left(\hat{\boldsymbol{H}}_i^{\mathrm{eff},1}\right)_{\alpha_{i-1},\alpha_i,\alpha_{i-1}',\alpha_i'}^{\sigma_i,\sigma_i'} M_{\alpha_{i-1}',\alpha_i'}^{\sigma_i'} - \lambda M_{\alpha_{i-1},\alpha_i}^{\sigma_i} = \hat{\boldsymbol{H}}_i^{\mathrm{eff},1} \boldsymbol{M}_i - \lambda \boldsymbol{M}_i. \quad (9.21)$$

i.e., the local tensor component $\boldsymbol{M}_i$ of the target ground state can be calculated by diagonalizing the one-site effective Hamiltonian $\hat{\boldsymbol{H}}_i^{\mathrm{eff},1}$. And the target ground state tends to be converged by sweeping the calculation of $M_{\alpha_{i-1},\alpha_i}^{\sigma_i}$ with i from 1 to n and backward. This algorithm of TISE within the MPS/MPO framework is the one-site density matrix renormalization group (DMRG) approach.

In summary, the processes for one sweep (from left to right, for instance) of one-site DMRG are listed below (for more detailed algorithm descriptions as well as implements, readers can find them somewhere):

(1) Initializing a guess MPS (random number or a reasonable state, e.g., Hartree-Fock configuration for electronic structure calculation) and building its right canonical form.
(2) Setting $i = 0$.
(3) Calculating $\hat{\boldsymbol{H}}_i^{\text{eff},1}$ and diagonalizing. The first eigenvector is set as new $\boldsymbol{M}_i$, whose eigenvalue is storing for the convergence criterion.
(4) Building the mix canonical MPS whose active site is $i + 1$ with the updated $\boldsymbol{M}_i$.
(5) Setting $i = i + 1$ and repeat step 3 and 4 until converge or $i = n$.

9.3 Time-Dependent Density Matrix Renormalization Group

Besides the TISE, the MPS/MPO framework can also be used to solve the time-dependent Schrodinger equation (TDSE).

Generally, the propagation operator $\hat{U}\,(t, t + \delta t)$ with a small time-step δt can be defined for the solution of TDSE and time-evolution of a state $|\psi\,(t)\rangle$ at time t,

$$\hat{U}\,(t, t + \delta t) = e^{-i\hat{H}(t)\delta t/\hbar}. \tag{9.22}$$

Then, the state is updated formulaically,

$$|\psi\,(t + \delta t)\rangle = \hat{U}\,(t, t + \delta t)\,|\psi\,(t)\rangle\,. \tag{9.23}$$

For complex systems, it is nearly impossible to construct the extract propagation operator. There are several ways to solve the TDSE by constructing an approximated propagation operator (e.g., a truncated Taylor expansion, diagonalization of the Hamiltonian via subspace methods, the Chebyshev method and so on) [28] or by updating the state without the explicit construction of $\hat{U}$ (e.g., the Runge–Kutta methods, split operator methods). Taking advantage of the MPS/MPO framework (for instance, effective compress of quantum state, local representation and treatment of states and systems), the above approaches for time-evolution can be implemented within the framework, namely time-dependent density matrix renormalization group (tDMRG) methods.

Based on the implements of these time integrators for MPS/MPO, there are several tDMRG approaches.

9.3.1 The Runge–Kutta Approaches

Firstly, we only consider the convenience of the effective information compress of state with MPS form, several time integration solvers can be performed directly and globally to the MPS. For example, the MPS could be updated by using the fourth-order Runge–Kutta methods (here, a time-independent Hamiltonian $\hat{H}$ is used for convenience),

$$\begin{aligned}|K_1\rangle &= -\frac{i\delta t}{\hbar}\hat{H}\,|\psi(t)\rangle \\ |K_2\rangle &= -\frac{i\delta t}{\hbar}\hat{H}\left(|\psi(t)\rangle + \frac{1}{2}|K_1\rangle\right) \\ |K_3\rangle &= -\frac{i\delta t}{\hbar}\hat{H}\left(|\psi(t)\rangle + \frac{1}{2}|K_2\rangle\right) \\ |K_4\rangle &= -\frac{i\delta t}{\hbar}\hat{H}\left(|\psi(t)\rangle + |K_3\rangle\right).\end{aligned} \tag{9.24}$$

$$|\psi(t+\delta t)\rangle = |\psi(t)\rangle + \frac{1}{6}\left(|K_1\rangle + 2|K_2\rangle + 2|K_3\rangle + |K_4\rangle\right). \tag{9.25}$$

Here, the application operations could be practically done within the MPS/MPO framework. For the addition operation of MPS, it follows the rule of tensor train addition, the block diagonal tensors constructed by local tensor components of the two operands result to the local tensor components of resulting MPS, i.e., given two TT-format coefficient sets of two states,

$$c_A = \boldsymbol{A}_1^{\sigma_1}\boldsymbol{A}_2^{\sigma_2}\cdots\boldsymbol{A}_n^{\sigma_n} \text{ and } c_B = \boldsymbol{B}_1^{\sigma_1}\boldsymbol{B}_2^{\sigma_2}\cdots\boldsymbol{B}_n^{\sigma_n}, \tag{9.26}$$

and the coefficient set of the sum of two states $c_C = c_A + c_B$,

$$c_C = \boldsymbol{C}_1^{\sigma_1}\boldsymbol{C}_2^{\sigma_2}\cdots\boldsymbol{C}_n^{\sigma_n}. \tag{9.27}$$

Then, the local tensor components of c_C are defined naively as

$$\boldsymbol{C}_1^{\sigma_1} = \begin{pmatrix} \boldsymbol{A}_1^{\sigma_1} & \boldsymbol{B}_1^{\sigma_1} \end{pmatrix}$$
$$\boldsymbol{C}_i^{\sigma_i} = \begin{pmatrix} \boldsymbol{A}_i^{\sigma_i} & 0 \\ 0 & \boldsymbol{B}_i^{\sigma_i} \end{pmatrix} \text{ with } i \neq 0, n$$
$$\boldsymbol{C}_n^{\sigma_n} = \begin{pmatrix} \boldsymbol{A}_n^{\sigma_n} \\ \boldsymbol{B}_n^{\sigma_n} \end{pmatrix}. \tag{9.28}$$

Further SVD processes would be applied to compress the resulting MPS.

9.3.2 The Krylov Subspace Approach

Similarly, the Krylov subspace approach could also be implemented for the time-evolution of MPS, namely, the global Krylov method. For an order-r Krylov subspace spanned by the images of the state $|\psi(t)\rangle$ under the first r powers of Hamiltonian $\hat{H}$,

$$\boldsymbol{K}_r(\hat{H}, |\psi\rangle) = \text{span}\{|\psi\rangle, \hat{H}|\psi\rangle, \cdots, \hat{H}^r|\psi\rangle\}. \tag{9.29}$$

The orthonormalization and construction of the exact propagator $\hat{U}'$ of this subspace can be done efficiently with the TT-format states and Hamiltonian. Given the projector of total space onto the orthonormalized $\boldsymbol{K}_r(\hat{H}, |\psi\rangle)$ space $\hat{P}_r$, the MPS is updated approximately by

$$|\psi(t+\delta t)\rangle \simeq \hat{P}_r^{\dagger}\hat{U}'\hat{P}_r|\psi(t)\rangle \tag{9.30}$$

Both Runge–Kutta and global Krylov method treat the state as whole part. Taking advantage of the near-exactly representing of states, the results of high-order $\hat{H}^r|\psi\rangle$ remain highly accurate. However, it would become expensive to keep the high accuracy of such intermediate states (e.g., $|K_i\rangle$ of Runge–Kutta approach and $\hat{H}^r|\psi\rangle$ of global Krylov method) when a long-time simulation is performing, because of the nature of larger entanglement of these states. To overcome this bottleneck, the advantage of local representation of MPS/MPO structure need to be considered.

9.3.3 The Time-Evolving Block Decimation Methods

Based on the split operator methods, the first-order time-evolving block decimation (TEBD) [6, 42, 44, 46, 47] separates the total

Hamiltonian into some commutative local terms $\hat{h}_j$. Then, the propagator can be approximated as

$$\hat{U}(t, t+\delta t) = e^{-i\sum_j \hat{h}_j(t)\delta t/\hbar} \simeq \prod_j e^{-i\hat{h}_j(t)\delta t/\hbar}. \tag{9.31}$$

Due to the localization feature of each term $\hat{h}_j$, the MPO form of local propagator $e^{-i\hat{h}_j(t)\delta t/\hbar}$ can be constructed directly from its tensor product formulate.

It would be highly efficient for systems with only short-range interaction to perform TEBD methods, since it only requires a small set of local terms. While it would be not a good choice to use original TEBD approaches for systems with long-range interaction, such as quantum chemistry systems, which we will discuss below.

9.3.4 The Time-Dependent Variational Principle Method

The time-dependent variational principle (TDVP) method [14, 15, 21] is another important local tDMRG approach. TDVP method is based on a basic idea of variational principle,

$$\min_{|\psi(t)\rangle} ||\hat{H}|\psi(t)\rangle - i\hbar\frac{\partial}{\partial t}|\psi(t)\rangle||^2. \tag{9.32}$$

i.e., minimizing the Dirac–Frenkel functional $||\hat{H}|\psi(t)\rangle - i\hbar\frac{\partial}{\partial t}|\psi(t)\rangle||^2$.

Similar to the DMRG method, TDVP aims to update the state $|\psi(t)\rangle$ by optimizing local tensor components based on the minimization in Eq. 9.29. For such optimization, the projector $\hat{P}_{T,|\psi(t)\rangle}$ on the tangent space of $|\psi(t)\rangle$ is defined as

$$\hat{P}_{T,|\psi(t)\rangle} = \sum_{i=1}^{n} \hat{P}_{i-1}^{L} \otimes \hat{I}_i \otimes \hat{P}_{i+1}^{R} - \sum_{i=1}^{n-1} \hat{P}_{i}^{L} \otimes \hat{P}_{i+1}^{R}, \tag{9.33}$$

where the block projection operators are given by

$$\hat{P}_i^L = \sum_{\alpha_i} \left|\mathcal{I}_{\alpha_i}^{[1:i]}\right\rangle \left\langle \mathcal{I}_{\alpha_i}^{[1:i]}\right|$$

$$\hat{P}_i^R = \sum_{\alpha_i} \left|\mathcal{R}_{\alpha_i}^{[i:n]}\right\rangle \left\langle \mathcal{R}_{\alpha_i}^{[i:n]}\right|. \tag{9.34}$$

By inserting the projector into the two sides of TDSE,

$$i\hbar\frac{\partial}{\partial t}|\psi(t)\rangle = \hat{P}_{T,|\psi(t)\rangle}\hat{H}|\psi(t)\rangle, \tag{9.35}$$

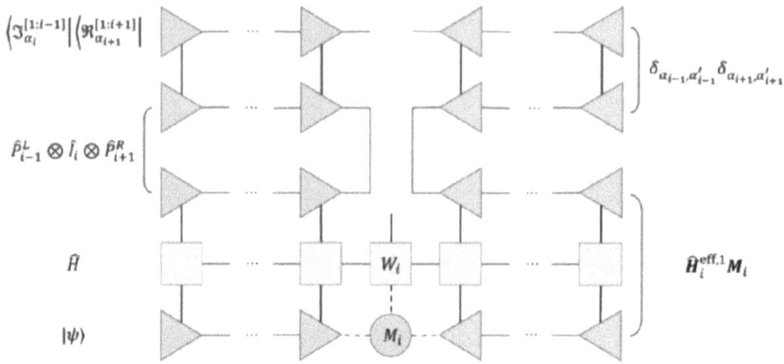

Figure 9.7 Illustration for the right-hand side of forward evolution of the one-site tensor component (from Eq. 9.33 to Eq. 9.35).

The TDSE can be separated approximately as n forward-evolving terms,

$$i\hbar \frac{\partial}{\partial t} |\psi(t)\rangle = \sum_{i=1}^{n} \hat{P}^L_{i-1} \otimes \hat{I}_i \otimes \hat{P}^R_{i+1} \hat{H} |\psi(t)\rangle. \tag{9.36}$$

And $n-1$ backward-evolving terms,

$$i\hbar \frac{\partial}{\partial t} |\psi(t)\rangle = -\sum_{i=1}^{n-1} \hat{P}^L_i \otimes \hat{P}^R_{i+1} \hat{H} |\psi(t)\rangle. \tag{9.37}$$

For each term in Eq. 9.33, it is equivalent the time-evolution of local tensor component by multiply the single-site map $\langle \mathcal{I}^{[1:i-1]}_{\alpha_{i-1}} | \langle \mathcal{R}^{[1:i+1]}_{\alpha_{i+1}} |$ (see Fig. 9.7), instead of performing the updating of full MPS.

$$i\hbar \frac{\partial}{\partial t} M^{\sigma_i}_{\alpha_{i-1},\alpha_i} = \sum_{\sigma'_i,\alpha'_{i-1},\alpha'_i} \left(\hat{\boldsymbol{H}}^{\text{eff},1}_i \right)^{\sigma_i,\sigma'_i}_{\alpha_{i-1},\alpha_i,\alpha'_{i-1},\alpha'_i} M^{\sigma'_i}_{\alpha'_{i-1},\alpha'_i}. \tag{9.38}$$

Or a shorter notation,

$$i\hbar \frac{\partial}{\partial t} \boldsymbol{M}_i = \hat{\boldsymbol{H}}^{\text{eff},1}_i \boldsymbol{M}_i. \tag{9.39}$$

For terms in Eq. 9.34, zero-site form of MPS is defined as

$$|\psi\rangle = \sum_{\alpha_i,\alpha'_i} C^i_{\alpha_i,\alpha'_i} \left| \mathcal{I}^{[1:i]}_{\alpha_i} \right\rangle \left| \mathcal{R}^{[i+1:n]}_{\alpha'_i} \right\rangle \tag{9.40}$$

After the multiplication with the map $\langle \mathcal{I}^{[1:i]}_{\alpha_i} | \langle \mathcal{R}^{[1:i+1]}_{\alpha_{i+1}} |$, these equations can be regarded as the time-evolution of the local zero-site

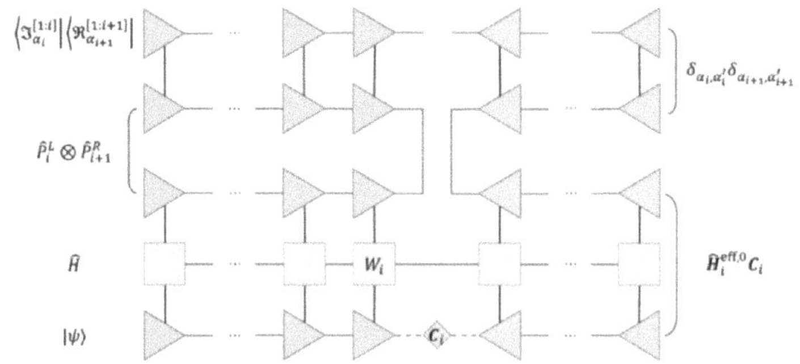

Figure 9.8 Illustration for the right-hand side of backward evolution of the zero-site tensor component (from Eq. 9.34 to Eq. 9.38).

components $\boldsymbol{C}_i$ by introducing the zero-site effective Hamiltonian (see Fig. 9.8),

$$i\hbar\frac{\partial}{\partial t}\boldsymbol{C}_i = -\hat{\boldsymbol{H}}_i^{\text{eff},0}\boldsymbol{C}_i, \tag{9.41}$$

where the effective Hamiltonian $\hat{\boldsymbol{H}}_i^{\text{eff},0}$ (see Fig. 9.9a) is given as,

$$\left(\hat{\boldsymbol{H}}_i^{\text{eff},0}\right)_{\alpha_i'',\alpha_i,\alpha_i''',\alpha_i'} = \sum_{\substack{\sigma_i,\,\alpha_{i-1},\\ \sigma_i',\alpha_{i-1}'}} \left(L^{\sigma_i}_{\alpha_{i-1},\alpha_i''}\right)^* \left(\hat{\boldsymbol{H}}_i^{\text{eff},1}\right)^{\sigma_i,\sigma_i'}_{\alpha_{i-1},\alpha_i,\alpha_{i-1}',\alpha_i'} L^{\sigma_i'}_{\alpha_{i-1}',\alpha_i'''} \tag{9.42}$$

In practice, Equations in 9.36 and 9.38 are solved in sequence defined by the order of the local sites of the system. In detail, the processes of (one-site) TDVP (i.e., 1TDVP) are listed below (one sweep from left to right for self-time updating),

(1) Constructing the right canonical form of state at time t.
(2) Setting $i = 1$.
(3) Forward evolving the local tensor component $\boldsymbol{M}_i$ based on Eq. 9.36 with time-step $\delta t/2$.
(4) Obtaining the zero-site component $\boldsymbol{C}_i$ by left-orthonormalizing $\boldsymbol{M}_i$, $\boldsymbol{M}_i = \boldsymbol{L}_i\boldsymbol{C}_i$.
(5) Backward evolving the local component $\boldsymbol{C}_i$ based on Eq. 9.38 with time-step $\delta t/2$.
(6) Absorbing the time-evolved $\boldsymbol{C}_i$ into the right-orthonormal component $\boldsymbol{R}_{i+1}$ and forming $\boldsymbol{M}_{i+1}$, $\boldsymbol{M}_{i+1} = \boldsymbol{C}_i\boldsymbol{R}_{i+1}$.
(7) Setting $i = i + 1$ and repeating steps 3-6 until $i = n - 1$.

After this sweep, right-to-left sweep for another self-time updating would complete the time-evolution of state $|\psi(t)\rangle$ to $|\psi(t+\delta t)\rangle$. It is easy to find that the bond dimension of state does not increase after time- updating via 1TDVP method, which is contradictory to the fact that the quantum entanglement of state is growing during the time evolution. Therefore, the error of 1TDVP calculation of the long-time dynamic simulation for an initial state with small bond dimension (e.g., linear combination of few local configurations) would be incredibly large. One way to solve this problem is to preform global methods (e.g., Range–Kutta method, global Krylov method) for the initial several steps to prepare a state with larger bond dimension for 1TDVP calculation.

Another way is to use a two-site approach, namely, 2TDVP method. For such method, the two-site effective Hamiltonian $\hat{\boldsymbol{H}}_{i,i+1}^{\text{eff},2}$ (see Fig. 9.9b) is defined for the forward-evolution terms,

$$\left(\hat{\boldsymbol{H}}_{i,i+1}^{\text{eff},2}\right)_{\alpha_{i-1},\alpha_{i+1},\alpha'_{i-1},\alpha'_{i+1}}^{\sigma_i,\sigma_{i+1},\sigma'_i,\sigma'_{i+1}} = \sum_{w_{i-1},w_i,w_{i+1}} \hat{H}^{L,[1:i-1]}_{w_{i-1},\alpha_{i-1},\alpha'_{i-1}} W^{\sigma_i,\sigma'_i}_{w_{i-1},w_i} W^{\sigma_{i+1},\sigma'_{i+1}}_{w_i,w_{i+1}} \hat{H}^{R,[i+2:n]}_{w_{i+1},\alpha_{i+1},\alpha'_{i+1}}. \quad (9.43)$$

Since the information of the MPO bond between site i and $i+1$ is included in the two-site effective Hamiltonian, the bond dimension of MPS should be enlarged after the forward time-evolving. Besides, truncation via SVD could be applied after the two-site updating for the consideration of computational efficiency.

Consequently, the backward-evolution is worked for one-site local components. And the total steps of 2TDVP are listed below:

(1) Constructing the right canonical form of state at time t.
(2) Setting $i = 1$.
(3) Forward evolving the two-site tensor component $\boldsymbol{M}_{i,i+1}$ using two-site effective Hamiltonian $\hat{\boldsymbol{H}}_{i,i+1}^{\text{eff},2}$ with time-step $\delta t/2$. $\boldsymbol{M}_{i,i+1} = \boldsymbol{M}_i \boldsymbol{R}_{i+1}$.
(4) Obtaining new one-site component $\boldsymbol{M}_{i+1}$ by left-orthonormalizing the time-updated $\boldsymbol{M}_{i,i+1}$. $\boldsymbol{M}_{i,i+1} = \boldsymbol{L}_i \boldsymbol{M}_{i+1}$.
(5) Backward evolving the local component $\boldsymbol{M}_{i+1}$ using one-site effective Hamiltonian $\hat{\boldsymbol{H}}_{i+1}^{\text{eff},1}$ with time-step $\delta t/2$.
(6) Setting $i = i + 1$ and repeating steps 3-5 until $i = n - 1$.

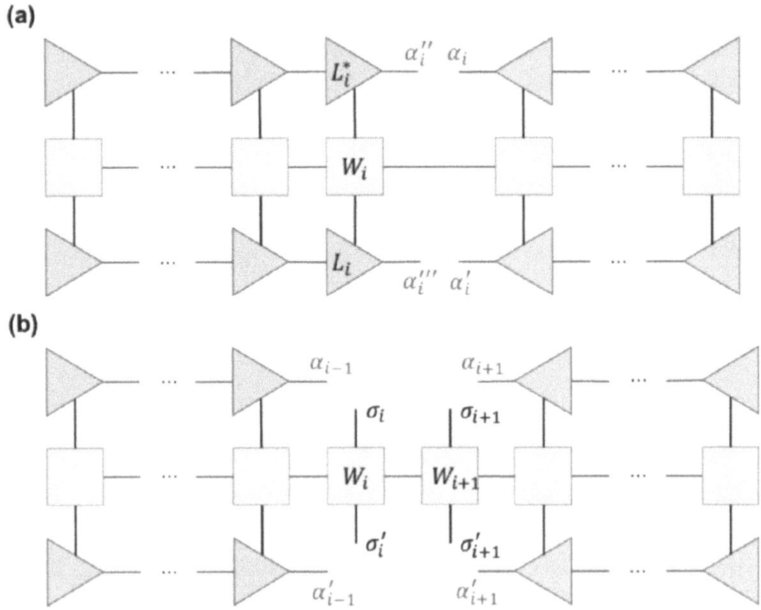

Figure 9.9 Illustration of (a) zero-site effective Hamiltonian $\hat{\boldsymbol{H}}_{\mathbf{i}}^{\mathbf{eff,0}}$ at site $\boldsymbol{i}$ with its four legs and (b) two-site effective Hamiltonian $\hat{\boldsymbol{H}}_{\mathbf{i,i+1}}^{\mathbf{eff,2}}$ at site $\boldsymbol{i}$ and $\boldsymbol{i}$ $+\ \mathbf{1}$ with its eight legs.

(7) Right sweep with i from $n - 1$ to 1 for another $\delta t/2$ time-updating.

Additionally, in order to avoid the case in which a state is trapped in local minima via 2TDVP when the initial state is near eigenstates of system Hamiltonian, one can prepare a better start point for 2TDVP calculation by simulating initial states with several time steps via global approaches firstly.

9.4 Examples

9.4.1 A General Exciton-Vibration Model for Chemistry Systems

Before discussing the applications of tDMRG for specific chemistry system, we would like to introduce a general exciton-vibration (ex-vib) model, which is widely used to study ultrafast nonadiabatic dynamics in organic systems with strong electron-vibration interaction.

With a second-order truncation, the Hamiltonian of an ex-vib coupled system is written as

$$\hat{H} = \hat{H}_{ex} + \hat{H}_{vib} + \hat{H}_{ex-vib}, \tag{9.44}$$

with

$$\hat{H}_{ex} = \sum_i e_i \left|\psi_i\right\rangle \left\langle\psi_i\right| + \sum_{i \neq j} V_{ij} \left|\psi_i\right\rangle \left\langle\psi_j\right|, \tag{9.45}$$

$$\hat{H}_{vib} = \sum_I \frac{1}{2} \hbar\omega_I \left(-\frac{\partial^2}{\partial Q_I^2} + Q_I^2\right), \tag{9.46}$$

$$\hat{H}_{el-vib} = \sum_{i,j,I} g_{ij}^I Q_I \left|\psi_i\right\rangle \left\langle\psi_i\right| + \sum_{i,j,I,J} g_{ij}^{IJ} Q_I Q_J \left|\psi_i\right\rangle \left\langle\psi_j\right|. \tag{9.47}$$

Here, e_i is the energy of the electronic state $|\psi_i\rangle$ and V_{ij} is the electronic coupling between states $|\psi_i\rangle$ and $|\psi_j\rangle$ of a reference geometry (commonly, the equilibrium geometry of initial state). For the vibrational term $\hat{H}_{vib}$, harmonic oscillator approximation is applied with ω_I and Q_I represent the frequency and the dimensionless displacement of the vibration mode I, respectively. g_{ij}^I are linear ex-vib couplings with g_{ii}^I being the local ex-vib coupling and $g_{ij}^I (i \neq j)$ being nonlocal ones. g_{ij}^{IJ} are second-order terms.

Considering the discreteness feature of basis set of states and matrix representation of Hamiltonian operator used in MPS/MPO language, the second quantization is more convenient for the tDMRG simulations, i.e., occupation representation is used for both electronic and vibrational parts. And the three terms of Hamiltonian can be reformulated as

$$\hat{H}_{ex} = \sum_i e_i \hat{a}_i^\dagger \hat{a}_i + \sum_{i \neq j} V_{ij} \hat{a}_i^\dagger \hat{a}_j, \tag{9.48}$$

$$\hat{H}_{vib} = \sum_I \hbar\omega_I \left(\hat{b}_I^\dagger \hat{b}_I + \frac{1}{2}\right), \tag{9.49}$$

$$\hat{H}_{el-vib} = \sum_{i,j,I} g_{ij}^I \hat{q}_I \hat{a}_i^\dagger \hat{a}_j + \sum_{i,j,I,J} g_{ij}^{IJ} \hat{q}_I \hat{q}_J \hat{a}_i^\dagger \hat{a}_j. \tag{9.50}$$

Here, $\hat{a}_i^\dagger$ ($\hat{a}_i$) and $\hat{b}_i^\dagger$ ($\hat{b}_i$) are the creators (annihilators) of the electronic state $|\psi_i\rangle$ and the vibration mode I, respectively. $\hat{q}_I = (\hat{b}_I^\dagger + \hat{b}_I)/\sqrt{2}$ is the dimensionless displacement operator of mode I.

The corresponding occupation representation set is $\{0, 1\}$ for each local electronic site (Fermion) and the total occupation number for all electronic site should be fixed as 1 due to the $U(1)$ symmetry of electronic part in the total Hamiltonian, while the complete basis set for each vibrational site (Bosonic) should be theoretically infinite. In practice, truncation of basis set needs to be applied to vibrational sites for MPS/MPO construction and tDMRG simulation by setting maximal occupation numbers for vibration modes, e.g., the basis set for mode I is $\{|0\rangle_I, \cdots, |n_{\max}^I\rangle_I\}$.

9.4.2 Charge Carrier Dynamics in Polymer Chain

The polymer can be described by a one-dimensional (1D) lattice chain with each site representing a unit group such as C-H unit in polyacetylene. Due to the special topology, a nonlinear elementary excitation, so-called polaron, serves as the essential candidate of charge carrier in the polymer chain [23–25, 55]. The polaron is formed by an electron or a hole surrounding by lattice distortions, which is named as self-trapping effect. This nonlinear effect leads to the remarkable stability of polarons which can freely drift on the polymer chain and hop between chains. In this perspective, the conductivity of polymers is then strongly dependent on both the drift velocity of polarons in one chain and the hopping rate between them. The latter one could be investigated by kinetic Monte Carlo simulations, and herein we will mainly consider the former case.

Let us first introduce the celebrated model Hamiltonian that was proposed by Su, Schrieffer and Heeger (SSH) to describe the topological elementary excitations in polymers. That is, the V_{ij} in Eq. 9.42 is modified as $V_0 - \alpha u_i$, with V_0 being the hopping integral, α the vibronic coupling strength and u_i the displacement of i-th site.

SSH model refers to a noninteracting case, but in the realistic polymers there are numerous π electrons on carbon atoms which are conjugated and interacting with each other. The Coulomb

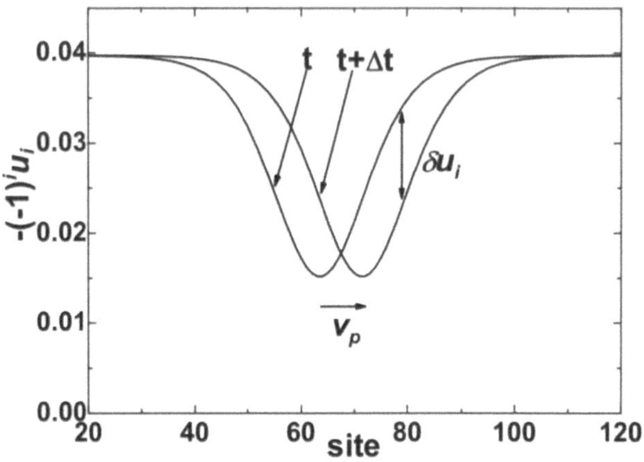

Figure 9.10 Time evolution of the lattice distortion on each site. Reproduced from Ref. [55] with permission from American Physical Society.

repulsive interaction between electrons could enlarge the energy gap and then weaken the conductivity due to the strong scattering. Actually, many polymers with poor conductivity suffer from the Coulomb interactions. To this end, the interaction term, namely Hubbard U term, should be added into the model, i.e.,

$$\hat{H}_U = U \sum_i \hat{a}_{i\uparrow}^\dagger \hat{a}_{i\uparrow} \hat{a}_{i\downarrow}^\dagger \hat{a}_{i\downarrow}, \tag{9.51}$$

where U is the on-site Coulomb interaction. In terms of Hubbard U term, the system becomes many-body system and the usual dynamical approach does not work, so we can use tDMRG to simulate the dynamics of polarons.

Figure 9.10 displays a schematic picture of the motion of polaron. There is a distorted potential valley for the trapped electron which will move on the site as an entire quasi-particle. The size of the polaron is around 30 sites, so that it can be called large polaron. The velocity of the polaron can be defined by the movement of its center.

In order to study the velocity of polarons on the chain, we should first locate the central position of polaron. As usual, we record the site of maximum value of charge density as the polaron charge center X_c and then calculate the velocity accordingly as shown in Fig. 9.11. It is clear that following U increasing the velocity decreases significantly, consistent with our expectation. This is easy

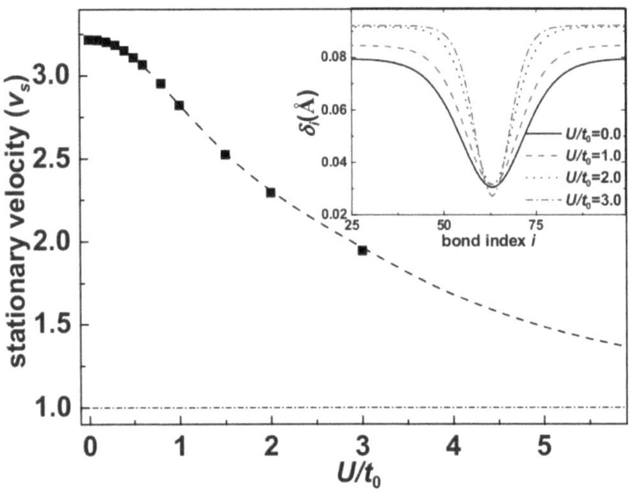

Figure 9.11 Velocity of polaron versus the Hubbard $\boldsymbol{U}$. The inset shows the change of the shape of polarons for various $\boldsymbol{U}$. Reproduced from Ref. [49] with permission from American Physical Society.

to understand since, as shown in the inset of Fig. 9.11, the size of the polaron shrinks and the binding energy becomes larger than that of noninteracting system.

9.4.3 Exciton Dissociation at Donor/Acceptor Interface

While talking about organic solar cells, one always asks this question: what is explicitly the difference between organic and inorganic cells? Although, we can argue that there is not a well-defined energy band in organics and the transportation mechanism of carriers turns out to be incoherent hopping, similar effects for these issues can be still observed in the two systems. The sole difference stems from the quantumness of charge photogeneration process [51].

For solar cells, there is a famous Shockley–Queisser (S-Q) limit of the power conversion efficiency. They regard the sun light as a high-temperature heat reservoir and the ambient circumstance as the low-temperature reservoir, and the cell itself matters as the working substance to converse the solar energy to electric energy. Neglecting

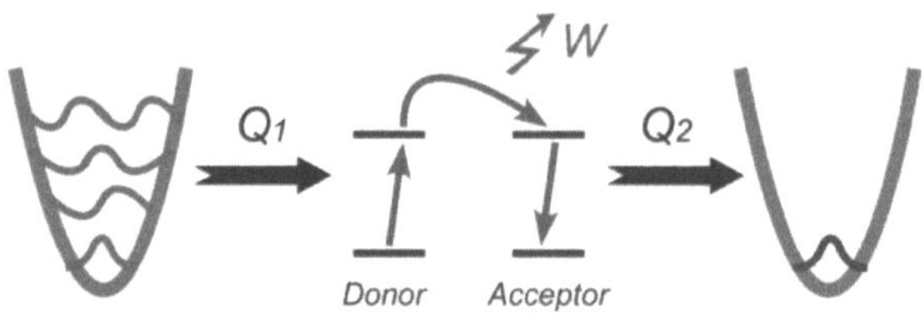

Figure 9.12 Schematic of our quantum heat engine model for organic donor/acceptor heterojunction. Reproduced from Ref. [51] with permission from American Physical Society.

all the details of the cell, they consider an ideal band structure and the electrons obey the distribution at equilibrium and reversible thermodynamic processes. The S-Q limit can then be derived and set as the upper limit of efficiency. This ideal heat engine model is valid for majority kinds of cells.

How to exceed the S-Q limit? The answer should be utilizing the quantum coherence. As it is well known, the efficiency of a heat engine is pretty large in the early working stage, and afterward the efficiency soon falls down due to the rapid increase of entropy following by the decoherence processes and then gets a steady state once achieving the maximum entropy. In order to increase the efficiency, one has to decrease the entropy at steady state, and the sole way is to keep the quantum coherence alive as long as possible. This turns out to be the very motivation of quantum heat engine.

Organic molecules possess separate orbitals and strong vibronic couplings. The former feature allows us to freeze some redundant orbitals so that the available states are limited and the maximum entropy can be kept small. The vibronic coupling can be also modified to be friendly to the quantum coherence. In this context, we can establish a quantum heat engine model taking the orbitals in donor and acceptor into account. The schematic of the model is sketched in Fig. 9.12. We can use tDMRG to calculate the efficiency of the model.

Figure 9.13 displays the efficiency of the quantum heat engine for various J values. J refers to the interaction between donor and acceptor. As usual, the initial efficiency is almost 100% and then decays following time evolving until reaching a steady value which

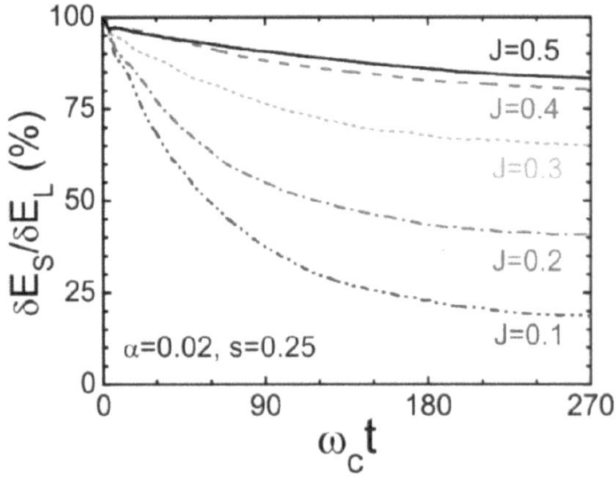

Figure 9.13 Efficiency of quantum heat engine model for various J. Reproduced from Ref. [45] with permission from American Physical Society.

can be regarded as the ultimate efficiency of the quantum heat engine. It is found that, when $J = 0.5$ the maximum efficiency can be as high as 80%, much higher than S-Q limit for the efficiency. This result can be applied to the practical organic solar cells, that is, one has to optimize the packing of donor and acceptor to enhance their interaction in order to increase the power conversion efficiency.

9.4.4 Excited State Charge Transfer

Charge transfer (CT) state is a special excited state in organic materials [50, 54]. Different from the extensive feature in the systems with periodic symmetry, the wavefunction of electrons in amorphous organic molecules is completely localized in certain moieties. This implies that the electron and hole could be distinguishable in organics, so that we can safely argue there are two different kinds of excited states: One is Frenkel exciton and the other is CT exciton. The latter can be either the direct yield of photo-excitation or the intermediate state of charge photogeneration.

In normal cases, the binding energy of Frenkel exciton is quite larger than the thermal energy at room temperature, implying the dissociation of exciton serves as the main bottleneck of photon-charge conversion efficiency. The CT state, on the other hand, holds much smaller binding energy than Frenkel exciton. No matter if it is the main yield of photo-excitation or it is formed coherently from Frenkel excitons, the CT state has always to be the key to resolve the problems in organic solar cells. Herein, we consider the ultrafast dynamics of the conversion from local exciton to CT state.

Our model is following the general form of ex-vib coupling Hamiltonian 1.41. To describe the binding energy of excitons, we additionally introduce a Coulomb attraction between electron and hole. Due to the presence of the strong Coulomb interaction and the ubiquitous vibronic couplings, the exciton dissociation is always a many-body problem that has to be dealt with tDMRG algorithm.

Figure 9.14 displays the time evolution of charge density in both donor and acceptor. We set the initial state as the singlet Frenkel exciton. It is observed that for intermediate coupling the wavepacket of charges splits into two and spread to the ends of the system until being bounced back by the boundaries. This process can be well analogized with the traditional CT process at the D/A interface. We thus demonstrated that the ultrafast CT dynamics can be simulated by tDMRG giving the appropriate interactions.

Through our calculations, the off-diagonal vibronic coupling manifests remarkable importance. In common sense, the warm and wet environments do not favor the quantum coherence, so why should we usually highlight the essential role of it? The answer could be the competition between diagonal and off-diagonal couplings. The former prefers to the localization and decoherence while the latter loves the opposite. In normal materials, the off-diagonal coupling is smaller than the diagonal one, but if there are free radicals that are likely delocalized in the backbone, the off-diagonal coupling will largely increase and even dominate the diagonal one. As a result, radicals should be the essential item for the presence of quantumness.

9.4.5 Photo-Dynamics and Absorption Spectrum for Pyrazine

As a well-defined benchmark model system, the internal conversion between S_1 and S_2 electronic states of pyrazine molecule system has been investigated via several quantum dynamics methods [49]. In

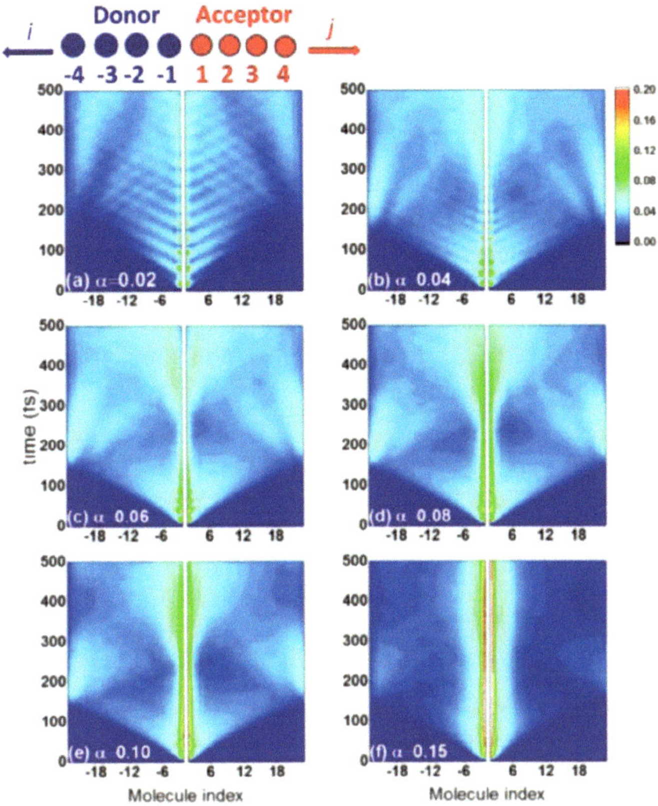

Figure 9.14 Time evolution of charge density hole and electron in their own layers for various vibronic coupling strength. Reproduced from Ref. [54] with permission from American Chemical Society.

detail, this model consisting of 2 electronic states and 24 vibrational modes, which contain 4 modes with strong ex-vib coupling and 20 modes weakly coupled with electronic states. And the ex-vib Hamiltonian is written as

$$\hat{H} = \begin{pmatrix} -\Delta & 0 \\ 0 & \Delta \end{pmatrix} + \sum_{I=1}^{24} \hbar\omega_I \left(\hat{b}_I^\dagger \hat{b}_I + \frac{1}{2} \right) + \sum_{I=1}^{24} \hat{q}_I \begin{pmatrix} g_1^I & g_{12}^I \\ g_{12}^I & g_2^I \end{pmatrix} + \sum_{I,J=1}^{24} \hat{q}_I \hat{q}_J \begin{pmatrix} g_1^{IJ} & g_{12}^{IJ} \\ g_{12}^{IJ} & g_2^{IJ} \end{pmatrix}. \quad (9.52)$$

Here, the electronic terms with two basis states are expressed with matrix representation. $\Delta = \left(E_{S_2} - E_{S_1}\right)/2$ represents energy terms of electronic states, while the electronic coupling is zero since two electronic states are two adiabatic states under the equilibrium geometry.

All the parameters in Eq. 9.48 have been calculated by computing potential energy surfaces of S_1 and S_2 states via ab initio quantum chemistry methods, such as the configuration interaction (CI) method and the complete active space self-consistent field (CASSCF) method. And the parameters in References [32, 48] are used to construct the MPO form of Hamiltonian and simulate the internal conversion process via tDMRG.

First, we test the performance of three tDMRG approaches by using the simplest 4-mode model (namely $v6a$, $v1$, $v9a$ and $v10$, discarding other 20 weak-coupled modes) with different maximal occupation number $n_{\max}^{l} = n_{\max}$. The initial state is $|\psi(0)\rangle = |S_2\rangle \otimes |0\rangle_{v6a} \otimes |0\rangle_{v1} \otimes |0\rangle_{v9a} \otimes |0\rangle_{v10}$, and the total simulation time is 120 fs with 0.12 fs per step. The truncation threshold of SVD processes during the MPS construction and tDMRG simulation is fixed as 1×10^{-15}. Results of S2 populations and absolute value of autocorrelation function $c = \langle\psi(0)|\psi(t)\rangle$ are displayed in Fig. 9.15.

In Fig. 9.15a,b, the results of second-order Taylor expansion show crashed lines for the system with $n_{\max} = 10$ for 4 modes after about 40 fs, which indicates the invalidation of this approach for large systems due to the large truncation error of propagator. On the contrary, both global Krylov and TDVP methods give consistent results, while the TDVP method is cheaper and more stable than global Krylov method. Therefore, the TDVP method is used for following calculations. On the other hand, the convergence test of basis set for vibrational sites shows that the model with smaller vibrational basis set give incorrect dynamics results at earlier time, suggesting that basis states with high occupation number become important for a long-time evolution and the basis set of vibration sites should be careful tested to balance the accuracy and efficiency for tDMRG simulation. We fix the suitable maximal occupation numbers for each mode by testing the convergence of time-evolution results, and the values are 24, 18, 10 and 18 for mode $v6a$, $v1$, $v9a$ and $v10$, respectively, for further time simulations.

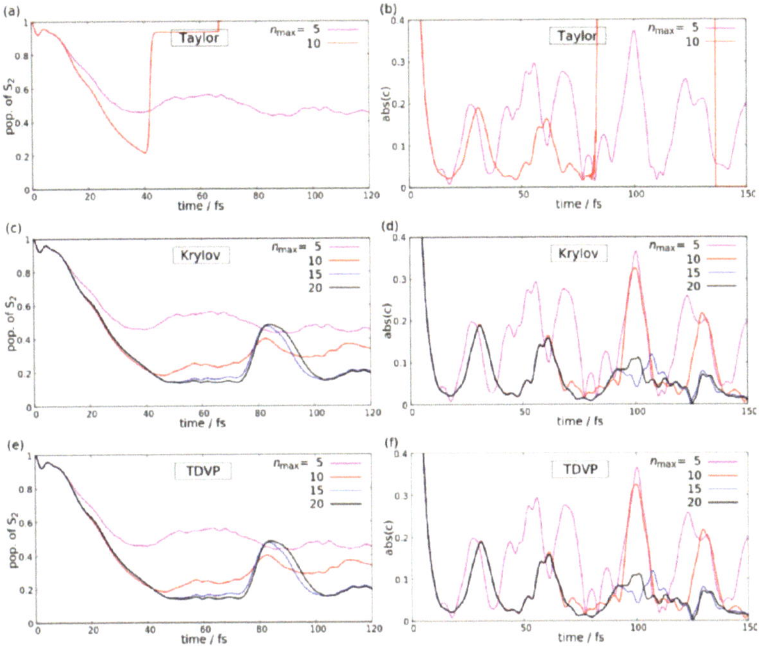

Figure 9.15 The dynamics simulation results of 4-modes systems with different maximal occupation number for vibration sites via (a, b) the 2nd-order Taylor expansion, (c, d) the global Krylov method and (e, f) the 2TDVP method. (a, c, d) show time evolution results of S_2 population and (b, d, f) give results of the absolute value of the autocorrelation function with respect to the simulation time. Reproduced from Ref. [49] with permission from American Institute of Physics.

After the decision of tDMRG approach (i.e., TDVP), we test some other parameters which may affect the performance of simulation, such as time interval, truncation threshold as well as the difference between the one-site/two-site TDVP variant. The results are given in Fig. 9.16. The tests of time interval in Fig. 9.16a,b show that simulations with $\delta t < 20$ a.u. (about 0.48 fs) give almost the same results for both population evolution and autocorrelation function, while the trajectories display an evident difference after about 50 fs due to the larger error of the time-integrator. Thus, the simulation time interval for pyrazine system should not be larger than 20 a.u.

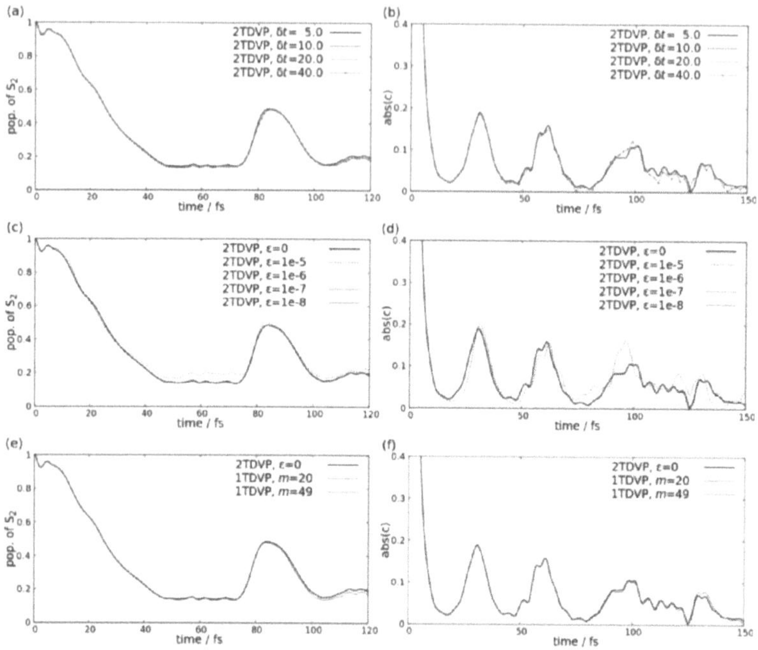

Figure 9.16 The tests of parameters for the dynamics simulation of the 4-mode pyrazine S_1/S_2 systems via TDVP methods. (a, b). The time evolution results of the S_2 population and absolute value of the autocorrelation function with different time intervals ($\delta t = 5.0$, 10.0, 20.0 and 40.0 a.u., respectively). (c, d). The time evolution results with different truncations in 2TDVP. (e, f). The results for difference initial bond dimension via 1TDVP method comparing to the benchmark 2TDVP calculation. Reproduced from Ref. [49] with permission from American Institute of Physics.

The truncation threshold for SVD during MPS construction and tDMRG simulation is a key parameter to cut off redundancy information of states effectively and reduce the computational cost, individually. It is worthwhile to test this parameter to balance accuracy and efficiency. As shown in Fig. 9.16c,d, the convergence of the population of an electronic state requires a truncation threshold $\varepsilon < 1 \times 10^{-7}$ while the condition of convergence for the autocorrelation function is $\varepsilon < 1 \times 10^{-8}$, which implies that the population of state is easier to get converge that other observables (e.g., the autocorrelation function).

As mentioned before, the performance of 1TDVP is limited by the fact of unchanged bond dimension of state during time evolution, especially for large systems. In Fig. 9.16e,f, the 1TDVP is tested by preparing initial states with larger bond dimension. Comparing to the benchmark 2TDVP results, 1TDVP gives correct simulation results when the bond dimension of initial state is 49. Though the computational cost of 1TDVP is much cheaper than 2TDVP for the 4-mode system, it should be noted that a sufficiently large bond dimension is required for the 1TDVP simulation of large systems, which make it hard to apply for such systems. Because the bond dimension grows automatically during 2TDVP simulation, it reveals a much better applicability of 2TDVP approach for systems with different sizes.

Based on the above tests, parameters of 2TDVP method are optimized and applied to other 4-mode as well as 24-mode model systems. As an application, the spectrum of pyrazine is calculated by calculating the Fourier transform of the product of the autocorrelation function $c(t)$ and a damping function $f(t) = e^{-|t|/\tau}$ with τ being the time decay constant. The results of spectrums are shown in Fig. 9.17 comparing to simulation results of MCTDH as well as experimental measurements. The nice agreement for both comparisons suggests that tDMRG, especially the 2TDVP approach is a feasible and accurate method to perform quantum dynamics simulation of realistic chemical systems.

9.4.6 Singlet Fission

Singlet fission is a special effect in organic solar cells [26, 52]. Due to the large binding energy, excitons in organics possess long lifetime and do not instantaneously dissociate into free electrons and holes. This makes another reaction pathway available, that is, a singlet exciton changes to be two triplet excitons. Benefiting from the special molecular structure of acenes, there are strong spin exchange interaction that largely lowers the energy of triplet, such that the energy of one singlet is close to that of two triplets enabling the singlet fission.

In a physical viewpoint, it is not easy to comprehend the simultaneity of exciton splitting and spin flipping. There has to

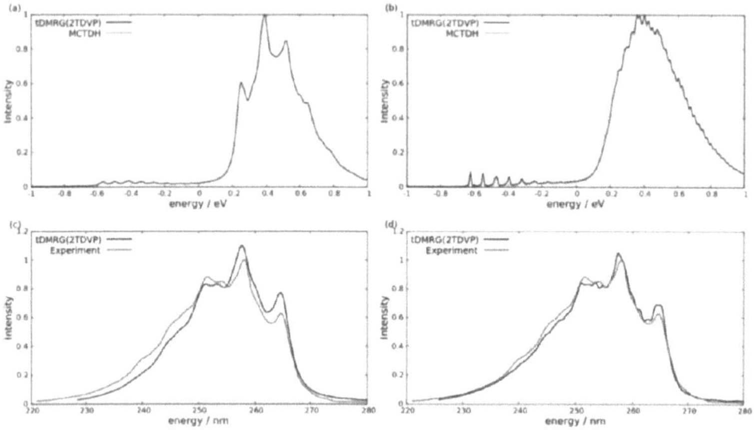

Figure 9.17 Spectrum results of the molecule pyrazine. (a) Spectrum (black line) of the 4-mode model with parameters in reference comparing to the simulation result (red line) of MCTDH ($\tau = 30$fs). (b) Spectrum (black line) of the 24-mode model with parameters as given in reference, compared to a MCTDH simulation (red line) result ($\tau = 30$ fs). (c) Spectrum (black line) of the 4-mode model with parameters as in reference compared to experimental results (red line) ($\tau = 30$ fs). (d) Spectrum (black line) of the 24-mode model with parameters as in reference compared to experimental results (red line) result ($\tau = 50$ fs). Reproduced from Ref. [49] with permission from American Institute of Physics.

be an intermediate state that separates the two steps of singlet fission. This state is the so-called correlated triplet-triplet pairing state, or TT state. The total spin of TT state is zero, so that from singlet to TT state one exciton splits to two without changing of spin configuration. Then the two triplets lose the quantum correlation and thus the spin coherence, and the TT state is changed to be two separated triplets, or T...T state. In the second step, only the spin state is changed due to the decoherence process. Subsequently, the role of TT state emerges to be an essential issue in the study of singlet fission.

To this end, we establish a diabatic Hamiltonian based on a space involving one singlet state, two charge-transfer (CT) states and two TT states, so that e_i and V_{ij} form a 5×5 matrix. For convenience, we set the energy of singlet to be zero, and the other two are 0.6 eV for CT state and 0.1 eV for TT state. The other parameters are calculated from quantum chemistry simulations: $V_{13} = V_{31} = -0.051$ eV, $V_{14} = V_{41} = -0.074$ eV, $V_{23} = V_{32} = -0.118$ eV, $V_{24} = V_{42} = -0.111$ eV, $V_{35} = V_{53} = -0.081$ eV, $V_{45} = V_{54} = 0.056$ eV. The CT state is virtually considered to bridge the singlet and TT state, since the direct coupling between them is not sufficiently strong to enable an efficient conversion. In order to investigate the decoherence process, two phonon baths, locally coupling to the excitons and nonlocally coupling to the CT states, are taken into account as well. As this is a complicated many-body system, we have to utilize the tDMRG to calculate the entire dynamical processes.

Figure 9.18 shows the population of singlet, triplet and CT state with the local vibronic coupling being 0.1. It is found that the singlet population goes down from 1 to a lower value and the triplet population increases from 0, manifesting the singlet fission takes place. The quantum beating mode is observed as well. More importantly, the increase of triplet is slower than the decay of singlet, suggesting the indirect conversion through the bridging role of CT state dominates the dynamical process of singlet fission.

9.5 Summary and Outlook

In this chapter, basic knowledge of MPS/MPO language as well as the general idea of DMRG algorithm are introduced. For the quantum dynamics field, various tDMRG approaches, including several global methods and local methods, within the MPS/MPO framework are listed with their basic ideas. In particular, TDVP methods, whose procedures are similar with the sweep DMRG algorithm, take advantage of both efficient compression of quantum states and localized representation of states and Hamiltonian operator.

By adopting the electron-vibration interaction Hamiltonian, tDMRG approaches can be applied to study the dynamics behaviors of realistic organic chemical systems with moderate and complicated electron-electron or electron-vibration interactions. Several application examples in organic systems are provided in this chapter, including charge transport in polymers, charge transfer and separation on the organic donor/acceptor interface, non-adiabatic excitonic dynamics of molecular pyrazine and singlet

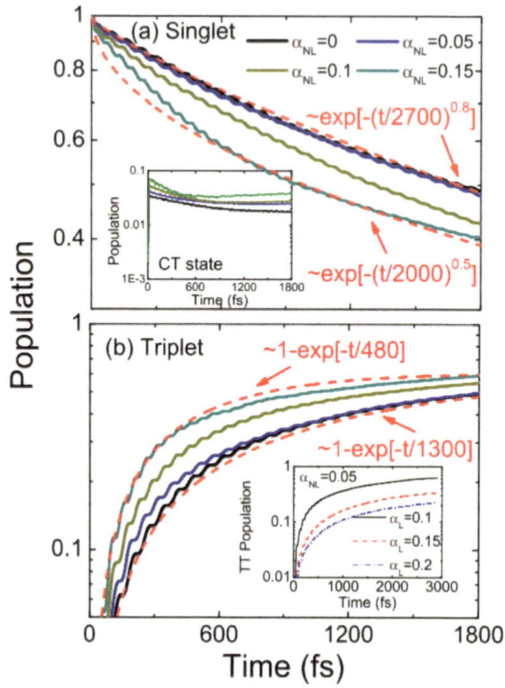

Figure 9.18 The time evolution of population of singlet, triplet and CT state for various nonlocal vibronic coupling strength. Reproduced from Ref. [52] with permission from American Physical Society.

fission processes in organic systems. These cases demonstrate that tDMRG methods are powerful tools to simulate detailed dynamic behaviors and study the microscopic mechanism of chemical phenomena, especially for ultrafast processes in organic systems.

Compared to other dynamic approaches, tDMRG manifests its advantage in such complex systems consisting of multiple (orbital, charge, and spin) degrees of freedom and with multiple (electron-vibration, electron-electron, and spin exchange) interactions, all of which are moderate and competing with each other. In a word, tDMRG can be applied to non-perturbative but normalizable systems. The word "non-perturbative" means there is not a dominating degree of freedom or interaction in the system, and the "normalizable" implies the local entropy will not increase without limit. Fortunately, many chemical systems fulfill these two conditions, thus we can safely extend the applicable subjects of tDMRG in chemical systems in the future.

However, there are still limitation of the applications of tDMRG for excessively large systems or large time-scale simulation due to the full quantum treatment of both electronic and vibrational part of systems. For such cases, the bond dimension of time-evolution state within MPS format grows fast with the increasing time, which makes the computational cost to increase quickly. Several improvements may be developed and implemented to tackle such bottlenecks. For example, by separating electron-vibration interaction into strong electron-vibration couplings and weak system-both interactions, quantum dissipation methods based on open quantum theory can be introduced and implemented within the MPS/MPO framework, i.e., solving quantum Liouville equations instead of TDSE using MPS/MPO representations. Another way is to adopt more general TNS frameworks since the realistic chemistry systems have more complicated topologic structures.

On the other hand, the low-order electron-vibration interaction model may not be sufficiently accurate for realistic systems with much more complex potential energy surfaces. The tDMRG approaches can be combined with quantum chemistry methods such as density functional theory or wavefunction methods such that the dynamics processes can be simulated via so called on-the-fly tDMRG approaches.

Acknowledgments

This work was supported by the National Natural Science Foundation of China (Grant Number 22073045), and we thank Prof. Ulrich Schollwöck and Prof. Zhigang Shuai for stimulating discussions.

References

1. Baiardi, A. (2021). Electron dynamics with the time-dependent density matrix renormalization group. *J. Chem. Theory Comput.* **17**, 3320–3334.

2. Baiardi, A., and Reiher, M. (2019). Large-scale quantum-dynamics with matrix product states. *J. Chem. Theory Comput.* **15**, 3481–3498.
3. Beck, M. H., Jäckle, A., Worth, G. A., and Meyer, H.-D. (2000). The multi-configuration time-dependent Hartree (MCTDH) method: a highly efficient algorithm for propagating wavepackets. *Phys. Rep.* **324**, 1–105.
4. Borrelli, R. (2019). Density matrix dynamics in twin-formulation: an efficient methodology based on tensor-train representation of reduced equations of motion. *J. Chem. Phys.* **150**, 234102.
5. Borrelli, R., and Gelin, M. F. (2017). Simulation of quantum dynamics of excitonic systems at finite temperature: an efficient method based on thermo field dynamics. *Sci. Rep.* **7**, 9127.
6. Daley, A. J., Kollath, C., Schollwöck, U., and Vidal, G. (2004). Time-dependent density-matrix renormalization-group using adaptive effective Hilbert spaces. *J. Stat. Mech.: Theory Exp.* **2004**, P04005.
7. De, L., De Moor, B., and Vandewalle, J. (2000). On the best rank-1 and rank-(r_1, r_2, ..., r_n) approximation of higher-order tensors. *SIAM J. Matrix Anal. Appl.* **21**, 1324–1342.
8. De Lathauwer, L., De Moor, B., and Vandewalle, J. (2000). A multilinear singular value decomposition. *SIAM J. Matrix Anal. Appl.* **21**, 1253–1278.
9. Dolgov, S. (2014). *Tensor Product Methods in Numerical Simulation of High-Dimensional Dynamical Problems,* PhD thesis, University of Leipzig.
10. Frahm, L.-H., and Pfannkuche, D. (2019). Ultrafast ab initio quantum chemistry using matrix product states. *J. Chem. Theory Comput.* 15, 2154–2165.
11. García-Ripoll, J. J. (2006). Time evolution of matrix product states. *New J. Phys.* **8**, 305.
12. Greene, S. M., and Batista, V. S. (2017). Tensor-train split-operator Fourier trans-form (TT-SOFT) method: multidimensional nonadiabatic quantum dynamics. *J. Chem. Theory Comput.* **13**, 4034–4042.
13. Gunst, K., Verstraete, F., Wouters, S., Legeza, Ö., and Van Neck, D. (2018). T3NS: three-legged tree tensor network states. *J. Chem. Theory Comput.* **14**, 2026–2033.
14. Haegeman, J., Cirac, J. I., Osborne, T. J., Pižorn, I., Verschelde, H., and Verstraete, F. (2011). Time-dependent variational principle for quantum lattices. *Phys. Rev. Lett.* **107**, 070601.
15. Haegeman, J., Lubich, C., Oseledets, I., Vandereycken, B., and Verstraete, F. (2016). Unifying time evolution and optimization with matrix product states. *Phys. Rev. B* **94**, 165116.

16. Kloss, B., Reichman, D. R., and Tempelaar, R. (2019). Multiset matrix product state calculations reveal mobile Franck-Condon excitations under strong Holstein-type coupling. *Phys. Rev. Lett.* **123**, 126601.
17. Kolda, T. G., and Bader, B. W. (2009). Tensor decompositions and applications. *SIAM Rev.* **51**, 455–500.
18. Kurashige, Y. (2018). Matrix product state formulation of the multiconfiguration time-dependent Hartree theory. *J. Chem. Phys.* **149**, 194114.
19. Legeza, Ö., Röder, J., and Hess, B. (2003). Controlling the accuracy of the density matrix renormalization-group method: the dynamical block state selection approach. *Physical Review B* **67**, 125114.
20. Li, W., Ren, J., and Shuai Z. (2021). Theory and applications of time dependent density matrix renormalization group. *Chem. J. Chinese U.* **42**, 2085–2102.
21. Lubich, C., Oseledets, I. V., and Vandereycken, B. (2015). Time integration of tensor trains. *SIAM J. Numer. Anal.* **53**, 917–941.
22. Ma, H., Luo, Z., and Yao, Y. (2018). The time-dependent density matrix renormalisation group method. *Mol. Phys.* **116**, 854–868.
23. Ma, H., and Schollwöck, U. (2008). Dynamical simulations of charged soliton transport in conjugated polymers with the inclusion of electron-electron interactions. *J. Chem. Phys.* **129**, 244705.
24. Ma, H., and Schollwöck, U. (2009). Dynamical simulations of polaron transport in conjugated polymers with the inclusion of electron-electron interactions. *J. Phys. Chem. A* **113**, 1360–1367.
25. Ma, H., and Schollwöck, U. (2010). Effect of electron-electron interactions on the charge carrier transitions in trans-polyacetylene. *J. Phys. Chem. A* **114**, 5439–5444.
26. Mardazad, S., Xu, Y., Yang, X., Grundner, M., Schollwöck, U., Ma, H., and Paecke, S. (2021). Quantum dynamics simulation of intramolecular singlet fission in covalently linked tetracene dimer. To be submitted.
27. Meyer, H.-D., Manthe, U., and Cederbaum, L. S. (1990). The multi-configurational time-dependent Hartree approach. *Chem. Phys. Lett.* **165**, 73–78.
28. Moler, C., and Van Loan, C. (2003). Nineteen dubious ways to compute the exponential of a matrix, twenty-five years later. *SIAM Rev.* **45**, 3–49.
29. Nakatani, N., and Chan, G. K.-L. (2013). Efficient tree tensor network states (TTNS) for quantum chemistry: generalizations of the density matrix renormalization group algorithm. *J. Chem. Phys.* **138**, 134113.

30. Oseledets, I. V. (2011). Tensor-train decomposition. *SIAM J. Sci. Comput.* **33**, 2295–2317.
31. Paeckel, S., Köhler, T., Swoboda, A., Manmana, S. R., Schollwöck, U., and Hubig, C. (2019). Time-evolution methods for matrix-product states. *Ann. Phys.* 167998.
32. Raab, A., Worth, G. A., Meyer, H.-D., and Cederbaum, L. (1999). Molecular dynamics of pyrazine after excitation to the S_2 electronic state using a realistic 24-mode model Hamiltonian. *J. Chem. Phys.* **110**, 936–946.
33. Ren, J., Shuai, Z., and Chan, G. K.-L. (2018). Time-dependent density matrix renormalization group algorithms for nearly exact absorption and fluorescence spectra of molecular aggregates at both zero and finite temperature. *J. Chem. Theory Comput.* **14**, 5027–5039.
34. Ronca, E., Li, Z., Jimenez-Hoyos, C. A., and Chan, G. K.-L. (2017). Time-step targeting time-dependent and dynamical density matrix renormalization group algorithms with ab initio Hamiltonians. *J. Chem. Theory Comput.* **13**, 5560–5571.
35. Schollwöck, U. (2011). The density-matrix renormalization group in the age of matrix product states. *Ann. Phys.* **326**, 96–192.
36. Schröder, F. A., Turban, D. H., Musser, A. J., Hine, N. D., and Chin, A. W. (2019). Tensor network simulation of multi-environmental open quantum dynamics via machine learning and entanglement renormalisation. *Nat. Commun.* **10**, 1062.
37. Shi, Q., Xu, Y., Yan, Y., and Xu, M. (2018). Efficient propagation of the hierarchical equations of motion using the matrix product state method. *J. Chem. Phys.* **148**, 174102.
38. Shi, Y.-Y., Duan, L.-M., and Vidal, G. (2006). Classical simulation of quantum many-body systems with a tree tensor network. *Phys. Rev. A* **74**, 022320.
39. Suzuki, M. (1976). Generalized Trotter's formula and systematic approximants of exponential operators and inner derivations with applications to many-body problems. *Commun. Math. Phys.* **51**, 183–190.
40. Tucker, L. R. (1996). Some mathematical notes on three-mode factor analysis. *Psychometrika* **31**, 279–311.
41. Verstraete, F., and Cirac, J. I. (2004). Renormalization algorithms for quantum-many body systems in two and higher dimensions. arXiv preprint condmat/0407066.
42. Verstraete, F., Garcia-Ripoll, J. J., and Cirac, J. I. (2004). Matrix product density operators: simulation of finite-temperature and dissipative systems. *Phys. Rev. Lett.* **93**, 207204.

43. Verstraete, F., Murg, V., and Cirac, J. I. (2008). Matrix product states, projected entangled pair states, and variational renormalization group methods for quantum spin systems. *Adv. Phys.* **57,** 143–224.

44. Vidal, G. (2003). Efficient classical simulation of slightly entangled quantum computations. *Phys. Rev. Lett.* **91**, 147902.

45. Vidal, G. (2004). Efficient simulation of one-dimensional quantum many-body systems. *Phys. Rev. Lett.* **93**, 040502.

46. Wang, H., and Thoss, M. (2003). Multilayer formulation of the multi-configuration time-dependent Hartree theory. *J. Chem. Phys.* **119**, 1289–1299.

47. White, S. R., and Feiguin, A. E. (2004). Real-time evolution using the density matrix renormalization group. *Phys. Rev. Lett.* **93**, 076401.

48. Worth, G. A., Meyer, H.-D., and Cederbaum, L. (1996). The effect of a model environment on the S_2 absorption spectrum of pyrazine: a wave packet study treating all 24 vibrational modes. *J. Chem. Phys.* **105**, 4412–4426.

49. Xie, X., Liu, Y., Yao, Y., Schollwöck, U., Liu, C., and Ma, H. (2019). Time-dependent density matrix renormalization group quantum dynamics for realistic chemical systems. *J. Chem. Phys.* **151**, 224101.

50. Xie, X., Zhang, C., and Ma, H. (2020). Charge transfer via deep hole in the J51/N2200 blend. *J. Chem. Phys.* **153**, 054705.

51. Yao, Y. (2015). Spin-boson theory for charge photogeneration in organic molecules: role of quantum coherence. *Phys. Rev. B* **91**, 045421.

52. Yao, Y. (2016). Coherent dynamics of singlet fission controlled by nonlocal electron-phonon coupling. *Phys. Rev. B* **93**, 115426.

53. Yao, Y., Sun, K.-W., Luo, Z., and Ma, H. (2018). Full quantum dynamics simulation of a realistic molecular system using the adaptive time-dependent density matrix renormalization group method. *J. Phys. Chem. Lett.* **9**, 413–419.

54. Yao, Y., Xie, X., and Ma, H. (2016). Ultrafast long-range charge separation in organic photovoltaics: promotion by off-diagonal vibronic couplings and entropy increase. *J. Phys. Chem. Lett.* **7**, 4830–4835.

55. Zhao, H., Yao, Y., An, Z., and Wu, C.-Q. (2008). Dynamics of polarons in conjugated polymers: an adaptive time-dependent density-matrix renormalization group study. *Phys. Rev. B* **78**, 035209.

Chapter 10

Spin-Flip TDDFT for Photochemistry

John M. Herbert and Aniket Mandal
Department of Chemistry and Biochemistry, The Ohio State University, Columbus, Ohio USA
herbert@chemistry.ohio-state.edu

10.1 Computational Photochemistry

10.1.1 Conical Intersections

Conical intersections (CXs) play a central role in photophysics and photochemistry [74, 97], serving as the "funnels" for nonadiabatic dynamics between coupled potential energy surfaces representing different electronic states [1, 71, 125]. For the special case of two coupled electronic states, the CX is really a "conical seam" of dimension $N_{\text{int}} - 2$ where N_{int} is the number of internal (vibrational) degrees of freedom. Within this $(N_{\text{int}} - 2)$-dimensional subspace the two states in question are degenerate, and when plotted in two dimensions this seam collapses to a point and the topology is that of a double cone, hence "CX." For points within the seam space, even an infinitesimal displacement outside of this space will lift the degeneracy, and this two-dimensional "branching space" is spanned

Time-Dependent Density Functional Theory: Nonadiabatic Molecular Dynamics
Edited by Chaoyuan Zhu

ISBN 978-981-4968-42-3 (Hardcover), 978-1-003-31921-4 (eBook)
www.jennystanford.com

by a pair of nonorthogonal vectors that are usually labeled **g** and **h**. For a two-state intersection between Born–Oppenheimer electronic potential energy surfaces $E_J(\mathbf{R})$ and $E_K(\mathbf{R})$, these branching-plane vectors consist of the *gradient-difference vector*,

$$\mathbf{g}_{JK}(\mathbf{R}) = \hat{\nabla}_{\mathbf{R}}\left[E_J(\mathbf{R}) - E_K(\mathbf{R})\right] , \tag{10.1}$$

which is simply the difference in slopes between the two surfaces, and the *nonadiabatic coupling vector*,

$$\mathbf{h}_{JK}(\mathbf{R}) = \left\langle \Psi_J(\mathbf{R}) \middle| \hat{\nabla}_{\mathbf{R}} \hat{H}(\mathbf{R}) \middle| \Psi_K(\mathbf{R}) \right\rangle . \tag{10.2}$$

In both equations, **R** represents the nuclear coordinates that define the potential energy surface and $\hat{\nabla}_{\mathbf{R}}$ indicates the vector-valued operator consisting of derivatives with respect to these coordinates. The quantities $|\Psi_J(\mathbf{R})\rangle$ and $|\Psi_K(\mathbf{R})\rangle$ are the Born–Oppenheimer electronic wavefunctions for electronic states J and K, defined by

$$\hat{H}|\Psi_K\rangle = E_K|\Psi_K\rangle \tag{10.3}$$

but written as $|\Psi_K(\mathbf{R})\rangle$ in Eq. (10.2) to remind the reader that the Born–Oppenheimer wavefunctions depend upon the geometry **R** at which Eq. (10.3) is solved.

As a simple example, two different CXs for the ethylene molecule are plotted in Fig. 10.1. As indicated by the molecular structures, these photochemical funnels are often associated with bond breaking and access very different geometries as compared to those sampled in ground-state dynamics. To access the "twisted-pyramidalized" CX in Fig. 10.1a, $\pi \rightarrow \pi^*$ excitation from the ground state leads to dissolution of the C=C double bond followed by rotation around the intact σ bond. In the "ethylidene" CX of Fig. 10.1b, a hydrogen atom has transferred from one carbon atom to the other.

The dimension of the branching-plane vectors $\mathbf{g}_{JK}$ and $\mathbf{h}_{JK}$ is that of **R**. For a single nuclear coordinate x, a simplified notation is

$$g_{JK}^{[x]} = \frac{\partial E_J(\mathbf{R})}{\partial x} - \frac{\partial E_K(\mathbf{R})}{\partial x} \tag{10.4a}$$

$$h_{JK}^{[x]} = \left\langle \Psi_J(\mathbf{R}) \middle| (\partial \hat{H}/\partial x) \middle| \Psi_K(\mathbf{R}) \right\rangle . \tag{10.4b}$$

The nonadiabatic coupling vector $\mathbf{h}_{JK}$ is related to the (first-order) derivative coupling vector, defined as

$$\mathbf{d}_{JK}(\mathbf{R}) = \left\langle \Psi_J(\mathbf{R}) \middle| \hat{\nabla}_{\mathbf{R}} \Psi_K(\mathbf{R}) \right\rangle , \tag{10.5}$$

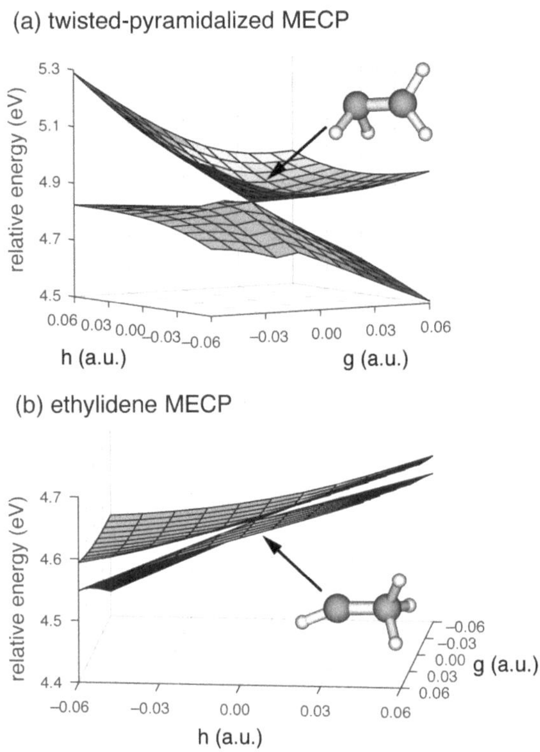

Figure 10.1 Minimum-energy crossing points (MECPs) along two different S_0/S_1 conical seams for the ethylene molecule, plotted as a function of orthogonalized **g** and **h** coordinates. These calculations were performed using SF-TDDFT and the resulting potential surfaces exhibit the correct double-cone topology of a CX. The intersection in (a) is strongly peaked whereas the one in (b) is more sloped. Energy is measured relative to the minimum-energy S_0 geometry. Reprinted from Ref. [37]; copyright 2016 American Chemical Society.

and the relationship is

$$\mathbf{d}_{JK}(\mathbf{R}) = \frac{\mathbf{h}_{JK}(\mathbf{R})}{E_K(\mathbf{R}) - E_J(\mathbf{R})} . \tag{10.6}$$

The derivative coupling describes the fact that the Born–Oppenheimer potential surfaces $E_J(\mathbf{R})$ and $E_K(\mathbf{R})$ are coupled due to nuclear motion, represented by the operator $\hat{\nabla}_{\mathbf{R}}$. The general solution to the coupled nuclear–electronic problem involves a *vibronic* Schrödinger equation [11], rather than the purely electronic one in Eq. (10.3), the latter of which describes the uncoupled Born–Oppenheimer approximation. Unless $\partial \hat{H}/\partial x$ happens to vanish by symmetry along a particular coordinate x, then electronic states J and K will be strongly coupled in regions where the energy gap $|E_J(\mathbf{R}) - E_K(\mathbf{R})|$ is small.

The topography around a given CX (peaked versus sloped, and symmetrical versus ellipsoidal) can be characterized using parameters related to $\mathbf{g}_{JK}$, $\mathbf{h}_{JK}$, and a *seam coordinate*

$$\mathbf{s}_{JK} = \frac{1}{2}\hat{\nabla}_{\mathbf{R}}\left[E_J(\mathbf{R}) + E_K(\mathbf{R})\right]. \tag{10.7}$$

Omitting the subscript JK in cases where there is no ambiguity, and denoting $g = \|\mathbf{g}\|$ and $h = \|\mathbf{h}\|$, Yarkony [125] defines parameters

$$s^x = (\mathbf{s} \cdot \mathbf{g})/g^2 \tag{10.8a}$$

$$s^y = (\mathbf{s} \cdot \mathbf{h})/h^2 \tag{10.8b}$$

that describe the "tilt" of the CX. When s^x and s^y are small, the CX is strongly peaked or hourglass-shaped, as in Fig. 10.1a. Larger values of either s^x or s^y correspond to a more sloped CX in either the $\mathbf{g}$ or the $\mathbf{h}$ direction, respectively, as in Fig. 10.1b. Peaked CXs are generally understood to be better funnels [1], leading to more efficient nonadiabatic transitions. Note also that it is common to minimize the energy along the seam coordinate $\mathbf{s}_{JK}$, maintaining the degeneracy $E_J(\mathbf{R}) = E_K(\mathbf{R})$. This affords the *minimum-energy crossing point* (MECP) along the conical seam, which is sometimes characterized as "the" CX. Both of the intersection points that are plotted in Fig. 10.1 are MECPs along a particular conical seam.

10.1.2 Time-Dependent DFT

CXs can be appropriately described using a variety of (mostly multireference) electronic structure methods [73], but the present chapter focuses on their description using time-dependent density functional theory (TDDFT) [19, 32, 33] in its linear-response (LR) formulation [8, 18, 23]. The frequency-domain formulation of LR-TDDFT involves an eigenvalue problem for the excitation energies and no explicit time dependence, nevertheless that method is nowadays largely synonymous with simply "TDDFT." Explicitly time-dependent approaches do exist and their use is increasing [61], but these are not discussed in the present chapter. LR-TDDFT represents something of a "sweet spot" in terms of its price-to-performance ratio, with an accuracy of $\sim$0.3 eV for many classes of vertical excitation energies [52, 91, 103], and a cost that scales as $n_{\text{states}} \times \mathcal{O}(n_{\text{basis}}^4)$, i.e., fourth-order scaling with a prefactor that reflects the number of desired states [24]. For many photochemical problems, only a few low-lying electronic states are required and the cost is not significantly greater than that of the ground-state DFT calculation.

That is the good news, and it has made LR-TDDFT absolutely ubiquitous for the calculation of vertical excitation spectra. The bad news, from the standpoint of photochemical investigations, is that the topology of any CX that involves the ground state is fundamentally flawed in LR-TDDFT, with seams that are $(N_{\text{int}} - 1)$-dimensional rather that $(N_{\text{int}} - 2)$-dimensional [37, 60]. This can be seen clearly from the LR-TDDFT potential energy surfaces for ethylene that are plotted in Fig. 10.2, around the same two geometries used to generate the plots in Fig. 10.1. Using LR-TDDFT, no intersection with the correct topology can be found. Numerical values of the tilt parameters in Eq. (10.8) also fail to correlate with the observed potential energy landscape [37].

This incorrect behavior originates in a lack of proper coupling between the ground (reference) state and the response (excited) states. At some level, this is a problem with response theory rather than with TDDFT *per se*, and from a different point of view the issue is that LR-TDDFT affords an unbalanced treatment of ground versus excited states. The ground state is variationally optimized via solution of the Kohn–Sham equations but then the excited states are obtained from a separate eigenvalue problem, and therefore do not satisfy a variational principle with respect to the ground state. (The excited states are, however, variational with respect to one another. As a result, CXs between two excited states are free of the aforementioned topology problems.) The predicament is brought to the forefront by Jahn–Teller problems involving symmetry-required degeneracy of the ground state, and the simplest such example is the H_3 radical in its D_{3h} geometry. As a result of the unbalanced manner in which ground and excited states are described, conventional LR-

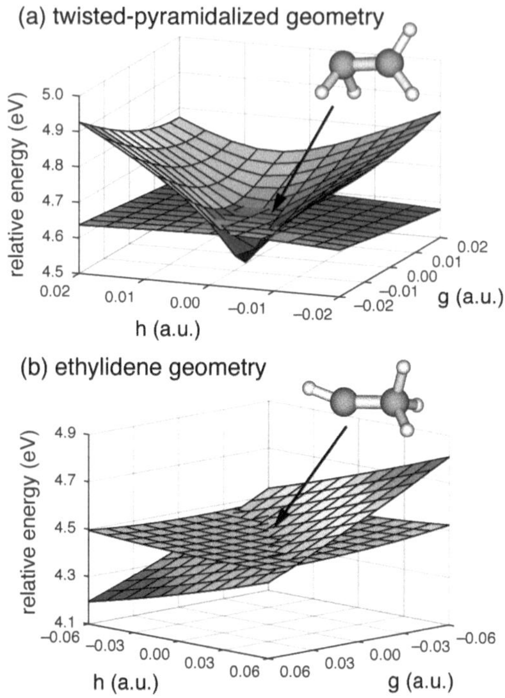

Figure 10.2 Potential surfaces for the S_0 and S_1 states of C_2H_4, computed around the same two geometries that are depicted in Fig. 10.1 but this time using conventional LR-TDDFT. Nowhere in the vicinity of the known MECP structures does a CX (i.e., a zero-dimensional point of intersection) manifest, because LR-TDDFT exhibits fundamentally incorrect topology for any CX that involves the reference state, which is S_0 for these calculations. Reprinted from Ref. [37]; copyright 2016 American Chemical Society.

TDDFT struggles to describe the degeneracy [130], as can be seen clearly in Fig. 10.3b. To emphasize that this is not a DFT problem *per se*, Fig. 10.3a shows that the configuration interaction singles (CIS) method [18] also struggles, even in a restricted open-shell formulation.

Despite these drawbacks, LR-TDDFT has been formulated for use with nonadiabatic molecular dynamics methods [3, 16, 46, 108–110, 113]. It is sometimes found that the electronic structure calculation is difficult to converge in regions of the potential surface where the ground state is quasi-degenerate with the first excited

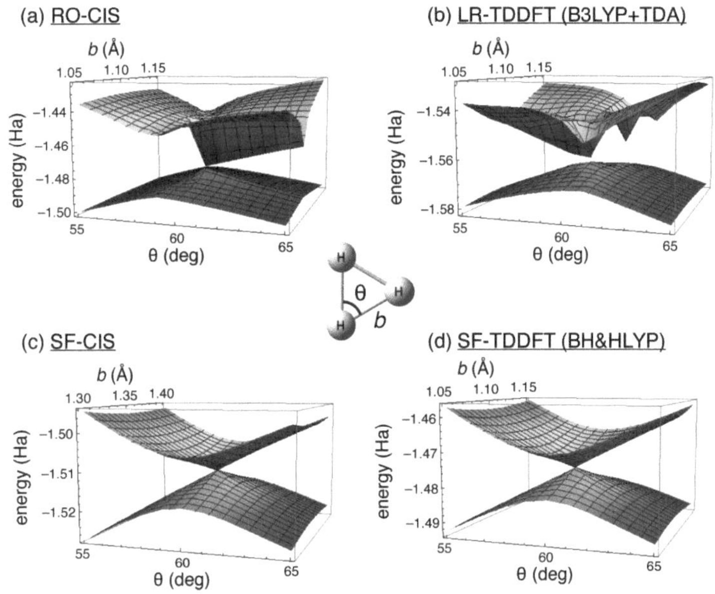

Figure 10.3 Potential energy surfaces for the lowest two doublet electronic states of H_3 radical, centered at the D_{3h} geometry with displacements along one bond-length coordinate (b) and one bond-angle coordinate (θ), as defined in the center diagram, which reduce the symmetry to C_{2v}. The methods considered are: (a) restricted open-shell (RO-)CIS, (b) unrestricted LR-B3LYP within the Tamm–Dancoff approximation (TDA), (c) SF-CIS, and (d) SF-BH&HLYP. Energies are in Hartree. Adapted from Ref. [130]; copyright 2014 American Institute of Physics.

state [128, 130], which is not altogether surprising given the warped topography of the potential surfaces for H_3 that is documented in Fig. 10.3a and Fig. 10.3b. A practical workaround is to halt any excited-state dynamics simulation before it can reach a ground-state CX [88, 134], using LR-TDDFT only to simulate excited-state dynamics and nonadiabatic transitions between excited states. That procedure, however, is at odds with the desire to perform first-principles simulations of the crucial excited-state deactivation step, leading to the onset of vibrational cooling on the ground state.

The present chapter describes a solution to these dilemmas based on the "spin-flip" (SF) approach to TDDFT [7, 37, 100]. The formalism of SF-TDDFT is described in Section 10.2 but the essential idea is rather simple: we recognize that the shortcomings described above are limited to CXs that involve the reference state that is used in LR theory, which is typically the ground state but it need not be. SF-TDDFT uses a sacrificial reference state with a different spin multiplicity as compared to the target states of interest, e.g., a triplet reference is used to simulate singlet photochemistry or a quartet reference for the doublet H_3 radical. The change in multiplicity is accomplished via single excitations in conjunction with a single $\alpha \rightarrow \beta$ spin flip, so that states having the target multiplicity (including the ground state) appear in the excitation manifold and are described in a variational manner with respect to one another. This cures the topology problem [130]. (In fact, the double cones in Fig. 10.1 were generated using SF-TDDFT calculations.) In Fig. 10.3c, the SF-CIS approach repairs the incorrect RO-CIS description of the conical intersection in H_3. Adding dynamical correlation, SF-TDDFT recovers a double cone as well; see Fig. 10.3d.

The SF approach is not a panacea and some problems do remain, most notably that it exacerbates spin contamination, which can make it difficult to identify the multiplicities of interest. These problems are described in Section 10.2 and some recently emergent solutions, based on augmented SF-type approaches, are discussed in Section 10.3.

10.2 Spin-Flip TDDFT Approach

10.2.1 Theory

We first provide a conceptual overview of SF-TDDFT (Section 10.2.1.1) before briefly reviewing the formalism of LR-TDDFT as a means to introduce the SF version in more detail (Section 10.2.1.2). A discussion of LR- and SF-TDDFT derivative couplings appears in Section 10.2.1.3.

10.2.1.1 Conceptual overview

As with other SF approaches [7], the basic idea behind SF-TDDFT is to use a high-spin ($M_S = S$) reference state whose total spin S is one unit larger than the states of interest, whose spin quantum number is $S-1$. A simple example for a (4e,4o) model is shown in Fig. 10.4. There, a triplet ($S=1$) reference configuration is used to generate a singlet ($S=0$) excitation manifold. For the proper qualitative description of a S_0/S_1 intersection, the minimal excitation space is labeled "o-o" in Fig. 10.4 and consists of excitations within the half-filled orbitals of the reference configuration, including de-excitations. (Notably, if the molecule in question has a singlet ground state then the lowest SF-TDDFT excitation energy starting from a triplet reference is negative.) The o-o space contains a closed-shell configuration that resembles S_0. It also contains two open-shell determinants with $M_S=0$, linear combinations of which will generate either an open-shell singlet (S_1) or else the $M_S=0$ component of the triplet state, depending on whether these determinants are added (triplet) or subtracted (singlet). Lastly, the o-o space contains a determinant that is doubly excited with respect to S_0, which provides coupling between S_0 and S_1. The four determinants in the o-o space thus constitute a minimalist model that can describe the S_0/S_1 intersection correctly [60]. Notably, the doubly excited determinant that is required is not contained in the usual LR-TDDFT excitation manifold starting from a closed-shell S_0 reference state, because that excitation manifold consists of single excitations only. From another point of view, SF-TDDFT thus represents a cost-effective way to augment the LR-TDDFT excitation manifold with a certain subset of doubly excited determinants.

The SF-TDDFT excitation manifold also consists of determinants involving the virtual orbitals ("o-v" and "c-v" subspaces in Fig. 10.4), along with the "c-o" space of excitations between doubly occupied and singly occupied orbitals. Unlike the o-o space, none these other subspaces is spin-complete [133], meaning that they are missing one or more determinants needed to construct $\hat{S}^2$ eigenstates but which cannot be generated from the reference state via a single excitation combined with a single spin-flip. These are the subspaces that are responsible for spin contamination. We will return to this point later in this chapter.

Operationally, a SF-TDDFT calculation consists of a configuration-interaction (CI) calculation within the extended manifold of determinants generated by single excitations combined with $\alpha \rightarrow \beta$

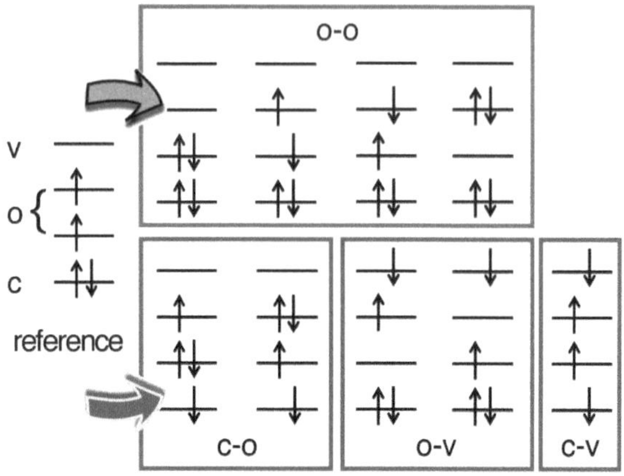

Figure 10.4 Illustration of the SF-TDDFT excitation space for a (4e,4o) model (i.e., four electrons in four orbitals), starting from a high-spin triplet reference at left. Single excitations combined with a single $\alpha \to \beta$ spin flip generate the determinants that are shown. The "o-o" set of determinants is spin-complete and represents a minimalist model of S_0 and S_1 along with the doubly excited determinant that couples them. Subspaces labeled c-o, o-v, and c-v are each missing some of the determinants needed to form spin-pure $\hat{S}^2$ eigenstates. Adapted from Ref. [133]; copyright 2015 American Institute of Physics.

spin flip. Unlike the SF-CIS method [49], however, the use of Kohn–Sham orbitals captures some dynamical correlation effects. Because the double excitations are limited, this CI problem does not incur the full $\mathcal{O}(N^6)$ cost of a singles-doubles (CISD) calculation, but rather is only a few times more expensive than conventional LR-TDDFT. The number of determinants in the excitation space remains $\mathcal{O}(N_{\text{occ}}N_{\text{virt}})$ [98], and does not grow in a way that increases with system size any faster than that of LR-TDDFT.

Crucially, this extension of the excitation manifold cures the topology problem around conical intersections involving any of the states of the target multiplicity, because the appropriate double excitation to couple them is included in the excitation manifold. Problems with symmetry-required (Jahn–Teller) degeneracy are mitigated, and the warped topography around such intersections is avoided (see Fig. 10.3), because both of the quasi-degenerate electronic states are treated on the same footing and emerge as solutions to a common eigenvalue problem [130, 134]. Convergence problems around such regions of near-degeneracy are mitigated as well.

10.2.1.2 Formalism

To proceed let us recapitulate the form of the ground-state (Kohn–Sham DFT) and excited-state (LR-TDDFT) eigenvalue problems. The former is

$$\hat{F}\,\psi_{p\sigma}(\mathbf{r}) = \varepsilon_{p\sigma}\,\psi_{p\sigma}(\mathbf{r})\,, \tag{10.9}$$

which defines the molecular orbitals (MOs) $\{\psi_{p\sigma}\}$, where σ is a spin index. The excited-state eigenvalue problem is [18, 23]

$$\begin{pmatrix} \mathbf{A} & \mathbf{B} \\ \mathbf{B}^* & \mathbf{A}^* \end{pmatrix} \begin{pmatrix} \mathbf{x} \\ \mathbf{y} \end{pmatrix} = \omega \begin{pmatrix} \mathbf{1} & \mathbf{0} \\ \mathbf{0} & -\mathbf{1} \end{pmatrix} \begin{pmatrix} \mathbf{x} \\ \mathbf{y} \end{pmatrix} \tag{10.10}$$

where the matrices **A** and **B** involve derivatives of the ground-state Fock matrix **F** with respect to the ground-state density matrix **P** [18]:

$$A_{ia\sigma,jb\sigma'} = (\varepsilon_{a\sigma} - \varepsilon_{i\sigma})\delta_{ij}\delta_{ab}\delta_{\sigma\sigma'} + \frac{\partial F_{ia\sigma}}{\partial P_{jb\sigma'}} \tag{10.11a}$$

$$B_{ia\sigma,jb\sigma'} = \frac{\partial F_{ai\sigma}}{\partial P_{jb\sigma'}}. \tag{10.11b}$$

The term $\partial F_{ia\sigma}/\partial P_{jb\sigma'}$ in **A**, along with the entirety of **B**, has the form of a coupling matrix that modifies the zeroth-order (independent-particle) excitation energies, $\varepsilon_{a\sigma} - \varepsilon_{i\sigma}$.

The solution of Eq. (10.10), for the Ith excited state, consists of an excitation energy (eigenvalue) $\omega_I = E_I - E_0$, along with vectors $\mathbf{x}^I$ and $\mathbf{y}^I$ consisting of excitation amplitudes and de-excitation amplitudes, respectively. We will use indices $i, j, \ldots$ to denote occupied MOs and $a, b, \ldots$ for virtual MOs, and $p, q, \ldots$ for arbitrary MOs, as in the p that appears in Eq. (10.9). The solution vector $(\mathbf{x}, \mathbf{y})$ to Eq. (10.10) parameterizes the transition density $\rho_{0I}(\mathbf{r}, \mathbf{r}')$ of the excited state in question [23, 24, 105]:

$$\rho_{0I}(\mathbf{r}, \mathbf{r}') = \sum_{ia\sigma} \left[x_{ia\sigma}\,\psi_{a\sigma}(\mathbf{r})\,\psi^*_{i\sigma}(\mathbf{r}') + y_{ia\sigma}\,\psi_{i\sigma}(\mathbf{r})\,\psi^*_{a\sigma}(\mathbf{r}') \right]. \tag{10.12}$$

The de-excitation amplitudes y_{ia}, which appear in the random phase approximation as a correlation contribution to the ground state [75, 92], and are typically $\sim$100$\times$ smaller than the excitation amplitudes x_{ia}. Neglecting the $\{y_{ia}\}$, affords the so-called Tamm–Dancoff approximation (TDA) to Eq. (10.10) [18, 38]:

$$\mathbf{Ax} = \omega\mathbf{x} \,. \tag{10.13}$$

Equation (10.13) has the same form as the CIS eigenvalue equation except that the matrix elements contain a contribution from the exchange-correlation kernel, $f_{\text{xc}} = \delta^2 E_{\text{xc}}/\delta\rho^2$. In the LR-TDDFT case, the matrix elements are [18]

$$\begin{aligned} A_{ia\sigma,jb\sigma'} = {} & (\varepsilon_a - \varepsilon_i)\delta_{ia}\delta_{jb}\delta_{\sigma\sigma'} + (ia|jb) - C_{\text{HFX}}(ij|ab)\delta_{\sigma\sigma'} \\ & + (1 - C_{\text{HFX}})(ia\sigma|f_{\text{xc}}|jb\sigma') \,. \end{aligned} \tag{10.14}$$

The quantity C_{HFX} is the fraction of Hartree–Fock exchange (HFX), in the case that the functional is a hybrid.

The TDA is generally a very good approximation, but for photochemical applications its utility goes deeper than that. To understand why, consider that the full LR-TDDFT eigenvalue problem in Eq. (10.10) can be rewritten as

$$(\mathbf{A} - \mathbf{B})(\mathbf{A} + \mathbf{B})(\mathbf{x} + \mathbf{y}) = \omega^2(\mathbf{x} + \mathbf{y}) \,, \tag{10.15}$$

where the matrices $\mathbf{A} \pm \mathbf{B}$ are orbital Hessians, i.e., stability matrices [123]. Solution of this equation for excitation energies ω will fail if the triplet instability matrix $\mathbf{A} + \mathbf{B}$ exhibits any negative eigenvalues. Such instabilities are found to be widespread once one an excited-state trajectory moves away from the Franck–Condon excitation point and begins to explore bond-breaking regions of the potential energy surface [14], leading to the suggestion that the TDA is effectively mandatory for photochemical applications.

The preceding discussion describes the conventional LR-TDDFT formalism, in which $x_{ia\sigma}$ has a single spin index because the $\psi_{i\sigma} \rightarrow \psi_{a\sigma}$ excitation is spin-conserving. For a spin-flipping transition, and under the usual assumption that f_{xc} comes from a semilocal exchange-correlation functional, the coupling matrix $\partial F_{ia\sigma}/\partial P_{jb\sigma'}$ in Eq. (10.11a) vanishes except for the HFX contribution [100], because only HFX can couple $\sigma = \alpha$ to $\sigma' = \beta$. As noted later [43], the fact that the SF-TDDFT coupling matrix involves only the HFX term (and not the Coulomb term or the semilocal kernel f_{xc}) explains the observation, made already in the very first SF-TDDFT study of singlet–triplet gaps [100], that a hybrid functional with 50% HFX significantly outperforms B3LYP with its 20% HFX. This observation was later confirmed in application of SF-TDDFT to excitation energies [64]. As a result, the "Becke half-and-half" functional BH&HLYP, consisting of 50% HFX and 50% semilocal B88 [5] exchange ("BH&H"), in conjunction with LYP correlation [53], has become the standard choice for SF-TDDFT calculations.

The formulation of SF-TDDFT that is outlined above is sometimes called the "collinear" approach. This was later generalized to an alternative "noncollinear" method, which does bring the semilocal part of the functional into the coupling matrix [119, 120]. (See Refs. [43] and [89] for additional discussion of the noncollinear approach. The name comes from generalized Hartree–Fock theory [104], where a noncollinear formalism is one that allows α and β spin to mix, as used for example in relativistic DFT [25, 26, 118].) The noncollinear version of SF-TDDFT is found to improve the performance for some problematic diradicals [6, 120], and is more accurate for excitation energies when functionals with a low percentage of HFX are used [45, 124]. A significant drawback, however, is that the noncollinear exchange-correlation kernel involves a numerically problematic factor of spin density $(\rho_\alpha - \rho_\beta)$ in the denominator [6, 89]. Likely for this reason, the noncollinear formulation is less widely used and is not discussed further in this chapter.

10.2.1.3 Nonadiabatic (derivative) couplings

To fully explore nonadiabatic photochemistry with TDDFT (or any other electronic structure method) it is necessary to possess not only an analytic gradient but also nonadiabatic coupling vectors $\mathbf{h}_{JK}$ [Eq. (10.2)]. As compared to the gradient, the couplings are more complicated to derive and implement, and as a result they are available only in selected quantum chemistry programs and only for a limited number of electronic structure models, which includes TDDFT. A survey of implementations can be found in Ref. [3], although it omits the implementation in the Q-Chem program [50]. The latter is the joint product of the Herbert group [130, 132] and the Subotnik group [21, 46, 84]. Semi-empirical implementations have also been reported recently [68, 81].

Out of a variety of formalisms for computing nonadiabatic couplings within TDDFT [12, 99, 112, 130, 132], the one that is conceptually and computationally simplest is the "pseudo-wavefunction" approach [21, 84, 130]. Working within the TDA, both for conceptual simplicity and for the practical reasons discussed above, one might write the wavefunction for the Kth excited state as a linear combination of singly excited Slater determinants, as in the CIS method:

$$|\Psi_K\rangle = \sum_{ia\sigma} x_{ia\sigma}^{K} |\Phi_{ia\sigma}\rangle \,. \tag{10.16}$$

Here, $|\Phi_{ia\sigma}\rangle$ is obtained from the ground state configuration by $\psi_{i\sigma} \to \psi_{a\sigma}$ excitation. Strictly speaking, LR-TDDFT affords only the transition density $\rho_{0K}(\mathbf{r}, \mathbf{r}')$ [Eq. (10.12)], not the excited-state wavefunction *per se*, but analogies along the lines of Eq. (10.16) have been used since the earliest days of the theory [8–10, 96]. In condensed-matter physics, it is also common to identify the product functions $\psi_a(\mathbf{r}_{\text{elec}})\,\psi_i^*(\mathbf{r}_{\text{hole}})$ as a quasiparticle basis for electron–hole pairs ("excitons"), and to consider $\rho_{0K}(\mathbf{r}_{\text{elec}}, \mathbf{r}_{\text{hole}})$ to be an "exciton wavefunction" [90, 101]. This is despite the fact that the Green's function methods that are often used in that context afford transition densities rather than proper wavefunctions, just like LR-TDDFT.

Taking Eq. (10.16) as an *ansatz* for the excited-state wavefunction, analytic expressions can then be derived for $\mathbf{h}_{JK}$ using standard analytic gradient theory, which essentially means careful application of the chain rule [21, 84, 130]. The relevant expressions will not be repeated here; see Zhang and Herbert for a concise derivation [130] and for a generalization beyond the TDA [132]. The results are in good agreement with derivative couplings computed via quadratic response theory [132], which is the formally correct way to obtain $\mathbf{h}_{JK}$ within TDDFT [83, 86], at least when quadratic response is well-behaved. That method, however, is vulnerable to spurious divergences [67, 83, 86, 132], a fact that was actually noted long ago in the context of time-dependent Hartree–Fock theory [17]. As explained later by Furche and co-workers [87], these are artifacts of approximate response theory in general, not limited to TDDFT, and they complicate the application of quadratic response to the point that the pseudo-wavefunction approach is the much more widely used method. Similarly, the topology problem around a CX is not unique to LR-TDDFT and is also an artifact of response theory, and in particular the unbalanced (and nonvariational) manner in which it treats the reference and response states. Correlated Green's functions methods also afford wrong topology at any CX involving the reference state, whereas SF versions of those methods restore the correct topology [58].

Derivative couplings computed within the pseudo-wavefunction formalism show good agreement with exact (full CI) results for small molecules [132], good agreement with LR theory for derivative couplings $\mathbf{d}_{0K}$ that involve the ground state [83], and (for couplings between two excited states) good agreement with quadratic response theory [83, 132], provided that the latter is well-behaved. Some exemplary data are shown in Fig. 10.5, where derivative couplings $\mathbf{d}_{JK}$ computed within the pseudo-wavefunction approach are compared to results from quadratic response theory for a set of small molecules, and also to exact (full CI) results for linear H_3 as a function of internuclear distance. Derivative couplings based on the pseudo-wavefunction formalism also correctly reproduce the branching plane and the Berry phase around a CX [82], for which the orbital response contributions to $\mathbf{d}_{JK}$ prove to be crucial. Results presented below demonstrate that MECPs located with the aid of pseudo-wavefunction $\mathbf{h}_{JK}$ vectors afford good agreement with benchmark results from multireference electronic structure methods. Within the TDA, the pseudo-wavefunction formalism also affords the same result for $\mathbf{d}_{JK}$ as does the equation-of-motion formalism developed by Li and Liu [65]. All of these features argue in favor of using the pseudo-wavefunction approach, to the exclusion of quadratic response theory.

A subtle aspect of the derivative couplings is that of translational invariance. As noted by Fatehi et al. [21], expressions for $\mathbf{h}_{JK}$ obtained from straightforward application of the pseudo-wavefunction approach are not translationally invariant, which is problematic for nonadiabatic molecular dynamics simulations [22], and certain "electronic translation factors" were introduced to restore translational invariance [21]. It was noted that the terms

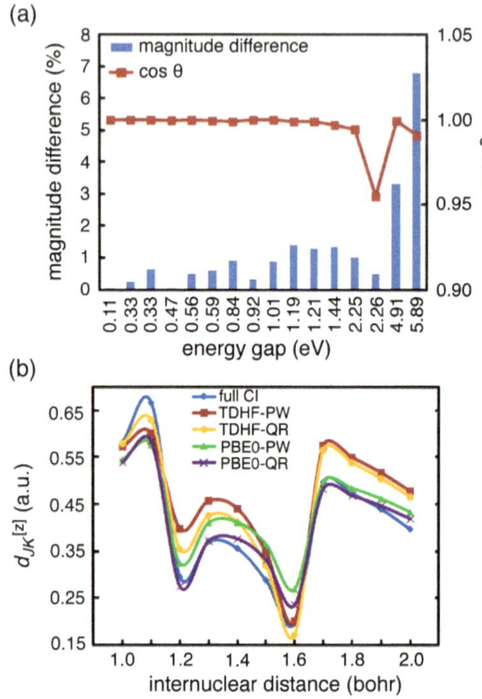

Figure 10.5 (a) Differences between $\mathbf{d}_{JK}$ computed using the pseudo-wavefunction (PW) approach versus quadratic response (QR) theory, expressed both in terms of magnitude $\|\mathbf{d}_{JK}^{\mathrm{PW}} - \mathbf{d}_{JK}^{\mathrm{QR}}\|$ and also direction ($\cos\theta$, where θ is the angle between $\mathbf{d}_{JK}^{\mathrm{PW}}$ and $\mathbf{d}_{JK}^{\mathrm{QR}}$). Results correspond to S_1/S_2 and S_1/S_3 couplings for formaldehyde, ethene, benzene, adenine, thymine, uracil, cytosine, and azulene. The energy gap axis consists of the individual gaps for this set of molecules and states. (b) Derivative coupling between the lowest ${}^2\Sigma_g^+$ states of linear H_3 radical ($D_{\infty h}$ symmetry) as a function of internuclear distance. Reproduced from Ref. [132]; copyright 2015 American Institute of Physics.

violating translational invariance arise from a kind of "Pulay force" [36], meaning derivatives of the overlap matrix in an atom-centered Gaussian basis set. It was later demonstrated by Zhang and Herbert [130] that the electronic translation factors introduced in Ref. [21] precisely annihilate the non-Hellmann–Feynman parts of $\mathbf{h}_{JK}$. The translationally invariant Hellmann–Feynman expression is easy to state, at least in schematic form [130]. Within the TDA, it is

$$h_{JK}^{[x]} = \sum_{iajb} \sum_{\sigma,\sigma'} x_{ia\sigma}^{J} A_{ia\sigma,jb\sigma'}^{[x]} x_{jb\sigma'}^{K}, \tag{10.17}$$

where the superscript $[x]$ indicates a derivative with respect to the nuclear coordinate x, as in Eq. (10.4b). The quantity $A_{ia\sigma,jb\sigma'}^{[x]}$ involves derivatives of the integrals in Eq. (10.14).

As will be discussed below, analytic derivative coupling vectors are useful for nonadiabatic molecular dynamics simulations. The $\mathbf{h}_{JK}$ vectors are also very useful for locating MECPs, because they facilitate the use of more efficient branching-plane updating methods [4, 102], in which the branching space (spanned by $\mathbf{g}_{JK}$ and $\mathbf{h}_{JK}$) is projected out of the gradient so that optimization proceeds entirely within the seam space. As compared to MECP-optimization algorithms that do not require $\mathbf{h}_{JK}$ vectors, such as penalty-function methods [59, 78] or approximate branching-plane updating [70, 131], the use of an *exact* branching-plane updating method requires far fewer optimization steps [48, 130, 131]. When programmed efficiently, evaluation of $\mathbf{h}_{JK}$ requires only modest overhead on top of the TDDFT gradient calculation that is already required to compute $\hat{\nabla}_{\mathbf{R}} E_J$ and $\hat{\nabla}_{\mathbf{R}} E_K$, because solution of coupled-perturbed equations is required in either case. Timing data confirm that the overhead for computing $\mathbf{h}_{JK}$ amounts to no more than 10–20% on top of the cost of TDDFT analytic gradients [99, 130].

10.2.2 Photochemical Applications

Although it can be tempting to jump right into nonadiabatic molecular dynamics simulations (which are discussed in Section 10.2.2.2), those simulations can be quite expensive because energy conservation typically requires time steps $\Delta t = 0.5$–1.0 fs, and accurate integration of the couplings (which vary rapidly in regions where energy gaps are small) may require even smaller time steps. Moreover, the results of such simulations are generally meaningful only in the aggregate, i.e., when averaged over an ensemble of trajectories. It may therefore be useful to first locate critical points on the potential energy surface, such as excited-state local minima and MECPs between electronic states, the latter of which not only

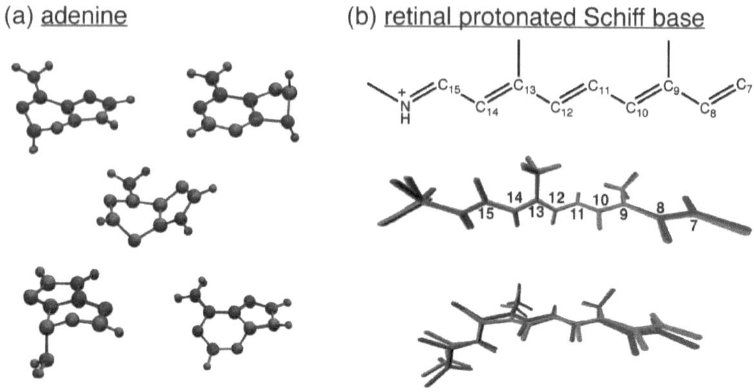

Figure 10.6 Comparison of optimized MECP structures obtained using SF-BH&HLYP to benchmark results from multireference electronic structure methods. (a) Comparison against MR-CIS for five different MECPs of adenine. (b) Comparison against CASSCF results for two MECPs of a truncated model of the rhodopsin chromophore, corresponding to *trans* → *cis* isomerization around either C_{11}=C_{12} or C_{13}=C_{14}. Panel (a) is adapted from Ref. [131]; copyright 2014 American Chemical Society. Panel (b) is adapted from Ref. [37]; copyright 2016 American Chemical Society.

indicate where the photochemical funnels are found in coordinate space, but also which vibrational modes provide the strongest coupling between the electronic states in question. Examples of MECP optimization are presented in Section 10.2.2.1.

10.2.2.1 Exploring excited-state potential surfaces

Figure 10.6 presents some examples of MECPs optimized using SF-TDDFT (with the BH&HLYP functional) in comparison to results from multireference electronic structure methods. In each case, the MECP obtained using SF-TDDFT is nearly indistinguishable from the benchmark multireference result. The test set includes five MECPs for adenine and two MECPs for a truncated model of the retinal protonated Schiff base that functions as the chromophore in the rhodopsin photoreceptor protein [20]. In retinal, *trans* → *cis* photoisomerization occurs around a different double bond in solution [117] (C_{13}=C_{14}) than it does in the protein environment

(C_{11}=C_{12}) [20], and the structures in Fig. 10.6b include MECPs for both isomerization reactions. Additional comparisons in Ref. [44] also indicate that SF-TDDFT accurately predicts MECP geometries.

Although applications of SF-TDDFT have been somewhat limited to date, the method has been used to map out potential energy surfaces and/or locate critical points for a variety of small molecules, including both uracil [131] and thymine [77], where the goal in both cases was to rationalize observed differences between the excited-state lifetime in the gas phase versus that in aqueous solution. Detailed photoisomerization pathways have also been mapped out for *cis* → *trans* photoisomerization in stilbene [78], and for the photochemical ring-opening of cyclohexadiene [95]. The method has been combined with exhaustive search algorithms in an attempt to elucidate all of the critical points (minima, transition states, and MECPs) for the S_0 and S_1 states of ethylene, 1,3-butadiene, thymine, and a coumarin dye [35, 69].

10.2.2.2 Trajectory surface hopping

Excited-state, nonadiabatic molecular dynamics simulations are mostly based on trajectory surface hopping (TSH) methods [2, 3, 47, 80, 106, 121, 122], most of which derive from Tully's seminal "fewest switches" algorithm [114], although modern variants incorporate some level of decoherence effects that are absent in Tully's original method [106, 121]. A notable exception (not based on Tully's algorithm) is the "multiple spawning" approach of Martínez and co-workers [15], which is an approximation to quantum wavepacket dynamics. The discussion below regarding the need for derivative coupling vectors is relevant to all of these methods. An example of a nonadiabatic dynamics algorithm that does *not* require derivative couplings is Zhu's "global switching" algorithm [126, 129], which derives from Landau-Zener theory. It is worth recalling (as discussed above) that properly coded TDDFT derivative couplings require only 10–20% additional computational overhead on top of a TDDFT analytic gradient calculation [99, 130]. Furche and co-workers report that TDDFT-based TSH simulations of the photodynamics of thymine, including the lowest three electronic states, require only 5× the cost of ground-state Born–Oppenheimer molecular dynamics, even with explicit calculation of derivative couplings [88]. As such, the discussion here focuses on the more popular TSH-based approaches and it is assumed that vectors $\mathbf{d}_{J\,K}$ are available for all pairs $J\,K$ of energetically accessible states.

In one way or another, TSH methods derive from the time-dependent Schrödinger equation of motion for the coefficients C_K of the electronic states in a superposition $|\Psi(t)\rangle = \sum_K C_K(t)|\Psi_K\rangle$. In the adiabatic basis $\{|\Psi_K\rangle\}$ of Born–Oppenheimer electronic states, the foundational equation of motion is usually [114, 121]

$$\mathrm{i}\hbar\frac{dC_J}{dt} = E_J(\mathbf{R})C_J - \mathrm{i}\hbar\sum_{K\neq J} C_K\dot{\mathbf{R}}\cdot\mathbf{d}_{J\,K} \qquad (10.18)$$

where $\dot{\mathbf{R}}$ is the nuclear velocity vector. Clearly, it is the derivative coupling $\mathbf{d}_{J\,K}$ that is responsible for stimulating transitions between electronic states. See Ref. [27] for a discussion regarding the absence of second-order derivative couplings in Eq. (10.18).

Derivative couplings are defined as

$$d_{J\,K}^{[x]} = \langle\Psi_J|(\partial/\partial x)|\Psi_K\rangle \qquad (10.19)$$

and these are only available in a limited number of quantum chemistry programs and only at selected levels of theory. More generally, the term that includes $\mathbf{d}_{J\,K}$ in Eq. (10.18) can be recast as

$$\dot{\mathbf{R}}\cdot\mathbf{d}_{J\,K} = \left\langle\Psi_J\left|\frac{d\Psi_K}{dt}\right.\right\rangle \approx \frac{\langle\Psi_J(t)|\Psi_K(t+\Delta t)\rangle}{\Delta t}, \qquad (10.20)$$

where the first equality is exact but the second represents a finite-difference approximation that allows the coupling term in Eq. (10.18) to be evaluated by determining the overlap (in time) of the adiabatic basis functions [94]. While the approximation in Eq. (10.20) is numerically advantageous as compared to approximating Eq. (10.19) via finite differences involving coordinate displacements, even the more efficient finite-difference in time is less desirable as compared to analytic evaluation of $\mathbf{d}_{J\,K}$.

There have so far been only a few TSH studies using SF-TDDFT. Yue et al. [127] used this approach to make a first-principles investigation of the mechanism of firefly bioluminescence, specifically, the photoinitiated decomposition reaction to form the anionic emitter species, oxyluciferin. Nakai and co-workers recently implemented SF-TDDFT at the level of semi-empirical density-functional tight-binding ("DFTB") [115, 116], and have used the method to perform TSH simulations of the archetypal azobenzene photoisomerization, in explicit solvent [116]. Lastly, Minezawa and Nakajima have simulated S_1 lifetimes for several small molecules [79], demonstrating good agreement with previous TSH simulations employing multireference electronic structure methods. That study recognizes that because LR-TDDFT may skew the topology of CXs that involve the ground state, TSH simulations that employ LR-TDDFT ought to be terminated prior to the final internal conversion step that returns the molecule to its ground state. If this paradigm is followed then S_1 lifetimes are inaccessible to LR-TDDFT simulations, whereas SF-TDDFT suffers no such limitation.

Along these lines, two recent TSH studies have made a side-by-side comparison of S_1 lifetimes predicted by LR- versus SF-TDDFT. Zhu and co-workers [128] examined *cis* → *trans* photoisomerization of azobenzene and found little difference between either the predicted S_1 lifetime or the branching ratio between *cis* and *trans* photoproducts, at odds with the notion that topology of the S_0/S_1 CX should lead to differences in measurable observables. It is certainly true that most nonadiabatic transitions happen *near* rather than *at* a CX, if only because the volume of configuration space is much larger if one admits quasi-degenerate geometries. It is therefore possible that in some cases the topological details may have no observable impact on the outcome of a particular photochemical reaction. A contrasting example, however, comes from a recent TSH study of the *trans* → *cis* photoisomerization of the protonated Schiff base $C_5H_6NH_2^+$, for which LR- and SF-TDDFT methods predict significantly different S_1 lifetimes [134]. The lifetime predicted using SF-TDDFT agrees with that obtained in CASPT2-based TSH simulations, whereas the LR-TDDFT lifetime is much longer. The reason is that topological problems in LR-TDDFT warp the S_1 potential surface in the vicinity of the photochemically relevant S_0/S_1 intersection, leading to the presence of a barrier that is not present in either the SF-TDDFT calculations or the multireference benchmarks [134]. LR-TDDFT's topology problem thus manifests as a *topography* problem, and trapping behind an artificial barrier leads to slower dynamics. Such problems appear to manifest in some but not all cases, e.g., LR-TDDFT potential energy surfaces for $CH_2NH_2^+$ are reported to be in good agreement with multireference results in the region of an S_0/S_1 intersection [111]. Results for $C_5H_6NH_2^+$ as well as theoretical considerations, however, demonstrate that there is a potential for serious problems.

It is also notable that in the azobenzene simulations reported in Ref. [128], a significant fraction of the LR-TDDFT trajectories had to be terminated and discarded due to convergence failure in the region of quasi-degeneracy, whereas none of the SF-TDDFT trajectories suffered this problem. This is likely a direct result of the more balanced treatment of the two states in the SF approach. Furthermore, the time-dependent energy gap $E_1 - E_0$ is observed to be much more oscillatory in the LR-TDDFT simulations, leading to much more frequent hopping events and concomitant oscillations in the populations of the two electronic states [128]. At the SF-TDDFT level, the energy gap is less oscillatory, the population dynamics are smoother, and hopping events are relatively infrequent [128].

10.2.2.3 Spin contamination and state tracking

While SF-TDDFT rigorously cures the topology problem around CXs, it tends to significantly worsen the spin contamination as compared to LR-TDDFT, because some double excitations are admitted into the excitation space but not in a manner that introduces all of the complementary determinants needed to form spin-pure $\hat{S}^2$ eigenstates. In the (4e,4o) example of Fig. 10.4, the o-o subspace in spin-complete and contains a minimalist description of the S_0/S_1 intersection, however the remaining subspaces lack some of the determinants needed to form $\hat{S}^2$ eigenstates. Any excited state that contains significant contributions from outside the o-o subspace is likely to exhibit significant spin contamination. It is possible to minimally augment the excitation space to restore spin symmetry, and this approach has been used to construct a spin-complete version of SF-CIS [98]. Notably, this can be done without increasing the formal complexity of the method; the number of determinants remains $\mathcal{O}(N_{\text{occ}}N_{\text{virt}})$, albeit with a somewhat larger prefactor as compared to conventional CIS. More automated versions of this approach are discussed in Section 10.3, including methods that incorporate a DFT treatment of dynamical correlation.

For practical application of the standard SF-TDDFT approach, especially for TSH simulations, it is necessary to use some kind of "state tracking" procedure that attempts to identify the best approximate multiplicity in regions of heavy spin contamination. Spin contamination tends to be relatively small near the ground-state geometry (Franck–Condon excitation point), whereas it becomes severe in bond-breaking regions of the potential surface where singlet and triplet states may be quasi-degenerate and therefore extensively mixed, in the absence of proper spin symmetry. A practical workaround is to categorize states based on their similarity to the wavefunctions computed in the previous time step, with the idea that spin multiplicities are identifiable at $t = 0$ so that the multiplicity of $|\Psi_K(t + \Delta t)\rangle$ can be assigned based on its overlap with the states $\{|\Psi_J(t)\rangle\}$. Several methods along these lines have been suggested [13, 34, 133]. The use of such algorithms is not guaranteed to lead to an unambiguous identification of the states in question, but it has been used successfully in TSH simulations [34].

10.3 Augmented Spin-Flip Methods

In the final section of this chapter, we introduce some methods that go beyond the simple SF *ansatz* in an attempt to correct some of its problems, most notably spin contamination. One of these is a spin-complete version of SF-TDDDFT [133], which minimally augments that method's excitation space in order to obtain a set of determinants from which spin-pure $\hat{S}^2$ eigenfunctions can be constructed. This is accomplished in an automated way using a tensor equation-of-motion (TEOM) formalism, as described in Section 10.3.1. An alternative approach is the "mixed reference spin-flip" (MRSF) procedure developed recently by Filatov, Choi, and coworkers [54], which is discussed in Section 10.3.2.

10.3.1 Spin-Adapted Spin-Flip Approach

10.3.1.1 Formalism

Originally developed from a wavefunction point of view [98], spin-complete SF-CIS lacks dynamical correlation but that shortcoming can be overcome using a TEOM framework to re-derive this method within the framework of TDDFT [133]. The TEOM formalism itself was originally developed in nuclear physics [93], but later extended to molecular systems by Li and Liu [62, 63, 66]. As applied to TDDFT, the result is a "spin-adapted" (SA-)SF-TDDFT method that maintains SF-TDDFT's correct treatment of the topology around CXs but restores spin multiplicity as a good quantum number.

In SA-SF-TDDFT, excitation operators in the TEOM formalism are truncated at the single excitation level. The single-excitation operators can be grouped into two categories: those with zero rank (singlet coupling) and those with rank one (triplet coupling). Because SF methods use a target state of lower spin angular momentum as compared to the reference state, only the triplet-coupled excitation operators are useful in this case. These can be represented in the MO basis as

$$\hat{O}^{\dagger}_{pq}(1,1) = -\hat{a}^{\dagger}_{p}\hat{a}_{\bar{q}} \tag{10.21a}$$

$$\hat{O}^{\dagger}_{pq}(1,0) = (\hat{a}^{\dagger}_{p}\hat{a}_{q} - \hat{a}^{\dagger}_{\bar{p}}\hat{a}_{\bar{q}})/\sqrt{2} \tag{10.21b}$$

$$\hat{O}^{\dagger}_{pq}(1,-1) = \hat{a}^{\dagger}_{\bar{p}}\hat{a}_{q} \tag{10.21c}$$

where $\hat{a}^{\dagger}_{p}$ creates an α-spin electron in orbital ψ_p and $\hat{a}_{\bar{q}}$ annihilates a β-spin electron in orbital ψ_q. Previous work by Li and Liu [62] uses a different working equation as a TEOM as compared to that used in SA-SF-CIS and SA-SF-DFT [133], and that equation generates spurious solutions due to the presence of redundant terms in the excitation space. The working equation used by Zhang and Herbert for SA-SF-TDDFT avoids this problem. The overall tensor operator can be written as

$$\begin{aligned}\hat{O}^{\dagger}(1) = &\sum_{ia}\hat{O}^{\dagger}_{ia}(1)x^{\text{CV}}_{ia} + \sum_{iu}\hat{O}^{\dagger}_{iu}(1)x^{\text{CO}}_{iu} + \sum_{ua}\hat{O}^{\dagger}_{ua}(1)x^{\text{OV}}_{ua}\\ &+\sum_{tu}\hat{O}^{\dagger}_{tu}(1)x^{\text{OO}}_{tu} + \sum_{ai}\hat{O}^{\dagger}_{ai}(1)y^{\text{CV}}_{ia}\\ &+\sum_{ui}\hat{O}^{\dagger}_{ui}(1)y^{\text{CO}}_{iu} + \sum_{au}\hat{O}^{\dagger}_{au}(1)y^{\text{OV}}_{ua}.\end{aligned} \tag{10.22}$$

As in conventional TDDFT, x_{ia} and y_{ia} are excitation and de-excitation amplitudes, respectively, with superscripts corresponding to the various subspaces in Fig. 10.4. For example, x_{ia}^{cv} is a coefficient from doubly occupied MO ψ_i to empty (virtual) MO ψ_a. Indices t and u in Eq. (10.22) are used to indicate singly occupied MOs.

The above SA-SF formalism can be used to generate a spin-complete version of SF-CIS. To extend this to DFT in a formally exact manner, it would be necessary to introduce a Hamiltonian $\hat{H}^{\text{DFT}}$ that generates the exact ground-state energy as the expectation value of the single-determinant reference state. No such Hamiltonian is known in analytical form, so dynamical correlation is instead introduced as an *ad hoc* DFT correction to the SA-SF-CIS matrix elements, in a manner that is reminiscent of the *ad hoc* correction introduced in the DFT/MRCI method [31, 72]. The final matrix elements for SA-SF-DFT have the form

$$\begin{aligned} \langle \Phi_{pq} | \hat{H}^{\text{DFT}} - E_0^{\text{DFT}} | \Phi_{rs} \rangle = \delta_{qs} F_{pr} - \delta_{pr} F_{qs} + \langle pq || qp \rangle \\ + (1 - C_{\text{HFX}}) \langle pq | pq \rangle \, . \end{aligned} \tag{10.23}$$

This ends up being precisely the same matrix element as that used in collinear SF-TDDFT [130], which rationalizes the good performance of the method. In conjunction with Kohn–Sham orbitals determined via ground-state DFT, the extra determinants that are introduced to make the excitation manifold spin-complete do incur some double-counting of electron correlation effects, although one may argue this is likely to be small because the number of additional determinants is rather limited. In fact, the same issue arises in DFT/MRCI, where the overcounting is potentially more severe, and in that approach an empirical damping factor is introduced for the off-diagonal matrix elements in order to limit the scope of the double counting. A similar modification may help the SA-SF-DFT method, but has not yet been tested.

10.3.1.2 Applications

Eliminating the need for failure-prone state-tracking algorithms is crucial in order to make SF-TDDFT into a robust, black-box engine for TSH simulations, but elimination of spin contamination has practical consequences for excited-state optimizations and MECP

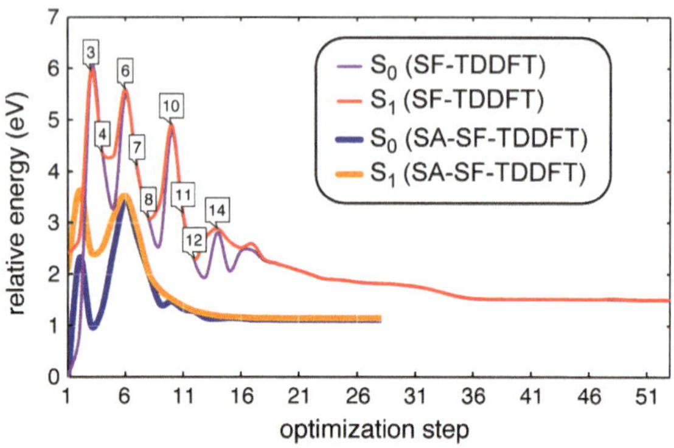

Figure 10.7 Optimization trajectories for a S_0/S_1 MECP of C_2H_4, starting from the ground-state geometry and using either SF-TDDFT or SA-SF-TDDFT with the BH&HLYP functional. In the SF-TDDFT case, flags at steps 3, 4, 6, 7, ... indicate geometries at which the assignment of spin multiplicities becomes ambiguous due to spin contamination. Data are taken from Ref. [133].

searches. An example of the latter, for an S_0/S_1 MECP of ethylene, is presented in Fig. 10.7 in the form of an "optimization trajectory," i.e., a plot of energy (for both S_0 and S_1) versus optimization steps. Using conventional SF-TDDFT, singlet and triplet states quickly become mixed so that already by the third optimization step it is difficult to assign spin multiplicities based on the value of $\langle \hat{S}^2 \rangle$ alone. At the third step, the two lowest-energy states have $\langle \hat{S}^2 \rangle = 1.14$ and $\langle \hat{S}^2 \rangle = 0.90$ (in units of $\hbar^2$), indicating nearly equal mixing of singlet and triplet [133]. Flags on steps 3, 4, 6, 7, 8, ... in Fig. 10.7 indicate points at which the state assignment is ambiguous. Energies of the two selected adiabats are strongly oscillatory, suggesting that the algorithm may be switching between states with differing amounts of singlet character, confounding the optimization algorithm. In contrast, the SA-SF-TDDFT optimization proceeds relatively smoothly and converges in far fewer steps. It is worth noting, however, that no analytic gradient has yet been reported for SA-SF-TDDFT, so MECP optimizations at that level of theory are proof-of-concept exercises based on finite-difference gradients [133].

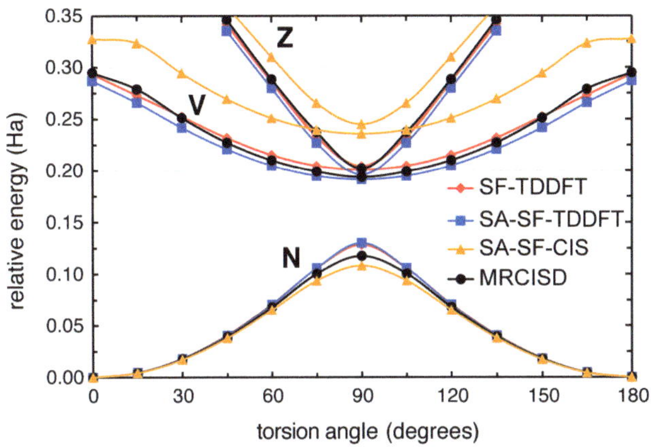

Figure 10.8 Potential energy curves along the C–C twisting coordinate for the lowest three singlet states of ethylene: N for the "normal" ground state or $(\pi)^2$ configuration, V for "valence" $(\pi)^1(\pi^*)^1$ state, and Z for "zwitterionic" $(\pi^*)^2$ state. (The notation is historical [76].) Reprinted from Ref. [133]; copyright 2015 American Institute of Physics.

As a second example of SA-SF-TDDFT, we consider the classic strong correlation problem of twisting C_2H_4 about its C–C axis. The $(\pi)^2$ and $(\pi^*)^2$ electron configurations become degenerate at a twist angle of 90°, causing single-reference methods based on spin-restricted orbitals to exhibit an unphysical cusp [51, 100]. Potential energy surfaces for the lowest three singlet states are plotted in Fig. 10.8, computed using several different SF methods and compared to a multireference benchmark. Both the simple SF-TDDFT method as well as SA-SF-TDDFT afford reasonably good agreement with the benchmarks for all three of the lowest singlet states, with no unphysical cusps. SA-SF-CIS results are also free of cusps although the absence of dynamical correlation causes the S_1 and S_2 surfaces to deviate significantly from the benchmarks.

10.3.2 Mixed-Reference Spin-Flip Approach

The MRSF approach to TDDFT has emerged recently as an alternative way to eliminate spin contamination in SF-TDDFT [54], if not rigorously and exactly (as in SA-SF-TDDFT) then at least to good numerical accuracy, without significantly increasing the complexity with respect to the original method SF-TDDFT method. In Section 10.3.2.1, we briefly explain how MRSF-TDDFT works in the case where states of singlet multiplicity are targeted starting from a triplet reference state, and then a few exemplary applications are discussed in Section 10.3.2.2.

10.3.2.1 Formalism

MRSF-TDDFT is based on a mixed-reference (MR) state whose density matrix is a linear combination of the $M_S = 1$ and $M_S = -1$ components of a triplet reference state:

$$\rho_0^{\mathrm{MR}}(\mathbf{r}, \mathbf{r}') = \frac{1}{2}\left[\rho_0^{M_S=+1}(\mathbf{r}, \mathbf{r}') + \rho_0^{M_S=-1}(\mathbf{r}, \mathbf{r}')\right] . \tag{10.24}$$

The two components of this density matrix are subjected to separate spin-flipping operations, $\alpha \rightarrow \beta$ for the $M_S = 1$ component (as in the basic SF-TDDFT method discussed above) and $\beta \rightarrow \alpha$ for the $M_S = -1$ component. This is shown schematically in Fig. 10.9 for the same (4e,4o) model that was used to introduce the basic method, using the same notation is used for the various subspaces of Slater determinants. Comparing Fig. 10.4 to Fig. 10.9, it is evident that MRSF-TDDFT introduces many more electronic configurations as compared to SF-TDDFT. Operationally, this removes the majority of the spin contamination even thought the excitation space is not formally spin-complete because some of the necessary o-v determinants (shown in gray in Fig. 10.9) cannot be generated by this procedure. However, these are higher-energy configurations and thus play little role in low-lying excited states, so that $\langle \hat{S}^2 \rangle$ is close to its spin-pure value in practice [54].

Derivation of MRSF-TDDFT is based on the density matrix formulation of the time-dependent Kohn–Sham methodology [32, 33], which means that the density matrix needs to be idempotent. This can be demonstrated for the density matrix in Eq. (10.24) based on a complex rotation of the spin functions [54]. Separation of the $M_S = 1$ and $M_S = -1$ triplet density matrices leads to two independent eigenvalue equations,

$$\mathbf{A}_{\mathrm{t}}\mathbf{x}_{\mathrm{t}} = \omega_{\mathrm{t}}\mathbf{x}_{\mathrm{t}} \tag{10.25a}$$

$$\mathbf{A}_{\mathrm{s}}\mathbf{x}_{\mathrm{s}} = \omega_{\mathrm{s}}\mathbf{x}_{\mathrm{s}} . \tag{10.25b}$$

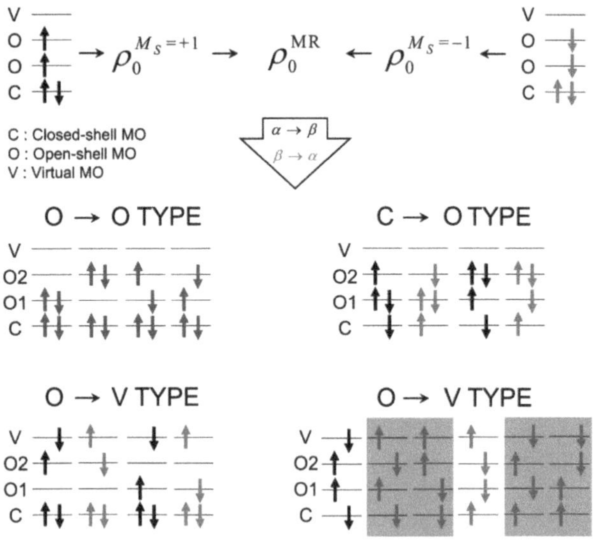

Figure 10.9 Schematic illustration of MRSF-TDDFT for a (4e,4o) model. The excitation space for singlet states is generated starting from both high- and low-spin triplet configurations, via single excitations combined with either $\alpha \to \beta$ spin flip (for $M_S = 1$) or $\beta \to \alpha$ spin flip (for $M_S = -1$). The four determinants shown in gray in the o-v subspace are missing from the excitation manifold because they cannot be generated in this manner. Adapted from Ref. [41]; copyright 2021 American Chemical Society.

However, it is necessary to incorporate coupling matrix elements between determinants from the $M_S = 1$ and $M_S = -1$ parts of the calculation on an *ad hoc* basis [54]. This amounts to new contributions to the orbital Hessian **A**, along the lines of [56]

$$A'_{pq,rs} = C_{\mathrm{HFX}} \left\langle \Phi^{M_S=+1}_{p\alpha q\beta} \middle| \hat{H} \middle| \Phi^{M_S=-1}_{r\beta s\alpha} \right\rangle . \tag{10.26}$$

The orbital Hessian that is used in practice is then $\mathbf{A}_k + \mathbf{A}'$, for k = t or s in Eq. (10.25).

10.3.2.2 Applications

MRSF-TDDFT has several advantages over SF-TDDFT, the most significant being that spin contamination is drastically reduced, simplifying the identification of relevant states. This is important for excited-state geometry optimization, reaction-path following, and

TSH simulations. (Notably, the analytic gradient has been formulated for MRSF-TDDFT [56], along with approximate nonadiabatic couplings [55], whereas no analytic gradient for SA-SF-TDDFT is yet available.)

MECP geometries are found to be in good agreement with multireference benchmarks but also in good agreement with the simple SF-TDDFT method [57], whereas vertical excitation energies are comparable in accuracy or slightly better with MRSF [39]. Both of these comparisons are based on the BH&HLYP functional, which is also used in nearly all of the SF-TDDFT results quoted in this chapter. Interestingly, when the fraction of exact exchange (C_{HFX}) is optimized in order to minimize errors in vertical excitation energies of polyenes, the MRSF-TDDFT approach lands on $C_{\text{HFX}} = 0.5$ as the optimal value, corresponding precisely to the BH&HLYP functional. The same optimization procedure applied to LR-TDDFT lands on $C_{\text{HFX}} = 0.2$ as the optimal value, corresponding to B3LYP [40]. This observation further corroborates the idea that 50% HFX is close to optimal for collinear spin-flip methods.

TSH simulations of thymine photodynamics have been reported using MRSF-TDDFT [85], which support trapping on S_1 as the explanation for a long-lived ($\sim$5 ps) decay component time-resolved experiments. This is consistent with multireference results that include dynamical electron correlation [107], but it is at odds with CASSCF predictions that indicate trapping on S_2 [42]. The discrepancy appears to originate in an excited-state barrier that is too high in the absence of dynamical electron correlation, supporting other results indicating that CASSCF is not a quantitative method for excited-state dynamics [28–30].

10.4 Summary and Outlook

This chapter has surveyed systemic problems with the description of ground-state CXs in conventional LR-TDDFT. Although the focus here has been on singlet photochemistry (S_0/S_1 intersections), similar problems can be expected for any CX that involves the ground state of the molecule, when that state is used as the reference state for LR-TDDFT. Numerous (apparently successful) applications of conventional LR-TDDFT to nonadiabatic molecular dynamics can be found in the literature, yet there are reasons to worry about that method's ability to describe the final internal conversion step that returns the molecule to its ground state. Therefore we must conclude that, at present, the situation remains unclear as to how pervasive these problem are, and the extent to which they effect real observables. There are undeniably cases where LR-TDDFT warps the topography around a S_0/S_1 intersection seam to such an extent that it has a significant effect on the simulated S_1 lifetime and thus the internal conversion timescale [134]. The unbalanced treatment of a quasi-degenerate ground and first excited state in LR-TDDFT can also lead to convergence problems that stymie attempts to use this method as a black-box engine for nonadiabatic molecular dynamics.

A "spin-flip" modification rectifies these problems, within a tractable computational formalism for which nonadiabatic (derivative) coupling vectors have been formulated and implemented based on analytic gradient theory [130, 132]. SF-TDDFT can be used in surface hopping calculations, as well as simply to locate MECPs with correct topology, including those that involve the ground state. For reasons discussed herein, SF-TDDFT calculations are typically performed using the BH&HLYP functional, and MECPs computed in this way show excellent agreement with multireference benchmarks.

A drawback of the simple SF-TDDFT approach is that it often leads to significant spin contamination in bond-breaking regions of the potential surface. For singlet photochemistry, the practical upshot is that it can be difficult to distinguish singlets from triplets in the photochemically relevant regions of the surface. State-tracking algorithms (based on wavefunction overlap) are used in practice to overcome this problem [13, 34, 133], although this feels to us like a stopgap solution. A "spin-adapted" version of the theory (SA-SF-TDDFT) has been reported [133], which solves this problem in an elegant way by introducing the minimal number of additional determinants necessary to obtain spin-pure $\hat{S}^2$ eigenstates, however at present its analytic gradient is not available. This method can be used to spot-check the state assignments of the simple SF-TDDFT approach. Alternatively, a "mixed reference" version of SF-

TDDFT has also been developed [54], which eliminates the spin contamination to a useful degree for low-lying singlet states, even if it does not rigorously afford spin eigenstates.

Acknowledgements

The Herbert group's work on TDDFT has been sponsored for many years by the U.S. National Science Foundation, at present under grant no. CHE-1955282.

References

1. Atchity, G. J., Xantheas, S. S., and Ruedenberg, K. (1991). Potential energy surfaces near intersections, *J. Chem. Phys.* **95**, pp. 1862–1876.
2. Barbatti, M. (2011). Nonadiabatic dynamics with trajectory surface hopping method, *WIREs Comput. Mol. Sci.* **1**, pp. 620–633.
3. Barbatti, M., and Crespo-Otero, R. (2016). Surface hopping dynamics with DFT excited states, in N. Ferré, M. Filatov, and M. Huix-Rotllant (eds.), *Density-Functional Methods for Excited States, Topics in Current Chemistry*, Vol. 368 (Springer International Publishing, Switzerland), pp. 415–444.
4. Bearpark, M. J., Robb, M. A., and Schlegel, H. B. (1994). A direct method for the location of the lowest energy point on a potential surface crossing, *Chem. Phys. Lett.* **223**, pp. 269–274.
5. Becke, A. D. (1988). Density-functional exchange-energy approximation with correct asymptotic behavior, *Phys. Rev. A* **38**, pp. 3098–3100.
6. Bernard, Y. A., Shao, Y., and Krylov, A. I. (2012). General formulation of spin-flip time-dependent density functional theory using non-collinear kernels: Theory, implementation, and benchmarks, *J. Chem. Phys.* **136**, pp. 204103:1–17.
7. Casanova, D., and Krylov, A. I. (2020). Spin-flip methods in quantum chemistry, *Phys. Chem. Chem. Phys.* **22**, pp. 4326–4342.
8. Casida, M. E. (1995). Time-dependent density functional response theory for molecules, in D. P. Chong (ed.), *Recent Advances in Density Functional Methods, Part I, Recent Advances in Computational Chemistry*, Vol. I, chap. 5 (World Scientific, River Edge, NJ), pp. 155–192.

9. Casida, M. E. (1996). Time-dependent density functional response theory of molecular systems: Theory, computational methods, and functionals, in J. M. Seminario (ed.), *Recent Developments and Applications of Modern Density Functional Theory, Theoretical and Computational Chemistry*, Vol. 4 (Elsevier, Amsterdam), pp. 391–439.

10. Casida, M. E., Gutierrez, F., Guan, J., Gadea, F.-X., Salahub, D., and Daudey, J.-P. (2000). Charge-transfer correction for improved time-dependent local density approximation excited-state potential energy curves: Analysis within the two-level model with illustration for H_2 and LiH, *J. Chem. Phys.* **114**, pp. 7062–7071.

11. Cederbaum, L. S. (2004). Born–Oppenheimer approximation and beyond, in W. Domcke, D. R. Yarkony, and H. Köppel (eds.), *Conical Intersections: Electronic Structure, Dynamics & Spectroscopy, Advanced Series in Physical Chemistry*, Vol. 15, chap. 1 (World Scientific, Singapore), pp. 3–40.

12. Chernyak, V., and Mukamel, S. (2000). Density-matrix representation of nonadiabatic couplings in time-dependent density functional (TDDFT) theories, *J. Chem. Phys.* **112**, pp. 3572–3579.

13. Closser, K. D., Gessner, O., and Head-Gordon, M. (2014). Simulations of the dissociation of small helium clusters with *ab initio* molecular dynamics in electronically excited states, *J. Chem. Phys.* **140**, pp. 134306:1–12.

14. Cordova, F., Doriol, L. J., Ipatov, A., Casida, M. E., Filippi, C., and Vela, A. (2007). Troubleshooting time-dependent density-functional theory for photochemical applications: Oxirane, *J. Chem. Phys.* **127**, pp. 164111:1–18.

15. Curchod, B. F. E., and Martínez, T. J. (2018). Ab initio nonadiabatic quantum molecular dynamics, *Chem. Rev.* **118**, pp. 3305–3336.

16. Curchod, B. F. E., Rothlisberger, U., and Tavernelli, I. (2013). Trajectory-based nonadiabatic dynamics with time-dependent density functional theory, *ChemPhysChem* **14**, pp. 1314–1340.

17. Dalgaard, E. (1982). Quadratic response functions within the time-dependent Hartree-Fock approximation, *Phys. Rev. A* **26**, pp. 42–52.

18. Dreuw, A., and Head-Gordon, M. (2005). Single-reference ab initio methods for the calculation of excited states of large molecules, *Chem. Rev.* **105**, pp. 4009–4037.

19. Elliott, P., Furche, F., and Burke, K. (2009). Excited states from time-dependent density functional theory, in K. B. Lipkowitz and T. R. Cundari (eds.), *Reviews in Computational Chemistry*, Vol. 26, chap. 3 (Wiley-VCH, New York), pp. 91–165.

20. Ernst, O. P., Lodowski, D. T., Elstner, M., Hegemann, P., Brown, L. S., and Kandori, H. (2014). Microbial and animal rhodopsins: Structures, functions, and molecular mechanisms, *Chem. Rev.* **114**, pp. 126–163.
21. Fatehi, S., Alguire, E., Shao, Y., and Subotnik, J. E. (2011). Analytic derivative couplings between configuration-interaction-singles states with built-in electron-transation factors for translational invariance, *J. Chem. Phys.* **135**, pp. 234105:1–21.
22. Fatehi, S., and Subotnik, J. E. (2012). Derivative couplings with built-in electron-translation factors: Application to benzene, *J. Phys. Chem. Lett.* **3**, pp. 2039–2043.
23. Furche, F. (2001). On the density matrix based approach to time-dependent density functional response theory, *J. Chem. Phys.* **114**, pp. 5982–5992.
24. Furche, F., and Rappoport, D. (2005). Density functional methods for excited states: Equilibrium structure and electronic spectra, in M. Olivucci (ed.), *Computational Photochemistry, Theoretical and Computational Chemistry*, Vol. 16, chap. 3 (Elsevier, Amsterdam), pp. 93–128.
25. Gao, J., Liu, W., Song, B., and Liu, C. (2004). Time-dependent four-component relativistic density functional theory for excitation energies, *J. Chem. Phys.* **121**, pp. 6658–6666.
26. Gao, J., Zou, W., Liu, W., Xiao, Y., Peng, D., Song, B., and Liu, C. (2005). Time-dependent four-component relativistic density-functional theory for excitation energies. II. The exchange-correlation kernel, *J. Chem. Phys.* **123**, pp. 054102:1–13.
27. Gherib, R., Ye, L., Ryabinkin, I. G., and Izmaylov, A. F. (2016). On the inclusion of the diagonal Born-Oppenheimer correction in surface hopping methods, *J. Chem. Phys.* **144**, pp. 154103:1–10.
28. Gozem, S., Huntress, M., Schapiro, I., Lindh, R., Granovsky, A. A., Angeli, C. and Olivucci, M. (2012). Dynamic electron correlation effects on the ground state potential energy surface of a retinal chromophore model, *J. Chem. Theory Comput.* **8**, pp. 4069–4080.
29. Gozem, S., Melaccio, F., Lindh, R., Krylov, A. I., Granovsky, A. A., Angeli, C. and Olivucci, M. (2013). Mapping the excited state potential energy surface of a retinal chromophore with multireference and equation-of-motion coupled-cluster methods, *J. Chem. Theory Comput.* **9**, pp. 4495–4506.
30. Gozem, S., Melaccio, F., Valentini, A., Filatov, M., Huix-Rotllant, M., Ferré, N., Frutos, L. M., Angeli, C., Krylov, A. I., Granovsky, A. A., Lindh,

R., and Olivucci, M. (2014). Shape of multireference, equation-of-motion coupled-cluster, and density functional theory potential energy surfaces at a conical intersection, *J. Chem. Theory Comput.* **10**, pp. 3074–3084.

31. Grimme, S., and Waletzke, M. (1999). A combination of Kohn–Sham density functional theory and multi-reference configuration interaction methods, *J. Chem. Phys.* **111**, pp. 5645–5655.
32. Gross, E. K. U., and Kohn, W. (1990). Time-dependent density functional theory, *Adv. Quantum Chem.* **21**, pp. 255–291.
33. Gross, E. K. U., and Maitra, N. T. (2012). Introduction to TDDFT, in M. A. L. Marques, N. T. Maitra, F. M. S. Nogueira, E. K. U. Gross and A. Rubio (eds.), *Fundamentals of Time-Dependent Density Functional Theory, Lecture Notes in Physics*, Vol. 837, chap. 1 (Springer-Verlag, Heidelberg), pp. 53–97.
34. Harabuchi, Y., Keipert, K., Zahariev, F., Taketsugu, T., and Gordon, M. S. (2014). Dynamics simulations with spin-flip time-dependent density functional theory: Photoisomerization and photocyclization mechanisms of *cis*-stilbene in $\pi\pi^*$ states, *J. Phys. Chem. A* **118**, pp. 11987–11998.
35. Harabuchi, Y., Maeda, S., Taketsugu, T., Minezawa, N., and Morokuma, K. (2013). Automated search for minimum energy conical intersection geometries between the lowest two singlet states S_0/S_1-MECIs by the spin-flip TDDFT method, *J. Chem. Theory Comput.* **9**, pp. 4116–4123.
36. Herbert, J. M., and Head-Gordon, M. (2004). Curvy-steps approach to constraint-free extended-Lagrangian *ab initio* molecular dynamics, using atom-centered basis functions: Convergence toward Born-Oppenheimer trajectories, *J. Chem. Phys.* **121**, pp. 11542–11556.
37. Herbert, J. M., Zhang, X., Morrison, A. F., and Liu, J. (2016). Beyond time-dependent density functional theory using only single excitations: Methods for computational studies of excited states in complex systems, *Acc. Chem. Res.* **49**, pp. 931–941.
38. Hirata, S., and Head-Gordon, M. (1999). Time-dependent density functional theory for radicals. An improved description of excited states with substantial double excitation character, *Chem. Phys. Lett.* **302**, pp. 375–382.
39. Horbatenko, Y., Lee, S., Filatov, M., and Choi, C. H. (2019). Performance analysis and optimization of mixed-reference spin-flip time-dependent density functional theory (MRSF-TDDFT) for vertical excitation energies and singlet–triplet energy gaps, *J. Phys. Chem. A* **123**, pp. 7991–8000.

40. Horbatenko, Y., Lee, S., Filatov, M., and Choi, C. H. (2021). How beneficial is the *explicit* account of doubly-excited configurations in linear response theory? *J. Chem. Theory Comput.* **17**, pp. 975–984.
41. Horbatenko, Y., Sadiq, S., Lee, S., Filatov, M., and Choi, C. H. (2021). Mixed-reference spin-flip time-dependent density functional theory (MRSF-TDDFT) as a simple yet accurate method for diradicals and diradicaloids, *J. Chem. Theory Comput.* **17**, pp. 848–859.
42. Hudock, H. R., Levine, B. G., Thompson, A. L., Satzger, H., Townsend, D., Gador, N., Ullrich, S., Stolow, A., and Martinez, T. J. (2007). Ab initio molecular dynamics and time-resolved photoelectron spectroscopy of electronically excited uracil and thymine, *J. Phys. Chem. A* **111**, pp. 8500–8508.
43. Huix-Rotllant, M., Natarajan, B., Ipatov, A., Wawire, C. M., Deutsch, T., and Casida, M. E. (2010). Assessment of noncollinear spin-flip Tamm–Dancoff approximation time-dependent density-functional theory for the photochemical ring-opening of oxirane, *Phys. Chem. Chem. Phys.* **12**, pp. 12811–12825.
44. Huix-Rotllant, M., Nikiforov, A., Thiel, W., and Filatov, M. (2016). Description of conical intersections with density functional methods, in N. Ferré, M. Filatov, and M. Huix-Rotllant (eds.), *Density-Functional Methods for Excited States, Topics in Current Chemistry*, Vol. 368 (Springer International Publishing), pp. 445–476.
45. Isegawa, M., and Truhlar, D. G. (2013). Valence excitation energies of alkenes, carbonyl compounds, and azabenzenes by time-dependent density functional theory: Linear response of the ground state compared to collinear and noncollinear spin-flip TDDFT with the Tamm-Dancoff approximation, *J. Chem. Phys.* **138**, pp. 134111:1–13.
46. Jain, A., Alguire, E., and Subotnik, J. E. (2016). An efficient, augmented surface hopping algorithm that includes decoherence for use in large-scale simulations, *J. Chem. Theory Comput.* **12**, pp. 5256–5268.
47. Jasper, A. W., and Truhlar, D. G. (2004). Non-Born–Oppenheimer molecular dynamics for conical intersections, avoided crossings, and weak interactions, in W. Domcke, D. R. Yarkony, and H. Köppel (eds.), *Conical Intersections: Theory, Computation and Experiment, Advanced Series in Physical Chemistry*, Vol. 17, chap. 10 (World Scientific, Singapore), pp. 375–414.
48. Keal, T. W., Koslowski, A., and Thiel, W. (2007). Comparison of algorithms for conical intersection optimisation using semiempirical methods, *Theor. Chem. Acc.* **118**, pp. 837–844.

49. Krylov, A. I. (2001). Spin-flip configuration interaction: An electronic structure model that is both variational and size-consistent, *Chem. Phys. Lett.* **350**, pp. 522–530.

50. Krylov, A. I., and Gill, P. M. W. (2013). Q-Chem: An engine for innovation, *WIREs Comput. Mol. Sci.* **3**, pp. 317–326.

51. Krylov, A. I., Sherrill, C. D., Byrd, E. F. C., and Head-Gordon, M. (1998). Size-consistent wave functions for nondynamical correlation energy: The valence active space optimized orbital coupled-cluster doubles model, *J. Chem. Phys.* **109**, pp. 10669–10678.

52. Laurent, A. D., and Jacquemin, D. (2013). TD-DFT benchmarks: A review, *Int. J. Quantum Chem.* **113**, pp. 2019–2039.

53. Lee, C., Yang, W., and Parr, R. G. (1988). Development of the Colle-Salvetti correlation-energy formula into a functional of the electron density, *Phys. Rev. B* **37**, pp. 785–789.

54. Lee, S., Filatov, M., Lee, S., and Choi, C. H. (2018). Eliminating spin-contamination of spin-flip time dependent density functional theory within linear response formalism by the use of zeroth-order mixed-reference (MR) reduced density matrix, *J. Chem. Phys.* **149**, pp. 104101:1–11.

55. Lee, S., Horbatenko, Y., Filatov, M., and Choi, C. H. (2021). Fast and accurate computation of nonadiabatic coupling matrix elements using the truncated Leibniz formula and mixed-reference spin-flip time-dependent density functional theory, *J. Phys. Chem. Lett.* **12**, pp. 4722–4728.

56. Lee, S., Kim, E. E., Nakata, H., Lee, S., and Choi, C. H. (2019). Efficient implementations of analytic energy gradient for mixed-reference spin-flip time-dependent density functional theory (MRSF-TDDFT), *J. Chem. Phys.* **150**, pp. 184111:1–11.

57. Lee, S., Shostak, S., Filatov, M., and Choi, C. H. (2019). Conical intersections in organic molecules: Benchmarking mixed-reference spin–flip time-dependent DFT (MSRF-TD-DFT) vs spin–flip TD-DFT, *J. Phys. Chem. A* **123**, pp. 6455–6462.

58. Lefrancois, D., Tuna, D., Martínez, T. J., and Dreuw, A. (2017). The spin-flip variant of the algebraic-diagrammatic construction yields the correct topology of S_1/S_0 conical intersections, *J. Chem. Theory Comput.* **13**, pp. 4436–4441.

59. Levine, B. G., Coe, J. D., and Martínez, T. J. (2008). Optimizing conical intersections without derivative coupling vectors: Application to multistate multireference second-order perturbation theory (MS-CASPT2), *J. Phys. Chem. B* **112**, pp. 405–413.

60. Levine, B. G., Ko, C., Quenneville, J., and Martínez, T. J. (2006). Conical intersections and double excitations in time-dependent density functional theory, *Mol. Phys.* **104**, pp. 1039–1051.
61. Li, X., Govind, N., Isborn, C., DePrince III, A. E., and Lopata, K. (2020). Real-time time-dependent electronic structure theory, *Chem. Rev.* **120**, pp. 9951–9993.
62. Li, Z., and Liu, W. (2010). Spin-adapted open-shell random phase approximation and time-dependent density functional theory. I. Theory, *J. Chem. Phys.* **133**, pp. 064106:1–22.
63. Li, Z., and Liu, W. (2011). Spin-adapted open-shell time-dependent density functional theory. III. An even better and simpler formulation, *J. Chem. Phys.* **135**, pp. 194106:1–14, erratum: *ibid.* **138**, 029904 (2013).
64. Li, Z., and Liu, W. (2012). Theoretical and numerical assessments of spin-flip time-dependent density functional theory, *J. Chem. Phys.* **136**, pp. 024107:1–14.
65. Li, Z., and Liu, W. (2014). First-order nonadiabatic coupling matrix elements between excited states: A Lagrangian formulation at the CIS, RPA, TD-HF, and TD-DFT levels, *J. Chem. Phys.* **141**, pp. 014110:1–10.
66. Li, Z., Liu, W., Zhang, Y., and Suo, B. (2011). Spin-adapted open-shell time-dependent density functional theory. II. Theory and pilot application, *J. Chem. Phys.* **134**, pp. 134101:1–22.
67. Li, Z., Suo, B., and Liu, W. (2014). First-order nonadiabatic coupling matrix elements between excited states: Implementation and application at the TD-DFT and pp-TDA levels, *J. Chem. Phys.* **141**, pp. 244105:1–16.
68. Liu, J., Koslowski, A., and Thiel, W. (2018). Analytic gradient and derivative couplings for the spin-flip extended configuration interaction singles method: Theory, implementation, and application to proton transfer, *J. Chem. Phys.* **148**, pp. 244108:1–11.
69. Maeda, S., Harabuchi, Y., Taketsugu, T., and Morokuma, K. (2014). Systematic exploration of minimum energy conical intersection structures near the Franck–Condon region, *J. Phys. Chem. A* **118**, pp. 12050–12058.
70. Maeda, S., Ohno, K., and Morokuma, K. (2010). Updated branching plane for finding conical intersections without coupling derivative vectors, *J. Chem. Theory Comput.* **6**, pp. 1538–1545.
71. Malhado, J. P., Bearpark, M. J., and Hynes, J. T. (2014). Non-adiabatic dynamics close to conical intersections and the surface hopping perspective, *Front. Chem.* **2**, pp. 97:1–21.

72. Marian, C. M., Heil, A., and Kleinschmidt, M. (2018). The DFT/MRCI method, *WIREs Comput. Mol. Sci.* **9**, pp. e1394:1–31, erratum: *ibid.* **9**, p. e1437 (2019).
73. Matsika, S. (2021). Electronic structure methods for the description of nonadiabatic effects and conical intersections, *Chem. Rev.* **121**, pp. 9407–9449.
74. Matsika, S., and Krause, P. (2011). Nonadiabatic events and conical intersections, *Annu. Rev. Phys. Chem.* **62**, pp. 621–643.
75. McLachlan, A., and Ball, M. (1964). Time-dependent Hartree-Fock theory for molecules, *Rev. Mod. Phys.* **36**, pp. 844–855.
76. Merer, A. J., and Mulliken, R. S. (1969). Ultraviolet spectra and excited states of ethylene and its alkyl derivatives, *Chem. Rev.* **69**, pp. 639–656.
77. Minezawa, N. (2014). Optimizing minimum free-energy crossing points in solution: Linear-response free energy/spin-flip density functional theory approach, *J. Chem. Phys.* **141**, pp. 164118:1–9.
78. Minezawa, N., and Gordon, M. S. (2011). Photoisomerization of stilbene: A spin-flip density functional theory study, *J. Phys. Chem. A* **115**, pp. 7901–7911.
79. Minezawa, N., and Nakajima, T. (2019). Trajectory surface hopping molecular dynamics simulation by spin-flip time-dependent density functional theory, *J. Chem. Phys.* **150**, pp. 204120:1–10.
80. Nelson, T. R., White, A. J., Bjorgaard, J. A., Sifain, A. E., Zhang, Y., Nebgen, B., Fernandez-Alberti, S., Mozyrsky, D., Roitberg, A. E., and Tretiak, S. (2020). Non-adiabatic excited-state molecular dynamics: Theory and applications or modeling photophysics in extended molecular materials, *Chem. Rev.* **120**, pp. 2215–2287.
81. Niehaus, T. A. (2021). Ground-to-excited derivative couplings for the density functional-based tight-binding method: Semi-local and long-range corrected formulations, *Theor. Chem. Acc.* **140**, pp. 34:1–21.
82. Ou, Q., Alguire, E. C., and Subotnik, J. E. (2015). Derivative couplings between time-dependent density functional theory excited states in the random-phase approximation based on pseudo-wavefunctions: Behavior around conical intersections, *J. Phys. Chem. B* **119**, pp. 7150–7161.
83. Ou, Q., Bellchambers, G. D., Furche, F., and Subotnik, J. E. (2015). First-order derivative couplings between excited states from adiabatic TDDFT response theory, *J. Chem. Phys.* **142**, pp. 064114:1–14.
84. Ou, Q., Fatehi, S., Alguire, E., Shao, Y., and Subotnik, J. E. (2014). Derivative couplings between TD-DFT excited states obtained by direct

differentiation in the Tamm-Dancoff approximation, *J. Chem. Phys.* **141**, pp. 024114:1–13.

85. Park, W., Lee, S., Huix-Rotllant, M., Filatov, M., and Choi, C. H. (2021). Impact of the dynamic electron correlation on the unusually long excited-state lifetime of thymine, *J. Phys. Chem. Lett.* **12**, pp. 4339–4346.
86. Parker, S. M., Rappoport, D., and Furche, F. (2018). Quadratic response properties from TDDFT: Trials and tribulations, *J. Chem. Theory Comput.* **14**, pp. 807–819.
87. Parker, S. M., Roy, S., and Furche, F. (2016). Unphysical divergences in response theory, *J. Chem. Phys.* **145**, pp. 134105:1–12.
88. Parker, S. M., Roy, S., and Furche, F. (2019). Multistate hybrid time-dependent density functional theory with surface hopping accurately captures ultrafast thymine photodeactivation, *Phys. Chem. Chem. Phys.* **21**, pp. 18999–19010.
89. Rinkevicius, Z., Vahtras, O., and Ågren, H. (2010). Spin-flip time dependent density functional theory applied to excited states with single, double, or mixed electron excitation character, *J. Chem. Phys.* **133**, pp. 114104:1–12.
90. Rohlfing, M., and Louie, S. G. (2000). Electron-hole excitations and optical spectra from first principles, *Phys. Rev. B* **62**, pp. 4927–4944.
91. Rohrdanz, M. A., Martins, K. M., and Herbert, J. M. (2009). A long-range-corrected density functional that performs well for both ground-state properties and time-dependent density functional theory excitation energies, including charge-transfer excited states, *J. Chem. Phys.* **130**, pp. 054112:1–8.
92. Rowe, D. J. (1968). Equations-of-motion method and the extended shell model, *Rev. Mod. Phys.* **40**, pp. 153–166.
93. Rowe, D. J., and Ngo-Trong, C. (1975). Tensor equations of motion for the excitations of rotationally invariant or charge-independent systems, *Rev. Mod. Phys.* **47**, pp. 471–485.
94. Ryabinkin, I. G., Nagesh, J., and Izmaylov, A. F. (2015). Fast numerical evaluation of time-derivative nonadiabatic couplings for mixed quantum-classical methods, *J. Phys. Chem. Lett.* **6**, pp. 4200–4203.
95. Salazar, E., and Faraji, S. (2020). Theoretical study of cyclohexadiene/hexatriene photochemical interconversion using spin-flip time-dependent density functional theory, *Mol. Phys.* **118**, pp. e1764120:1–12.

96. Savin, A., Umrigar, C. J., and Gonze, X. (1998). Relationship of Kohn–Sham eigenvalues to excitation energies, *Chem. Phys. Lett.* **288**, pp. 391–395.
97. Schuurman, M. S., and Stolow, A. (2018). Dynamics at conical intersections, *Annu. Rev. Phys. Chem.* **69**, pp. 427–450.
98. Sears, J. S., Sherrill, C. D., and Krylov, A. I. (2003). A spin-complete version of the spin-flip approach to bond breaking: What is the impact of obtaining spin eigenfunctions? *J. Chem. Phys.* **118**, pp. 9084–9094.
99. Send, R., and Furche, F. (2010). First-order nonadiabatic couplings from time-dependent hybrid density functional response theory: Consistent formalism, implementation, and performance, *J. Chem. Phys.* **132**, pp. 044107:1–12.
100. Shao, Y., Head-Gordon, M., and Krylov, A. I. (2003). The spin-flip approach within time-dependent density functional theory: Theory and applications to diradicals, *J. Chem. Phys.* **118**, pp. 4807–4818.
101. Sharifzadeh, S. (2018). Many-body perturbation theory for understanding optical excitations in organic molecules and solids, *J. Phys.: Condens. Matt.* **30**, pp. 153002:1–12.
102. Sicilia, F., Blancafort, L., Bearpark, M. J., and Robb, M. A. (2008). New algorithms for optimizing and linking conical intersection points, *J. Chem. Theory Comput.* **4**, pp. 257–266.
103. Silva-Junior, M. R., Schreiber, M., Sauer, S. P. A., and Thiel, W. (2008). Benchmarks for electronically excited states: Time-dependent density functional theory and density functional theory based multireference configuration interaction, *J. Chem. Phys.* **129**, pp. 104103:1–14.
104. Small, D. W., Sundstrom, E. J., and Head-Gordon, M. (2015). A simple way to test for collinearity in spin symmetry broken wave functions: General theory and application to generalized Hartree Fock, *J. Chem. Phys.* **142**, pp. 094112:1–9.
105. Stratmann, R. E., Scuseria, G. E., and Frisch, M. J. (1998). An efficient implementation of time-dependent density-functional theory for the calculation of excitation energies of large molecules, *J. Chem. Phys.* **109**, pp. 8218–8224.
106. Subotnik, J. E., Jain, A., Landry, B., Petit, A., Ouyang, W., and Bellonzi, N. (2016). Understanding the surface hopping view of electronic transitions and decoherence, *Annu. Rev. Phys. Chem.* **67**, pp. 387–417.
107. Szymczak, J. J., Barbatti, M., Hoo, J. T. S., Adkins, J. A., Windus, T. L., Nachtigallová, D., and Lischka, H. (2009). Photodynamics simulations

of thymine: Relaxation into the first excited singlet state, *J. Phys. Chem. A* **113**, pp. 12686–12693.

108. Tapavicza, E., Bellchambers, G. D., Vincent, J. C., and Furche, F. (2013). *Ab initio* non-adiabatic molecular dynamics, *Phys. Chem. Chem. Phys.* **15**, pp. 18336–18348.
109. Tapavicza, E., Tavernelli, I., and Rothlisberger, U. (2007). Trajectory surface hopping within linear response time-dependent density functional theory, *Phys. Rev. Lett.* **98**, pp. 023001:1–4.
110. Tavernelli, I., Röhrig, U. F., and Rothlisberger, U. (2005). Molecular dynamics in electronically excited states using time-dependent density functional theory, *Mol. Phys.* **103**, pp. 963–981.
111. Tavernelli, I., Tapavicza, E., and Rothlisberger, U. (2009). Non-adiabatic dynamics using time-dependent density functional theory: Assessing the coupling strengths, *J. Mol. Struct. (Theochem)* **914**, pp. 22–29.
112. Tavernelli, I., Tapavicza, E., and Rothlisberger, U. (2009). Nonadiabatic coupling vectors within linear response time-dependent density functional theory, *J. Chem. Phys.* **130**, pp. 124107:1–10.
113. Tretiak, S., and Mukamel, S. (2002). Density matrix analysis and simulation of electronic excitations in conjugated and aggregated molecules, *Chem. Rev.* **102**, pp. 3171–3212.
114. Tully, J. C. (1990). Molecular dynamics with electronic transitions, *J. Chem. Phys.* **93**, pp. 1061–1071.
115. Uratani, H., Morioka, T., Yoshikawa, T., and Nakai, H. (2020). Fast nonadiabatic molecular dynamics via spin-flip time-dependent density-functional tight-binding approach: Application to nonradiative relaxation of tetraphenylethylene with locked aromatic rings, *J. Chem. Theory Comput.* **16**, pp. 7299–7313.
116. Uratani, H., Yoshikawa, T., and Nakai, H. (2021). Trajectory surface hopping approach to condensed-phase nonradiative relaxation dynamics using divide-and-conquer spin-flip time-dependent density-functional tight binding, *J. Chem. Theory Comput.* **17**, pp. 1290–1300.
117. Wand, A., Gdor, I., Zhu, J., Sheves, M., and Ruhman, S. (2013). Shedding new light on retinal protein photochemistry, *Annu. Rev. Phys. Chem.* **64**, pp. 437–458.
118. Wang, F., and Liu, W. (2003). Comparison of different polarization schemes in open-shell relativistic density functional calculations, *J. Chin. Chem. Soc. (Taipei)* **50**, pp. 597–606.

119. Wang, F., and Ziegler, T. (2004). Time-dependent density functional theory based on a noncollinear formulation of the exchange-correlation potential, *J. Chem. Phys.* **121**, pp. 12191:1–6.

120. Wang, F., and Ziegler, T. (2005). The performance of time-dependent density functional theory based on a noncollinear exchange- correlation potential in the calculations of excitation energies, *J. Chem. Phys.* **122**, pp. 074109:1–9.

121. Wang, L., Akimov, A., and Prezhdo, O. V. (2016). Recent progress in surface hopping: 2011–2015, *J. Phys. Chem. Lett.* **7**, pp. 2100–2112.

122. Wang, L., Qiu, J., Bai, X., and Xu, J. (2020). Surface hopping methods for nonadiabatic dynamics in extended systems, *WIREs Comput. Mol. Sci.* **10**, pp. e1435:1–24.

123. Weiss, H., Ahlrichs, R., and Häser, M. (1993). A direct algorithm for self-consistent-field linear response theory and application to C_{60}: Excitation energies, oscillator strengths, and frequency-dependent polarizabilities, *J. Chem. Phys.* **99**, pp. 1262–1270.

124. Xu, X., Yang, K. R., and Truhlar, D. G. (2014). Testing noncollinear spin-flip, collinear spin-flip, and conventional time-dependent density functional theory for predicting electronic excitation energies of closed-shell atoms, *J. Chem. Theory Comput.* **10**, pp. 2070–2084.

125. Yarkony, D. R. (2001). Nuclear dynamics near conical intersections in the adiabatic representation: I. The effects of local topography on interstate transitions, *J. Chem. Phys.* **114**, pp. 2601–2613.

126. Yu, L., Xu, C., Lei, Y., Zhu, C., and Wen, Z. (2014). Trajectory-based nonadiabatic molecular dynamics without calculating nonadiabatic coupling in the avoided crossing case: *Trans* ↔ *cis* photoisomerization in azobenzene, *Phys. Chem. Chem. Phys.* **16**, pp. 25883–25895.

127. Yue, L., Lan, Z., and Liu, Y.-J. (2014). The theoretical estimation of the bioluminescent efficiency of the firefly via a nonadiabatic molecular dynamics simulation, *J. Phys. Chem. Lett.* **6**, pp. 540–548.

128. Yue, L., Liu, Y., and Zhu, C. (2018). Performance of TDDFT with and without spin-flip in trajectory surface hopping dynamics: *Cis–trans* azobenzene photoisomerization, *Phys. Chem. Chem. Phys.* **20**, pp. 24123–24139.

129. Yue, L., Yu, L., Xu, C., Lei, Y., Liu, Y., and Zhu, C. (2017). Benchmark performance of global switching versus local switching for trajectory surface hopping molecular dynamics simulation: *Cis* ↔ *trans* azobenzene photoisomerization, *Chem. Phys. Chem.* **18**, pp. 1274–1287.

130. Zhang, X., and Herbert, J. M. (2014). Analytic derivative couplings for spin-flip configuration interaction singles and spin-flip time-dependent density functional theory, *J. Chem. Phys.* **141**, pp. 064104:1–9.

131. Zhang, X., and Herbert, J. M. (2014). Excited-state relaxation pathways in uracil versus hydrated uracil: Solvatochromatic shift in the $^1n\pi^*$ state is the key, *J. Phys. Chem. B* **118**, pp. 7806–7817.

132. Zhang, X., and Herbert, J. M. (2015). Analytic derivative couplings in time-dependent density functional theory: Quadratic response theory versus pseudo-wavefunction approach, *J. Chem. Phys.* **142**, pp. 064109:1–12.

133. Zhang, X., and Herbert, J. M. (2015). Spin-flip, tensor equation-of-motion configuration interaction with a density-functional correction: A spin-complete method for exploring excited-state potential energy surfaces, *J. Chem. Phys.* **143**, pp. 234107:1–10.

134. Zhang, X., and Herbert, J. M. (2021). Nonadiabatic dynamics with spin-flip versus linear-response time-dependent density functional theory: A case study for the protonated Schiff base $C_5H_6NH_2^+$, *J. Chem. Phys.* **155**, pp. 124111:1–15.

Chapter 11

Phase Space Mapping Theory for Nonadiabatic Quantum Molecular Dynamics

Baihua Wu,* Xin He,* and Jian Liu

Beijing National Laboratory for Molecular Sciences,
Institute of Theoretical and Computational Chemistry,
College of Chemistry and Molecular Engineering, Peking University,
Haidian District, Beijing, China
jianliupku@pku.edu.cn

The chapter reviews our recent progress in phase space mapping theory for nonadiabatic systems. We first show a unified phase space formulation, where continuous nuclear degrees of freedom are mapped onto phase space with infinite boundaries in the conventional way, but discrete electronic-state degrees of freedom should be mapped onto constraint phase space. The well-known Meyer–Miller mapping Hamiltonian is intrinsically a special case of the comprehensive mapping phase function for the molecular Hamiltonian. Trajectory-based phase space approaches are then developed, which outperform both Ehrenfest dynamics and surface hopping in several challenging models.

*Both authors contributed equally.

Time-Dependent Density Functional Theory: Nonadiabatic Molecular Dynamics
Edited by Chaoyuan Zhu

ISBN 978-981-4968-42-3 (Hardcover), 978-1-003-31921-4 (eBook)
www.jennystanford.com

11.1 Introduction

11.1.1 Nonadiabatic Dynamics in the Wavefunction Picture

The whole chemistry as well as molecular sciences depend on behavior of both electrons and nuclei which are quantum particles in nature [1]. Consider a molecular system that is described by the Hamiltonian operator in the nonrelativistic scheme,

$$\begin{aligned}\hat{H}(\mathbf{R},\mathbf{r}) &= \hat{T}_{\text{nuc}} + \hat{H}_{\text{el}}(\mathbf{r},\mathbf{R}) \\ &= -\sum_{I=1}^{N_a}\frac{\hbar^2\nabla_I^2}{2M_I} - \sum_{i=1}^{N_e}\frac{\hbar^2\nabla_i^2}{2m_e} - \sum_{I=1}^{N_a}\sum_{i=1}^{N_e}\frac{Z_I e^2}{4\pi\varepsilon_0|\mathbf{R}_I-\mathbf{r}_i|} \\ &\quad + \sum_{i>j}^{N_e}\frac{e^2}{4\pi\varepsilon_0|\mathbf{r}_i-\mathbf{r}_j|} + \sum_{I>J}^{N_a}\frac{Z_I Z_J e^2}{4\pi\varepsilon_0|\mathbf{R}_I-\mathbf{R}_J|},\end{aligned} \tag{11.1}$$

where $\hat{T}_{\text{nuc}} = -\sum_{I=1}^{N_a}\hbar^2\nabla_I^2/2M_I$ denotes the nuclear kinetic energy term and the electronic Hamiltonian $H_{\text{el}}(\mathbf{r},\mathbf{R})$ includes the electronic kinetic energy and the Coulombic interactions among electrons and nuclei. In the chapter $\mathbf{R}$ and $\mathbf{r}$ stand for the nuclear and electronic coordinates, respectively.

We first consider the wavefunction representation. The evolution of the molecular system satisfies the time-dependent Schrödinger equation (TDSE), should be written into,

$$i\hbar\frac{\partial}{\partial t}\Psi(\mathbf{R},\mathbf{r},t) = \hat{H}(\mathbf{R},\mathbf{r})\Psi(\mathbf{R},\mathbf{r},t), \tag{11.2}$$

where $\Psi(\mathbf{R},\mathbf{r},t)$ is the total wavefunction for both electrons and nuclei. Because the mass of a nucleus is at least three orders of magnitude heavier than that of an electron, it is natural to expand the total wavefunction $\Psi(\mathbf{R},\mathbf{r},t)$ in the stationary orthonormal electronic basis set $\{\phi_n(\mathbf{r};\mathbf{R})\}$ for each nuclear coordinate $\mathbf{R}$, i.e., as the Born–Huang expansion suggests

$$\Psi(\mathbf{R},\mathbf{r},t) = \sum_n \chi_n(\mathbf{R},t)\phi_n(\mathbf{r};\mathbf{R}). \tag{11.3}$$

The orthonormal electronic basis set $\{\phi_n(\mathbf{r};\mathbf{R})\}$ can be diabatic, adiabatic, or in any other representations. The time-dependent coefficients $\{\chi_n(\mathbf{R},t)\}$ for each nuclear coordinate $\mathbf{R}$ in Eq. (11.3) denote nuclear wavefunctions.

Prevailing electronic structure theory solves the eigen-system problem of the electronic Hamiltonian $\hat{H}_{\rm el}(\mathbf{r}, \mathbf{R})$ with the frozen nuclei approximation (that is, the Born–Oppenheimer approximation) where nuclear coordinate $\mathbf{R}$ is fixed. That is,

$$H_{\rm el}(\mathbf{r}; \mathbf{R})\, \phi_n(\mathbf{r}; \mathbf{R}) = E_n(\mathbf{R})\, \phi_n(\mathbf{r}; \mathbf{R}), \tag{11.4}$$

where n denotes the index of the adiabatic electronic state, and $E_n(\mathbf{R})$ represents the corresponding adiabatic potential energy surface (PES). When the stationary adiabatic electronic basis set is used, substitution of Eq. (11.3) into Eq. (11.2) yields the coupled equations of motion for nuclear wavefunctions

$$\begin{aligned} i\hbar\frac{\partial}{\partial t}\chi_m(\mathbf{R}, t) = & \left[-\sum_{I=1}^{N_a}\frac{\hbar^2\nabla_I^2}{2M_I} + E_m(\mathbf{R})\right]\chi_m(\mathbf{R}, t) \\ & - \sum_n\sum_{I=1}^{N_a}\frac{\hbar^2}{2M_I}\left[2\mathbf{d}_{mn}^{(I)}(\mathbf{R})\cdot\nabla_I + D_{mn}^{(I)}(\mathbf{R})\right]\chi_n(\mathbf{R}, t), \end{aligned} \tag{11.5}$$

where

$$\mathbf{d}_{mn}^{(I)}(\mathbf{R}) = \langle\phi_m|\,\nabla_I\phi_n\rangle = \int d\mathbf{r}\phi_m(\mathbf{R}, \mathbf{r})\,\nabla_I\phi_n(\mathbf{R}, \mathbf{r}), \tag{11.6}$$

and

$$D_{mn}^{(I)}(\mathbf{R}) = \langle\phi_m|\,\nabla_I^2\phi_n\rangle = \int d\mathbf{r}\phi_m(\mathbf{R}, \mathbf{r})\,\nabla_I^2\phi_n(\mathbf{R}, \mathbf{r}), \tag{11.7}$$

represent the first and second order nonadiabatic coupling between the m-th and n-th electronic states for the I-th nucleus, respectively. The bracket inner product stands for the integration over electronic coordinate $\mathbf{r}$.

Use the notation of the momentum operator, $\hat{P}_I = -i\hbar\nabla_I$. The matrix element of the total Hamiltonian (Eq. (11.1)) in the adiabatic electronic basis set, as expressed on the right-handed side of Eq.(11.5), is

$$\begin{aligned} \left[\hat{H}(\mathbf{R}, \mathbf{P})\right]_{mn} = & \left[\sum_{I=1}^{N_a}\frac{\hat{P}_I^2}{2M_I} + E_m(\hat{\mathrm{R}})\right]\delta_{mn} \\ & - i\hbar\sum_{I=1}^{N_a}\mathbf{d}_{mn}^{(I)}(\hat{\mathrm{R}})\cdot\frac{\hat{P}_I}{M_I} - \sum_{I=1}^{N_a}\frac{\hbar^2 D_{mn}^{(I)}(\hat{\mathrm{R}})}{2M_I} \\ = & \;\delta_{mn}\sum_{I=1}^{N_a}\frac{\hat{P}_I^2}{2M_I} + V_{mn}(\hat{\mathrm{R}}, \hat{\mathrm{P}}), \end{aligned} \tag{11.8}$$

The number of electronic states under investigation, F, is often finite. Recent advance in the Kohn–Sham density functional theory (DFT) [2] as well as time-dependent DFT (TDDFT) [3] offers useful tools for studying nonadiabatic chemistry for (large) molecular systems.

Equation (11.5) indicates that the nonadiabatic coupling terms entangle stationary electronic wavefunctions with nuclear equations of motion. This leads to the population transfer and energy exchange, as well as branching and coherence among nuclear wavefunctions.

11.1.2 Nonadiabatic Dynamics with the Density Operator

The eigen energy levels of nuclear degrees of freedom (DOFs) are dense, especially in large molecular systems. The difference between two neighbor vibrational/rotational energy levels is typically on the order of wavenumber (cm^{-1}) or less. The initial condition of nuclear DOFs is often a mixed state rather than a pure state. The density operator is more convenient than the wavefunction representation to describe behavior of nuclei in large molecular systems.

The equations of motion of the total density operator, $\hat{\rho}(t)$, including both nuclear and electronic DOFs, are governed by the quantum Liouville theorem,

$$\frac{\partial}{\partial t}\hat{\rho}(t) = -\frac{1}{i\hbar}[\hat{\rho}(t), \hat{H}(\hat{\mathrm{R}}, \hat{\mathrm{P}})]. \tag{11.9}$$

It is often difficult, if not impossible, to utilize numerically exact methods for solving Eq. (11.5) or Eq. (11.9) for experiment-related anharmonic systems. A great deal of effort for developing practical methodologies employs the concept of the trajectory.

A common strategy is to use Wigner phase space only for continuous nuclear DOFs while keeping electronic state DOFs discrete in Eq. (11.9). When the second order term of $\hbar$ as well as higher-order ones for nuclear DOFs are ignored, such an approximation leads to the so-called mixed quantum-classical Liouville equation (MQCL) [4]. However, faithful implementation of MQCL often meets numerical problems [5]. Below we show a substantially different strategy to tackle the problem.

11.2 Unified Phase Space Formulation for both Nuclear and Electronic Freedoms

11.2.1 Meyer–Miller Mapping Hamiltonian Model

The Meyer–Miller mapping Hamiltonian model as a classical analogy proposed in 1979 suggests that it is possible to treat both nuclear and electronic DOFs in the same fashion in dynamics [6]. The model was shortly successfully applied to electronic-to-rotational or electronic-to-vibrational resonance energy transfer in nonadiabatic collision reactions. In 1997 Stock and Thoss [7] further showed that the Meyer–Miller model is an exact Hamiltonian model by utilizing Schwinger's angular momentum theory of mapping the electronic basis onto the singly excited oscillator (SEO), i.e., when the mapping

$$|\phi_n\rangle\langle\phi_m| \mapsto \hat{a}_n^{\dagger}\hat{a}_m = \frac{1}{2}(\hat{x}^{(n)} - i\hat{p}^{(n)})(\hat{x}^{(m)} + i\hat{p}^{(m)}) \tag{11.10}$$

is introduced, Eq. (11.8) becomes

$$\begin{aligned} H_{map}(\hat{\mathrm{R}}, \hat{\mathrm{P}}, \hat{\mathbf{x}}, \hat{\mathbf{p}}) &= \sum_{I=1}^{N_a} \frac{\hat{P}_I^2}{2M_I} + \sum_{n,m=1}^{F} \frac{1}{2}(\hat{x}^{(n)} - i\hat{p}^{(n)})(\hat{x}^{(m)} + i\hat{p}^{(m)})V_{nm}\left(\hat{\mathbf{R}}, \hat{\mathbf{P}}\right) \\ &= \sum_{I=1}^{N_a} \frac{\hat{P}_I^2}{2M_I} + \sum_{n,m=1}^{F} \left[\frac{1}{2}(\hat{x}^{(n)}\hat{x}^{(m)} + \hat{p}^{(n)}\hat{p}^{(m)}) - \gamma\right] V_{nm}\left(\hat{\mathbf{R}}, \hat{\mathbf{P}}\right). \end{aligned} \tag{11.11}$$

Here $\gamma = 1/2$ comes from the conventional commutation relation

$$\left[\hat{x}^{(m)}, \hat{p}^{(n)}\right] = i\delta_{mn}. \tag{11.12}$$

Dynamics generated by the quantum Hamiltonian (Eq. (11.11)) is then in principle exact. Parameter γ of Eq. (11.11) is often viewed as a parameter for the zero point energy (ZPE) of the underlying SEO. In practice, parameter $\gamma = 1/3$ or $(\sqrt{F+1} - 1)/F$ is chosen in semi-classical/quasi-classical trajectory applications. More recently, Cotton and Miller have proposed the symmetric quasi-classical dynamics with the window function and demonstrated a few useful applications [8].

11.2.2 Unified Formulation of Mapping Phase Space

It is straightforward to implement traditional phase space theory for the Hamiltonian operator (Eq. (11.11)) such that both nuclear and electronic state DOFs are expressed by phase space variables with no boundaries. This, however, is inconsistent with the constraint implied in the conventional Meyer–Miller mapping Hamiltonian (Eq. (11.11)),

$$\mathcal{S}(\mathbf{x},\mathbf{p}) = \delta\left(\sum_{n=1}^{F}\frac{\left(x^{(n)}\right)^2+\left(p^{(n)}\right)^2}{2} - F\gamma - 1\right). \tag{11.13}$$

The subtle point was first explicitly illustrated by He and Liu in Ref. [9]. When a fundamentally different viewpoint is introduced by Liu for constructing the mapping Hamiltonian in Cartesian phase space [10, 11], the model reminiscent of the conventional Meyer–Miller model reads

$$H_{map}(\hat{\mathbf{R}},\hat{\mathbf{P}},\hat{\mathbf{x}},\hat{\mathbf{p}}) = \sum_{n,m=1}^{F}\left[\frac{1}{2}(\hat{x}^{(n)}\hat{x}^{(m)}+\hat{p}^{(n)}\hat{p}^{(m)})-\gamma\right]H_{nm}\left(\hat{\mathbf{R}},\hat{\mathbf{P}}\right). \tag{11.14}$$

Comparison of Eq. (11.14) to Eq. (11.11) suggests that Eq. (11.13) is the constraint condition implicitly used for the kinetic energy term $\hat{T}_{\mathrm{nuc}} = \sum_{I=1}^{N_a}\frac{\hat{P}_I^2}{2M_I}$ [9, 12, 13].

The simplest way is to use the phase space defined by the constraint condition. That is, the mapping phase space for discrete electronic-state DOFs is constructed on constraint phase space $\mathcal{S}(\mathbf{x},\mathbf{p})$ of Eq. (11.13). Below we introduce a general formulation of mapping both continuous nuclear DOFs and discrete electronic state DOFs onto phase space.

We first define the mapping kernel (and inverse kernel) from quantum operators to phase space functions [12], where the one-to-one correspondence is

$$\mathrm{Tr}\left[\hat{A}\hat{B}\right] = \int\frac{d\mathbf{R}d\mathbf{P}}{(2\pi\hbar)^N}\int_{\mathcal{S}(\mathbf{x},\mathbf{p})}F\,d\mathbf{x}d\mathbf{p}\,A\left(\mathbf{R},\mathbf{P},\mathbf{x},\mathbf{p}\right)\tilde{B}\left(\mathbf{R},\mathbf{P},\mathbf{x},\mathbf{p}\right). \tag{11.15}$$

with

$$\begin{aligned}A\left(\mathbf{R},\mathbf{P};\mathbf{x},\mathbf{p}\right) &= \mathrm{Tr}_{n,e}\left[\hat{A}\hat{K}_{nuc}\left(\mathbf{R},\mathbf{P}\right)\otimes\hat{K}_{ele}\left(\mathbf{x},\mathbf{p}\right)\right]\\ \tilde{B}\left(\mathbf{R},\mathbf{P};\mathbf{x},\mathbf{p}\right) &= \mathrm{Tr}_{n,e}\left[\hat{K}_{nuc}^{-1}\left(\mathbf{R},\mathbf{P}\right)\otimes\hat{K}_{ele}^{-1}\left(\mathbf{x},\mathbf{p}\right)\hat{B}\right].\end{aligned} \tag{11.16}$$

The one-to-one correspondence inverse mapping from $A(\mathbf{R}, \mathbf{P}; \mathbf{x}, \mathbf{p})$ (or $\tilde{B}(\mathbf{R}, \mathbf{P}; \mathbf{x}, \mathbf{p})$) to operator $\hat{A}$ (or $\hat{B}$) is

$$\hat{A} = \int (2\pi\hbar)^{-N} \mathrm{d}\mathbf{R}\mathrm{d}\mathbf{P} \int_{\mathcal{S}(\mathbf{x},\mathbf{p})} F\mathrm{d}\mathbf{x}\mathrm{d}\mathbf{p}\ A(\mathbf{R}, \mathbf{P}; \mathbf{x}, \mathbf{p})\, \hat{K}_{nuc}^{-1}(\mathbf{R}, \mathbf{P}) \otimes \hat{K}_{ele}^{-1}(\mathbf{x}, \mathbf{p})$$
$$\hat{B} = \int (2\pi\hbar)^{-N} \mathrm{d}\mathbf{R}\mathrm{d}\mathbf{P} \int_{\mathcal{S}(\mathbf{x},\mathbf{p})} F\mathrm{d}\mathbf{x}\mathrm{d}\mathbf{p}\ \tilde{B}(\mathbf{R}, \mathbf{P}; \mathbf{x}, \mathbf{p})\, \hat{K}_{nuc}(\mathbf{R}, \mathbf{P}) \otimes \hat{K}_{ele}(\mathbf{x}, \mathbf{p}). \tag{11.17}$$

Here, $(2\pi\hbar)^{-N} d\mathbf{R}d\mathbf{P} \otimes F d\mathbf{x}d\mathbf{p}$ represents the invariant measure on the mapping phase space for the nuclear and electronic state DOFs, and Tr_n and Tr_e stand for the trace over the nuclear DOFs and that over the F electronic states, respectively. Provided that Eq. (11.16) and Eq. (11.17) hold, the one-to-one correspondence can be established between quantum Hilbert space and mapping phase space, including both nuclear DOFs $(\mathbf{R}, \mathbf{P})$ and electronic DOFs $(\mathbf{x}, \mathbf{p})$.

Because nuclear space is continuous in the Hamiltonian operator defined by Eq. (11.8), it is routine to employ the unified scheme [14, 15], in which mapping kernel $\hat{K}_{nuc}(\mathbf{R}, \mathbf{P})$ is

$$\hat{K}_{nuc}(\mathbf{R}, \mathbf{P}) = \left(\frac{\hbar}{2\pi}\right)^N \int \mathrm{d}\boldsymbol{\zeta} \int \mathrm{d}\boldsymbol{\eta}\ e^{i\boldsymbol{\zeta}\cdot(\hat{\mathbf{R}}-\mathbf{R})+i\boldsymbol{\eta}\cdot(\hat{\mathbf{P}}-\mathbf{P})} f(\boldsymbol{\zeta}, \boldsymbol{\eta}), \tag{11.18}$$

and the corresponding inverse kernel is

$$\hat{K}_{nuc}^{-1}(\mathbf{R}, \mathbf{P}) = \left(\frac{\hbar}{2\pi}\right)^N \int \mathrm{d}\boldsymbol{\zeta} \int \mathrm{d}\boldsymbol{\eta}\ e^{i\boldsymbol{\zeta}\cdot(\hat{\mathbf{R}}-\mathbf{R})+i\boldsymbol{\eta}\cdot(\hat{\mathbf{P}}-\mathbf{P})} \left[f(-\boldsymbol{\zeta}, -\boldsymbol{\eta})\right]^{-1}. \tag{11.19}$$

When scalar function $f(\boldsymbol{\zeta}, \boldsymbol{\eta}) = 1$, Wigner phase space [16] is produced [12]. Regardless of which scalar function is used in Eqs. (11.18) and (11.19), nuclear phase space has infinite boundaries.

In comparison, the integral for the mapping phase variables for the F electronic states is over the hypersurface function (or weighting function), $\mathcal{S}(\mathbf{x}, \mathbf{p})$. The integral for electronic-state mapping phase space in Eq. (11.17) is performed as

$$\int_{\mathcal{S}(\mathbf{x},\mathbf{p})} F\mathrm{d}\mathbf{x}\mathrm{d}\mathbf{p}\ g(\mathbf{x}, \mathbf{p}) = \frac{\int F\mathrm{d}\mathbf{x}\mathrm{d}\mathbf{p}\mathcal{S}(\mathbf{x}, \mathbf{p})\, g(\mathbf{x}, \mathbf{p})}{\int \mathrm{d}\mathbf{x}\mathrm{d}\mathbf{p}\mathcal{S}(\mathbf{x}, \mathbf{p})}. \tag{11.20}$$

It is evident that a different definition of $\mathcal{S}(\mathbf{x}, \mathbf{p})$ leads to a different mapping phase space approach. In the chapter, we use Eq. (11.13) for constraint space $\mathcal{S}(\mathbf{x}, \mathbf{p})$. As proved in Appendix A of Ref.

[9] and the Supporting Information of Ref. [12], the form of the mapping Hamiltonian (Eq. (11.14)) hints that the mapping kernel and corresponding inverse kernel for the F discrete electronic states are

$$\hat{K}_{ele}(\mathbf{x},\mathbf{p}) = \sum_{n,m=1}^{F}\left[\frac{1}{2}\left(x^{(n)}+ip^{(n)}\right)\left(x^{(m)}-ip^{(m)}\right)-\gamma\delta_{nm}\right]|n\rangle\langle m|$$
$$\hat{K}_{ele}^{-1}(\mathbf{x},\mathbf{p}) = \sum_{n,m=1}^{F}\left[\frac{1+F}{2(1+F\gamma)^2}\left(x^{(n)}+ip^{(n)}\right)\left(x^{(m)}-ip^{(m)}\right)-\frac{1-\gamma}{1+F\gamma}\delta_{nm}\right]|n\rangle\langle m|, \tag{11.21}$$

which meet the normalized conditions,

$$\mathrm{Tr}_e\left[\hat{K}_{ele}(\mathbf{x},\mathbf{p})\right] = \mathrm{Tr}_e\left[\hat{K}_{ele}^{-1}(\mathbf{x},\mathbf{p})\right] = 1,$$
$$\int_{\mathcal{S}(\mathbf{x},\mathbf{p})} F\mathrm{d}\mathbf{x}\mathrm{d}\mathbf{p}\hat{K}_{ele}(\mathbf{x},\mathbf{p}) = \int_{\mathcal{S}(\mathbf{x},\mathbf{p})} F\mathrm{d}\mathbf{x}\mathrm{d}\mathbf{p}\hat{K}_{ele}^{-1}(\mathbf{x},\mathbf{p}) = \hat{I}_{ele} \tag{11.22}$$

and

$$\mathrm{Tr}\left[|\rangle\langle m|\,k\rangle\langle l|\right] = \delta_{mk}\delta_{nl}$$
$$= \int_{\mathcal{S}(\mathbf{x},\mathbf{p})} F\,\mathrm{d}\mathbf{x}\mathrm{d}\mathbf{p}\left[\hat{K}_{ele}(\mathbf{x},\mathbf{p})\right]_{nm}\left[\hat{K}_{ele}^{-1}(\mathbf{x},\mathbf{p})\right]_{kl}. \tag{11.23}$$

The value of parameter γ lies in $\left(-\frac{1}{F},+\infty\right)$, which means that parameter γ can be negative and that it is *not* consistent at all to view γ as the zero-point-energy parameter as done in the conventional way [10, 12].

11.3 Trajectory-Based Approaches

11.3.1 Extended Classical Mapping Model

Most dynamical properties can be expressed by the correlation function, $C_{AB}(t) = \mathrm{Tr}\left[\hat{A}e^{i\hat{H}t/\hbar}\hat{B}e^{-i\hat{H}t/\hbar}\right] = \mathrm{Tr}\left[\hat{A}(0)\hat{B}(t)\right]$. When both nuclear and electronic DOFs are involved, the correlation function can be reformulated as

$$C_{AB}(t) = \int\frac{\mathrm{d}\mathbf{R}\mathrm{d}\mathbf{P}}{(2\pi\hbar)^N}\int_{\mathcal{S}(\mathbf{x},\mathbf{p})} F\mathrm{d}\mathbf{x}\mathrm{d}\mathbf{p}A(\mathbf{R},\mathbf{P},\mathbf{x},\mathbf{p};0)\tilde{B}(\mathbf{R},\mathbf{P},\mathbf{x},\mathbf{p};t). \tag{11.24}$$

Here, $A(\mathbf{R}, \mathbf{P}, \mathbf{x}, \mathbf{p})$ for $\hat{A}(0)$ and $\tilde{B}(\mathbf{R}, \mathbf{P}, \mathbf{x}, \mathbf{p}; t)$ for $\hat{B}(t)$ are given by Eq. (11.16). Equation (11.24) is an exact formulation. It is, however, difficult to obtain the exact solution of nuclear dynamics for general molecular systems. For instance, the quantum Liouville theorem, Eq. (11.9), can be expressed in mapping phase space for both nuclear and electronic-state DOFs, but its numerically exact solution is often intractable.

An approximation is to replace $\tilde{B}(\mathbf{R}, \mathbf{P}, \mathbf{x}, \mathbf{p}; t)$ by $\tilde{B}(\mathbf{R}_t, \mathbf{P}_t, \mathbf{x}_t, \mathbf{p}_t)$ in Eq. (11.24), where trajectory $\{\mathbf{R}_t, \mathbf{P}_t, \mathbf{x}_t, \mathbf{p}_t\}$ is generated to conserve the corresponding mapping Hamiltonian. When Eq. (11.14) is used, the approach is denoted as the extended classical mapping model (eCMM) method. This is closely related to the linearized semi-classical initial value representation (LSC-IVR) approach [17, 18] for treating *only* the nuclear DOFs.

As suggested by Liu and coworkers in Refs. [10, 13, 19], a more comprehensive mapping Hamiltonian takes the form,

$$H_{map} = \sum_{n,m=1}^{F} H_{nm}(\mathbf{R}, \mathbf{P}) \left(\frac{1}{2} \left(x^{(n)} x^{(m)} + p^{(n)} p^{(m)} \right) - \Gamma_{nm} \right), \tag{11.25}$$

where Γ_{nm} is the element of matrix $\mathbf{\Gamma}$ and comes from the general commutation relation

$$\left[\hat{x}^{(n)}, \hat{p}^{(m)} \right] + \left[\hat{x}^{(m)}, \hat{p}^{(n)} \right] = 2i \left(\Gamma_{nm} + \Gamma_{mn} \right), \tag{11.26}$$

Matrix $\mathbf{\Gamma}$ denotes the commutator matrix. It is trivial to see that Eq. (11.12) is a special case of Eq. (11.26) and that commutator matrix $\mathbf{\Gamma}$ is Hermitian. Similar to Eq. (11.13), the constraint applied to the kinetic energy term $\hat{T}_{\text{nuc}} = \sum_{I=1}^{N_a} \frac{\hat{P}_I^2}{2M_I}$ satisfies

$$\sum_{n=1}^{F} \left(\frac{1}{2} \left(x^{(n)} x^{(n)} + p^{(n)} p^{(n)} \right) - \Gamma_{nn} \right) = 1. \tag{11.27}$$

Commutator matrix $\mathbf{\Gamma}$ can be represented as

$$\mathbf{\Gamma} = \sum_{k=1}^{F} \lambda_k \mathbf{c}_k \mathbf{c}_k^{\dagger}, \tag{11.28}$$

where λ_k and $\mathbf{c}_k$ are k-th (real) eigenvalue and corresponding (column) eigenvector, respectively. By defining $\mathbf{c}_k = \left(\tilde{\mathbf{x}}_k + i \tilde{\mathbf{p}}_k \right) / \sqrt{2|\lambda_k|}$, we express Eq. (11.28) as

$$\mathbf{\Gamma} = \sum_{k=1}^{F} \frac{s_k}{2} \left(\tilde{\mathbf{x}}_k + i\tilde{\mathbf{p}}_k\right)\left(\tilde{\mathbf{x}}_k^T - i\tilde{\mathbf{p}}_k^T\right). \tag{11.29}$$

Here, $s_k = \operatorname{sgn}(\lambda_k)$, $\tilde{\mathbf{x}}_k = \left(\tilde{x}_k^{(1)}, \cdots, \tilde{x}_k^{(F)}\right)^T$ and $\tilde{\mathbf{p}}_k = \left(\tilde{p}_k^{(1)}, \cdots, \tilde{p}_k^{(F)}\right)^T$ are real-valued vectors. The comprehensive Hamiltonian Eq. (11.25) is then recast into

$$\begin{aligned} H_{map} = & \sum_{n,m=1}^{F} H_{nm}(\mathbf{R}, \mathbf{P}) \\ & \times \left(\frac{x^{(n)}x^{(m)} + p^{(n)}p^{(m)}}{2} - \sum_{k=1}^{F} \frac{s_k}{2}\left(\tilde{x}_k^{(n)}\tilde{x}_k^{(m)} + \tilde{p}_k^{(n)}\tilde{p}_k^{(m)}\right)\right). \end{aligned} \tag{11.30}$$

It can be used to generate trajectory-based dynamics for mapping variables $\{\mathbf{x}, \mathbf{p}\}$ as well as commutator variables $\{\tilde{\mathbf{x}}_k, \tilde{\mathbf{p}}_k\}$ for electronic-state DOFs. In many cases, only one electronic state of the system is occupied before the coupling region is encountered. Short-time dynamics then approaches the Born–Oppenheimer limit because the coupling is effectively zero around the beginning [13, 20]. The Born–Oppenheimer limit is satisfied at time zero when the initial condition of commutator variables satisfies

$$\begin{aligned} \xi\left(\tilde{\mathbf{x}}(0), \tilde{\mathbf{p}}(0)\right): \prod_{k,n=1}^{F} \delta & \left(\frac{s_k}{2}\left[\left(\tilde{x}_k^{(n)}(0)\right)^2 + \left(\tilde{p}_k^{(n)}(0)\right)^2\right]\right. \\ & \left. - \left[\frac{\left(x^{(n)}(0)\right)^2 + \left(p^{(n)}(0)\right)^2}{2} - \delta_{n,j_{\mathrm{occ}}}\right]\delta_{nk}\right), \end{aligned} \tag{11.31}$$

where j_{occ} denotes the index of initially occupied state. It should be noted that we adopt the same mapping kernel that is irrelevant to $\mathbf{\Gamma}$, for phase space variables for the electronic-state DOFs. That is, the initial condition for electronic-state mapping variables is sampled from Eq. (11.13), and the formulation, Eq. (11.24), is used for calculation of the correlation function. The only approximation is to replace $\tilde{B}(\mathbf{R}, \mathbf{P}, \mathbf{x}, \mathbf{p}; t)$ by $\tilde{B}(\mathbf{R}_t, \mathbf{P}_t, \mathbf{x}_t, \mathbf{p}_t)$ where trajectory $\{\mathbf{R}_t, \mathbf{P}_t, \mathbf{x}_t, \mathbf{p}_t\}$ is generated to preserve the comprehensive mapping

Hamiltonian, Eq. (11.25). Such trajectory-based dynamics is an approximation to the quantum Liouville theorem in mapping phase space. We denote such an approach the extended classical mapping model with commutator variables (eCMMcv).

11.3.2 Equations of Motion in the Adiabatic Representation

It is trivial to show $D_{nm}^{(I)}(\mathbf{R}) = \sum_{k=1}^{F} \mathbf{d}_{nk}^{(I)} \cdot \mathbf{d}_{km}^{(I)} + \nabla_{\mathbf{R}_I} \cdot \mathbf{d}_{nm}^{(I)}$. The effective potential matrix element, $V_{mn}(\hat{\mathbf{R}}, \hat{\mathbf{P}})$, of Eq. (11.8) is recast into

$$V_{mn}(\hat{\mathbf{R}}, \hat{\mathbf{P}}) = E_n\left(\hat{\mathbf{R}}\right)\delta_{mn} - i\hbar\sum_{I=1}^{N_a} \mathbf{d}_{mn}^{(I)}\left(\hat{\mathbf{R}}\right) \cdot \frac{\hat{\mathbf{P}}_I}{2M_I} - i\hbar\sum_{I=1}^{N_a} \frac{\hat{\mathbf{P}}_I}{2M_I} \cdot \mathbf{d}_{mn}^{(I)}\left(\hat{\mathbf{R}}\right) - \sum_{I=1}^{N_a}\sum_{k=1}^{F} \frac{\hbar^2 \mathbf{d}_{mk}^{(I)}\left(\hat{\mathbf{R}}\right) \cdot \mathbf{d}_{kn}^{(I)}\left(\hat{\mathbf{R}}\right)}{2M_I}. \tag{11.32}$$

The Wigner–Weyl correspondence (in Wigner phase space for nuclear DOFs) leads to

$$i\hbar\sum_{I=1}^{N_a} \mathbf{d}_{mn}^{(I)}\left(\hat{\mathbf{R}}\right) \cdot \frac{\hat{\mathbf{P}}_I}{2M_I} + i\hbar\sum_{I=1}^{N_a} \frac{\hat{\mathbf{P}}_I}{2M_I} \cdot \mathbf{d}_{mn}^{(I)}\left(\hat{\mathbf{R}}\right) \mapsto i\hbar\sum_{I=1}^{N_a} \frac{\mathbf{P}_I}{M_I} \cdot \mathbf{d}_{mn}^{(I)}(\mathbf{R}). \tag{11.33}$$

The potential matrix element is then de-quantized to

$$V_{mn}(\mathbf{R}, \mathbf{P}) = \sum_{I=1}^{N_a}\left(-\frac{i\hbar}{M_I}\mathbf{P}_I \cdot \mathbf{d}_{mn}^{(I)}(\mathbf{R}) - \frac{\hbar^2}{2M_I}\sum_{k=1}^{F} \mathbf{d}_{mk}^{(I)} \cdot \mathbf{d}_{kn}^{(I)}(\mathbf{R})\right) + E_n(\mathbf{R})\delta_{mn}. \tag{11.34}$$

It is convenient to use the matrix form, e.g.,

$$\mathbf{d}^{(I)}(\mathbf{R}) = \sum_{n,m=1}^{F} \mathbf{d}_{mn}^{(I)}(\mathbf{R}) \left|\phi_m\right\rangle \left\langle\phi_n\right|, \tag{11.35}$$

$$\mathbf{I}_{ele} = \sum_{n=1}^{F} \left|\phi_n\right\rangle \left\langle\phi_n\right|, \tag{11.36}$$

and

$$\mathbf{E}(\mathbf{R}) = \sum_{n=1}^{F} E_n(\mathbf{R}) \left|\phi_n\right\rangle \left\langle\phi_n\right|. \tag{11.37}$$

Define the (Hermitian) kinematic momentum matrix

$$\mathbf{\Pi}_I\left(\mathbf{R},\mathbf{P}\right) \triangleq \mathbf{P}_I \mathbf{I}_{ele} - i\hbar \mathbf{d}^{(I)}\left(\mathbf{R}\right). \tag{11.38}$$

The matrix form of Eq. (11.34) is then

$$\mathbf{V}(\mathbf{R},\mathbf{P}) = \sum_{m,n=1}^{F} V_{mn}(\mathbf{R},\mathbf{P}) \left|\phi_m\right\rangle \left\langle\phi_n\right|. \tag{11.39}$$

Finally, the matrix form of the Hamiltonian of Eq. (11.8) becomes

$$\mathbf{H}_{map}(\mathbf{R},\mathbf{P}) = \sum_{I=1}^{N_a} \frac{\mathbf{P}_I \cdot \mathbf{P}_I}{2M_I} + \mathbf{V}(\mathbf{R},\mathbf{P}). \tag{11.40}$$

or equivalently

$$\mathbf{H}_{map}(\mathbf{R},\mathbf{P}) = \sum_{I=1}^{N_a} \frac{\mathbf{\Pi}_I \cdot \mathbf{\Pi}_I}{2M_I} + \mathbf{E}\left(\mathbf{R}\right). \tag{11.41}$$

Note that only the first-order nonadiabatic coupling term is requested in Eq. (11.34) or Eq. (11.39). We adopt the strategy of using the mapping form of the kinematic momentum $\mathbf{P}_{kin,I}\left(\mathbf{R},\mathbf{P},\mathbf{x},\mathbf{p},\mathbf{\Gamma}\right) = \mathrm{Tr}\left[\mathbf{\Pi}_I\left(\mathbf{R},\mathbf{P}\right)\left(\frac{1}{2}\left(\mathbf{x}+i\mathbf{p}\right)\left(\mathbf{x}^T - i\mathbf{p}^T\right) - \mathbf{\Gamma}\right)\right]$ as proposed in Ref. [13, 21, 40].

The equations of motion read

$$\begin{aligned}
\dot{\mathbf{R}}_I &= \frac{\mathbf{P}_{kin,I}}{M_I} \\
\dot{\mathbf{P}}_{kin,I} &= -\sum_{n,m=1}^{F}\left[\frac{1}{2}\left(x^{(n)}x^{(m)} + p^{(n)}p^{(m)}\right) - \sum_{k=1}^{F}\frac{s_k}{2}\left(\tilde{x}_k^{(n)}\tilde{x}_k^{(m)} + \tilde{p}_k^{(n)}\tilde{p}_k^{(m)}\right)\right] \\
&\quad \times \left[\nabla_{\mathbf{R}_I} E_n(\mathbf{R})\delta_{mn} + \left(E_n(\mathbf{R}) - E_m(\mathbf{R})\right)\mathbf{d}_{mn}^{(I)}(\mathbf{R})\right]
\end{aligned} \tag{11.42}$$

for nuclear DOFs,

$$\begin{aligned}
\dot{x}^{(m)} &= \hbar^{-1} E_m\left(\mathbf{R}\right) p^{(m)} - \sum_{n=1}^{F}\sum_{I=1}^{N_a} \frac{\mathbf{P}_{kin,I}}{M_I} \cdot \mathbf{d}_{mn}^{(I)}\left(\mathbf{R}\right) x^{(n)} \\
\dot{p}^{(m)} &= -\hbar^{-1} E_m\left(\mathbf{R}\right) x^{(m)} - \sum_{n=1}^{F}\sum_{I=1}^{N_a} \frac{\mathbf{P}_{kin,I}}{M_I} \cdot \mathbf{d}_{mn}^{(I)}\left(\mathbf{R}\right) p^{(n)}
\end{aligned} \tag{11.43}$$

for mapping phase variables for electronic-state DOFs, and

$$\begin{aligned}
\dot{\tilde{x}}_k^{(m)} &= -s_k \hbar^{-1} E_m\left(\mathbf{R}\right) \tilde{p}_k^{(m)} + s_k \sum_{n=1}^{F}\sum_{I=1}^{N_a} \frac{\mathbf{P}_{kin,I}}{M_I} \cdot \mathbf{d}_{mn}^{(I)}\left(\mathbf{R}\right) \tilde{x}_k^{(n)} \\
\dot{\tilde{p}}_k^{(m)} &= s_k \hbar^{-1} E_m\left(\mathbf{R}\right) \tilde{x}_k^{(m)} + s_k \sum_{n=1}^{F}\sum_{I=1}^{N_a} \frac{\mathbf{P}_{kin,I}}{M_I} \cdot \mathbf{d}_{mn}^{(I)}\left(\mathbf{R}\right) \tilde{p}_k^{(n)}
\end{aligned} \tag{11.44}$$

for auxiliary mapping variables for the commutator matrix.

Define $\mathbf{g}(t) = \mathbf{x}(t) + i\mathbf{p}(t)$ and the electronic propagator

$$\mathbf{U}(\mathbf{R}, \mathbf{P}_{kin}; \Delta t) = \exp\left[-i\Delta t \mathbf{V}^{\text{eff}}(\mathbf{R}, \mathbf{P}_{kin})/\hbar\right] \quad (11.45)$$

in each step, where Δt is the time step size and the effective potential element defined as $\mathbf{V}^{\text{eff}}_{mn}(\mathbf{R}, \mathbf{P}_{kin}) = E_n(\mathbf{R})\delta_{mn} - i\hbar \sum_{I=1}^{N_a} \frac{\mathbf{P}_{kin,I}}{M_I} \cdot \mathbf{d}^{(I)}_{mn}(\mathbf{R})$. Equations (11.43)—(11.44) can be reformulated in a more compact form

$$\begin{aligned} \mathbf{g}(t+\Delta t) &= \mathbf{U}(\mathbf{R}, \mathbf{P}_{kin}; \Delta t)\,\mathbf{g}(t) \\ \boldsymbol{\Gamma}(t+\Delta t) &= \mathbf{U}(\mathbf{R}, \mathbf{P}_{kin}; \Delta t)\,\boldsymbol{\Gamma}(t)\,\mathbf{U}^{\dagger}(\mathbf{R}, \mathbf{P}_{kin}; \Delta t)\,. \end{aligned} \quad (11.46)$$

In the eCMMcv approach, the initial condition of commutator matrix is

$$\Gamma_{nm}(0) = (|g_n(0)|^2 - \delta_{n,j_{occ}})\delta_{nm}, \quad (11.47)$$

where $\boldsymbol{\Gamma}(t)$ varies with time.

We note that Eqs. (11.42)–(11.44) are identical to the classical Hamilton equations of Eq. (11.30) in the adiabatic representation [40]. Equation (11.42) for nuclear dynamics requires only the first-order nonadiabatic coupling term, as well as the adiabatic electronic potential energy and its first-order derivative. These elements are often available in standard quantum chemistry packages. So the phase space mapping approach is in principle practical for the on-the-fly simulation for nonadiabatic processes of realistic molecular systems.

Regardless of which representation of the electronic basis is employed in nonadiabatic quantum dynamics, once the initial condition is provided, results for dynamical observables remain the same. In a consistent approximate nonadiabatic method, the equations of motion for such as nuclei should be independent of the representation of choice. It is not difficult to show that the important criterion is satisfied in the phase space mapping approach of the chapter.

11.3.3 Ehrenfest Dynamics and Surface Hopping

Perhaps the simplest approach for nonadiabatic dynamics is Ehrenfest dynamics, which is also called mean field dynamics and whose essential idea can be traced back to the Ehrenfest theorem in quantum mechanics [22, 23]. Ehrenfest dynamics employs the ansatz of total wavefunction

$$\Psi(\mathbf{R},\mathbf{r},t)=\sum_n c_n(t)\,\phi_n(\mathbf{R}(t),\mathbf{r}), \tag{11.48}$$

where complex-valued $c_n(t)$ is the electronic amplitude of the nth state, of which the evolution is determined by the TDSE projection,

$$i\hbar\dot{c}_n=\sum_m\left(E_m(\mathbf{R})\,\delta_{nm}-i\hbar\sum_I\frac{\mathbf{d}_{nm}^{(I)}(\mathbf{R})\cdot\mathbf{P}_I}{M_I}\right)c_m. \tag{11.49}$$

The equations of motion of nuclear DOFs in Ehrenfest dynamics read

$$\begin{aligned}\dot{\mathbf{R}}_I&=M_I^{-1}\mathbf{P}_I\\ \dot{\mathbf{P}}_I&=\langle\Psi(\mathbf{R},t)|\,\nabla_{\mathbf{R}_I}H_{\text{ele}}(\mathbf{R},\hat{\mathbf{r}})\,|\Psi^*(\mathbf{R},t)\rangle\\ &=-\int d\mathbf{r}\Psi^*(\mathbf{R},\mathbf{r},t)\,\nabla_{\mathbf{R}_I}H_{\text{ele}}(\mathbf{R},\mathbf{r})\,\Psi(\mathbf{R},\mathbf{r},t)\\ &=-\sum_n|c_n|^2\nabla_{\mathbf{R}_I}E_n(\mathbf{R})+\sum_{n,m}c_n^*c_m\left[E_n(\mathbf{R})-E_m(\mathbf{R})\right]\mathbf{d}_{mn}^{(I)}.\end{aligned} \tag{11.50}$$

The electronic reduced density matrix is $\rho_{\text{ele}}(t)=\mathbf{c}(t)\,\mathbf{c}^\dagger(t)$.

Another popular nonadiabatic dynamics method is the fewest switch algorithm of surface hopping (FSSH) of Tully [24]. In FSSH the evolution of the electronic amplitude is governed by Eq. (11.43), but nuclei incessantly hop between different adiabatic potential surfaces and the nuclear motion follows Born–Oppenheimer dynamics on the active state. The hopping rate is referred to a net population transfer probability rate during the time interval dt,

$$\omega_{i\to j}=\begin{cases}0, & j\neq i\\ \max\left(-\frac{2\mathrm{Im}\left[c_jc_i^*\left(E_i\delta_{ij}-i\hbar\mathbf{d}_{ij}\cdot\mathbf{M}^{-1}\mathbf{P}\right)\right]}{c_ic_i^*}dt,0\right), & j\neq i\end{cases}, \tag{11.51}$$

which, if necessary, is truncated to fall into [0, 1]. Hopping events can be made by the Monte Carlo algorithm on each step, by generating a uniform random number between [0, 1] and determining which subsection the random number lies in. For example, if the random number falls in $\left[\sum_{j=1}^{k-1}\omega_j,\sum_{j=1}^{k}\omega_j\right]$, then the hopping to the k-th state will be made.

When hopping to a new active state occurs, nuclear momenta should be corrected along the direction of the nonadiabatic coupling vector,

$$\frac{1}{2}\mathbf{P}^T\mathbf{M}^{-1}\mathbf{P}+E_i=\frac{1}{2}\left(\mathbf{P}+\lambda\mathbf{d}_{ij}\right)^T\mathbf{M}^{-1}\left(\mathbf{P}+\lambda\mathbf{d}_{ij}\right)+E_j, \tag{11.52}$$

such that the energy conservation is satisfied. If λ of Eq. (11.52) has no physical solution, such a hopping event is called frustrated hopping and should be abandoned in the original FSSH algorithm. There exist other variants of surface hopping algorithms that have different definitions of the hopping probability, or different ways to deal with the frustrated hopping. In the chapter, we compare eCMMcv to the original FSSH algorithm of Tully as well as the conventional Ehrenfest dynamics to demonstrate the numerical performance. We use Wigner phase space to include nuclear quantum effects in the initial condition for all these trajectory-based methods.

11.4 Applications

Below we present several illustrative applications of phase space mapping approaches in the adiabatic and diabatic representations.

11.4.1 Spin-Boson Model in Condensed Phase

The spin-boson model is a standard model for such as electron transfer processes in condensed phase, which describes a two-level system coupled with a harmonic bath environment. The interaction between the bath and system takes the bilinear form. The spin-boson model includes most key aspects of dissipative nonadiabatic systems.

Most numerically exact methods, including quasi-adiabatic propagator path integral (QuAPI) approach [25, 26], (extend) hierarchical equations of motion [(e)HEOM] [27, 28] and (multi-layer)-multiconfiguration time-dependent Hartree [(ML-)MCTDH] approach [29, 30] employ the diabatic representation to study the spin-boson model. In comparison, it is convenient for trajectory-based approaches to use either the diabatic or adiabatic picture. Because on-the-fly nonadiabatic dynamics involves the adiabatic representation, benchmark tests for the spin-boson model in the adiabatic representation will be useful to shed light on the numerical performance of the trajectory-based method for realistic nonadiabatic systems in condensed phase.

In the diabatic basis, the spin-boson model Hamiltonian includes three parts,

$$\hat{H} = \hat{H}_s + \hat{H}_b + \hat{H}_{sb}, \tag{11.53}$$

with the system part $\hat{H}_s = \varepsilon\hat{\sigma}_z + \Delta\hat{\sigma}_x = \begin{pmatrix} \varepsilon & \Delta \\ \Delta & -\varepsilon \end{pmatrix}$, bath part $\hat{H}_b = \sum_{j=1}^{N_b} \frac{1}{2}\left(\hat{P}_j^2 + \omega_j^2 \hat{R}_j^2\right)$, and the interaction part $\hat{H}_{sb} = \sum_{j=1}^{N_b} c_j \hat{R}_j \hat{\sigma}_z$. Here $\hat{\sigma}_{\mu=x,y,z}$ denote Pauli matrices for spin 1/2. The bath is described by the Ohmic spectrum density with the exponential cutoff,

$$J(\omega) = \frac{\pi}{2}\alpha\omega e^{-\omega/\omega_c}, \tag{11.54}$$

where α denotes the Kondo parameter. It can be represented by a set of discretized bath modes with [38, 39]

$$\begin{aligned} \omega_j &= -\omega_c \ln\left(1 - \frac{j}{N_b + 1}\right), \; j = 1, \ldots, N_b \\ c_j &= \sqrt{\frac{\alpha\omega_c}{N_b + 1}}\omega_j, \; j = 1, \ldots, N_b. \end{aligned} \tag{11.55}$$

We use $N_b = 300$ discretized modes to converge the results.

The corresponding adiabatic potential energy surfaces are

$$E_\pm(\mathbf{R}) = \sum_{j=1}^{N_b} \frac{1}{2}\omega_j^2 R_j^2 \pm \sqrt{\left(\varepsilon + \sum_{j=1}^{N_b} c_j R_j\right)^2 + \Delta^2}. \tag{11.56}$$

Adiabatic states $|\phi_\pm\rangle$ for $E_\pm(\mathbf{R})$ are

$$\begin{pmatrix} |\phi_-\rangle \\ |\phi_+\rangle \end{pmatrix} = \begin{pmatrix} \cos\theta & \sin\theta \\ -\sin\theta & \cos\theta \end{pmatrix} \begin{pmatrix} |1\rangle \\ |2\rangle \end{pmatrix}. \tag{11.57}$$

Here, the mixed angle $\theta(\mathbf{R}) = -\frac{1}{2}\arctan\left(\frac{2V_{12}(\mathbf{R})}{V_{11}(\mathbf{R}) - V_{22}(\mathbf{R})}\right)$ valued in $[0, \pi/2)$ is introduced for simplicity. It is easy to show that the first-order nonadiabatic coupling vector is $\mathbf{d}_{mn}(\mathbf{R}) = i\left(\sigma_y\right)_{mn} \partial_{\mathbf{R}}\theta(\mathbf{R})$.

We choose four challenging cases at low-temperature for numerical simulations. In consistence with the benchmark tests in Ref. [12], we set that the first diabatic state is initially occupied. It is projected onto the adiabatic basis. We calculate the population dynamics via the eCMM approach, Ehrenfest dynamics and FSSH. All simulations are performed in the adiabatic representation, and we transform the population results back to the diabatic basis at time t. The results produced by eHEOM in the diabatic basis are used for comparison. As expected, all results for either Ehrenfest dynamics or eCMM yielded in the adiabatic representation reproduce those obtained in the diabatic representation as reported in Ref. [12].

Figure 11.1 shows the population difference for the four spin-boson models. Ehrenfest dynamics performs most poorly in all cases. FSSH slightly improves the performance, but its asymptotic results still significantly deviate from the exact data for those cases with strong non-adiabatic coupling or high characteristic frequency. In contrast, eCMM generates much more faithful results in all the cases. More results in Ref. [13] also demonstrate that eCMM or eCMMcv is more reasonable as well as useful in studying nonadiabatic processes of condensed-phase systems.

11.4.2 Tully's Gas Phase Scattering Models

Tully's gas phase scattering models [24] are often used as benchmark tests for nonadiabatic dynamics methods. The set of models are described by a two-state Hamiltonian,

$$\hat{H} = \frac{\hat{P}^2}{2m} + \begin{pmatrix} V_{11}(\hat{R}) & V_{12}(\hat{R}) \\ V_{21}(\hat{R}) & V_{22}(\hat{R}) \end{pmatrix}, \tag{11.58}$$

with mass $m = 2000$ a. u. Atomic units are used for this set of models. The models involve plateau regions where $V_{nn}(R \to \infty)$ and $V_{nn}(R \to -\infty)$ are flat.

In the single avoided crossing (SAC) model, the diagonal elements of the potential matrix are $V_{11}(R) = -V_{22}(R) = A(1 - e^{-B|R|})\text{sgn}(R)$ and off-diagonal ones are $V_{12}(R) = V_{21}(R) = Ce^{-DR^2}$, with parameters $A = 0.01$, $B = 1.6$, $C = 0.005$, and $D = 1.0$. State 1 is initially occupied, and the initial nuclear wavepacket is $\Psi(R; t = 0) \propto \exp[-\alpha(R - R_0)^2/2 + i(R - R_0)P_0/\hbar]$ with Gaussian width parameter $\alpha = 1$ a. u., initial average coordinate $R_0 = -3.8$ a. u., and initial average momentum P_0. The initial Wigner distribution for the nuclear phase variables is then $\rho_{\text{W}}^{\text{nuc}}(R, P) \propto \exp[-\alpha(R - R_0)^2 - (P - P_0)^2/(\alpha\hbar^2)]$.

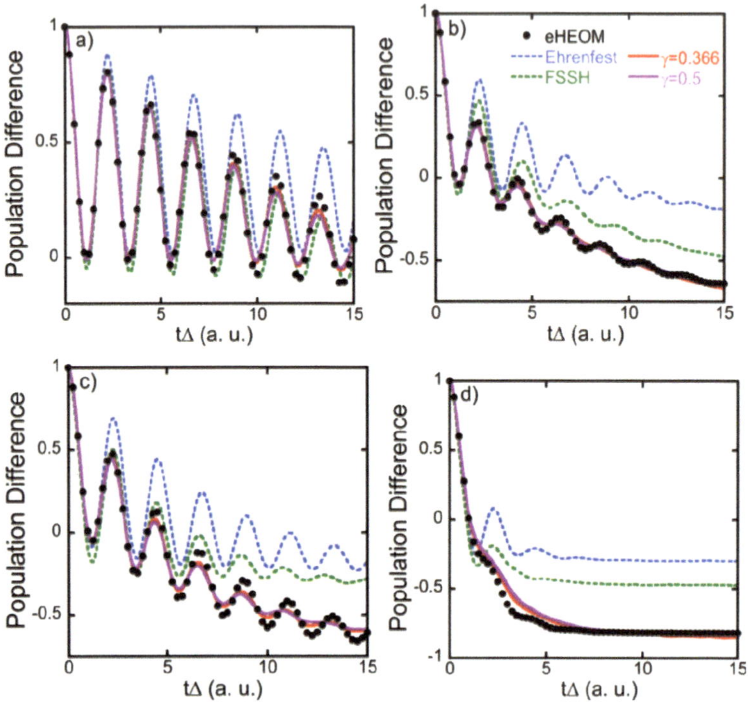

Figure 11.1 Panel (a) reports the population dynamics for the spin-boson model with the Ohmic bath, with parameter $\varepsilon = \Delta = 1$, $\beta = 5$, $\omega_c = 1$, $\alpha = 0.1$. Panel (b) is similar to panel (a) but for $\omega_c = 2.5$. Panel (c) is similar to panel (a) but for $\alpha = 0.4$. Panel (d) is similar to panel (a) but for $\omega_c = 2.5$, $\alpha = 0.4$. Solid circles: eHEOM. Cyan dashed lines: Ehrenfest dynamics. Green dashed lines: FSSH. Red and purple solid lines: eCMM with $\gamma = 0.366$ and 0.5, respectively.

In the dual avoid crossing (DAC) model, the diagonal elements of the potential matrix are $V_{11}(R) = 0$ and $V_{22}(R) = -Ae^{-BR^2} + E_0$, and the off-diagonal ones are $V_{12}(R) = V_{21}(R) = Ce^{-DR^2}$, with parameters $A = 0.10$, $B = 0.28$, $E_0 = 0.05$, $C = 0.015$ and $D = 0.06$. Two crossing points appear in the diagonal potential energy surfaces in the DAC model. Similar to the SAC model, state 1 is initially occupied, and the initial nuclear wavepacket is $\Psi(R; t = 0) \propto \exp[-\alpha(R - R_0)^2/2 + i(R - R_0)P_0/\hbar]$ with Gaussian width parameter $\alpha = 1$ a. u., initial average coordinate $R_0 = -10$ a. u., and initial average momentum P_0.

We investigate the transmission and reflection coefficients of each state. In the two models, the initial average momentum, P_0, varies from 2 a. u. to 50 a. u.. Similar to the spin boson model, Tully's SAC or DAC model, though originally described in the diabatic basis, can be presented in the adiabatic representation. The corresponding adiabatic potential energy surfaces and nonadiabatic coupling vectors can be obtained by the diagonalization of the potential matrix in Eq. (11.58).

We use three types of trajectory-based methods, which include Ehrenfest dynamics, FSSH, and eCMMcv. Their results are compared with the exact data obtained by the discrete variable representation (DVR) [31].

Figure 11.2 shows the transmission probability for the SAC and DAC models. In panel (a) and panel (c) all approaches perform well for the SAC model. For the DAC model shown in panel (b) and panel (d), the eCMMcv approach delivers the most accurate description of the peak near $\mathrm{p}_0 = 32$ a. u., either with $\gamma = 0.366$ or 0.5. Although the performance of eCMMcv should be improved in the low momentum region of the DAC model, results between [15 a.u., 50 a.u.] in panel (b) and panel (d) demonstrate that eCMMcv is superior to FSSH or Ehrenfest dynamics in studying fast nonadiabatic processes.

11.4.3 Atom-in-Cavity Models

Optical cavity-modified chemical dynamics has recently attracted considerable interest. Nonadiabatic couplings between different electronic states can be induced by the interaction between the (strong) external field and the molecular system. Such induced nonadiabatic couplings can alter and control a wide variety of chemical dynamical processes.

We study a benchmark system where an atom with a stationary position interacts with a one-dimensional lossless cavity. A two-state model and a three-state one are simulated. Four hundred field modes are used to produce converged data. More details of

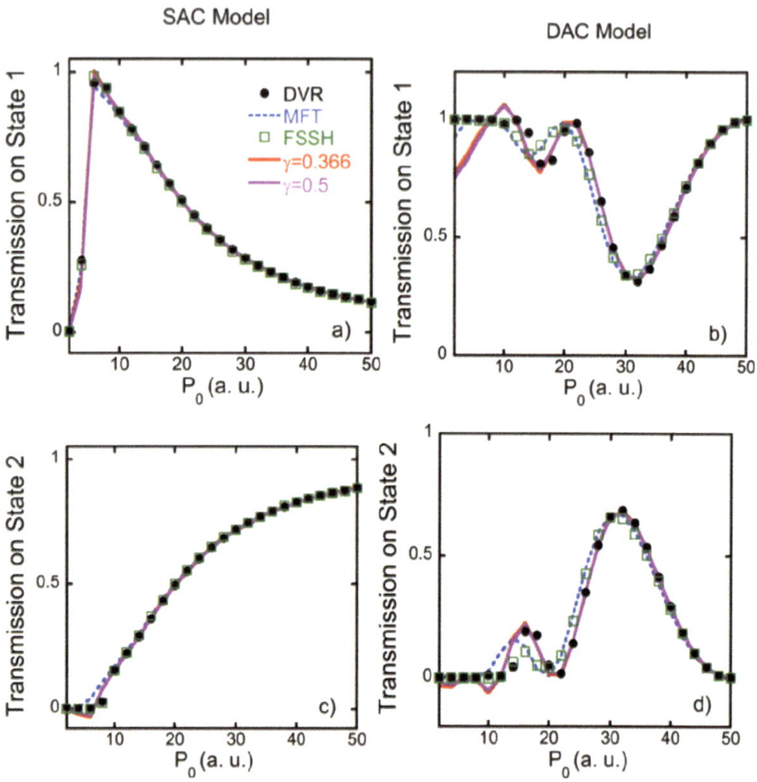

Figure 11.2 Panel (a) presents transmission coefficients on adiabatic state 1 for the SAC model. Panel (b) is similar to panel (a) but for the DAC model. Panel (c) is the same as panel (a) but shows transmission coefficients on adiabatic state 2 instead. Panel (d) is similar to panel (c) but for the DAC model. Solid circles: DVR. Blue dashed lines: Ehrenfest dynamics. Green hollow squares: FSSH. Red and purple solid lines: eCMMcv with $\gamma = 0.366$ and 0.5, respectively. (The eCMMcv and Ehrenfest dynamics results in the adiabatic representation have been reported in Ref. [13].)

the two atom-in-cavity models and the eCMMcv implementation are available in Refs. [13, 32–34]. The diabatic representation is used for all calculations, except that FSSH employs the adiabatic representation. The eCMMcv approach is compared with Ehrenfest dynamics and FSSH. Exact data from Refs. [32, 33] are used as the reference.

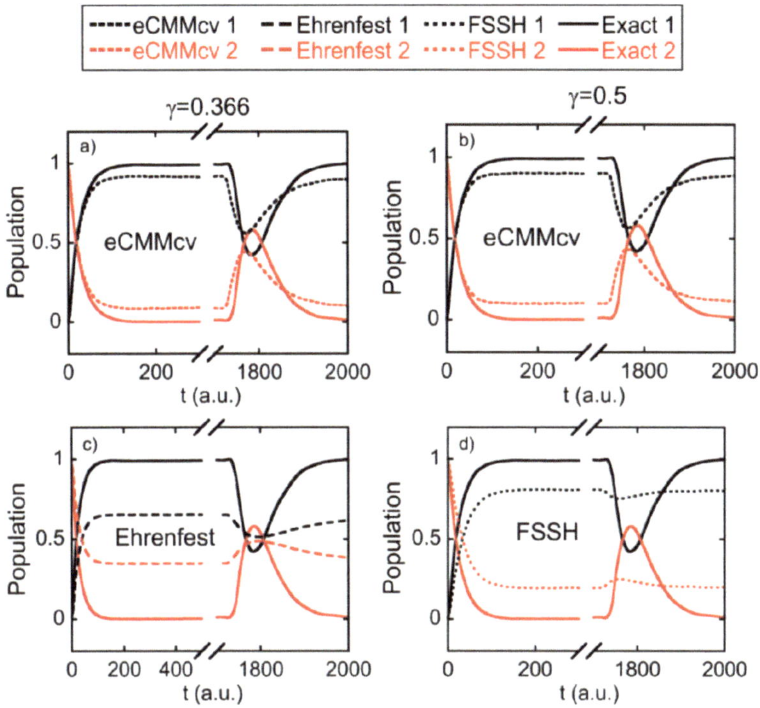

Figure 11.3 Population dynamics for the two-state atom-in-cavity model. Black and red curves stand for population of state 1 and 2, respectively. Short-dashed lines in Panel (a): eCMMcv with $\gamma = \left(\sqrt{F+1} - 1\right)/F \approx 0.366$. Panel (b): eCMMcv with $\gamma = 0.5$. Long-dashed lines in Panel (c): Ehrenfest dynamics. Dotted lines in Panel (d): FSSH. Solid lines in all panels: Exact results. (The eCMMcv and Ehrenfest dynamics results in the diabatic representation have been reported in Ref. [13].)

Figures 11.3 and 11.4 show the population dynamics of the two models. It is evident that Ehrenfest dynamics fails to describe the population transfer even at short time. More interestingly, eCMMcv semi-quantitatively captures the positive (negative) spike in the excited (ground) electronic state, while FSSH or Ehrenfest dynamics does a poor job. The spike at about $t = 1800$ a.u. is related to the re-absorption and re-emission process of the earlier emitted photon by the atom. Neither FSSH nor Ehrenfest dynamics is capable of depicting the phenomenon.

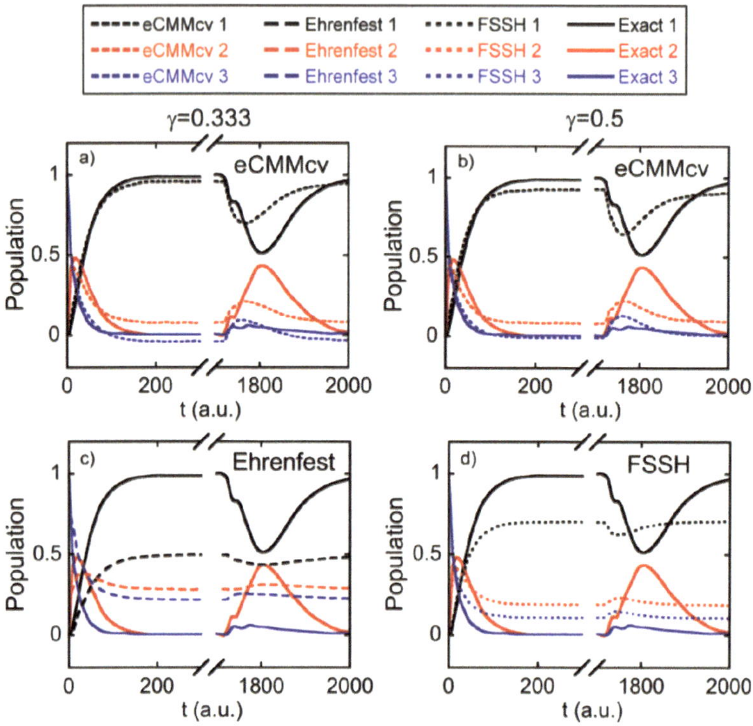

Figure 11.4 Population dynamics for the three-state atom-in-cavity model. Black, red and blue curves stand for population of state 1, 2 and 3, respectively. Short-dashed lines in Panel (a): eCMMcv with $\gamma = \left(\sqrt{F+1}-1\right)/F \approx 0.333$. Panel (b): eCMMcv with $\gamma = 0.5$. Long-dashed lines in Panel (c): Ehrenfest dynamics. Dotted lines in Panel (d): FSSH. Solid lines in all panels: Exact results. (The eCMMcv and Ehrenfest dynamics results in the diabatic representation have been reported in Ref. [13].)

11.5 Concluding Remarks

In this chapter, we report our recent progress in phase space mapping theory for nonadiabatic quantum molecular dynamics. We construct the mapping kernel and its inverse for the one-to-one correspondence mapping between quantum Hilbert space and phase space such that both continuous nuclear DOFs and discrete electronic-state DOFs are treated in a unified formulation. In such a formulation, the constraint condition implicitly applied to the kinetic energy term defines constraint space for mapping phase variables. The conventional Meyer–Miller mapping Hamiltonian is intrinsically a special case of the mapping phase function for the multi-electronic-state Hamiltonian. Interestingly, there exists a more comprehensive mapping Hamiltonian with a commutation matrix in phase space, in which the conventional commutation relation between the coordinate and momentum operators is replaced by a more general form. As a trajectory-based approximation to the quantum Liouville theorem in mapping phase space, the eCMMcv approach meets the Born–Oppenheimer limit before the system enters the nonadiabatic coupling region. Although we demonstrate the formulation of phase space mapping theory in the adiabatic representation, it can be applied to any other representations [13, 19], including the diabatic or quasi-diabatic [35] representation. It is expected that phase space mapping approaches can be directly used for ab initio nonadiabatic quantum molecular dynamics, especially for realistic systems where TDDFT offers a practical quantum chemistry tool, in a way similar to Refs. [36, 37]. More developments and extensions will be warranted in the future.

Acknowledgments

This work was supported by the National Natural Science Foundation of China (NSFC) Grant No. 21961142017, and by the Ministry of Science and Technology of China (MOST) Grant No. 2017YFA0204901. We acknowledge the High-performance Computing Platform of Peking University, Beijing PARATERA Tech CO., Ltd., and Guangzhou supercomputer center for providing computational resources.

References

1. Domcke, W., Yarkony, D. R., and Koppel, H. (2010). *Conical Intersections: Theory, Computation and Experiment*, World Scientific, Singapore.
2. Koch, W., and Holthausen, M. C. (2001). *A Chemist's Guide to Density Functional Theory*, 2nd ed, WILEY-VCH Verlag GmbH, Germany.

3. Ullrich, C. A. (2011). *Time-Dependent Density-Functional Theory: Concepts and Applications*, Oxford University Press, New York.
4. Kapral, R., and Ciccotti, G. (1999). Mixed quantum-classical dynamics. *J. Chem. Phys.* **110**, 8919–8929.
5. Hsieh, C.-Y., and Kapral, R. (2013). Analysis of the forward-backward trajectory solution for the mixed quantum-classical liouville equation. *J. Chem. Phys.* **138**.
6. Meyer, H.-D., and Miller, W. H. (1979). A classical analog for electronic degrees of freedom in nonadiabatic collision processes. *J. Chem. Phys.* **70**, 3214–3223.
7. Stock, G., and Thoss, M. (1997). Semiclassical description of nonadiabatic quantum dynamics. *Phys. Rev. Lett.* **78**, 578–581.
8. Miller, W. H., and Cotton, S. J. (2016). Classical molecular dynamics simulation of electronically non-adiabatic processes. *Faraday Discuss.* **195**, 9–30.
9. He, X., and Liu, J. (2019). A new perspective for nonadiabatic dynamics with phase space mapping models. *J. Chem. Phys.* **151**, 024105.
10. Liu, J. (2016). A unified theoretical framework for mapping models for the multi-state hamiltonian. *J. Chem. Phys.* **145**, 204105.
11. Liu, J. (2017). Isomorphism between the multi-state Hamiltonian and the second-quantized many-electron Hamiltonian with only 1-electron interactions. *J. Chem. Phys.* **146**, 024110.
12. He, X., Gong, Z., Wu, B., and Liu, J. (2021). Negative zero-point-energy parameter in the Meyer–Miller mapping model for nonadiabatic dynamics. *J. Phys. Chem. Lett.* **12**, 2496–2501.
13. He, X., Wu, B., Gong, Z., and Liu, J. (2021). Commutator matrix in phase space mapping models for nonadiabatic quantum dynamics. *J. Phys. Chem. A* **125**, 6845–6863.
14. Cohen, L. (1966). Generalized phase-space distribution functions. *J. Math. Phys.* **7**, 781.
15. Liu, J., and Miller, W. H. (2011). An approach for generating trajectory-based dynamics which conserves the canonical distribution in the phase space formulation of quantum mechanics. I. Theories. *J. Chem. Phys.* **134**, 104101.
16. Wigner, E. (1932). On the quantum correction for thermodynamic equilibrium. *Phys. Rev.* **40**, 0749–0759.
17. Liu, J. (2015). Recent advances in the linearized semiclassical initial value representation/classical Wigner model for the thermal correlation function. *Int. J. Quantum Chem.* **115**, 657–670.

18. Sun, X., Wang, H. B., and Miller, W. H. (1998). Semiclassical theory of electronically nonadiabatic dynamics: results of a linearized approximation to the initial value representation. *J. Chem. Phys.* **109**, 7064–7074.
19. Liu, J., He, X., and Wu, B. (2021). Unified formulation of phase space mapping approaches for nonadiabatic quantum dynamics. *Acc. Chem. Res.* **54**, 4215–4228.
20. Cotton, S. J., and Miller, W. H. (2019). Trajectory-adjusted electronic zero point energy in classical Meyer–miller vibronic dynamics: symmetrical quasiclassical application to photodissociation. *J. Chem. Phys.* **150**, 194110.
21. Cotton, S. J., Liang, R., and Miller, W. H. (2017). On the adiabatic representation of Meyer–Miller electronic-nuclear dynamics. *J. Chem. Phys.* **147**, 064112.
22. Sakurai, J. J. (1994). *Modern Quantum Mechanics*, Addison-Wesley, New York.
23. Ehrenfest, P. (1927). Bemerkung über die angenäherte gültigkeit der klassischen mechanik innerhalb der quantenmechanik. *Z. Physik* **45**, 455–457.
24. Tully, J. C. (1990). Molecular-dynamics with electronic-transitions. *J. Chem. Phys.* **93**, 1061–1071.
25. Makarov, D. E., and Makri, N. (1994). Path-integrals for dissipative systems by tensor multiplication. Condensed-phase quantum dynamics for arbitrarily long-time. *Chem. Phys. Lett.* **221**, 482–491.
26. Makri, N., and Makarov, D. E. (1995). Tensor propagator for iterative quantum time evolution of reduced density matrices. I. Theory. *J. Chem. Phys.* **102**, 4600–4610.
27. Tang, Z. F., Ouyang, X. L., Gong, Z. H., Wang, H. B., and Wu, J. L. (2015). Extended hierarchy equation of motion for the spin-boson model. *J. Chem. Phys.* **143**, 224112.
28. Tanimura, Y., and Kubo, R. (1989). Time evolution of a quantum system in contact with a nearly Gaussian–Markoffian noise bath. *J. Phys. Soc. Jpn.* **58**, 101–114.
29. Meyer, H. D., Manthe, U., and Cederbaum, L. S. (1990). The multi-configurational time-dependent Hartree approach. *Chem. Phys. Lett.* **165**, 73–78.
30. Wang, H. B., and Thoss, M. (2003). Multilayer formulation of the multiconfiguration time-dependent Hartree theory. *J. Chem. Phys.* **119**, 1289–1299.

31. Colbert, D. T., and Miller, W. H. (1992). A novel discrete variable representation for quantum-mechanical reactive scattering via the s-matrix Kohn method. *J. Chem. Phys.* **96**, 1982–1991.
32. Hoffmann, N. M., Schafer, C., Rubio, A., Kelly, A., and Appel, H. (2019). Capturing vacuum fluctuations and photon correlations in cavity quantum electrodynamics with multitrajectory Ehrenfest dynamics. *Phys. Rev. A* **99**, 063819.
33. Hoffmann, N. M., Schafer, C., Sakkinen, N., Rubio, A., Appel, H., and Kelly, A. (2019). Benchmarking semiclassical and perturbative methods for real-time simulations of cavity-bound emission and interference. *J. Chem. Phys.* **151**, 244113.
34. Saller, M. A. C., Kelly, A., and Geva, E. (2021). Benchmarking quasi-classical mapping hamiltonian methods for simulating cavity-modified molecular dynamics. *J. Phys. Chem. Lett.* **12**, 3163–3170.
35. Mandal, A., Yamijala, S. S., and Huo, P. (2018). Quasi-diabatic representation for nonadiabatic dynamics propagation. *J. Chem. Theory. Comput.* **14**, 1828–1840.
36. Yue, L., Liu, Y., and Zhu, C. (2018). Performance of TDDFT with and without spin-flip in trajectory surface hopping dynamics: cis–trans azobenzene photoisomerization. *Phys. Chem. Chem. Phys.* **20**, 24123–24139.
37. Hu, Y., Xu, C., Ye, L., Gu, F. L., and Zhu, C. (2021). Nonadiabatic molecular dynamics simulation for the ultrafast photoisomerization of dMe-OMe-NAIP based on TDDFT on-the-fly potential energy surfaces. *Phys. Chem. Chem. Phys.* **23**, 5236–5243.
38. Makri, N. (1999). The linear response approximation and its lowest order corrections: an influence functional approach. *J. Phys. Chem. B* **103**, 2823–2829.
39. Craig, I. R., Thoss, M., and Wang, H. B. (2007). Proton transfer reactions in model condensed-phase environments: accurate quantum dynamics using the multilayer multiconfiguration time-dependent Hartree approach. *J. Chem. Phys.* **127**, 144503.
40. He, X., Wu, B., Shang, Y., Li, B., Cheng, X., and Liu, J. (2022). New Phase Space Formulations and Quantum Dynamics Approaches. Wiley Interdiscip. *Rev. Comput. Mol. Sci.*, e1619.

Chapter 12

Global Switch Trajectory Surface Hopping Dynamics in the Framework of Time-Dependent Density Functional Theory

Ling Yue[a] **and Chaoyuan Zhu**[b]
[a]*Key Laboratory for Non-Equilibrium Synthesis and Modulation of Condensed Matter, Ministry of Education, School of Chemistry, Xi'an Jiaotong University, Xi'an, Shaanxi 710049, P. R. China*
[b]*Department of Applied Chemistry and Institute of Molecular Science, National Chiao-Tung University, Hsinchu 30010, Taiwan*
heartstar@mail.xjtu.edu.cn, cyzhu@mail.nctu.edu.tw

12.1 Introduction

Photochemical and photophysical processes are involved in photoexcitation with suitable energy that promotes molecule from its ground state to electronically excited states and then undergoes deactivation by radiative and/or nonradiative photophysical processes. The conical intersection (CI) where the electronically excited-state potential energy surfaces (PESs) are degenerate

Time-Dependent Density Functional Theory: Nonadiabatic Molecular Dynamics
Edited by Chaoyuan Zhu

ISBN 978-981-4968-42-3 (Hardcover), 978-1-003-31921-4 (eBook)
www.jennystanford.com

(intersect) plays an important role in nonradiative photophysical processes leading to chemical reactions [1]. In fact, two electronic states round CI have the same spin-multiplicity [2] and the nonadiabatic couplings between the two states are strong enough leading to the breakdown of Born–Oppenheimer approximation. In the recent years, many theoretical approaches have been developed to investigate the nonadiabatic molecular dynamics involving in CI [2, 3]. Theoretically modeling for nonadiabatic photochemical and photophysical processes requires excited-state dynamics that considering the couplings between the nuclear and the electronic motions beyond the Born–Oppenheimer approximation regime. The complexity and high computational costs of such molecular dynamic simulations in the framework of full quantum mechanics are big challenge for modeling systems with more than three or four atoms. In general, the mixed quantum-classical molecular dynamics methods have been employed in nonadiabatic molecular dynamics simulations for relatively larger systems, and trajectory surface hopping (TSH) based molecular dynamics (MD) method is one of the most important strategy to study nonadiabatic photophysical and photochemical processes for large complex chemical systems [2, 4, 5].

In the TSH-MD method, the nuclear motion is treated as classical trajectories running on on-the-fly electronically adiabatic PESs governed by Newtonian equation, while PESs and its ingredients are obtained by quantum chemistry method from electronic structure calculation. The nonadiabatic couplings between two adiabatic PESs drive classical trajectories hopping or switching from one PES to another. Nonaidabatic transition or switching probabilities that guide classical trajectories stochastically switching from one PES to another are calculated based on quantum mechanics or mixed quantum-classical methods. This stochastic method attempts to make the electronic probability distribution averaged over an ensemble of trajectories equal to the probability distribution computed from the electronic density matrix. Thus, the nonaidabatic transition probabilities between two PESs are calculated to determine whether adiabatic PES must instantaneously be changed or not for running classical trajectory [2]. The TSH-MD methods mostly differ from how to calculate nonadiabatic switching probability, and time-independent and/or time-dependent electronically coupled Schrödinger equations must be solved for obtaining the nonadiabatic switching probability.

Regarding ab initio quantum chemistry method for computing adiabatic PES, the multireference methods like the state-averaged complete active space self-consistent field (SA-CASSCF) [6–8] and multi-reference configurational interaction (MRCI) approaches [9] are most high level ab initio methods, but only applying for relatively small systems for performing on-the-fly single-point energy calculations in TSH-MD simulation. The multiconfigurational wavefunction is proved to properly describe the electronic structure around conical intersections and correctly construct the topology of ground and excited-state PESs [10]. In addition, the analytical NACVs at CASSCF levels are now widely implemented in popular quantum chemistry software. However, the computational costs of multireference methods exponentially rise with the increase of molecular size and the number of active molecular orbitals, so that it is not applicable for large systems. On the other hand, the time-dependent density functional theory (TDDFT) methods have been greatly improved in both accuracy and efficiency over the past twenty years. The relatively low computational costs and including electron correlation in density function have made TDDFT become an alternative tool to perform reliable calculations for excited-state electronic structures and dynamics, especially for relative large systems [11–13]. TDDFT methods have been successfully applied not only for simulating the absorption and fluorescence spectra of molecules, but also for excited-state dynamics simulations. In recent decade, the wavefunction overlap evaluations [14–16] and analytical NACVs in the framework of linear-response TDDFT (LR-TDDFT) were derived and implemented in several quantum chemistry software packages [17–20], the TSH-MD simulation combined with LR-TDDFT are rapidly developed in recent years [15, 19, 21, 22]. The TSH-MD simulation associated with LR-TDDFT on-the-fly PES calculations has been widely applied to investigate photo-induced photochemical and photophysical processes in all areas of quantum chemistry studies from fundamentals to material and biological [2, 5, 20, 23–31]. However, it should be realized that the conventional LR-TDDFT methods can suffer the certain problems in dealing with

CI between electronically ground and excited states, especially of the topology of PESs in the vicinity of CI that might be wrong. The spin-flip scheme has been introduced in LR-TDDFT to correct the topology of PESs in the vicinity of CI, but the spin-flip TDDFT (SF-TDDFT) [32] or its "spin-complete" version (SA-SF-TDDFT) [13, 33] plus analytic NACVs of the these newest developed TDDFT variants [34] are still under the development or not widely implemented in most popular quantum chemistry programs.

Direct nonadiabatic TSH-MD requires calculating PESs and its gradients plus nonadiabatic coupling vectors (NACVs) $\boldsymbol{d}_{ji}$ on-the-fly for solving time-dependent electronically coupled Schrödinger equations, and running trajectories can in principle make hopping at any time of propagation. This scheme is called as a local switch (LS) algorithm and thus LS switch-probability is time stepsize dependent. Moreover, if a large number of excited-states, e.g., m states are involved in TSH-MD simulations, the number of NACVs required by solving electronic time-dependent Schrödinger equation can be very huge, i.e., $C_m^2 = m\,(m-1)/2$. Thus, these time-consuming computations hinder the further applications of TSH-MD with time-dependent switch probabilities in more complex photophysical and photochemical processes. On the other hand, time-independent scheme is an alternative solution to obtain switching probabilities for TSH-MD simulation. In this scheme, the running trajectories only make hopping in the vicinity of CI during propagation. This scheme is called as a global switch (GS) algorithm.

Starting from the Landau–Zener two-state linear curve-crossing model, a GS probability has been formulated as an analytical formula from time-independent Zhu–Nakamura nonadiabatic transition probability obtained from very rigorous semiclassical theory [35]. The GS algorithm from Zhu–Nakamura formula [36–41] does not require calculation of NACVs as required by LS algorithm. Thus, GS algorithm is computationally much faster than the LS one for performing on-the-fly TSH-MD simulation. The GS-TSH-MD simulation is much easier to be combined with presently existed and newly developed quantum chemistry methods, especially in the case where NACVs are not available in computation. This is also potentially useful to incorporate with force field excited-state molecular dynamic simulation in which only electronically adiabatic potential energy surfaces are available. In this chapter, we introduce the GS-TSH method based on Zhu–Nakamura formula for general two-state curve crossing problem [42, 43], and focus on its applications and performances especially in the framework of TDDFT. This method could yield an accurate description of the nonadiabatic molecular dynamics simulation at low computational cost. In particular, the use of well-defined classical trajectories and switching probabilities in the GS algorithm make it possible to employ direct dynamics, i.e., an on-the-fly approach where only PESs and its gradients are required to be calculated at each point of time along running trajectory. This method, which is very suitable to be incorporated with the TDDFT method, can be another useful tool for studying TSH-MD simulation for large-size systems and/or longtime processes. Especially, the GS algorithm might partially overcome the inherent limitations of traditional LR-TDDFT while TSH-MD simulations involve CI between electronically ground and excited states. Moreover, we take our recent application on GS-TSH-MD simulation for chemiluminescence [44].

12.2 Global Switch Trajectory Surface Hopping Dynamics

In order to understand how GS algorithm is formulated, we briefly introduce the time-dependent LS algorithm as comparison.

12.2.1 Time-Dependent Scheme and Local Switch Probability

The time-dependent coupled Schrödinger equations for electronic wavefunctions can be obtained in both diabatic and adiabatic representation. We just take an example in adiabatic representation by expanding the total electronic wavefunction using an adiabatic electronic basis sets ($\Phi = \sum_i c_i \varphi_i$) as

$$i\hbar \frac{\partial c_j}{\partial t} = c_i E_j(\boldsymbol{R}) - i\hbar \sum_i c_i D_{ji}, \tag{12.1}$$

where $\boldsymbol{R}$ is N-dimensional nuclei coordinate vector, and electronic adiabatic PES is defined by $E_j(\boldsymbol{R}) = \langle \varphi_j | \mathbf{H}_e | \varphi_j \rangle$ along running classical trajectory with on-the-fly electronic structure calculations. D_{ji} stands for the nonadiabatic coupling between electronic i and j states which might be calculated approximately from the time-dependent electronic wavefunction overlap [14, 15], but the required time stepsize Δt for D_{ji} calculation is usually much smaller than time stepsize for integration of classical trajectory in order to obtain an accurate LS probability. Thus, as long as an analytic nonadiabatic coupling vector ($\boldsymbol{d}_{ji}$) is available, we can calculate the nonadiabatic D_{ji} based on the following relationship,

$$D_{ji} = \left\langle \varphi_j \,|\, \frac{\partial}{\partial t} \,|\, \varphi_i \right\rangle = \left\langle \varphi_j \,|\, \frac{\partial}{\partial \boldsymbol{R}} \,|\, \varphi_i \right\rangle \cdot \frac{\partial \boldsymbol{R}}{\partial t} = \boldsymbol{d}_{ji}(\boldsymbol{R}) \cdot \boldsymbol{v} \quad (12.2)$$

where $\boldsymbol{v}$ is an N-dimensional velocity vector of nuclei along running classical trajectory. Once the expansion coefficients (c_i and c_j) corresponding electronic state i and j are solved from Eq. (12.1), various algorithms can be employed to calculate LS probabilities. One of the standard LS algorithms is based on Tully's fewest switches (TFS) [45] that is designed to minimize the number of switches as

$$\begin{aligned} p_{i\to j} &= \max\left(0, \frac{-2Re\left[c_j^* c_i D_{ij}\right]\Delta t}{|c_j(t)|^2}\right) \\ &= \max\left(0, \frac{-2Re\left[c_j^* c_i \boldsymbol{d}_{ij} \cdot \boldsymbol{v}\right]\Delta t}{|c_j(t)|^2}\right), \end{aligned} \quad (12.3)$$

which determines LS probability from the current i state to the target j state. The $p_{i\to j}$ is obviously dependent on stepsize of time (Δt).

12.2.2 Time-Independent Scheme and Global Switch Probability

In contrast to time-dependent approach based on the time-dependent coupled Schrödinger Eq. (12.1), the time-independent approach is based on two-state coupled time-independent Schrödinger equations starting from one-dimensional two-state curve crossing model in a diabatic representation ($\{\psi_i\}$) [46–49],

$$\left[-\frac{\hbar^2}{2\mu}\frac{d^2}{d\boldsymbol{R}^2}+\begin{pmatrix} V_{11}(\boldsymbol{R}) & V_{12}(\boldsymbol{R}) \\ V_{21}(\boldsymbol{R}) & V_{22}(\boldsymbol{R}) \end{pmatrix}\right]\begin{pmatrix} \psi_1(\boldsymbol{R}) \\ \psi_2(\boldsymbol{R}) \end{pmatrix} = (E-E_x)\begin{pmatrix} \psi_1(\boldsymbol{R}) \\ \psi_2(\boldsymbol{R}) \end{pmatrix} \tag{12.4}$$

in which $\boldsymbol{R}$ can be considered as a one-dimensional nuclear coordinate along running classical trajectory, μ is the reduced mass of the molecule, E is the collision energy and E_x is potential energy at the crossing point. Diabatic potential energy and coupling are defined by $V_{ij} = \langle \psi_i \mid \mathrm{H}_e \mid \psi_j \rangle$. By applying a linear crossing model, that is, $V_{11}(\boldsymbol{R}) = -F_1\boldsymbol{R}$, $V_{22}(\boldsymbol{R}) = -F_2\boldsymbol{R}$, and $V_{12}(\boldsymbol{R}) = V_{21}(\boldsymbol{R}) = V_0$ (constant coupling), Landau–Zener nonadiabatic switching probability can be analytically formulated in terms of two unitless parameters, namely, effective coupling (a^2) and effective collision energy (b^2) [36],

$$a^2 = \frac{\hbar^2}{2\mu}\frac{\sqrt{|\boldsymbol{F}_2\boldsymbol{F}_1|}|\boldsymbol{F}_2-\boldsymbol{F}_1|}{(2V_{12})^3} \tag{12.5A}$$

$$b^2 = (E-E_x)\frac{|\boldsymbol{F}_2-\boldsymbol{F}_1|}{\sqrt{|\boldsymbol{F}_2\boldsymbol{F}_1|}(2V_{12})} \tag{12.5B}$$

Original Landau–Zener switching probability completely breaks down at $E \to E_x$ and the more accurate Zhu–Nakamura switching probability is obtained as

$$p = \exp\left[-\frac{\pi}{4\sqrt{a}^2}\sqrt{\frac{2}{b^2+\sqrt{|b^4 \pm 1|}}}\right] \tag{12.6}$$

where $+$ and $-$ stand for $\boldsymbol{F}_1\boldsymbol{F}_2 > 0$ and $\boldsymbol{F}_1\boldsymbol{F}_2 < 0$, respectively. The $\boldsymbol{F}_1\boldsymbol{F}_2 > 0$ means the Landau–Zener model in the two-state curve crossing model while the $\boldsymbol{F}_1\boldsymbol{F}_2 < 0$ means the nonadiabatic tunneling case as shown in Fig. 12.1. The switching probability in Eq. (12.6) is regarded as GS probability calculated in the vicinity of CI during trajectory propagation.

The LS probability with TFS algorithm defined in Eq. (12.3) is calculated at each integration timestep along running trajectory and in principle the trajectory can make an attempted hop anytime, whereas the GS probability defined in Eq. (12.6) is calculated at an avoided crossing point and it represents an averaged

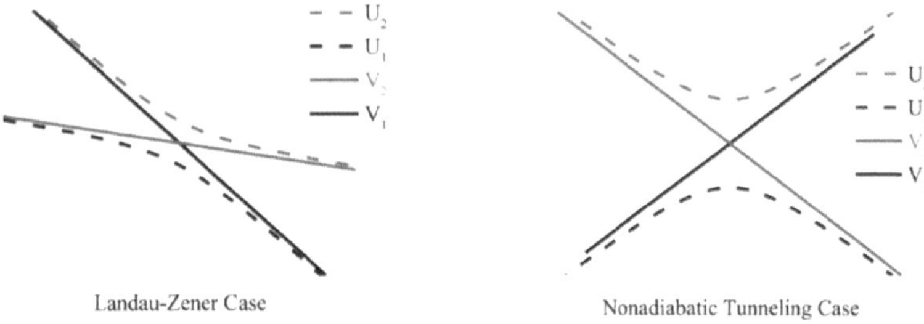

Figure 12.1 Landau–Zener and nonadiabatic tunneling linear crossing models. The U_1 (U_2) and V_1 (V_2) are the higher (lower) adiabatic and diabatic states, respectively.

nonadiabatic transition probability along running trajectory. The effective coupling (a^2) and effective collision energy (b^2) in Eq. (12.5) were extended to multidimensional nonadiabatic molecular dynamic simulations previously requiring calculation of seam surfaces, direction of nonadiabatic coupling vectors, or the Hessian matrix. In 2014, Zhu's group [37] has worked out the simplest way for calculating a^2 and b^2 in Eq. (12.5) by only needing two adiabatic potential energy surfaces and its gradients along running trajectory where avoided crossing occurs. One dimensional force parameters $|\boldsymbol{F}_2 - \boldsymbol{F}_1|$ and $|\boldsymbol{F}_2\boldsymbol{F}_1|$ in Eq. (12.6) are generalized into a multidimensional expression as

$$\frac{|\boldsymbol{F}_2 - \boldsymbol{F}_1|}{\sqrt{\mu}} = \sqrt{\left|\sum_{i=1}^{N} \frac{1}{m_i} \sum_{\alpha=x,y,z} \left(F_2^{i\alpha} - F_1^{i\alpha}\right)^2\right|} \tag{12.7A}$$

$$\frac{\sqrt{|\boldsymbol{F}_2\boldsymbol{F}_1|}}{\sqrt{\mu}} = \sqrt{\left|\sum_{i=1}^{N} \frac{1}{m_i} \sum_{\alpha=x,y,z} F_2^{i\alpha} F_1^{i\alpha}\right|}, \tag{12.7B}$$

where N is number of atoms in a molecule with atomic mass m_i ($i = 1, 2, \ldots, N$), and the diabatic forces $F_1^{i\alpha} = -\partial V_1/\partial R_{i\alpha}$ and $F_2^{i\alpha} = -\partial V_2/\partial R_{i\alpha}$ (where $R_{i\alpha}$ stands for the x, y, and z component of Cartesian coordinates for the i-th atom). Equation (12.6) for multidimensional-case can exactly go back to Eq. (12.5) for one-dimensional case when diatomic molecule is applied. The diabatic forces can be generalized by linear potential energy surface approximation from the adiabatic forces along running on-the-fly

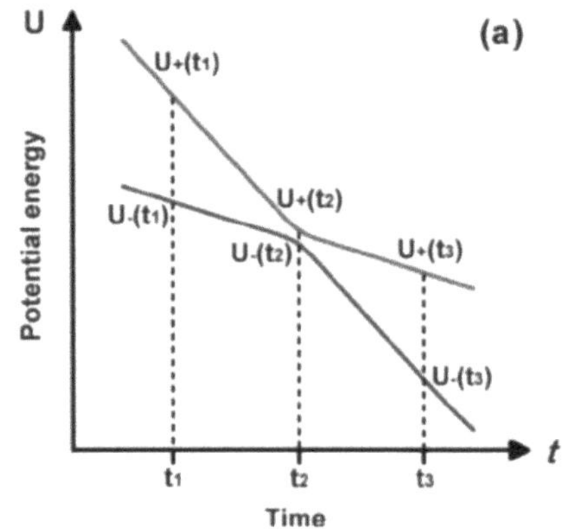

Figure 12.2 The hopping point at t_2 along running trajectory in GS algorithm, reprinted from Ref. [37], Copyright © 2014 Royal Society of Chemistry.

trajectory, and this linear approximation in the vicinity of CI is well hold. That is simple linear connection of an upper adiabatic force at t_1 with a lower adiabatic force at t_3, and another linear connection of a lower adiabatic force at t_1 with an upper adiabatic force at t_3 were made from three consecutive time steps $t_1 < t_2 < t_3$ as shown in Fig. 12.2.

The approximate diabatic forces $\boldsymbol{F}_1$ and $\boldsymbol{F}_2$ at time $t = t_2$ along running trajectory can be calculated based on only two adiabatic potential energy surfaces and its gradients:

$$F_1^{i\alpha}(t_2) = -\frac{\frac{\partial U_-}{\partial R_{i\alpha}(t_3)}\left(R_{i\alpha}(t_2) - R_{i\alpha}(t_1)\right) - \frac{\partial U_+}{\partial R(t_1)}\left(R_{i\alpha}(t_2) - R_{i\alpha}(t_3)\right)}{R_{i\alpha}(t_3) - R_{i\alpha}(t_1)} \tag{12.8A}$$

$$F_2^{i\alpha}(t_2) = -\frac{\frac{\partial U_+}{\partial R_{i\alpha}(t_3)}\left(R_{i\alpha}(t_2) - R_{i\alpha}(t_1)\right) - \frac{\partial U_-}{\partial R(t_1)}\left(R_{i\alpha}(t_2) - R_{i\alpha}(t_3)\right)}{R_{i\alpha}(t_3) - R_{i\alpha}(t_1)} \tag{12.8B}$$

in which U_+ (U_-) are the upper (lower) adiabatic potential energy surface along running trajectory. The collision energy E in this case is the potential energy plus the kinetic energy component in the hopping direction whose definition is described in the following subsections. For transition between states with the same spin, the energy at the crossing point is approximated as $E_x = [U_+(t_2) + U_-(t_2)]/2$, while diabatic coupling $V_{12}(\boldsymbol{R})$ in Eq. (12.5), is generalized by $V_{12} = [U_+(t_2) - U_-(t_2)]/2$. For transition between states with the different spins (intersystem crossings), e.g., $S_1 \leftrightarrow T_1$, the nonadiabatic coupling between these states are vanished due to the spin symmetry restriction. The singlets and triplets are spin-diabatic states, so $\boldsymbol{F}_1$ and $\boldsymbol{F}_2$ in Eq. (12.7) are the gradients of S_1 and T_1 themselves while the diabatic coupling in Eq. (12.5) is just the spin-orbit coupling (V_{SOC}) between S_1 and T_1 [40]. The p in Eq. (12.6) represents the switching probability between adiabatic states, so in the case of intersystem crossings, the true switching probabilities for $S_1 \leftrightarrow T_1$ is $p_{S \to T} = 1 - p$. Therefore, the two effective parameters a^2 and b^2 and diabatic coupling V_{12} (or V_{SOC}) hence can be calculated only from on-the-fly two adiabatic (or spin-diabatic) potential energy surfaces and its gradients along running trajectory, once the minimum potential energy gap at an avoided crossing or intersystem crossings was detected along running trajectory from three consecutive time steps.

The integration time stepsize in principle should be small enough to maintain good linear approximation of potential energy surfaces in order to have accurate diabatic forces in Eq. (12.8). However, in particle application to specific system, one can test converged time stepsize along sampling trajectories where the longer stepsize can be utilized in TSH-MD simulation to reduce computational cost for single-point energy calculation in quantum chemistry methods. Actually, the present linear approximation of the diabatic forces is consistent with the velocity Verlet method for on-the-fly trajectory calculation, in which the forces within the integration timestep are approximately considered as a constant. Once the timestep is small enough to converge calculation for on-the-fly trajectory, it converges multidimensional diabatic forces in Eq. (12.8). There is no additional small time stepsize requirement for nonadiabatic transitions in GS algorithm. This is different from LS algorithm in which the timestep for solving the time-dependent coupling equations and computing accurate nonadiabatic transitions is considerably smaller than the timestep for computing accurate velocity Verlet trajectory.

12.2.3 Velocity Adjustment

Once hopping occurs in a TSH-MD simulation, the total energy of the system is in general not conserved, since a discontinuous change occurs in the electronic energy at hooping point. This unphysical behavior is due to the fact that the hopping event is imposed from outside, through an ad hoc hopping algorithm, and acts as an external perturbation on the system. Its effect, of course, must be removed since the purpose is to simulate a closed system for which the total energy must be conserved. To this end, a velocity adjustment is necessary to be performed after each hopping event. When the system jumps from the i state to the j state, the electronic energy changes from $E_i(\boldsymbol{R})$ to $E_j(\boldsymbol{R})$. To compensate the potential energy change, velocity of the system is adjusted by

$$v'_X = v_X - \gamma \frac{\boldsymbol{x}_X}{M_X}, \tag{12.9}$$

where $\boldsymbol{x}$ is a $3N$-dimension vector, $\boldsymbol{v}_X$ and $\boldsymbol{v}'_X$ are the velocities of atom X before and after hopping, respectively. To maintain the energy conservation, the velocity adjustment induces a change in the nuclear kinetic energy, which is equal and opposite to the one in potential energy produced by surface hopping. Thus, the $\boldsymbol{v}'_X$ and $\boldsymbol{v}_X$ satisfied the following equation:

$$\Delta E_{\text{kin}} = \frac{1}{2}\sum_{X=1}^{N} M_X \boldsymbol{v}'^2_X - \frac{1}{2}\sum_{X=1}^{N} M_X \boldsymbol{v}^2_X = \gamma^2\alpha - \gamma\beta = \Delta E_{\text{pot}} = E_i - E_j, \tag{12.10}$$

where $\alpha$and β are defined as $\alpha = \frac{1}{2}\sum_{X=1}^{N} \frac{\boldsymbol{x}_X^2}{M_X}$ and $\beta = \sum_{X=1}^{N} \boldsymbol{v}_X \cdot \boldsymbol{x}_X$, respectively. By solving the Eq. (12.10), we obtained the scaling factor γ:

$$\gamma = \left(\beta \pm \sqrt{\beta^2 + 4\alpha \Delta E_{pot}}\right)/2a. \tag{12.11}$$

If the kinetic energy is too low to compensate for the energy variation caused by the electronic transition, the hopping is considered unphysical and is rejected in conventional TSH algorithm. In this case, the velocity of the system can be made toward reversed direction in analogy with elastic scattering.

In principle, the $\boldsymbol{x}$ vector can be chosen to be any $3N$-dimension vector. In LS, the velocity adjustment is usually performed in the direction of NACV $\boldsymbol{d}_{ji}$. An alternative choice for the scaling direction can be the regularized diabatic states introduced by Köppel et al. [50] or the gradient difference vector (GDVs, see Section 12.4.1 for its definition). Following these ideas, a normalized mass-weighted GDV called self-consistent hopping direction in which the nonadiabatic switching probability has the maximum at the avoided crossing are proposed for the velocity adjustment in GS algorithm:

$$n_i = \frac{1}{\sqrt{s_{ix}^2 + s_{iy}^2 + s_{iz}^2}} \left(s_{ix}, s_{iy}, s_{iz}\right) \tag{12.12}$$

where $s_{i\alpha} = \left(F_2^{i\alpha} - F_1^{i\alpha}\right) / \sqrt{m_i}$. In addition, the collision energy E in Eq. (12.5B) in this case can be expressed by E_t which is the potential energy plus the kinetic energy component in the hopping direction $\boldsymbol{n}$.

12.2.4 Implementation of Global Switch Algorithm

The implementation of GS-TSH is very similar with that of LS-TSH. The only difference is that the hopping in GS only occurs in the region of an avoided crossing and intersystem crossing. The PESs topology judgment must be evaluated for each timestep instead of the one-by-one computation of switching probability. Once the appropriate topology is detected, the global switching probabilities are calculated according to Eq. (12.6). Thus, the coordinates, velocities and gradients should be recorded for at least three consecutive timesteps ($t - \Delta t, t, t + \Delta t$). The general algorithm for the implementation of the GS-TSH-MD is summarized as follows:

(1) Initialization of velocity, gradients.
(2) Time propagation of coordinates ($\boldsymbol{x}_t^i$) and velocities ($\boldsymbol{v}_t^i$) on the PES of the selected electronic state i.
(3) Computation of energies of all relevant states ($\boldsymbol{E}_t^n$) and also the gradients of i state ($\boldsymbol{g}_t^i$) at new position and velocity.
(4) Judge the topology of the PESs close to dynamics state using the energies of three consecutive time steps ($\boldsymbol{E}_{t-\Delta t}^i$, $\boldsymbol{E}_{t-\Delta t}^j$, $\boldsymbol{E}_t^i$, $\boldsymbol{E}_t^j$, $\boldsymbol{E}_{t+\Delta t}^i$, $\boldsymbol{E}_{t+\Delta t}^j$). If the topology of relevant PESs is detected at an

avoided crossing or intersystem crossing, and then we calculate GS probability according to Eq. (12.6). If not, return back to step (2).

(5) Generate a random number and compare it with GS probability.
(6) If hopping is rejected, inverse the velocity. If hopping is performed, adjust the velocity according to the Eq. (12.9) and update the i state for GS-TSH-MD simulation.
(7) Calculate the electronic energies of all relevant states and the gradients of dynamics state (i state) at new position and velocity.
(8) If terminated condition of a GS-TSH-MD simulation is satisfied, e.g., maximum duration time of is reached, we collect product information, otherwise return back to step (2) continuing the simulation.

12.3 The Performance of Global Switch Versus Local Switch

The GS algorithm has successfully been applied to photoisomerization dynamics in relatively large organic molecules, and simulated quantum yields and lifetimes are in good agreement with experimental results [36, 38, 40, 41, 51–55]. In GS algorithm, the approximated diabatic coupling $V_{12} = [U_+(t_2) - U_-(t_2)]/2$ and self-consistent hopping direction $\boldsymbol{n}$ defined in Eq. (12.12) are employed in the switching probability calculations and velocity adjustments, instead of the utilization of NACVs. In addition, as a global switch, the hopping must not be judged at each time step but only appeared at conical intersections or avoid crossing, i.e., the hopping events far from the conical intersections or avoid crossing regions are ignored. To display the performance and accuracy of these approximation, it is necessary to perform a benchmark TSH simulation for a typical photochemical reaction by extensively comparing the results from GS algorithm with from conventional LS algorithm (Tully's fewest switches, TFS) simulation at the same platform of computational program.

12.3.1 Photoisomerization of Azobenzene on S_1-(n, π*) Excitation

We take photoisomerization of azobenzene as an example, and the *cis*- and *trans*-isomers of azobenzene can be interconverted by light of different wavelengths, the photoproducts having distinct spectral and geometric properties. These features open a variety of technical applications such as simple molecular switch [56], photoactuator [57], modification of surface properties in oriented films [58], or control of solution viscosity [59]. It hence aroused the interests of theoretical and experimental scientists [60, 61]. Therefore, this system was chosen to perform benchmark study for on-the-fly trajectory surface hopping molecular dynamics. More specifically, we focused on the *cis*-to-*trans* and *trans*-to-*cis* isomerization of azobenzene upon S_1-(n,π*) excitation as shown in Fig. 12.3.

The mechanism of azobenzene photoisomerization are investigated by many experimental and theoretical groups with different theoretical methods such as state-averaged CASSCF and semiempirical approach [61–65]. Both ab initio and semiempirical calculations showed alternative ground and excited-state reaction path for isomerization between *cis*- and *trans*-azobenzene. Our group also have performed mechanistic studies on azobenzene

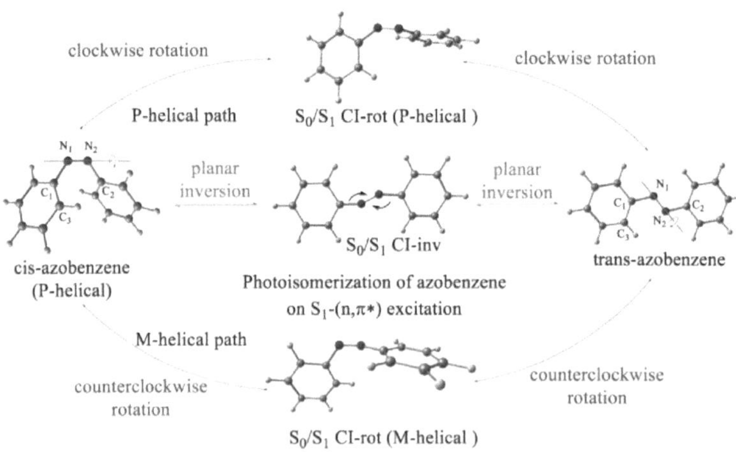

Figure 12.3 The *cis*-to-*trans* and *trans*-to-*cis* photoisomerization of azobenzene on S_1-(n, π*) excitation.

photoisomerization at SA2-CASSCF(6,6)/6-31G level [36] whose computational results are similar with previous CASSCF studies. The geometric optimization from SA2-CASSCF(6,6)/6-31G detected two conical intersections between the S_0 and S_1 states, namely rotation CI-rot and inversion CI-inv. At the SA2-CASSCF(6,6)/6-31G level, the CI-inv is usually hardly accessible in the TSH dynamic simulation starting from the S_1 state, because the energy of CI-inv is 0.42 eV higher than energy of CI-rot and 0.29 eV higher than the energy of *trans*-azobenzene in the S_1 state at SA2-CASSCF(6,6)/6-31G level (see Ref. [36] for details). It should be noted that two chiral structures (mirror images) exist for both *cis*-azobenzene and CI-rot shown in Fig. 12.3. The two chiral structures are called P-helical and M-helical, respectively. Two chiral rotation pathways are available for any one of the chiral *cis*-isomers. For example, P-helical *cis*-azobenzene might go through the P-helical S_0/S_1 CI-rot by clockwise rotation around the N–N bond while it might also isomerize via the M-helical S_0/S_1 CI-rot by counterclockwise rotation. Obviously, the P-helical S_0/S_1 CI-rot is energetically favorable for P-helical *cis*-azobenzene (vice versa). Thus, one or two chiral pathways might be observed in an actual TSH simulation which dependent on the activation barrier between P- and M-helical *cis*-azobenzene calculated at different theoretical levels. Besides the ab initio methods, some semiempirical methods were proved to provide a successful calculations on the relative energies and geometries for both ground and lowest excited-states for azobenzene system [60]. The geometric optimization and LS-TSH-MD simulation by OM2/MRCI [66, 67] is very similar with those measured by experiments [61] and SA-CASSCF calculations [36, 65], while the computational cost is much lower than the latter one. In order to perform fast TSH-MD simulations, the OM2/MRCI was utilized to compute electronic energies, nuclear gradients and nonadiabatic couplings.

12.3.2 Hopping Spots, Switching Probabilities and Velocity Adjustment

By applying the OM2/MRCI method with standard parameters, we performed 800 and 600 sampling trajectories starting from *cis*- and *trans*-azobenzene in Franck–Condon (FC) regions, respectively. The propagation time stepsize of the sampling trajectory was 0.1 fs. The nuclear coordinates and velocities were sampled by Wigner distribution with uncorrelated quantum harmonic oscillators [68, 69]. The sampled initial conditions are duplicated for both LS and GS algorithms, that means the theoretical levels and initial conditions are completely the same for both algorithms. The only differences between the two algorithms are hopping spots, switching probabilities and velocity adjustments. These features are extensively compared in the LS and GS-TSH dynamics simulations.

Hopping spots. Both SA-CASSCF and OM2/MRCI optimizations located a rotation CI-rot and CI-inv between the S_0 and S_1 states [36]. The two CIs aroused from the rotation and planar inversion of the CNNC dihedral angle (C_1–N_1–N_2–C_2 in Fig. 12.3), respectively. Figure 12.4 showed good agreement between LS and GS algorithms for hopping spot distributions. In *cis*-to-*trans* photoisomerization, both two chiral isomers of CI-rot respectively from clockwise and counterclockwise rotations around N–N bond are encountered at OM2/MRCI level as shown in Fig. 12.4. In contrast, only one chiral pathway was observed at SA2-CASSCF(6,6) level (Fig. 4 in Ref. [36]). The TSH simulations based on LS and GS algorithms show very similar hopping spots distributions with respect to the CNNC [–180°, 180°] and one of NNCC dihedral angles (e.g., N_2–N_1–C_1–C_3 in Fig. 12.3) [$-180°$, $180°$], in which two islands simulated from LS and GS show the same contour patterns. In the range of CNNC [75°, 105°] and NNCC [$-30°$, $30°$], which represent the P-helical (clockwise) pathway, there are 717 (LS) and 652 (GS) hopping points (some trajectories can hop more than twice), respectively. In the range of CNNC [$-75°$, $-105°$] and NNCC [75°, 135°], which represent the M-helical (counterclockwise) pathway, there are 239 (LS) and 203 (GS) hopping points, respectively. In addition, the ratio of the P-helical/M-helical channel $717/239 = 3$ from LS is very close to $652/203 = 3.2$ from GS algorithm. The more sampling trajectories decay from P-helical S_0/S_1 CI-rot is in line with the fact that P-helical channel is energetically favorable for P-helical *cis*-azobenzene. For *trans*-to-*cis* photoisomerization, two islands of the hopping spots around CI-rot simulated from LS and GS show the same contour

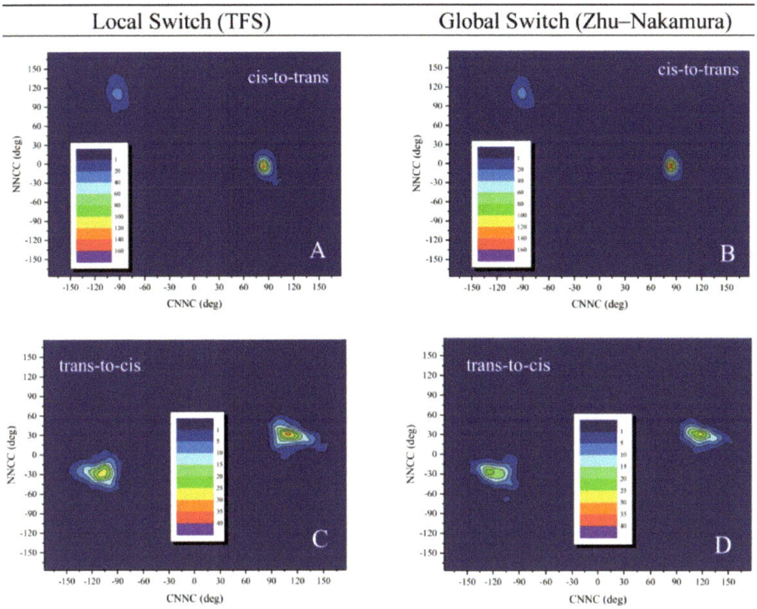

Figure 12.4 Simulated hopping spots in terms of CNNC and NNCC1 dihedral angles by LS (left panel) and GS (right panel) algorithm, reprinted from Ref. [70], Copyright © 2017 Wiley-VCH.

patterns as well. The ratio between clockwise and counterclockwise channels is 474/452 = 1.05 (LS) and 337/330 = 1.02 (GS), as expected because *trans*-azobenzene has a planar geometry. In *trans*-to-*cis* photoisomerization, neither LS nor GS shows the sampling trajectories decay from the CI-inv channel at OM2/MRCI level. This is the same as the GS-TSH at SA-CASSCF level.

Switching probability. Since the two algorithms show very similar hopping spots distribution, the ensemble statistical averages such as the averaged populations of electronic states and quantum yields of final products, which directly determined by the surface switching are good criterions to benchmark the switching probabilities calculated from LS and GS algorithms. The lifetimes of the photoisomerization (excited state) are estimated by $\tau + t_d$ according to the decay curves of the S_1 state (the time evolution of averaged S_1 state populations), where t_d is the onset time of the S_1 population loss,

Table 12.1 Comparison between LS and GS algorithms for simulated quantum yields (QY) and lifetimes (unit: fs) at OM2/MRCI level for reactive photoisomerization $S_0 \leftrightarrow S_1$ in azobenzene with respect to the effective number of trajectories N_{traj}

	N_{traj}	*cis*-to-*trans*		*trans*-to-*cis*	
Algorithm	**(*cis*/*trans*)**	**QY**	**Lifetime**	**QY**	**Lifetime**
LS	786/570	0.58 ± 0.03	82.1 ± 0.5	0.20 ± 0.08	0.318 ± 0.001
GS	797/575	0.57 ± 0.03	71.0 ± 0.5	0.15 ± 0.10	0.284 ± 0.001

and τ is the time constant fitted from the exponential decay function of S_1 state population $f(t) = \exp[(t - t_d)/\tau]$ [36]. The distribution of the CNNC dihedral angle shows significant difference between the *cis*- and *trans*-azobenzene, the final ground state products could be obviously determined by monitoring the CNNC torsion angle. Hence, the quantum yield (QY) is calculated by $\Phi_{QY} = N_r/N_{traj}$ with standard error $\sqrt{(N_{traj} - N_r)/(N_{traj} N_r)}$, where N_{traj} is total number of trajectories and N_r is number of the reactive trajectories. As shown in Fig. 12.5, LS and GS fit well with each other for the time evolution of averaged S_0 and S_1 state populations. Simulated quantum yields for the *cis*-to-*trans* $S_1(n,\pi^*)$ photoisomerization are 0.58 and 0.57, and the lifetimes are 82 and 71 fs, for LS and GS algorithms, respectively (0.20 and 0.15, and 318 and 284 fs for *trans*-to-*cis*), as shown in Table 12.1. Thus, the excellent agreement on both lifetimes and quantum yields confirms that the switching probability by GS is in line with that by LS.

Velocity adjustment. As trajectories run with the same initial conditions for LS and GS algorithms, simulated on-the-fly trajectories behave exactly the same before hopping. The different is that the trajectory simulated from LS varies according to the direction of the nonadiabatic vector coupling after hopping, whereas the trajectory simulated from GS changes according to the direction defined by Eq. (12.12). If two trajectories with completely the same initial condition respectively simulated hopped at the same (or very close) hopping spots (or hopping time), the following evolution of the two trajectories should be the same with each other for LS and GS. Thus, a series of typical trajectories were selected for the comparison. Figure 12.6 shows the time evolution of geometries

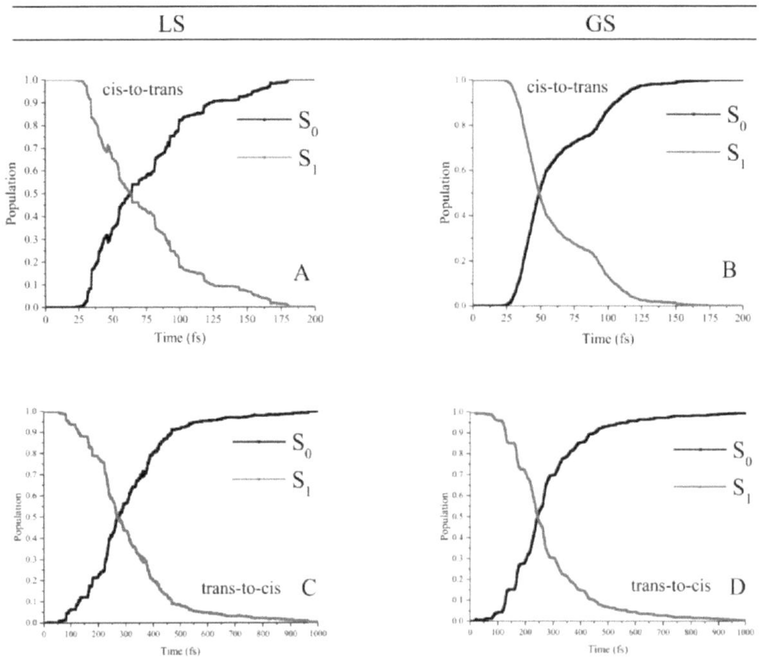

Figure 12.5 Simulated time evolution of average S_0 and S_1 populations from local (left panels) and global (right) switching algorithms, reprinted from Ref. [70], Copyright © 2017 Wiley-VCH.

in four types of trajectories for *cis*-to-*trans* simulations, including the clockwise reactive, clockwise nonreactive, counterclockwise reactive and counterclockwise nonreactive trajectories. The hopping times for the four types of trajectories, 31.5 (32.0), 51.6 (51.4), 97.8 (97.9), 131.7 (131.3) fs were found for LS (GS) algorithm. Before hopping, the important CNNC, two NNCC dihedral angles (NNCC1, and NCC2), and two NNC angles (NCC1 and NCC2) must show exactly the same variations as their initial coordinates and velocities are completely the same. Both algorithms indicate that the CNNC dihedral angle changes rapidly toward the clockwise or counterclockwise direction that corresponds to the P- and M-helical rotational channels. It is remarkable that after hopping occurs, the dihedral and bond angles still show nearly the same variations. This indicates that the self-consistent hopping direction in GS indeed agrees with the direction defined by nonadiabatic coupling vectors

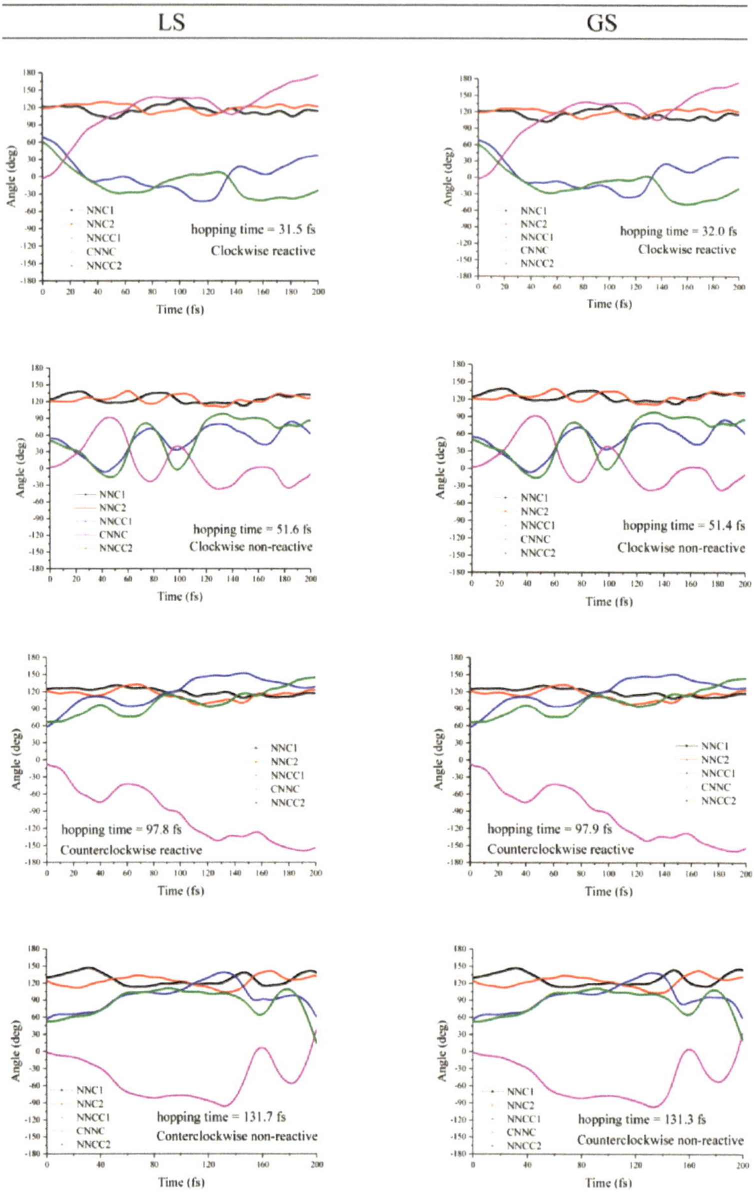

Figure 12.6 The selected clockwise reactive trajectories simulated from local (left panels) and global (right) switching starting at *cis*-azobenzene: the corresponding evolution of five important angles and dihedral angles, reprinted from Ref. [70], Copyright © 2017 Wiley-VCH.

in LS. For *trans*-to-*cis* simulations, similar results could be observed in the selected four types of trajectories (see Ref. [70] for details).

Concluding remarks. Excellent agreement between LS and GS algorithms was obtained not only for quantum yields and lifetimes, but also for hopping spots and hopping directions. The TSH benchmark at OM2/MRCI demonstrated that LS and GS algorithms are in excellent agreement for on-the-fly trajectory surface hopping molecular dynamics simulations involving conical intersections of *cis*-to-*trans* and *trans*-to-*cis* azobenzene photoisomerization. Not only are the quantum yields and lifetimes of highly averaged quantities similar, but the hopping spot distributions, and hopping directions are also highly consistent with each other. There has been a misunderstanding about GS, for which trajectory hopping spot was thought to only occur at conical intersection points. However, this is not true and in fact we showed hopping spots spread across the entire conical intersection area, in the same way as those simulated using LS algorithms (see Fig. 12.4). Moreover, it was confirmed that GS can perform trajectory surface hopping with a larger time stepsize than LS in order to maintain the same accuracy of convergent results [36]. This is because the GS approach requires only potential energies and their gradients, and its trajectory evolution as well as computational cost is similar to those of molecular dynamics on a single potential energy surface. For nonadiabatic molecular dynamic simulations involving conical intersections for larger complex systems and/or longer propagation time, the GS approach provides a powerful tool in excited-state molecular dynamic simulation of the photochemical and photophysical applications.

12.4 The Performance of Time-Dependent Density Functional Theory in Global Switch Algorithm

By extensively comparing LS with GS, the global switching (GS) algorithm in which nonadiabatic coupling calculation is not required has been proved to be a very powerful TSH method associated with on-the-fly TDDFT potential energy surfaces, especially as a suitable TSH-MD simulation tool for large size of systems. For TSH-MD simulations with LS algorithm, however, the PES topology around conical intersections is critical for the hopping judgment and switching probability calculations. A crucial question is how TDDFT method can correctly describe conical intersections—molecular geometries where the two electronic states are exactly degenerate. In traditional linear response TDDFT method, the reference state is usually a closed-shell Hartree–Fock configuration, so that strictly symmetry-adapted singlet response states could be constructed by a single electronic spin-preserve excitation while the closed-shell reference wavefunction computed by DFT is stable. In this case, the PES topology of conical intersections between two excited states is correct. However, traditional LR-TDDFT is usually not a good choice for describing S_0/S_1 CI even ignoring its failure on the systems with open-shell ground states. When dealing with traditional spin-preserve LR-TDDFT with closed-shell reference, the nonadiabatic coupling matrix elements between the reference S_0 state and all the excited states exactly vanish by virtue of Brillouin's theorem. This means that there is only one condition to be satisfied in order to obtain degeneracy, namely the branching space is one- rather than two-dimensional for any conical intersection that involves the ground state [10]. To correctly describe the S_0/S_1 CI and open-shell ground-state species, the spin-flip (SF) scheme was proposed for LR-TDDFT. In the spin-flip approach, the reference state is the lowest high-spin state, while the response states with low-spin are generated by the spin-flip of an electron from α to β spin. The low-spin target states can be treated as "excited states," and thus they hold equal footing with excited states of the same spin. The SF-TDDFT rigorously fixes the PES topology problem around CIs [19]. To show the performance of TDDFT simulations on the nonadiabatic process involving S_0/S_1 CIs, the TSH-MD simulations for the photoisomerization of azobenzene upon S_1-(n,π*) excitation was performed again at both traditional LR-TDDFT and SF-TDDFT level with the same functional and basis sets. The PES topologies of S_0 and S_1 around CI-rot are compared and discussed by running trajectories on both LR-TDDFT and SF-TDDFT on-the-fly PESs within GS-TSH-MD simulation. For convenience, in the following subsections, the traditional spin-conserving LR-TDDFT and spin-flip TDDFT are briefly noted as LR-TDDFT and SF-TDDFT, respectively.

12.4.1 Topology of S_0 and S_1 PESs Around Conical Intersections

In general, the two-dimensional subspaces constructed from directions of the gradient difference vector (GDV) $\boldsymbol{g}_{ij}$ and derivative coupling vector (DCV) $\boldsymbol{h}_{ij}$ (i.e., nonadiabatic coupling vectors, $\boldsymbol{d}_{ij}$) are utilized to describe PES topology around CI, and $g-h$ spaces are defined as [71]

$$\boldsymbol{g}_{ij} = \frac{\partial E_i}{\partial \boldsymbol{R}} - \frac{\partial E_j}{\partial \boldsymbol{R}} \tag{12.13A}$$

$$\boldsymbol{h}_{ij} = \left\langle \varphi_i \left| \frac{\partial}{\partial \boldsymbol{R}} \right| \varphi_j \right\rangle, \tag{12.13B}$$

where the E_i (φ_i) and E_j (φ_j) are the energy (wavefunction) corresponding to electronic state i and j, respectively. For azobenzene molecular, the direction $\boldsymbol{g}_{ij}$ and $\boldsymbol{h}_{ij}$ could be approximated as the two-dimensional subspaces described from two coordinates of CNNC and NNCC dihedral angles. Thus, we scanned two-dimensional PESs around S_0/S_1 CI-rot in terms of the CNNC and NNCC dihedral angles to show the PES topology. At a true CI, the degeneracy between two intersecting PESs can be lifted along $\boldsymbol{g}_{ij}$ or $\boldsymbol{h}_{ij}$. This is demonstrated by the two-dimensional PES topology of around S_0/S_1 CI-rot calculated at SA2-CASSCF(6,6) and SF-TDDFT (SF-BHHLYP/6-31G) levels as shown in Figs. 12.7A,B, respectively. PES topology at S_0/S_1 CI-rot is a normal crossing shape (see Fig. 12.7 upper panel) along the direction of the CNNC dihedral angle, and funnel shape (see Fig. 12.7, lower panel) along the direction of the NNCC dihedral angle, but inverted cone on S_0 PES. Two-dimensional PESs around S_0/S_1 CI-rot agree well with each other between two calculations from SA2-CASSCF(6,6) and SF-TDDFT. The PES topology of the conical intersections calculated by SF-TDDFT is also consistent with the CI loop energy profiles calculated by Olivucci et al. [72], and this means that the SF-TDDFT method can describe a correct topology of nonlinear conical intersection. In addition, the optimized geometries and excitation energies by SF-TDDFT also agree well

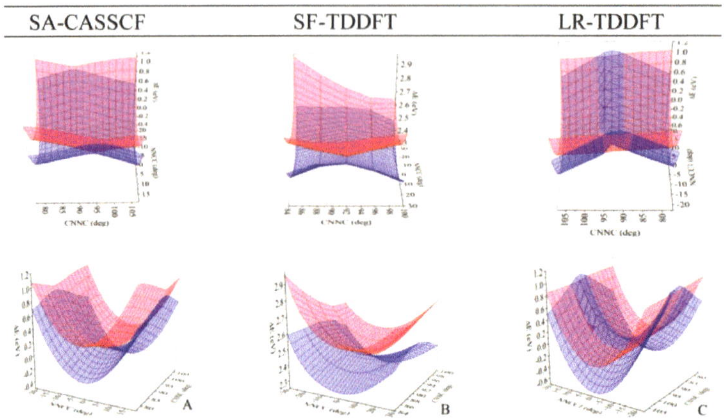

Figure 12.7 The two-dimensional PES topology in the vicinity of minimum energy conical intersections of S_0 and S_1 calculated from (A) SA2-CASSCF(6,6) and (B) SF-TDDFT or in vicinity of local minimum of S_1 PES calculated from (C) LR-TDDFT in different perspectives. BHHLYP functional and 6-31G basis set are used in both TDDFT calculations, reprinted from Ref. [27], Copyright © 2018 Royal Society of Chemistry. Upper and lower panels are same PESs viewed from different direction.

with the those obtained from multi-configurational methods, not only for electronic structures of key sudden points (see Ref. [27] for detailed comparison).

Due to a possible instability of closed-shell ground state around S_0/S_1 CI-rot based on LR-TDDFT calculation, the incorrect PES topology is actually shown in Fig. 12.7C for azobenzene system. Although the correct S_0/S_1 CI-rot cannot be located by LR-TDDFT (TD-BHHLP/6-31G) calculation, it was found that a local minimum on S_1 surface (marked as *cis*-min S_1) optimized by LR-TDDFT is close to geometry of the CI-rot S_0/S_1 optimized by SF-TDDFT. Thus, we scanned two-dimensional PES around *cis*-min S_1 (i.e., S_0/S_1 CI-rot) as shown in Fig. 12.7C. Outside range of $90^\circ \sim 100^\circ$ of CNNC dihedral angle, the S_0 and S_1 PESs scanned by LR-TDDFT are similar to those scanned by SF-TDDFT in Fig. 12.7B. However, due to the negative excitation energy computed from LR-TDDFT, there appeared significant dual crossings along the direction of CNNC dihedral angle in the range $90^\circ \sim 100^\circ$. The double-cone topology around *cis*-min S_1 scanned by LR-TDDFT is considered to be equivalent to single-cone S_0/S_1 CI-rot scanned by SF-TDDF. Dual crossings appearing in

Fig. 12.7C are purely artificial effect due to unstable closed-shell reference state around conical intersection in LR-TDDFT calculation. The artificial double-cone problem might be partially solved if a stable and unrestricted open-shell reference state is used in LR-TDDFT calculation in which the spin symmetry of alpha and beta orbitals is broken down. However, due to the heavy spin-contamination, the target spin states are difficult to be tracked and large quantitative errors might be observed especially for the S_0/S_1 energy gaps, which is critical for nonadiabatic transition. What's more, since LR-TDDFT does not include double excitations as well as configuration interaction truncated at single excitations (CIS), the derivative coupling $\boldsymbol{h}_{ij}$ (NNCC direction) vanishes and the intersecting states can be lifted only along $\boldsymbol{g}_{ij}$ (CNNC direction). In this case, only a one-dimension branching space is observed for CI-rot in LR-TDDFT scanning while an incorrect linear crossing with 3N-7 intersection space (IS) but rather than a true CI with 3N-6 IS is shown in Fig. 12.7C.

This artificial S_0/S_1 double-cone crossings with one-dimensional branching space is also observed in the PES of water molecule [10] and in the loop energy profile around the S_0/S_1 crossing points for penta-2,4-dieniminium-cation photoisomerization [72] from LR-TDDFT calculation. This means that the incorrect S_0/S_1 topology of the linear conical intersection could be quite general within LR-TDDFT calculation. Therefore, one should be aware that the TSH-MD simulation with LR-TDDFT method might produce a large error in the final statistics. However, it is curious to ask how much difference this artificial effect cause in the final statistics based on LR-TDDF on-the-fly potential energy surfaces in TSH-MD simulation. Especially, if we are only interested in highly averaged quantities like quantum yield and lifetime, GS algorithm can still show some meaningful simulation results. In other words, the artificial double-cone topology within GS-TSH-MD simulation could be an acceptable error.

12.4.2 GS-TSH-MD Simulations by Time-Dependent Density Functional Theory with and without Spin-Flip

We take the azobenzene photoisomerization as an example again to perform GS-TSH-MD simulation on on-the-fly PESs calculated from

TDDFT with and without spin-flip. Namely, we make extensively comparison between two simulated results from SF-TDDFT and LR-TDDFT methods in the same computational platform for *cis*-to-*trans* and *trans*-to-*cis* azobenzene photoisomerization up to excitation to the S_1-(n,π*) state. The both SF-TDDFT and LR-TDDFT methods utilize BHHLYP functional with 6-31G basis set for computing on-the-fly PESs. The propagation time stepsize is 0.5 fs and time limitation is 200 fs (1500 fs) for *cis*-to-*trans* (*trans*-to-*cis*) photoisomerization for all sampling trajectories. The nuclear coordinates and velocities are initially sampled by the Wigner distribution with uncorrelated quantum harmonic oscillators [68, 69] as performed in Section 12.3. For current SF-TDDFT method, the spin-contamination might introduce some undesired spin-mixed states in TSH-MD simulation. Thus, we utilize a practically simple scheme to handle undesired states by monitoring the expectation values of the spin operator $\langle \boldsymbol{S}^2 \rangle$ for on-the-fly PESs. For both *cis*-to-*trans* and *trans*-to-*cis* dynamics simulation together with total 300 sampling trajectories running on SF-TDDFT (LR-TDDFT) on-the-fly PESs, 300 (259) sampling trajectories successfully end on S_0 state for *cis*-to-*trans* isomerization and 226 (268) successfully end on S_0 state or end on S_1 state for *trans*-to-*cis* isomerization. During on-the-fly GS-TSH-MD simulations, there are SCF nonconvergent problems or/and discontinuous problems of PESs at on-the-fly electronic structure calculations in the vicinity of conical intersections for part of sampling trajectories and those trajectories are excluded in the final statistics. As a result, 41 sampling trajectories are excluded with use of LR-TDDFT method for *cis*-to-*trans* isomerization, while 74 (32) sampling trajectories are excluded for *trans*-to-*cis* isomerization with use of SF-TDDFT (LR-TDDFT) method.

***Cis*-to-*trans* photoisomerization.** Simulated quantum yield of 0.43 (lifetime 63 fs) from the SF-TDDFT method agrees well with the previous one 0.39 (53 fs) from the SA2-CASSCF(6,6) method (see Table 12.2), this is because PES topology around S_0/S_1 rot-CI is quite similar as shown in Figs. 12.7A,B for these two methods. LR-TDDFT method gives 0.34 (62 fs) that is not bad even with artificial double cone structure around S_0/S_1 CI-rot. One sampling trajectory within LR-TDDFT method can usually have many hops at artificial double-

Table 12.2 Simulated quantum yields and lifetimes from different theoretical methods and experimental measurements

Method	N_{traj} (*cis*/*trans*)	*cis*-to-*trans* QY	*cis*-to-*trans* Lifetime (fs)	*trans*-to-*cis* QY	*trans*-to-*cis* Lifetime (ps)
SF-TDDFT [27]	300/226	0.43 ± 0.07	63.1 ± 1.1	0.11~0.16	2.218 ± 0.010
LR-TDDFT [27]	259/268	0.34 ± 0.09	62.0 ± 0.9	0.13 ± 0.16	1.039 ± 0.009
SA2-CASSCF (6,6) [36]	800/600	0.39 ± 0.04	53.1 ± 3.0	0.33 ± 0.05	0.81 ± 0.10
Expt. [61]	-	0.41~0.56	100~170	0.23~ 0.35	0.9~1.4

cone region, while the sampling trajectory within SF-TDDFT method has usually one hop at real single-cone region. This means in average that overall nonadiabatic transition probability calculated from LR-TDDFT around double-cone is basically similar to that calculated from SF-TDDFT around single-cone, especially for highly averaged quantities of quantum yield and lifetime. We conclude that quantum yield is basically determined by PES topology around S_0/S_1 rot-CI, while lifetime is depend on global PES profiles because the main difference between the SF-TDDFT and TDDFT methods is in the S_0/S_1 rot-CI region.

Figure 12.8 shows average state populations of S_0 and S_1 as a function of time for the SF-TDDFT and TDDFT calculations. Both methods show the same tendency of rapid decay for S_1 population and onset time constant for S_1 population decreasing is 18.5 fs (18.0 fs) for the SF-TDDFT (TDDFT) calculations. However, the population decay curve with the SF-TDDFT calculation is much smoother than that with the LR-TDDFT calculation. The decay curve shows a zigzag pattern with LR-TDDFT due to the artificial effect of the double-cone induced many artificial back-and-forth hops between S_0 and S_1 states.

Figure 12.9 shows hopping spot distributions for the SF-TDDFT and LR-TDDFT calculations, in which LR-TDDFT gives larger area of hopping spots than those of SF-TDDFT since an artificial double-cone exists in LR-TDDFT PESs instead of real single-cone.

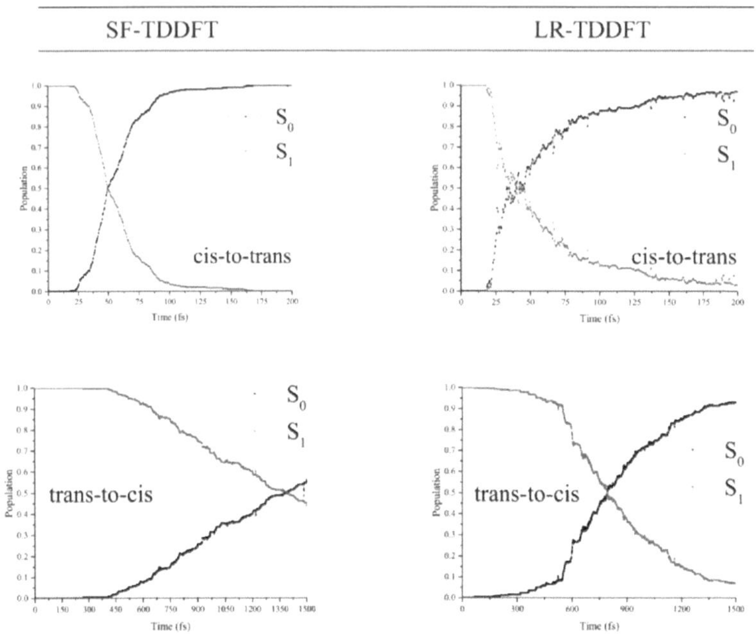

Figure 12.8 Simulated time evolution of average S_0 and S_1 populations with upper (lower) panel for *cis*-to-*trans* (*trans*-to-*cis*) isomerization, reprinted from Ref. [27], Copyright © 2018 Royal Society of Chemistry.

Nonetheless, both methods do show the same pattern of a hopping contour map around "CI" (it is a true CI corresponding to CI-rot S_0/S_1 for SF-TDDFT, while it represents *cis*-min S_1 in the case of LR-TDDFT). Actually, there are 330 (1452) hops from the SF-TDDFT (LR-TDDFT) calculation and all hops appear in the range [75°, 105°]/[−30°, 30°] of the CNNC/NNCC1 dihedral angles and correspond to the clockwise (P-helical) pathway for *cis*-to-*trans* isomerization. This is exactly the same as in the previous SA2-CASSCF(6,6) simulation where there are only clockwise pathway observed. However, both clockwise (P-helical) and counterclockwise (M-helical) pathways (ratio is about 3:1) are found with OM2/MRCI on-the-fly potential energy surfaces simulation. This means that potential energy surfaces in the Franck–Condon region are the same for three methods (SF-TDDFT, LR-TDDFT and SA-CASSCF), but are different from the OM2/MRCI method. The S_0 and S_1 PESs far from

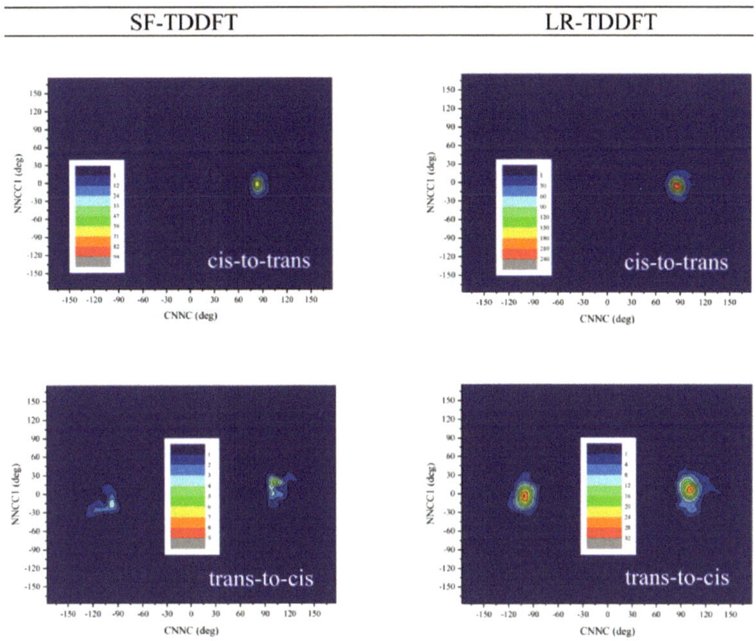

Figure 12.9 Simulated hopping spots in terms of CNNC and NNCC1 dihedral angles with upper (lower) panel for *cis*-to-*trans* (*trans*-to-*cis*) isomerization, reprinted from Ref. [27], Copyright © 2018 Royal Society of Chemistry.

the S_0/S_1 CI by LR-TDDFT is expected to be better than those by OM2/MRCI, although the former has incorrect topology at S_0/S_1 CI-rot.

The average hopping numbers per trajectory are 1.1 (5.6) for the SF-TDDFT (LR-TDDFT) simulation, in which most sampling trajectories hop in one of the double cone region from S_1 to S_0 and then return back in another to S_1, and after several back-and forth hops the trajectories finally depart from the CI region to the ground state. It is surprising that the number of switches (1.1 times per trajectory) in SF-TDDFT is much less than those (5.6 times per trajectory) in LR-TDDFT, but simulated quantum yields and lifetimes are quite similar in average. This reflects that the double-cone topology in LR-TDDFT makes some sense in comparison with the correct single-cone topology in SF-TDDFT, so that a one-time switching probability computed at the single cone can be approximately equal to the average of six-time switching probabilities computed at the double cones for one sampling trajectory. As a result, the overall nonadiabatic switching probability simulated from LR-TDDFT seems to be in reasonable agreement with that simulated from SF-TDDFT.

***Trans*-to-*cis* photoisomerization.** Simulated quantum yield 0.11 (by SF-TDDFT) agrees well with 0.13 (by LR-TDDFT), but simulated lifetime 2218 fs (by SF-TDDFT) is about twice of 1039 fs (by LR-TDDFT). Simulated quantum yields from TDDFT do not agree with the previous one 0.33 from SA2-CAS(6,6) method (see Table 12.2), but simulated lifetime 1039 fs (by LR-TDDFT) agrees with 890 fs (by SA2-CASSCF). The main difference happens at S_1/S_0 CI-inv that is conical intersection close to Franck–Condon region around *trans*-isomer. There is no sampling trajectory making hop at CI-inv from SA2-CASSCF simulation, while there are 74 (14) sampling trajectories (regarded as unsuccessful ones) making hops at CI-inv from SF-TDDFT (TDDFT) simulation and those trajectories are not taken into account in the final statistics in order to compare with the SA2-CAS(6,6) simulation. Although the PES topology is quite similar around CI-inv S_1/S_0 scanned from both SA2-CASSCF and SF-TDDFT as shown in Fig. 12.10.

The other reason is that *trans*-min S_1 is close to CI-inv in geometry from the SF-TDDFT simulation, and this leads to heavy spin contamination problem $\langle \boldsymbol{S}^2 \rangle \approx 1$ for sampling trajectories running

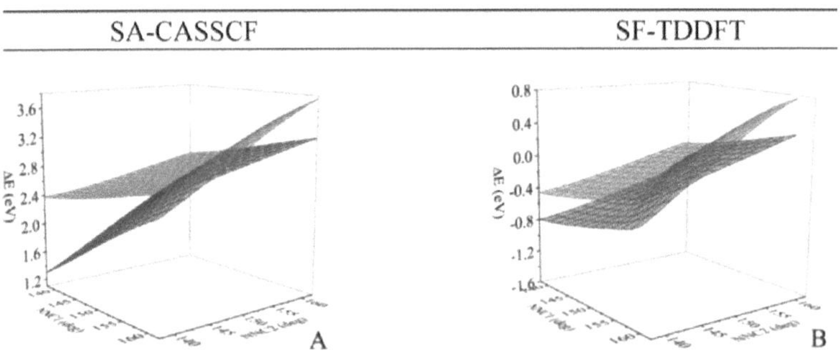

Figure 12.10 The two-dimensional PES topology of S_0 and S_1 in vicinity of minimum energy conical intersections simulated from (A) SA2-CASSCF and (B) SF-TDDFT, reprinted from Ref. [27], Copyright © 2018 Royal Society of Chemistry.

around *trans*-min S_1 for *trans*-to-*cis* isomerization. On the contrary, spin contamination is weak as $\langle \boldsymbol{S}^2 \rangle \approx 0.0 \sim 0.3$ for sampling trajectories starting from *cis*-to-*trans* isomerization. Besides, the sampling trajectories can trap in *trans*-min S_1 for a while, and thus longer lifetime is expected for *trans*-to-*cis* isomerization (this is true for SA2-CAS, SF-TDDFT, and LR-TDDFT simulation). Due to spin contamination accumulated in longtime sampling trajectories from ST-TDDFT simulation, number of resonance trajectories trapped on S_1 potential energy surface is quite different between SF-TDDFT and LR-TDDFT. LR-TDDFT may have problem around CI-inv, but it could be very similar to SA2-CAS in geometry around *trans*-min min S_1 and thus it leads to almost same lifetime ~1000 fs with SA2-CAS simulation. The spin-contamination problem in TSH-MD simulation is not easy to be solved in current SF-TDDFT implementation unless more rigorous methods like the state tracking [73] or spin-complete techniques [33] would be employed. Nevertheless, SF-TDDFT and LR-TDDFT methods can still agree each other for simulated quantum yield, but not for lifetime.

The number of sampling trajectories successfully ending on the S_0 state is 130 (249) with 132 (718) hopping points for the SF-TDDFT (TDDFT) simulations. As shown in Fig. 12.9 (lower panel), the two methods show similar hopping spot distribution in terms of the CNNC and NNCC1 dihedral angles; the left island with CNNC/NNCC1 dihedral angles in the range of [−130°, 90°]/[−30°, 30°] stands for the clockwise (P-helical) pathway of photoisomerization, while the right island in the range [90°, 130°]/[−30°, 30°] represents the counterclockwise (M-helical) pathway. The number of clockwise/counterclockwise hopping points is 61/71 (339/379) from SF-TDDFT (TDDFT) simulation, and thus the ratio for both methods is close to one. As the *trans*-isomer has a planar symmetric structure, the sampling trajectories can therefore have almost equal opportunity to follow the clockwise and counterclockwise rotation pathways.

Concluding remarks. The SF-TDDFT method can not only deal with the open-shell species like biradicals but also describe a correct single-cone topology of S_0/S_1 CI-rot, which is consistent with the CI loop energy profiles by multireference methods. For *cis*-to-*trans* and *trans*-to-*cis* azobenzene photoisomerization up to S_1-(n, π*) excitation, the overall performance of GS-TSH-MD simulation by SF-TDDFT is well agree with those by SA-CASSCF and experiments, especially for highly averaged quantities such as quantum yields, lifetimes, population decay distribution, and distributions of hopping spots. However, in order to track desired states in simulation with SF-TDDFT on-the-fly potential energy surfaces, one need additional computational costs to monitor spin properties and other type of data like MOs and CI coefficients to eliminate those undesired spin contaminated states. Moreover, in comparison with the traditional TDDFT which based on a closed-shell reference state, the unrestricted open-shell configuration used in SF-TDDFT unavoidably raises computational costs and slows down speed of the SCF convergence.

On the other hand, the conventional LR-TDDFT simulation may have error like artificial S_0/S_1 double-cone topology instead of a real single-cone CI. However, within GS-TSH-MD simulation, LR-TDDFT method presents quite similar results with SF-TDDFT method. The reason is that most of sampling trajectories do not make hops right at CI (where energy gap is zero), while make hops near CI where the energy gap can be small but still in valid region for LR-TDDFT method. Therefore, LR-TDDFT is very suitable method incorporated with GS-TSH-MD simulation for fairly large systems and/or long-time scale simulation. It should be emphasized that LR-TDDFT method might be system dependent for handing S_1/S_0 CI, some benchmark test might be necessary when it is applied for GS-TSH-MD simulation. We expect that LR-TDDFT associated with GS-TSH-MD simulation must work well for most of molecular systems since it works even in difficult situation of azobenzene photoisomerization in which the artificial double-cone CI shows up.

12.5 Time-Dependent Density Functional Theory Functional and Basis Set Dependence in GS-TSH-MD Simulation

The extensive comparisons between TDDFT with and without SF in the previous section clearly showed that simulated quantum yields, lifetime, hopping spot and population distributions agree fairly well with each other. It is very encouraging for further investigation of LR-TDDFT (hereafter referred to as TDDFT) functional and basis set dependence in GS-TSH-MD simulation. TDDFT is well-developed method that can be found in most popular computational programs and is quite friendly towards dynamic simulation for large systems. Although traditional TDDFT method might have problem to describe PES topology around S_0/S_1 CI due to a restricted closed-shell reference state, however, numerous publications which even involved transitions between ground and excited states are still available for TSH-MD simulations in the framework TDDFT [20, 22, 74, 75], and references therein. One reason is that closed-shell TDDFT can still work correctly as far as wavefunction of restricted closed-shell reference state is stable as S_0 and S_1 states are not completely degenerate, in which region most of on-the-fly running trajectories make hopping correctly while only very small portion of the trajectories really reach absolutely degenerate CI where TDDFT might fail. Another reason is fast computational efficiency of closed-shell TDDFT for excited-state calculations as many nicely developed techniques are available such as resolution-of-the-identity approximation [76–78], and the Car–Parrinello MD (CPMD) [79–83] which based on the plane-wave/pseudopotential basis and parallel acceleration by graphic process unit (GPU) are used to perform ab initio MD for very large size of systems and/or longtime simulation. Within these approaches, SF-TDDFT scheme is presently not available and implemented for the TDDFT (linear response, real time or imaginary time TDDFT) calculations. In many cases for large systems, TDDFT could be only choice for affordable accurate calculation in the on-the-fly TSH-MD simulations.

As is well known, functional and basis set dependence in DFT and TDDFT electronic structure calculations might be significant for obtaining accurate global potential energy surfaces which are required for on-the-fly TSH-MD simulation. Of course, there are systematic studies toward finding the best functional for all chemical problems under study as so-called Jacob's ladder of DFT functionals [84], while for specific excited-states such as Rydberg and charge transfer states [85], some functionals are designed and can have better performance in TDDFT calculation. For the photoinduced molecular dynamics mostly influenced by conical intersections, it is also necessary to choose an appropriate functional to make sure the PESs around crossing zones as smooth as possible. However, the more accurate functionals and bigger basis sets are implemented, the more computational costs are increased. How to balance calculation accuracy with less computational costs for affordable TSH-MD simulation is big question. If one is only interesting in estimating highly averaged quantities such as quantum yields, lifetimes, and population decay distribution in TSH-MD simulations, the functional and basis set dependences might be less sensitive. Practically speaking, on-the-fly TSH-MD simulation can be performed with benchmark test for suitable functional and affordable basis set for large complex systems under study. Therefore, we continue another benchmark study for on-the-fly GS-TSH-MD simulation within TDDFT method, and we focus on analyzing functional and basis set dependence. Three functionals, B3LYP, BHandHLYP (50% Hartree–Fock +50% local spin density approximation +50% Becke for exchange and LYP for correlation), and CAM-B3LYP [85] in Gaussian software package, are chosen for performing GS-TSH-MD simulation. B3LYP is the most popular functional in quantum chemistry calculations, BHandHLYP is quite popular for nonadiabatic dynamics simulations, and CAM-B3LYP is very suitable for excited-state calculations. All three functionals chosen here show smooth potential energy surfaces globally for preliminary trajectory surface hopping dynamic calculations with GS-TSH for *cis*-azobenzene photoisomerization. Four basis sets from small (3-21G) to large (cc-pVDZ), namely, 3-21G, 6-31G, 6-31G*, and cc-pVDZ are selected in performing on-the-fly GS-TSH-MD simulation.

12.5.1 Functional and Basis Set Dependence on Artificial Double Cone

As discussed in Section 12.4, TDDFT calculations with three functionals (B3LYP, CAM-B3LYP, and BHandHLYP) all show artificial double-cone CIs around true S_0/S_1 CI-rot (CNNC $= -92.0^{\circ}$) from SF-TDDFT. In order to clarify the influence of functionals and basis sets for the double-cone PES topology, the one-dimensional potential energy curves (PECs) that connect *cis*-, *trans*-azobenzene and two

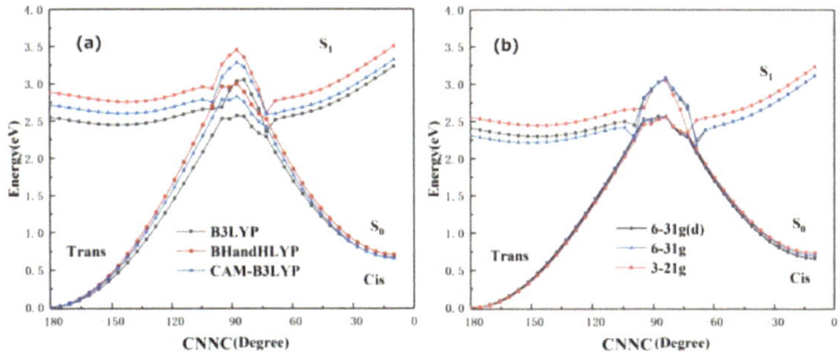

Figure 12.11 Potential energy profiles along the LIICs in terms of dihedral angle CNNC pairwise constructed by two local minima and two hopping points. (A) Three functionals with the same basis set 6-31G* and (B) three basis sets with the same functional B3LYP, reprinted from Ref. [29], Copyright © 2020 Wiley-VCH.

hopping points are scanned based on the linear interpolations of internal coordinates (LIICs) [86] as shown in Fig. 12.11. Two local minimum on S_0 surface for *cis*-(CNNC $= -9.8°$) and *trans*-azobenzene (CNNC $= -180.0°$) are optimized by B3LYP/6-31G* while the two hopping points with CNNC $= -72.9°$and $-94.7°$ for double cone structure come from GS-TSH-MD simulations. Based on the LIICs at B3LYP/6-31G(d) level, the S_0 and S_1 PECs are calculated by using five functional/basis set combinations.

Three functionals with the same basis set 6-31G* in Fig. 12.11A show that the vertical excitation energies ΔU of S_1 state (S_0/S_1 energy gaps) around *cis*- and *trans*-regions vary about 0.4 eV with order of ΔU (BHandHLYP) $>$ ΔU (CAM-B3LYP) $>$ ΔU (B3LYP), but differences of ΔU are getting smaller while approaching the CI seam surfaces where all ΔU are almost the same. Three basis sets with the same functional B3LYP in Fig. 12.11B show that the ΔU around *cis* and *trans* regions vary about 0.3 eV with order of ΔU (3-21G) $>$ ΔU (6-31G*) $>$ ΔU (6-31G), but potential energy profiles for both S_0 and S_1 around the double-cone CI seam surfaces almost overlap with each other. Moreover, it should be noted that for a potential energy curves, the basis set dependence is less sensitive than functional dependence for hopping spot distributions. Thus, it can be predicted that the total five functional/basis set combinations might provide very similar transition probability distributions in

GS-TSH-MD simulation, although they can have different vertical excitation energies around *cis*- and *trans*- Franck–Condon regions.

12.5.2 Functional and Basis Set Dependence on Dynamic Quantities

250, 400, and 600 sampling trajectories are performed in which about 20%, 50%, and 67% trajectories do not successfully end on the S_0 state, respectively, from B3LYP, BHandHLYP and CAM-B3LYP functionals within GS-TSH-MD simulation. In this way, we can ensure that about 200 successful trajectories are taken into account in final statistical analysis for all functional/basis set combinations. Except for those combinations including cc-pVDZ basis set, only 60, 100, 160 trajectories are performed to ensure about 50 successful sampling trajectories in final statistical analysis. Actually, total 12 functional/basis set combinations including three functionals (B3LYP, BHandHLYP, and CAM-B3LYP) plus four basis sets (3-21G, 6-31G, 6-31G*, and cc-pVDZ) are performed for *cis*-azobenzene photoisomerization GS-TSH-MD simulations.

Hopping spots. Figure 12.12 shows hopping spot distributions for energy gap as function of CNNC dihedral angle simulated from six combinations of functional/basis sets. All hopping points are taken place with energy gap in the range of 0.02~0.2 eV with respect to CNNC dihedral angle in the range of [$-100°$, $-92°$] and [$-77°$, $-68°$]. The number of hopping points is slightly less in [$-100°$, $-92°$] region than in [$-77°$, $-68°$] region. Since the hopping spots in [$-100°$, $-92°$] region is structurally closer to the *trans*-azobenzene than those in [$-77°$, $-68°$] region, most trajectories hopping in former region are expected to toward reactive photoisomerization, while those hopping in the latter region are mostly nonreactive as shown in Fig. 12.12. However, it should be noticed that for a given basis set 6-31G*, the number of nonreactive trajectories in [$-100°$, $-92°$] by TD-B3LYP (Fig. 12.12A) is obviously more than that by TD-CAM-B3LYP (Fig. 12.12B) and TD-BHandHLYP (Fig. 12.12C). This is not case for given functional B3LYP with three different basis sets, and the simulations with 3-21G, 6-31G, cc-pVDZ remain almost the same number of reactive trajectories in [$-100°$, $-92°$]

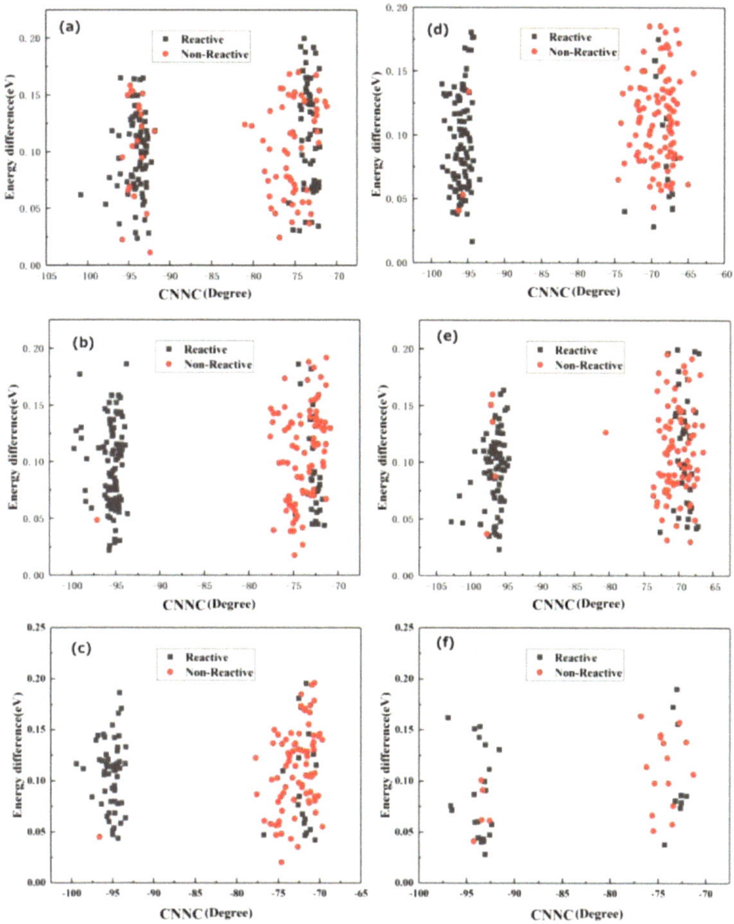

Figure 12.12 The hopping spot distributions for energy gap with respect to CNNC dihedral angle. The left panel is simulated from the same basis set 6-31G* with different functionals: (A) B3LYP, (B) BHandHLYP, and (C) CAM-B3LYP. The right panel is simulated from the same functional B3LYP with different basis sets: (D) 3-21G, (E) 6-31G, and (F) cc-pVDZ, reprinted from Ref. [29], Copyright © 2020 Wiley-VCH.

region as shown in Figs. 12.12D–F. Thus, we conclude that the functional dependence is more sensitive than basis set dependence for hopping spot distributions. This can be also seen from hopping spot distributions in terms of effective coupling parameter a^2 and

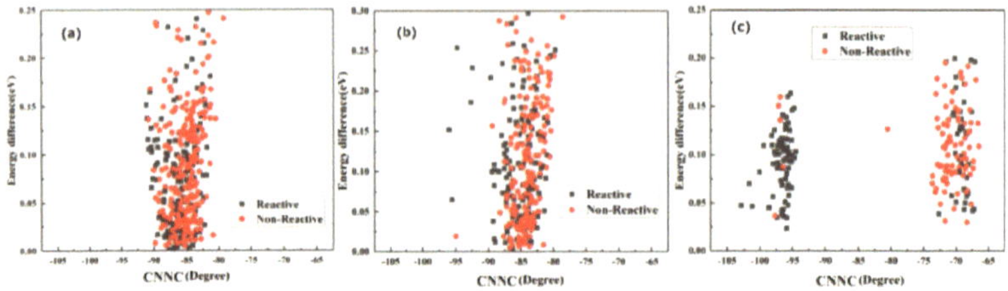

Figure 12.13 The hopping spot distributions for energy gap with respect to CNNC dihedral angle from (A) SA2-CAS(6,6)/6-31G from 400 sampling trajectories (CNNC = −92.6° at CI-rot) [36], (B) SF-BHHLYP/6-31g from 300 sampling trajectories (CNNC = −92° at CI-rot) [27], (C) TD-B3LYP/6-31G from 200 sampling trajectories, reprinted from Ref. [29], Copyright © 2020 Wiley-VCH.

effective collision energy b^2 (see Ref. [29] for detailed comparison and discussion).

The hopping spot distributions simulated from TDDFT are also compared with those simulated from SA2-CASSCF(6,6)/6-31G [36] and SF-BHHLYP/6-31G (Section 12.4) [27] methods. As shown in Figs. 12.13A,B, whether for reactive or nonreactive trajectories, the hopping spots simulated by SA2-CASSCF(6,6) and SF-BHHLYP mostly are taken place in region of CNNC dihedral angle [−91°, −80°] corresponding to the single-cone structure of S_0/S_1 CI-rot at CNNC = −92°. However, this single-cone hopping spots distribution is split into an artificial double-cone hopping spots distribution simulated by traditional TDDFT. For TD-B3LYP case as shown in Fig. 12.13C, one end of single-cone structure at CNNC = −91° is shifted into the range of (−100°, −92°) (dominant by reactive trajectories in this region) and the other end at CNNC = −80° is shifted into the range of [−77°, −68°] (dominant by nonreactive trajectories in this region). Nevertheless, this closed-shell TDDFT simulation can still agree with the previous SA2-CASSCF(6,6) and SF-TDDFT simulations for highly averaged quantities such as quantum yield and lifetime. This is probably because that around S_0/S_1 CI-rot, TDDFT double-cone topology shows similar a^2 and b^2 (Eq. 12.5) distributions in comparison with SA2-CAS (6,6)/6-31G single-cone topology, finally leading to similar distributions for nonadiabatic transition probabilities in GS algorithm.

Table 12.3 Quantum yields and lifetimes (in parenthesis, unit: fs) simulated by GS-TSH-MD with three functionals and four basis sets combinations for *cis*-to-*trans* azobenzene photoisomerization

	B3LYP	BHandHLYP	CAM-B3LYP
3-31G	0.51(35.3)	0.21(36.4)	0.10(36.8)
6-31G	0.57(37.3)	0.40(36.8)	0.40(37.3)
6-31G*	0.63(47.0)	0.59(34.6)	0.49(34.4)
cc-pVDZ	0.60(37.2)	0.52(34.4)	0.42(33.9)

Quantum yields and lifetimes. Figure 12.14 plots the average populations of S_0 and S_1 as function of time simulated from six combinations (Fig. 12.12) of functional/basis sets. These six combinations all show the same tendency of rapid decay for S_1 population and all have onset time constant 18.0 fs for S_1 population, especially all having the nearly same crossing point at time equal to 31.2 fs when S_1 population decreases by 50%. Most of the simulated quantum yields and lifetimes with three functionals are in the range from 0.40~0.63 and 33.9~47 fs, respectively, except for those with 3-21G basis set whose quantum yield is only 0.1 by TD-CAM-B3LYP as shown in Table 12.3. If the simulated results with 3-21G basis set in Table 12.3 are excluded, we can see that quantum yield is much more sensitive to the functional than basis sets. This tendency agrees with previous conclusion that the functional dependence is more sensitive than basis set dependence for hopping spot distributions. This means if we select appropriate TDDFT functional to perform GS-TSH-MD simulation, we could be able to use smaller basis set with keeping good accuracy. This is very useful when we deal large complex system, and we can save a lot computational time for electronic structure calculation on on-the-fly TDDFT PESs. For the present *cis*-azobenzene photoisomerization, it seems that TD-B3LYP method yields most stable quantum yield against change of basis sets as well as the highest rate for successful sampling trajectories, and even within 3-21G basis set quantum yield can be 0.51. Within average results from three basis sets (6-31G, 6-31G*, and cc-pVDZ), quantum yield (lifetime) is 0.6 with error ± 5% (40.5 fs with error ± 10%) by B3LYP, 0.5 with error ± 10% (35.5 fs with error ± 4%) by

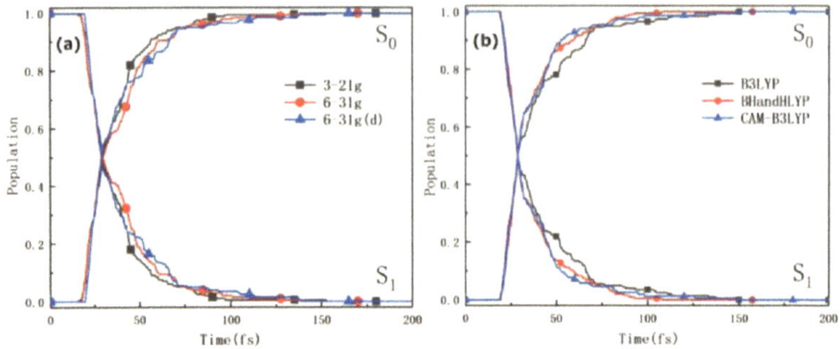

Figure 12.14 The average population distribution in S_1 and S_0 states as function of time. (A) Three basis sets with the same functional B3LYP and (B) Three functionals with the same basis set 6-31G*, reprinted from Ref. [29], Copyright © 2020 Wiley-VCH.

BHandHLYP, and 0.44 with error ± 9% (35.2 fs with error ± 10%) by CAM-B3LYP.

The quantum yields and lifetimes for all simulated results by GS algorithm with basis set 6-31G for the *cis*-azobenzene photoisomerization (summarized in Table 12.4) shows that quantum yields are all close to 0.40 (0.57 except for B3LYP) and lifetimes are about 40 fs from the present three TDDFT functionals, while 60 fs from the previous TDDFT functional with and without SF in comparison with 50 fs from SA2-CASSCF(6,6) method [36]. All simulated quantum yields are well fitted in the range of experimental value, but all simulated lifetimes are all shorter than experimental value, this might be due to gas phase model used in the GS-TSH-MD simulation while experimental measurements being all carried out in various solution environments.

Velocity adjustment. Figure 12.15 plots one typical reactive (nonreactive) trajectory with the same initial condition simulated by three functionals (TD-B3LYP, TD-BHandHLYP, and TD-CAM-B3LYP) under the same basis set 6-31G*. The time evolution of the three important dihedral angles (one CNNC and two NNCC) and two bond angles (NNC and CNN) are depicted in Fig. 12.15, in which upper (lower) panel shows hopping point happened almost at the same time 28, 31 and 32 fs (21, 23 and 24 fs) for the reactive (nonreactive)

Table 12.4 The quantum yields and lifetimes (fs) simulated by GS-TSH-MD from the same basis set 6-31G and experimental measurement for *cis*-to-*trans* azobenzene photoisomerization

Method	N_{traj}	QY	Lifetime (fs)
TD-B3LYP (using Gaussian 16) [29]	201	0.57	37.3
TD-BHandHLYP (using Gaussian 16) [29]	204	0.40	36.8
TD-CAM-B3LYP (using Gaussian 16) [29]	154	0.40	36.8
SF-BHHLYP (using GAMESS) [27]	300	0.43	63
TD-BHHLYP (using Turbomole) [27]	259	0.34	62
SA2-CASSCF(6,6) [36] (using Molpro) [36]	800	0.39	53
Expt. [61]		0.41–0.56	100–170

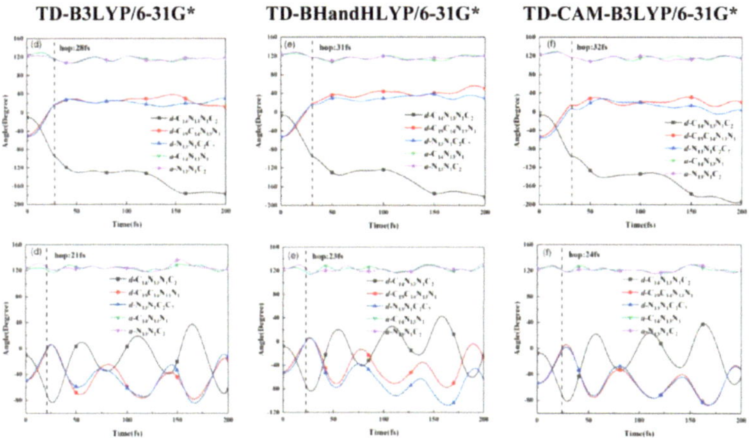

Figure 12.15 The time evolution of five important angles along with a typical reactive (nonreactive) trajectory shown in upper (lower) panel with the same initial condition and with the same basis set 6-31G* in GS-THS-MD simulation, reprinted from Ref. [29], Copyright © 2020 Wiley-VCH.

trajectory. Time-evolution profiles of the three dihedral angles and the two bond angles are quite similar, especially all three functionals showing that the CNNC dihedral angle follows quite same smooth oscillatory structures, while two NNCC dihedral angles change toward the opposite direction and the two NNC bond angles vibrate slowly around 120° for both reactive and nonreactive trajectory. This indicates that the hopping direction simulated by the three functionals is quite same after hopping.

Concluding remarks. Three functionals (TD-B3LYP, TD-BHandHLYP, and TD-CAM-B3LYP) in combination with four basis sets (3-21G, 6-31G, 6-31G, and cc-pVDZ) were used to perform on-the-fly GS-TSH-MD simulation for *cis*-to-*trans* azobenzene photoisomerization up to $S_1(n\pi^*)$ excitation. All three functionals in combination with four basis sets have shown artificial double cone structure of CI, but the simulated quantum yields, lifetimes and hopping directions are in good agreement with one another. The functional dependence for given basis set is slightly more sensitive than basis set dependence for given functional. This is demonstrated from population decay, hopping spot distribution and typical trajectory starting from the same initial condition. Among all functionals, the TD-B3LYP calculation shows the smallest sensitivity with respect to change of basis set, and even 3-21G basis set can work well. Thus, for highly averaged quantities such as quantum yield and lifetime, it might not be necessary to incorporate with large basis set, and the smaller basis set can perform well. This conclusion is very useful for the GS-TSH-MD simulation associated with TDDFT method for large complex system (in keeping good accuracy with small basis set).

12.6 GS-TSH-MD Simulation for Chemiluminescence

Chemiluminescence (CL) [1] is cold and visible light emissions from chemical reactions. The similar phenomenon observed in living organisms is usually referred to as bioluminescence. Most of currently discovered organic chemiluminescence in solution comes from the thermolysis of cyclic peroxides, especially four-membered cyclic peroxides like 1,2-dioxetane [87]. Following these reactions along with the path (elongation and breaking of the peroxide bond), the energy gaps between the ground and excited-states are gradually decreased into the surface crossings or degenerates of these electronic states. The S_0/S_1 conical intersections or S_0/T_1 intersystem crossings induced by the surface crossings can provide considerable probabilities for producing excited-state light emitters from nonadiabatic transitions. The chemiluminescence, to some extent, is the opposite of photochemical reactions, whose chemical transformations are initiated by light. In the former case, the energy of chemical bonds is converted into electronic excitation energy, whereas in photochemical processes the energy of the electromagnetic radiation is utilized to drive chemical transformations. Nonadiabatic transitions from ground state to excited-states are ignored sometimes in ultrafast photochemical or photophysical reaction. However, the nonadiabatic transitions from ground state to excited-states that is the origin of excited-state light emitter are obviously vital important for chemiluminescence. In this case, an open-shell ground state is inevitably occurred for most chemiluminescent reaction. Thus, to simulate a chemiluminescent reaction, all conical intersections, intersystem crossings and open-shell ground state have to be described correctly and balanced first. The multiconfigurational methods are hence the primary choice to perform the TSH-MD simulations for chemiluminescence. However, in comparison with ultrafast photochemical or photophysical reactions, chemiluminescence involves slow ground state chemical process, which requires long timescales of dynamics simulation. More efficient and accurate computational methods of electronic structure might be necessary, especially for LS algorithm. For GS algorithm, this might not be a problem as it is free from on-the-fly nonadiabatic couplings and spin-orbit couplings which are the most time-consuming calculations in LS-TSH-MD simulation. In addition, GS-TSH-MD simulation is proved to be less sensitive to the local PES topology around CI, and this can relax an accuracy requirement on the electronic structure calculation.

In comparison with the LS-TSH-MD simulation, the GS-TSH-MD simulation has advantages for nonadiabatic processes with longtime scale or involving S_0/S_1 CIs, it hence has been successfully applied to simulate the nonadiabatic processes like the chemiluminescence of organic cyclic peroxides in which the nonadiabatic transitions from ground state to excited-states are important.

12.6.1 Electron Transfer Catalyzed Chemiluminescence of Luminol

The luminol(5-amino-2,3-dihydro-1,4-phthalazinedione as shown in Fig. 12.16A) first discovered as man-made effective chemilumi-

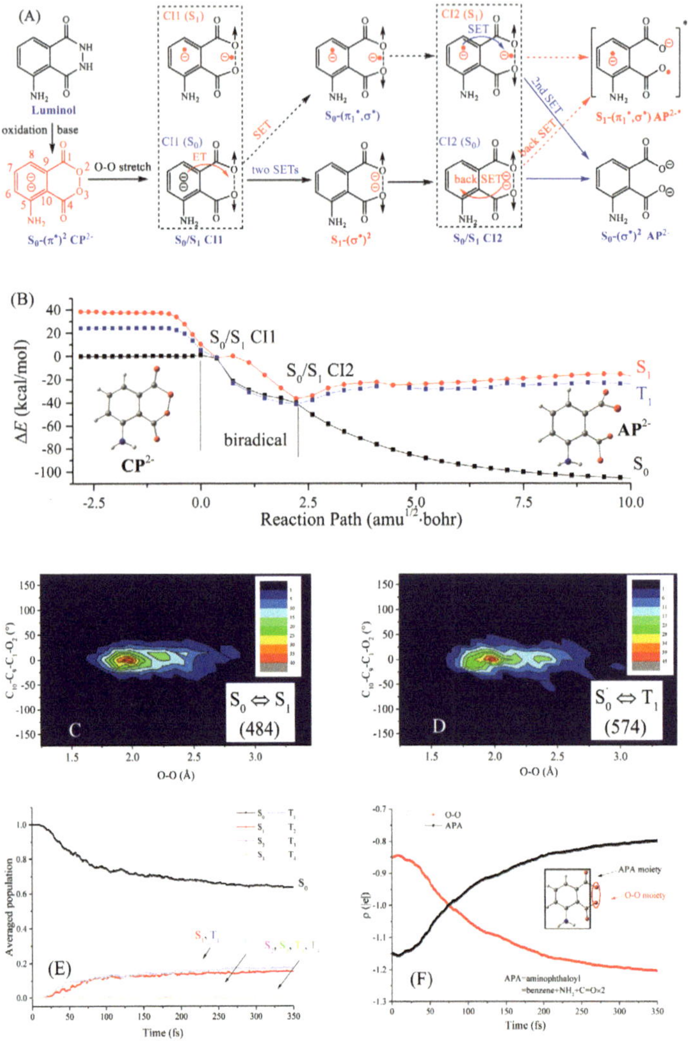

Figure 12.16 (A) The chemiexcitation mechanism of luminol; (B) the ground and excited PECs for the chemiexcitation; (C) and (D) Two-dimensional distribution of the O–O bond length and $C_{10}-C_9-C_1-O_2$ dihedral angle at $S_0 \leftrightarrow S_1$ hopping and $S_0 \leftrightarrow T_1$ hopping. The numbers of hopping spots are shown in parentheses. (E) The time evolution of the ground and excited states; (F) the time evolution of the Mülliken charge populations, reprinted with permission from Ref. [88] Copyright 2019 American Chemical Society.

nescence (CL) system is the best known as one of the most widely used CL materials. The chemiluminescent process of the luminol CL has not yet been fully elucidated, although the decomposition of 1,2-dioxane-3,6-dione dianion (**CP**$^{2-}$) is verified to be the key step to produce light emitter. The mechanisms of **CP**$^{2-}$ decomposition and effective singlet chemiexcitation are totally unknown, which is the outstanding obstacle to comprehending luminol CL. By means of the state-of-art multireference calculation and nonadiabatic molecular dynamics (NAMD) simulation, the decomposition mechanism of **CP**$^{2-}$ is investigated. A stepwise single electron transfer from the aminophthaloyl to the O–O bond initiates the decomposition of **CP**$^{2-}$, and the light emitter is produced via the two crossings of the potential energy surfaces between the ground state (S_0) and the first singlet excited state (S_1) as shown in Fig. 12.16B. The NAMD simulated quantum yield of the light emitter demonstrates that the located two conical intersections control an effective nonadiabatic pathway in luminol CL. The proposed mechanism of "two conical intersections" is not only suitable to luminol but also the other CL materials with cyclic peroxide as the chemical energy provider.

Recently, by means of the state-of-the-art multireference computation and the GS-TSH-MD simulation, the decomposition mechanism of **CP**$^{2-}$ is clearly revealed. Dynamic evolution picture how to initiate the decomposition of **CP**$^{2-}$ are observed for a stepwise single electron transfer from the aminophthaloyl to the O–O bond. The simulated quantum yield of the light emitter by GS-TSH-MD simulation demonstrates that the localized two conical intersections control an effective nonadiabatic pathway in luminol CL as shown in Figs. 12.16C–F (detailed discussions are given in Ref. [88]).

12.6.2 Uncatalyzed Chemiluminescence of Methylated 1,2-dioxetane

In contrast to the ET-catalyzed chemiluminescence, the 1,2-dioxetane derivatives can also yield chemiluminescence without catalysis by electron-withdrawing group. From the uncatalyzed chemiluminescence which often occurs in the decomposition of neutral cyclic peroxides, the triplet but not singlet states dominate the light emitter. The singlet chemiexcitations which cannot compete with the triplet chemiexcitation is usually very weak. The uncatalyzed chemiluminescence is difficult to be observed and measured by experiments due to very low phosphorescent efficiency in solution. In summary, the uncatalyzed and ET-catalyzed decompositions are the two main chemiexcitation paths for CL of cyclic peroxides. Although generous experimental and theoretical contributions were made, the chemiexcitation mechanism of cyclic peroxide is still in controversy up to now, because efficient nonadiabatic dynamics simulations especially involving singlet and triplet states simultaneously are still a challenge for theoretical investigation [21–28].

The decomposition of the methylated 1,2-dioxetanes are typical examples for the uncatalyzed chemiluminescence. According to the early experimental observation by Adam and Baade et al. in 1980s, the triplet chemiexcitation quantum yield increases significantly with the degree of methylation and varies from approximately 0.2% to 35%, while the quantum yield of singlet chemiexcitation slightly increases from 0.003% to 0.25%. According to previous high level multireference computations by Lindh et al. [17], in the biradical processes, the O–O bond is first broken down resulting in a flat biradical region (entropic trap) where the four singlet states and four triplet states are nearly degenerated, and then the C–C bond is cleaved resulting in the dissociation of the molecule into two carbonyl compounds as shown in Fig. 12.17.

Recently, we performed GS-TSH-MD simulation at the 8SA-CASSCF quantum level to estimate the quantum yield of chemiexcitation for the uncatalyzed decomposition reaction of the open-

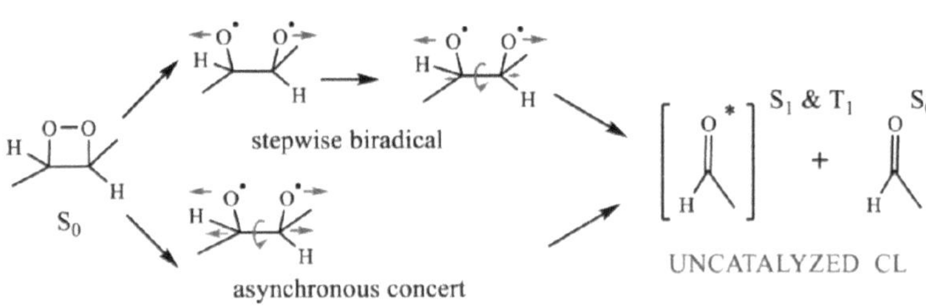

Figure 12.17 The stepwise and asynchronous concert biradical mechanism for uncatalyzed chemiluminescence of cyclic peroxides.

shell biradical *trans*-3,4-dimethyl-1,2-dioxetane system as shown in Fig. 12.16. The present ab initio nonadiabatic molecular dynamic simulation involving both conical intersection and intersystem crossing is to compute for the first time the population evolution of quantum yields at the four lowest singlet and four lowest triplet states. The simulated results demonstrate not only the stepwise dissociation of O–O and C–C bond breaking, but also confirm the existence of a biradical entropic trap which is responsible for chemiexcitation. The simulated quantum yield of the triplet chemiexcitation $\Phi_{T1} = 0.266 \pm 0.096$ agrees with the experimental value of 0.20 ± 0.04 very well [45]. This GS-TSH-MD simulation for dimethyl 1,2-dioxetanes provides a further advanced understanding and stepping stone for future studies on chemi- and bioluminescence.

Concluding remarks. The aforementioned two examples in GS-TSH-MD simulation are actually performed with multireference approaches but not by TDDFT. In principle, these chemiluminescent reactions are able to be simulated by GS-TSH-MD with spin-flip TDDFT which could correctly describe open-shell ground state and S_0/S_1 conical intersections, although some critical problems exist. One of the problems is the spin-contamination of SF-TDDFT. This could be solved by spin-adapted spin-flip TDDFT [33] approach. In SA-SF-TDDFT, additional configurations are included in the spin-flip CIS calculations based on a tensor equation-of-motion formalism to remove the spin-contamination, while the dynamic correlation from DFT correction to SA-SF-CIS is added in an ad hoc way. Another remaining problem is that current spin-flip scheme based on a triplet reference could not deal with the singlet and triplet states simultaneously at the same level, so the triplet states cannot be included in the TSH-MD simulations. Recently, Gordon et al. [89, 90] proposed a new and general spin-correct spin-flip configuration interaction (SF-CI) method by extending the occupation restricted multiple active spaces (ORMAS) CI method in GAMESS. In SF-ORMAS, not only triplet but also a higher spin determinant like quintette could be used as reference state and an arbitrary amount of spin-flipped excitations are carried out to generate a wave function of desired multiplicity involving singlet and triplets. The

SF-ORMAS method was not only used to successfully describe diradical transition states, single and multiple bonds breaking, but also the vertical excitations, and singlet–triplet gaps. Its accuracy is consistent along the PES, and the inclusion of dynamic correlation through SF-MRMP2 is comparable to methods such as CASPT2 and MRCI [89]. Based on this SF scheme, the singlet and triplet states are possible calculated simultaneously at the same level by future extension of SF-TDDFT.

For newly developed SF-TDDFT methods, an analytic calculation of the nonadiabatic couplings is probably not available or not implemented in a short time, thus the GS algorithm might be the only choice to perform the TSH-MD simulation. Moreover, the GS algorithm which does not need the NACVs is quite efficient, in the case of total eight electronic states (four singlet and four triplet states) are nearly degenerated at the biradical region during the decomposition of a 1,2-dioxetane derivative. These degenerated states have to be involved in the TSH-MD simulations calculated at the same theoretical levels, and this can be done with GS algorithm efficiently.

12.7 GS-TSH-MD Simulation for Photoisomerization of dMe-OMe-NAIP

The dMe-OMe-NAIP is one of biomimetic photoswitch species and display nice photo-switchable properties between two configurations (E and Z) as an ultrafast photoinduced isomerization. The quantum yields are experimentally observed as 0.26 and 0.24, respectively for E-to-Z and Z-to-E isomerization with relaxing times smaller than 500 fs [91]. Ultrafast dMe-OMe-NAIP photoisomerization has been governed by nonadiabatic dynamics via the conical intersections between the first excited (S_1) state and ground (S_0) state. Molecular size of dMe-OMe-NAIP as shown in Fig. 12.18 is to large too be affordable by TSH-MD simulation with multireference based on electronic structure calculation. On the other hand, quantum yields and lifetimes do not agree with experimental observations at all by performing THS-MD simulation based on the affordable

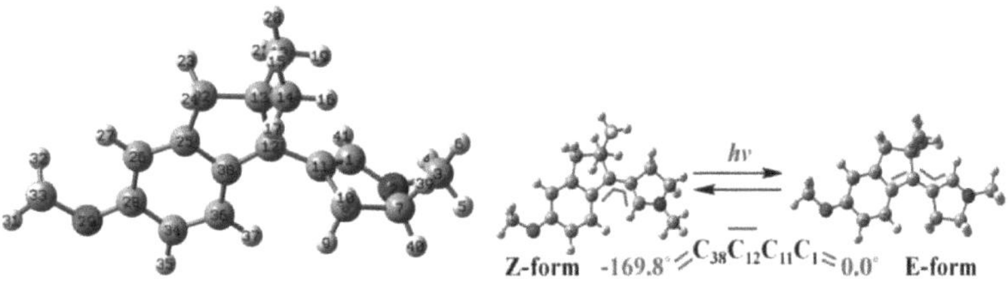

Figure 12.18 Left panel is atom numbering for dMe-OMe-NAIP (blue is for nitrogen, red for oxygen, dark grey for carbon and light grey for hydrogen atoms) in transition state structure on S_0. Right panel is for Z/E photoisomerization with Z-form and E-form electronic structures at ground-state minima.

semi-empirical OM2/MRCI electronic structure calculation for on-the-fly PESs [92]. Therefore, this is case requiring for GS-TSH-MD simulation associated with TDDFT electronic structure calculation for the on-the-fly PESs. As demonstrated in the previous sections for GS-TSH-MD simulation, the conventional on the-fly TDDFT potential energy surfaces can work successfully in involving with conical intersection between S_1 state and S_0 states.

12.7.1 Time-Dependent Density Functional Theory Calculations for Searching Conical Intersections

As dMe-OMe-NAIP has chiral symmetry in which the mirror image is found by changing signs of all dihedral angles for any configuration at potential energy surface. If dihedral angle $C_{38}C_{12}C_{11}C_8$ (see Fig. 12.18) is smaller (larger) than zero, it is defined as left (right)-handed enantiomer. E-form E-S_0 (Z-form Z-S_0) minimum has $C_{38}C_{12}C_{11}C_8$ equal to 0.0° (−169.8°) on ground S_0 state optimized by B3LYP/6-31G* method. Actually, (TD)-B3LYP/6-31G* is utilized for all electronic structure calculations as well as for on-the-fly PESs simulations. Optimized E-S_1 (Z-S_1) minimum on S_1 state has $C_{38}C_{12}C_{11}C_8$ equal to −1.3° (−176.2°) and this means that geometry of E-form (Z-form) minimum on S_0 state is very close to the corresponding geometry on S_1 state. By running a few tens of preliminary sampling trajectories starting from Franck–Condon region at both E-S_0 and Z-S_0 geometries with GS-TSH-MD simulation,

Table 12.5 Key geometries for four minima, one transition state, and three representative conical intersections with respect most important rotational angle $C_{38}C_{12}C_{11}C_8$ plus the other four associated dihedral angles for describing conical intersections

Geometry	$C_{38}C_{12}C_{11}C_8$	$C_{12}C_{11}C_1N_2$	$H_{41}C_1C_{11}C_{12}$	$N_2C_1C_{11}C_{14}$	$N_2C_1C_{11}C_{18}$
Z-S$_1$-L	−176.2	180	−6.6	−96.5	143.5
ZT-CI-L	−175.3	−106.3	112.1	−44.5	−59.4
Z-S$_0$-L	−169.8	−177.8	6.0	101.1	−127.3
Rot-CI-L	−93.9	166.9	2.2	105.2	145.1
TS-S$_0$-L	−92.9	174.6	−3.2	138.7	106.4
ET-CI-L	−11.4	96.7	−111.9	−100.1	−142.5
E-S$_1$-L	−1.3	−174.9	7.3	163.3	−159.4
E-S$_0$-L	0.0	180	0.0	156.4	−156.4

Note: Dihedral angles are given in degrees. The notation -L (-R) stands for left (right)-handed enantiomer.

we collect electronic geometry structures at all hopping points and we analyzed that those hopping points can be classified into three conical intersection zones; hopping points around $C_{38}C_{12}C_{11}C_8 \sim \pm 90^\circ$ is regarded as rotation CI (namely Rot-CI zone), around $C_{38}C_{12}C_{11}C_8 \sim \pm 160^\circ$ as Z-form twisting CI (namely ZT-CI zone) and around $C_{38}C_{12}C_{11}C_8 \sim \pm 20^\circ$ as E-form twisting CI (namely ET-CI zone). At each CI zone, we choose the hoping point with the minimum energy gap between S_1 and S_0 states as a representative conical intersection and its corresponding geometry parameters listed in Table 12.5, along with the other four important dihedral angles ($C_{12}C_{11}C_1C_2$, $H_{41}C_1C_{11}C_{12}$, $N_2C_1C_{11}C_{14}$, and $N_2C_1C_{11}C_{18}$), involving in motion for dMe-OMe-NAIP E/Z photoisomerization. The most important one is Rot-CI around $C_{38}C_{12}C_{11}C_8 = -93.9^\circ$ that uniquely determines reactive photoisomerization, and the other ZT-CI is around $C_{38}C_{12}C_{11}C_8 = -175.3^\circ$ (leading to Z-form nonreactive channel to Z-S_0 around $C_{38}C_{12}C_{11}C_8 = -169.8^\circ$) and ET-CI is around $C_{38}C_{12}C_{11}C_8 = -11.4^\circ$ (leading to E-form nonreactive channel to E-S_0 around $C_{38}C_{12}C_{11}C_8 = 0.0^\circ$).

Figure 12.19 shows potential energy profiles of all critical structure points in Table 12.5 arranged from L-form to R-form isomers (from -180° to $+180^\circ$ for $C_{38}C_{12}C_{11}C_8$). Figure 12.19 shows that Franck–Condon (FC) vertical excitation energies from E-S_0 and Z-S_0 are very close to its corresponding minimum energies

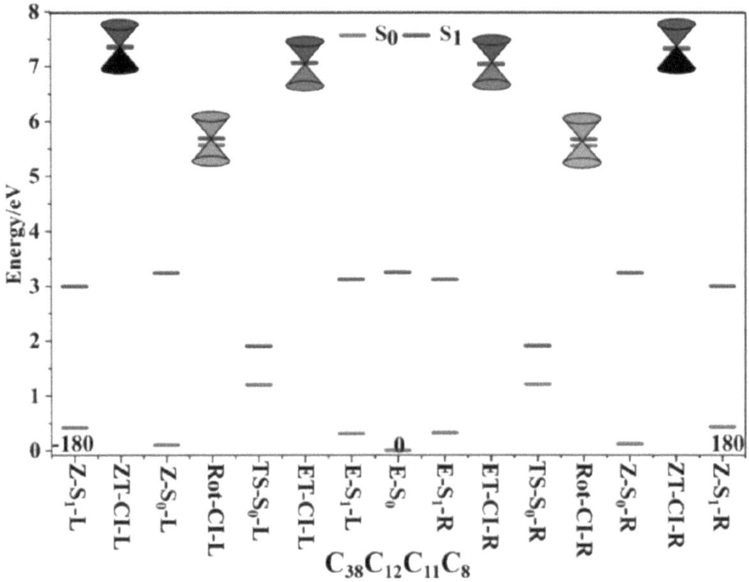

Figure 12.19 Potential energy profiles for four minima, one transition state and three conical intersections in Table 12.5 via $C_{38}C_{12}C_{11}C_8$ from $-180°$ to $+180°$ (but scale is arbitrary), reprinted from Ref. [93], Copyright © 2021 Royal Society of Chemistry.

of E-S_1 and Z-S_1, respectively, and furthermore the energies of all conical intersections are well above the Franck–Condon energy region. These results indicate that photoisomerization pathway for dMe-OMe-NAIP is uphill to conical intersections on potential energy surface S_1 from FC region, and thus it is very different from conventional cartoon picture downhill to conical intersections. Besides, this fact reflects that deep potential wells are surrounding excited-state minima E-S_1 and Z-S_1 around FC region, and thus it can lead to a quite few sampling trajectories trapping in the wells (undergoing fluorescence emission).

12.7.2 Both E-to-Z and Z-to-E Photoisomerization in GS-TSH-MD Simulation

A canonical ensemble with a temperature T = 300 K is utilized to specify initial conditions of the trajectories, and zero-point

Table 12.6 Simulated quantum yields and excited-state lifetimes for reactive photoisomerization for dMe-MeO-NAIP in comparison with experimental results

	From E-isomer		From Z-isomer	
	Quantum Yield	Lifetime (fs)	Quantum Yield	Lifetime (fs)
TD-B3LYP [93]	0.23	620	0.15	600
Exp. [94]	0.25	~480	–	–
Exp. [91]	0.26	~480	0.24	430

vibrational energy is utilized as the initial kinetic energy. The time stepsize is chosen to be 0.5 fs for both electronic and nuclear motions. All sampling trajectories are simulated for up to 1500 fs time constant where we collect dMe-OMe-NAIP reaction products. The 100 sampling trajectories for both E-to-Z and Z-to-E photoisomerization are performed. For E-to-Z case, 46 trajectories go through rotation Rot-CI (23 reactive trajectories ending to Z-S_0 and 23 nonreactive back to E-S_0), 7 trajectories go via twisting ET-CI back to E-S_0, and 47 trajectories trap on E-S_1 as resonances (undergoing fluorescence emission). For Z-to-E case, 24 trajectories go through rotation Rot-CI (15 reactive trajectories ending to E-S_0 and 9 nonreactive back to Z-S_0), 33 trajectories go via twisting ZT-CI back to Z-S_0, and 43 trajectories trap on Z-S_1 as resonances (undergoing fluorescence emission). Simulated E-to-Z (Z-to-E) quantum yield is about 0.23 (0.15) agrees well with experiment measurement of 0.25 (0.24) as summarized in Table 12.6.

Hopping spots. Figure 12.20 shows distributions of the number of hopping trajectories as function of time. 53 (57) hopping trajectories initially starting from E-isomer (Z-isomer) ranges from several tens fs to 1500 fs and the average hopping time (excited-state lifetime) is estimated as 620 fs (600 fs) in good agreement with experiment measurement of 480 fs (430 fs) in methanol solution. The sampling trajectories estimated 50% to 50% (38% to 62%) bifurcation to reactant and product at rotational conical intersection Rot-CI for E-to-Z (Z-to-E) case. However, a big difference is taken place at twisting conical intersection ZT-CI (33 trajectories go via), but ET-CI (only 7

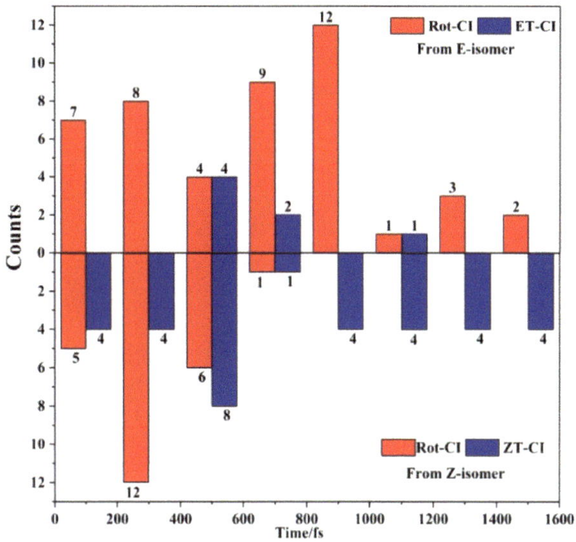

Figure 12.20 The number of hopping trajectories via conical intersections as function of time. The top (bottom) panel stands for sampling trajectories starting from FC region of E-S_0 (Z-S_0) initially, reprinted from Ref. [93], Copyright © 2021 Royal Society of Chemistry.

trajectories go via), and thus it results a big difference in quantum yield although excited-state lifetime is similar.

Among 53 hopping trajectories in the case of E-to-Z isomerization, hopping spots can be classified as shown in Fig. 12.21A, in which 25 sampling trajectories (13 of them are reactive) undergo the clockwise pathway via Rot-CI in the region of $C_{38}C_{12}C_{11}C_8/C_{12}C_{11}C_1N_2 = [60°, 110°]/\pm 180°$, and 21 sampling trajectories (10 of them are reactive) undergo the counterclockwise pathway via Rot-CI in the region of $C_{38}C_{12}C_{11}C_8/C_{12}C_{11}C_1N_2 = [-110°, -60°]/\pm 180°$, a number of clockwise/counterclockwise hopping points is 25/21 and its ratio is close to 1. This is because that E-S_0 is planar structure and the trajectories have almost equal opportunity to follow the clockwise and counterclockwise rotation pathways. However, 7 nonreactive sampling trajectories go via ET-CI are widely distributed in the region of $C_{38}C_{12}C_{11}C_8 = [-20°, 60°]$, in which only one goes counterclockwise pathway and 6 go clockwise pathway.

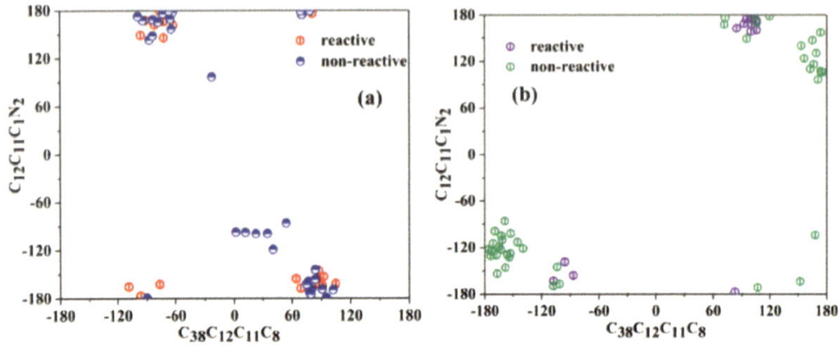

Figure 12.21 The distribution of trajectory hopping points in terms of two dihedral angles $C_{38}C_{12}C_{11}C_8$ and $C_{12}C_{11}C_1N_2$ starting from Frank-Condon region of (a) E-S_0 and (b) Z-S_0, reprinted from Ref. [93], Copyright © 2021 Royal Society of Chemistry.

Among 57 hopping trajectories in the case of Z-to-E isomerization, hopping spots can be classified as shown in Fig. 12.21B, in which 18 sampling trajectories (12 of them are reactive) undergo the clockwise pathway via Rot-CI in the region of $C_{38}C_{12}C_{11}C_8/C_{12}C_{11}C_1N_2 = [70°, 120°]/\pm 180°$, and 6 sampling trajectories (3 of them are reactive) undergo the counterclockwise pathway via Rot-CI in the region of $C_{38}C_{12}C_{11}C_8/C_{12}C_{11}C_1N_2 = [-110°, -80°]/180°$, and ratio of clockwise/counterclockwise hopping points is $18/6 = 3$ and this is because Z-S_0 is nonplanar geometry structure which is in favor to clockwise rotation related to Rot-CI. However, that 22 (11) nonreactive sampling trajectories go via ZT-CI are distributed in the region of $C_{38}C_{12}C_{11}C_8 = [-180°, -140°]$ ([160°, 180°]) in counterclockwise (clockwise) pathway.

The Franck–Condon vertical excitation energies estimated from S_0 minima are very close to corresponding minimum energies on S_1 state, and thus all sampling trajectories rapidly rise its potential energy (in 3 $\sim$4 eV) uphill to conical intersection zones within the first 10 fs as shown in Fig. 12.19. This makes both Z-to-E and E-to-Z ultrafast photoisomerization in the femtosecond regime. Besides, all sampling trajectories can move either clockwise ($C_{38}C_{12}C_{11}C_8$ >0) or counter-clockwise ($C_{38}C_{12}C_{11}C_8$ <0) rotation from Franck–Condon region to conical intersection zones starting from both Z-isomer and E-isomer.

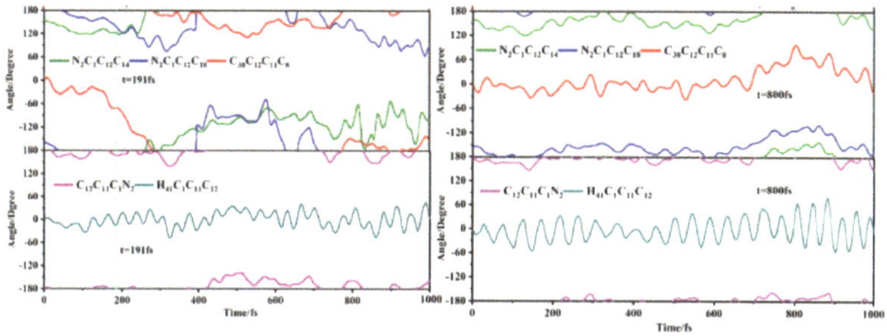

Figure 12.22 A selected counterclockwise reactive (clockwise nonreactive) trajectory goes via Rot-CI starting from E-isomer as function of time in left (right) panel in terms of evolution of five dihedral angles in Table 12.5, reprinted from Ref. [93], Copyright © 2021 Royal Society of Chemistry.

Photoisomerization mechanisms. Figure 12.22 (left panel) shows a typical counterclockwise reactive trajectory starting from E-isomer to hop down to S_0 state via Rot-CI (at $C_{38}C_{12}C_{11}C_8 \sim -90.0^\circ$) at time equal to 191 fs. At starting point, the $N_2C_1C_{11}C_{14}$ and $N_2C_1C_{11}C_{18}$ is 156.4° and -156.4°, respectively, but after 50 fs the $N_2C_1C_{11}C_{14}$ and $N_2C_1C_{11}C_{18}$ quickly turn to the same rotation direction making this reactive trajectory toward to Rot-CI smoothly with less influence from steric hindrance related to the rotation of two CH_3 groups (see Fig. 12.18). Figure 12.22 (right panel) shows a typical clockwise nonreactive trajectory running up to 720 fs when $N_2C_1C_{11}C_{14}$ and $N_2C_1C_{11}C_{18}$ move in the opposite rotation direction, the steric hindrance prevents necessary geometrical arrangement toward to Rot-CI. However, after 700 fs the two dihedral angles move in the same rotation direction and then this nonreactive trajectory could hop via Rot-CI at 800 fs ($C_{38}C_{12}C_{11}C_8 \sim 90.0^\circ$). In both reactive and nonreactive trajectories, $H_{41}C_1C_{11}C_{12}$ and $C_{12}C_{11}C_1C_2$ are just vibrating around their initial values all the time, respectively. It is concluded that if the $N_2C_1C_{11}C_{14}$ and $N_2C_1C_{11}C_{18}$ move in the opposite rotation direction, a considerable steric hindrance is encountered to prevent trajectory moving toward Rot-CI level by future extension of SF-TDDFT.

Figure 12.23 (left panel) shows a typical clockwise reactive trajectory starting from Z-isomer to hop down to S_0 state via Rot-CI ($C_{38}C_{12}C_{11}C_8 \sim 90.0^\circ$) at time equal to 203 fs. After 120 fs

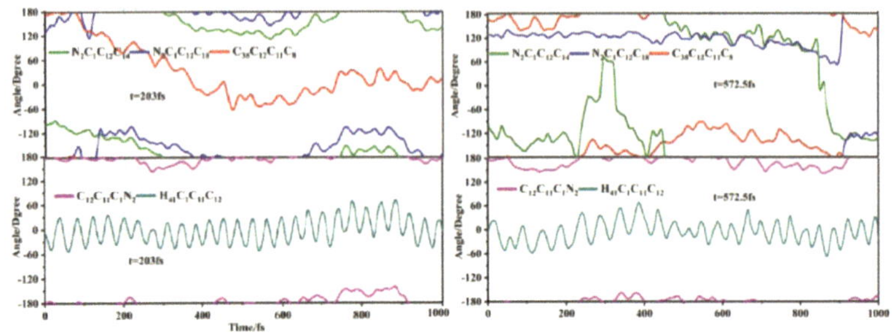

Figure 12.23 A selected clockwise reactive (counterclockwise nonreactive) trajectory goes via Rot-CI starting from Z-isomer as function of time in left (right) panel in terms of evolution of five dihedral angles in Table 12.5, reprinted from Ref. [93], Copyright © 2021 Royal Society of Chemistry.

the $N_2C_1C_{11}C_{14}$ and $N_2C_1C_{11}C_{18}$ quickly turn to the same rotation direction making this reactive trajectory toward to Rot-CI. On the other hand, $N_2C_1C_{11}C_{14}$ and $N_2C_1C_{11}C_{18}$ move in the opposite rotation direction as shown in Fig. 12.23 (right panel) up to 450 fs for a typical counterclockwise nonreactive trajectory, the steric hindrance again prevents trajectory moving toward to Rot-CI. Right after 450 fs the two dihedral angles move in the same rotation direction and then this nonreactive trajectory could hop via Rot-CI at 572 fs ($C_{38}C_{12}C_{11}C_8$ $\sim$90.0°).

Basically, photoisomerization mechanisms starting from both Z-isomer and E-isomer are almost the same and only difference is the steric hindrance having more influence in the pathway starting from Z-isomer than from E-isomer. Figure 12.20 shows that in a relatively short period (<600 fs) the both pathways have almost same hopping points to Rot-CI, but after 600 fs Z-isomer pathway suffers strong the steric hindrance that prevents $N_2C_1C_{11}C_{14}$ and $N_2C_1C_{11}C_{18}$ moving in the same rotation direction and thus hopping points all shift to ZT-CI. This steric hindrance is actually in close agreement with the discussion in Ref. [92].

Concluding remarks. For up to the $S_1(\pi\pi^*)$ excitation from Franck–Condon region for dMe-OMe-NAIP photoisomerization, the simulated quantum yield and lifetime 0.23 and 620 fs (0.15 and 600 fs) for E-to-Z (Z-to-E) isomerization, respectively, in good (relatively good) agreement with experiment observation of 0.25 and 480 fs

(0.24 and 430 fs). In the present GS-TSH-MD dynamic simulation, all sampling trajectories (including both Z-to-E and E-to-Z isomerization) make hops successfully where the conical intersections are well behavior without any problem. One of the reasons is that all sampling trajectories make hops before reaching exactly degenerate conical intersection points between S_1 and S_0 states. This means that the potential energy gaps at all hopping points are in the region where TDDFT method is still valid. This situation is very different from the previous dynamic simulation for azobenzene photoisomerization where small portions of sampling trajectories do reach degenerate CI point where TDDFT has negative excitation (double cone) problem between S_1 and S_0 states. The other reason could be system dependent whether or not TDDFT method produces certain erroneous behavior conical intersections. On the other hand, the present simulation revealed that photoisomerization pathways are initially uphill to conical intersection zones on S_1 potential energy surface and then downhill to product zones. This is very different from the conventional picture of downhill to conical intersection zones up to Franck–Condon excitation. Three types of representative conical intersections are found for determining dMe-OMe-NAIP photoisomerization mechanisms; one is the rotation type responsible for reactive isomerization and the other two are close to E and Z configurations, respectively only for nonreactive isomerization. This conclusion can be held in general for similar even larger NAIP systems of photoinduced isomerization based on E and Z two configurations where GS-TSH-MD simulation is very suitable to incorporate with on-the-fly TDDFT potential energy surfaces.

Acknowledgments

This work was partly supported by National Natural Science Foundation of P. R. China under grant nos. 21503156, 21773075 and 22073071. C. Z. thanks support from the Ministry of Science and Technology, Taiwan (grant no. MOST 109-2113-M-009-019).

References

1. Haas, Y. (2014). *Adv. Chem. Ser.*, **2014**, 419102.

2. Wang, L., Akimov, A., Prezhdo, O. V. (2016). *J. Phys. Chem. Lett.*, **7**, 2100–2112.
3. Crespo-Otero, R., Barbatti, M. (2018). *Chem. Rev.*, **118**, 7026–7068.
4. Curchod, B. F. E., Rothlisberger, U., Tavernelli, I. (2013). *ChemPhysChem*, **14**, 1314–1340.
5. Martens, C. C. (2016). *J. Phys. Chem. Lett.*, **7**, 2610–2615.
6. Per, S., Anders, H., Björn, R., Bernard, L. (1980). *Phys. Scr.*, **21**, 323–327.
7. Roos, B. O., Taylor, P. R., Siegbahn, P. E. M. (1980). *Chem. Phys.*, **48**, 157–173.
8. Malmqvist, P.-Å., Roos, B. O. (1989). *Chem. Phys. Lett.*, **155**, 189–194.
9. Hirao, K., *Recent Advances in Multireference Methods.*
10. Levine, B. G., Ko, C., Quenneville, J., MartÍnez, T. J. (2006). *Mol. Phys.*, **104**, 1039–1051.
11. Dreuw, A., Head-Gordon, M. (2005). *Chem. Rev.*, **105**, 4009–4037.
12. Casida, M. E. (2009). *J. Mol. Struct.: THEOCHEM*, **914**, 3–18.
13. Herbert, J. M., Zhang, X., Morrison, A. F., Liu, J. (2016). *Acc. Chem. Res.*, **49**, 931–941.
14. Pittner, J., Lischka, H., Barbatti, M. (2009). *Chem. Phys.*, **356**, 147–152.
15. Werner, U., Mitrić, R., Suzuki, T., Bonačić-Koutecký, V. (2008). *Chem. Phys.*, **349**, 319–324.
16. Lee, S., Kim, E., Lee, S., Choi, C. H. (2019). *J. Chem. Theory Comput.*, **15**, 882–891.
17. Ryabinkin, I. G., Nagesh, J., Izmaylov, A. F. (2015). *J. Phys. Chem. Lett.*, **6**, 4200–4203.
18. Ou, Q., Bellchambers, G. D., Furche, F., Subotnik, J. E. (2015). *J. Chem. Phys.*, **142**, 064114.
19. Zhang, X., Herbert, J. M. (2015). *J. Chem. Phys.*, **142**, 064109.
20. Parker, S. M., Roy, S., Furche, F. (2019). *Phys. Chem. Chem. Phys.*, **21**, 18999–19010.
21. Craig, C. F., Duncan, W. R., Prezhdo, O. V. (2005). *Phys. Rev. Lett.*, **95**, 163001.
22. Tapavicza, E., Tavernelli, I., Rothlisberger, U. (2007). *Phys. Rev. Lett.*, **98**, 023001.
23. Tapavicza, E., Tavernelli, I., Rothlisberger, U., Filippi, C., Casida, M. E. (2008). *J. Chem. Phys.*, **129**, 124108.
24. Tavernelli, I., Curchod, B. F. E., Rothlisberger, U. (2011). *Chem. Phys.*, **391**, 101–109.

25. Plasser, F., Crespo-Otero, R., Pederzoli, M., Pittner, J., Lischka, H., Barbatti, M. (2014). *J. Chem. Theory Comput.*, **10**, 1395–1405.
26. Yue, L., Lan, Z., Liu, Y.-J. (2015). *J. Phys. Chem. Lett.*, **6**, 540–548.
27. Yue, L., Liu, Y.-J., Zhu, C. Y. (2018). *Phys. Chem. Chem. Phys.*, **20**, 24123–24139.
28. Ding, B.-W., Liu, Y.-J. (2017). *J. Am. Chem. Soc.*, **139**, 1106–1119.
29. Ye, L., Xu, C., Gu, F. L., Zhu, C. (2020). *J. Comput. Chem.*, **41**, 635–645.
30. Fischer, S. A., Habenicht, B. F., Madrid, A. B., Duncan, W. R., Prezhdo, O. V. (2011). *J. Chem. Phys.*, **134**, 024102.
31. Menger, M. F. S. J., Plasser, F., Mennucci, B., González, L. (2018). *J. Chem. Theory Comput.*, **14**, 6139–6148.
32. Shao, Y., Head-Gordon, M., Krylov, A. I. (2003). *J. Chem. Phys.*, **118**, 4807–4818.
33. Zhang, X., Herbert, J. M. (2015). *J. Chem. Phys.*, **143**, 234107.
34. Zhang, X., Herbert, J. M. (2014). *J. Chem. Phys.*, **141**, 064104.
35. Child, M. S., Pechukas, P. (1974). *Molecular Collision Theory*, Academic Press, London.
36. Yu, L., Xu, C., Lei, Y., Zhu, C., Wen, Z. (2014). *Phys. Chem. Chem. Phys.*, **16**, 25883–25895.
37. Xu, C., Yu, L., Zhu, C., Yu, J. (2015). *J. Phys. Chem. A*, **119**, 10441–10450.
38. Yu, L., Xu, C., Zhu, C. (2015). *Phys. Chem. Chem. Phys.*, **17**, 17646–17660.
39. Li, C.-X., Guo, W.-W., Xie, B.-B., Cui, G. (2016). *J. Chem. Phys.*, **145**, 074308.
40. Xu, C., Yu, L., Zhu, C., Yu, J., Cao, Z. (2016). *Sci. Rep.*, **6**, 26768.
41. Xu, C., Yu, L., Gu, F. L., Zhu, C. (2018). *Phys. Chem. Chem. Phys.*, **20**, 23885–23897.
42. Zhu, C., Nakamura, H. (1994). *J. Chem. Phys.*, **101**, 10630–10647.
43. Zhu, C., Nakamura, H. (1995). *J. Chem. Phys.*, **102**, 7448–7461.
44. Yue, L., Yu, L., Xu, C., Zhu, C., Liu, Y. (2020). *Phys. Chem. Chem. Phys.*, **22**, 11440–11451.
45. Tully, J. C. (1990). *J. Chem. Phys.*, **93**, 1061–1071.
46. Zhu, C., Nakamura, H. (1992). *J. Chem. Phys.*, **97**, 8497–8514.
47. Zhu, C., Nakamura, H., Re, N., Aquilanti, V. (1992). *J. Chem. Phys.*, **97**, 1892–1904.
48. Zhu, C., Nakamura, H. (1993). *J. Chem. Phys.*, **98**, 6208–6222.
49. Zhu, C., Nakamura, H. (1994). *J. Chem. Phys.*, **101**, 4855–4866.
50. Köppel, H., Gronki, J., Mahapatra, S. (2001). *J. Chem. Phys.*, **115**, 2377–2388.

51. Gao, W., Yu, L., Zheng, X., Lei, Y., Zhu, C., Han, H. (2016). *RSC Adv.*, **6**, 39542–39552.
52. Lei, Y., Wu, H., Zheng, X., Zhai, G., Zhu, C. (2016). *J. Photochem. Photobiol. A: Chem.*, **317**, 39–49.
53. Gan, Y., Yue, L., Guo, X., Zhu, C., Cao, Z. (2017). *Phys. Chem. Chem. Phys.*, **19**, 12094–12106.
54. Yang, M., Huo, C., Li, A., Lei, Y., Yu, L., Zhu, C. (2017). *Phys. Chem. Chem. Phys.*, **19**, 12185–12198.
55. Fan, Y., Chen, J., Yu, L., Li, A., Zhai, G., Lei, Y., Zhu, C. (2018). *Phys. Chem. Chem. Phys.*, **20**, 2260–2273.
56. Yager, K. G., Barrett, C. J. (2006). *J. Photochem. Photobiol. A: Chem.*, **182**, 250–261.
57. Choi, H. J., Jeong, K.-U., Chien, L.-C., Lee, M.-H. (2009). *J. Mater. Chem.*, **19**, 7124–7129.
58. Pace, G., Ferri, V., Grave, C., Elbing, M., von Hänisch, C., Zharnikov, M., Mayor, M., Rampi, M. A., Samorì, P. (2007). *Proc. Natl. Acad. Sci.*, **104**, 9937–9942.
59. Tamesue, S., Takashima, Y., Yamaguchi, H., Shinkai, S., Harada, A. (2010). *Angew. Chem. Int. Ed.*, **49**, 7461–7464.
60. Weingart, O., Lan, Z., Koslowski, A., Thiel, W. (2011). *J. Phys. Chem. Lett.*, **2**, 1506–1509.
61. Bandara, H. M. D., Burdette, S. C. (2012). *Chem. Soc. Rev.*, **41**, 1809–1825.
62. Gao, A.-H., Li, B., Zhang, P.-Y., Han, K.-L. (2012). *J. Chem. Phys.*, **137**, 204305.
63. Harabuchi, Y., Ishii, M., Nakayama, A., Noro, T., Taketsugu, T. (2013). *J. Chem. Phys.*, **138**, 064305.
64. Cantatore, V., Granucci, G., Persico, M. (2014). *Comput. Theor. Chem.*, **1040–1041**, 126–135.
65. Pederzoli, M., Pittner, J., Barbatti, M., Lischka, H. (2011). *J. Phys. Chem. A*, **115**, 11136–11143.
66. Weber, W., Thiel, W. (2000). *Theor. Chem. Acc.*, **103**, 495–506.
67. Koslowski, A., Beck, M. E., Thiel, W. (2003). *J. Comput. Chem.*, **24**, 714–726.
68. Barbatti, M., Aquino, A. J. A., Lischka, H. (2010). *Phys. Chem. Chem. Phys.*, **12**, 4959–4967.
69. Dahl, J. P., Springborg, M. (1988). *J. Chem. Phys.*, **88**, 4535–4547.

70. Yue, L., Yu, L., Xu, C., Lei, Y., Liu, Y., Zhu, C. (2017). *ChemPhysChem*, **18**, 1274–1287.

71. Domcke, W., Yarkony, D. R., Köppel, H. (2004). *Conical Intersections Electronic Structure, Dynamics and Spectroscopy*, World Scientific, Singapore.

72. Gozem, S., Melaccio, F., Valentini, A., Filatov, M., Huix-Rotllant, M., Ferré, N., Frutos, L. M., Angeli, C., Krylov, A. I., Granovsky, A. A., Lindh, R., Olivucci, M. (2014). *J. Chem. Theory Comput.*, **10**, 3074–3084.

73. Harabuchi, Y., Keipert, K., Zahariev, F., Taketsugu, T., Gordon, M. S. (2014). *J. Phys. Chem. A*, **118**, 11987–11998.

74. Ibele, L. M., Nicolson, A., Curchod, B. F. E. (2020). *Mol. Phys.*, **118**, e1665199.

75. Tavernelli, I. (2015). *Acc. Chem. Res.*, **48**, 792–800.

76. Ren, X., Rinke, P., Blum, V., Wieferink, J., Tkatchenko, A., Sanfilippo, A., Reuter, K., Scheffler, M. (2012). *New J. Phys.*, **14**, 053020.

77. Bleiziffer, P., Heßelmann, A., Görling, A. (2012). *J. Chem. Phys.*, **136**, 134102.

78. Ihrig, A. C., Wieferink, J., Zhang, I. Y., Ropo, M., Ren, X., Rinke, P., Scheffler, M., Blum, V. (2015). *New J. Phys.*, **17**, 093020.

79. Hutter, J. (2012). *WIREs Comput. Mol. Sci.*, **2**, 604–612.

80. Kühne, T. D. (2014). *Wiley Interdiscip. Rev.: Comput. Mol. Sci.*, **4**, 391–406.

81. Ufimtsev, I. S., Martínez, T. J. (2008). *J. Chem. Theory Comput.*, **4**, 222–231.

82. Ufimtsev, I. S., Martinez, T. J. (2009). *J. Chem. Theory Comput.*, **5**, 1004–1015.

83. Ufimtsev, I. S., Martinez, T. J. (2009). *J. Chem. Theory Comput.*, **5**, 2619–2628.

84. Perdew, J. P., Ruzsinszky, A., Tao, J., Staroverov, V. N., Scuseria, G. E., Csonka, G. I. (2005). *J. Chem. Phys.*, **123**, 062201.

85. Yanai, T., Tew, D. P., Handy, N. C. (2004). *Chem. Phys. Lett.*, **393**, 51–57.

86. Merchán, M., Serrano-Andrés, L., Robb, M. A., Blancafort, L. (2005). *J. Am. Chem. Soc.*, **127**, 1820–1825.

87. Matsumoto, M. (2004). *J. Photochem. Photobiol. C: Photochem. Rev.*, **5**, 27–53.

88. Yue, L., Liu, Y.-J. (2019). *J. Chem. Theory Comput.*, **15**, 1798–1805.

89. Mato, J., Gordon, M. S. (2018). *Phys. Chem. Chem. Phys.*, **20**, 2615–2626.

90. Mato, J., Gordon, M. S. (2019). *J. Phys. Chem. A*, **123**, 1260–1272.

91. Gueye, M., Paolino, M., Gindensperger, E., Haacke, S., Olivucci, M., Léonard, J. (2020). *Faraday Discuss.*, **221**, 299–321.
92. Hu, D., Huang, J., Xie, Y., Yue, L., Zhuang, X., Lan, Z. (2015). *Chem. Phys.*, **463**, 95–105.
93. Hu, Y., Xu, C., Ye, L., Gu, F. L., Zhu, C. (2021). *Phys. Chem. Chem. Phys.*, **23**, 5236–5243.
94. Dunkelberger, A. D., Kieda, R. D., Shin, J. Y., Rossi Paccani, R., Fusi, S., Olivucci, M., Fleming Crim, F. (2012). *J. Phys. Chem. A*, **116**, 3527–3533.

Index